Teacher's Edition

by
Charles J. LaRue

AGS Publishing
Circle Pines, Minnesota 55014-1796
800-328-2560
www.agsnet.com

About the Author

Charles J. LaRue, Ph.D. The late Charles J. LaRue held a Ph.D. in science education, zoology, and botany from the University of Maryland, and an M.A. in zoology and botany from the University of Texas. He taught biology, zoology, and botany.

Photo and illustration credits for this textbook can be found on page 416.

The publisher wishes to thank the following consultants and educators for their helpful comments during the review process for *Biology*. Their assistance has been invaluable.

Susan B. Board, ESE Department Chair, Terry Parker High School (Duval County), Jacksonville, FL; **Brenda Bowman-Price,** Biology Teacher, Milby High School, Houston, TX; **Trish Duncan,** Special Education Teacher, Freedom High School, Tampa, FL; **Brian P. Johnson,** Science Teacher, Centennial High School, Circle Pines, MN; **Johnny McCarty,** Resource Science Teacher, Flour Bluff I.S.D. High School, Corpus Christi, TX; **Daniel A. McFarland,** Science Department Chair, Durant High School, Plant City, FL; **Helen M. Parke,** Ph.D., Meriwether Educational Designs, Greenville, NC; **Mary Schroder,** Science Department Head, St. Aloysius Gonzaga Secondary School, Mississauga, ON, Canada

Editorial and production services provided by Creative Services Associates, Inc.

Publisher's Project Staff

Vice President, Product Development: Kathleen T. Williams, Ph.D., NCSP; Associate Director, Product Development: Teri Mathews; Senior Editor: Julie Maas; Editor: Susan Weinlick; Development Assistant: Bev Johnson; Senior Designer: Diane McCarty; Creative Services Manager: Nancy Condon; Project Coordinator/Designer: Katie Sonmor; Purchasing Agent: Mary Kaye Kuzma; Senior Marketing Manager/Secondary Curriculum: Brian Holl

© 2004 AGS Publishing
4201 Woodland Road
Circle Pines, MN 55014-1796
800-328-2560 • www.agsnet.com

AGS Publishing is a trademark and trade name of American Guidance Service, Inc.

All rights reserved, including translation. No part of this publication may be reproduced or transmitted in any form or by any means without written permission from the publisher.

Printed in the United States of America

ISBN 0-7854-3614-6

Product Number 93902

A 0 9 8 7 6 5 4 3 2

Contents

Overview

Biology Overview T4
Skill Track Software T5
Biology Student Text Highlights T6
Biology Teacher's Edition Highlights T8
Biology Teacher's Resource Library Highlights T10
Other AGS Science Textbooks T11
Correlation of Biology to the National Science Education Standards T12
Skills Chart T14
Learning Styles T16

Lesson Plans

How to Use This Book: A Study Guide viii
The Nature of Science xiv
Chapter 1   The Basic Unit of Life xvi
Chapter 2   Organizing Living Things 24
Chapter 3   Classifying Animals 40
Chapter 4   Classifying Plant Groups 66
Chapter 5   Bacteria, Protists, and Fungi 84
Chapter 6   How Animals Stay Alive 114
Chapter 7   How Plants Live 142
Chapter 8   Human Body Systems 166
Chapter 9   Reproduction, Growth, and Development 208
Chapter 10   Staying Healthy 240
Chapter 11   Genetics 264
Chapter 12   Ecology 294
Chapter 13   The Behavior of Organisms 326
Chapter 14   Evolution 350
Appendix A: Animal Kingdom 385
Appendix B: Plant Kingdom 387
Appendix C: Body Systems 388
   The Skeletal System 388
   The Muscular System 389
   The Nervous System 390
   The Circulatory System 391
Appendix D: Measurement Conversion Factors 392
Appendix E: Alternative Energy Sources 394
Glossary 400
Index 410
Photo and Illustration Credits 416

Teacher's Resources

Midterm and Final Mastery Tests 417
Teacher's Resource Library Answer Key 420
   Workbook Activities 420
   Alternative Activities 426
   Lab Manual 431
   Community Connection 436
   Self-Study Guide 436
   Tests 437
Materials List for Biology Lab Manual 449
Some Suppliers of Science Education Materials 450

Biology

Biology is designed to help students and young adults learn about classification and organization; patterns of growth, development, and reproduction; the systems of the human body; ecological cycles, and other basic biological building blocks. Written to meet national standards, it offers students who read below grade level the opportunity to sharpen their abilities to interpret data, practice formulating hypotheses, and observe and record information. Throughout the text, comprehension is enhanced through the use of simple sentence structure and low-level vocabulary.

The textbook's short, concise lessons hold students' interest. Clearly stated objectives given at the beginning of each lesson outline what students will learn in the lesson.

Lesson Reviews and Chapter Reviews offer some open-ended questions to encourage students to use critical thinking skills. Hands-on Investigations lead students to apply the skills they are learning to everyday life. Full-color photographs and illustrations add interest and appeal as students learn key life science concepts.

Skill Track Software

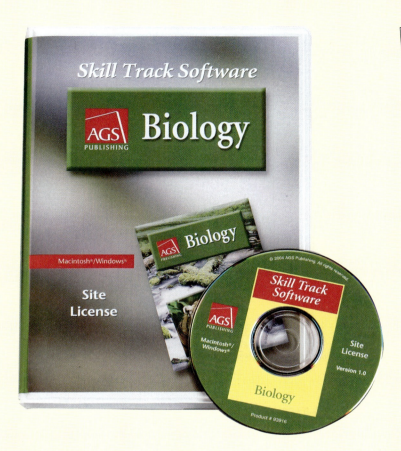

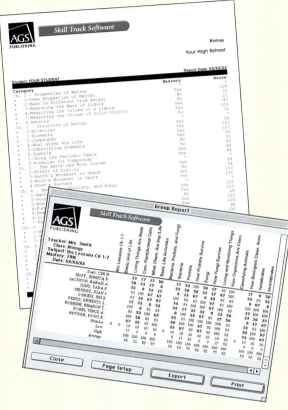

Skill Track Software The Skill Track software program allows students using AGS Publishing textbooks to be assessed for mastery of each chapter and lesson of the textbook. Students access the software on an individual basis and are assessed with multiple choice items.

Students can enter the program through two paths:

Lesson
Six items assess mastery of each lesson.

Chapter
Two parallel forms of chapter assessments are provided to determine chapter mastery. The two forms are equal in length and cover the same concepts with different items. The number of items in each chapter assessment varies by chapter, as the items are drawn from content of each lesson in the textbook.

The program includes high-interest graphics to accompany the items. Students are allowed to retake the chapter or lesson assessments over again at the instructor's discretion. The instructor has the ability to run and print out a variety of reports to track students' progress.

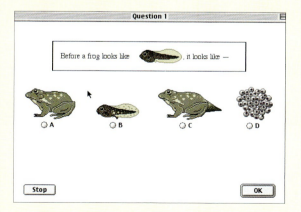

Biology **T5**

Student Text Highlights

- Each lesson is clearly labeled to help students focus on the skill or concept to be learned.

- Vocabulary terms are bold-faced and then defined in the margin and in the Glossary.

- Notes in the margin reinforce lesson content.

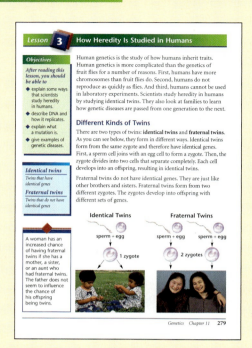

Identical twins
Twins that have identical genes

Fraternal twins
Twins that do not have identical genes

- Goals for Learning at the beginning of each chapter identify learner outcomes.

Goals for Learning

- To explain what a cell is and describe the organization of cells in living things
- To compare and contrast plant and animal cells
- To identify chemicals that are important for life and explain how living things use these chemicals
- To describe some basic life activities

Water in a nonvascular plant moves from one cell to another. Since water can't travel far this way, each cell must be near water. This limits the number of cells a nonvascular plant can have.

- Chapter Reviews allow students and teachers to check for skill mastery. Multiple-choice items are provided for practice in taking standardized tests.

- Test-Taking Tips at the end of each Chapter Review help reduce test anxiety and improve test scores.

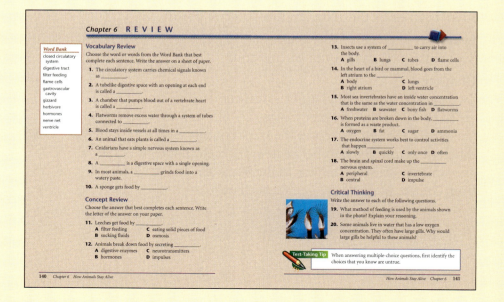

T6 Biology

Technology Note

To study some plant tissues, scientists use equipment to make very thin slices of a plant's stem. These thin slices are used to prepare microscope slides. When scientists first used microscopes, they had to draw what they saw. Today, they can use special cameras to take pictures of what they see. They can put the pictures on a computer and work with others.

Achievements in Science

Pasteurization Discovered

When people buy milk at a store, they can be assured that the milk does not contain pathogens, which could make them sick. Milk must be treated to kill pathogens before it can be sold. This treatment is known as pasteurization.

Pasteurization is named after the scientist Louis Pasteur. In the 1850s, he was asked to help solve a problem that local industry was having in Lille, France. During production, wine, beer, and vinegar would sometimes spoil. Pasteur discovered that microorganisms caused the spoilage and could cause disease. He also found that the microorganisms could be killed by gentle heating. This process of gentle heating is called pasteurization. It is still used today to make milk, orange juice, and other products safe to eat and drink.

Did You Know?

There are more than 500 species of bacteria that live in the human mouth. As much as 50 percent of the bacteria in the mouth live on the surface of the tongue.

Science Myth

Dinosaurs are the only extinct animals.

Fact: Many animal species have become extinct. Some scientists estimate that, at the current rate, half of all species now on Earth will be extinct within 100 years.

Science in Your Life

Are bacteria helpful or harmful?

Bacteria live everywhere. They are in the ocean, on top of mountains, in polar ice, on your hands, and even inside you. You cannot go anywhere without coming in contact with bacteria.

Most bacteria are helpful. Bacteria in soil break down plant and animal material and release nutrients. Bacteria take gases from the air, such as nitrogen. They change these gases into a form that plants and animals can use. Bacteria in your intestines make vitamin K. This vitamin helps blood to clot when you are cut.

Bacteria can also be harmful. Many bacteria cause disease in people. Bacteria cause diseases such as strep throat, tuberculosis, and cholera. Food that is contaminated with certain bacteria can cause illness.

Microbiologists are scientists who work with microorganisms. Some microbiologists study bacteria that cause disease. They grow the bacteria in the lab. Each cell multiplies until it forms millions of cells called a colony. It is impossible to see just one cell without a microscope. But it is easy to see a colony.

There are ways to get rid of most harmful bacteria. Antibiotics are drugs that kill bacteria in people and animals. Pasteurization, or rapid heating, kills harmful bacteria in food. Water is purified to remove bacteria and other organisms. Sewage is treated so that it will not pollute waterways. Some bacteria help to clean up sewage. Helpful bacteria eat the material in the sewage so that it does not pollute.

Science at Work

Tree Technician

Tree technicians need strength and balance. They need the ability to climb and to use equipment such as chainsaws and wood chippers. Tree technicians need knowledge of trees and their growth. Tree technicians need a high school diploma. A two-year degree is encouraged but not required.

Tree technicians trim trees, get rid of dead limbs, and remove trees. To do this, they often climb up into a tree and use ropes and pulleys to keep from falling. Other times they use trucks with buckets that lift them into the air. The work of tree technicians helps trees grow correctly and helps prevent problems.

Tree technicians might be hired by homeowners to trim trees on their property. They also might be hired by companies to trim trees that could damage electrical wires or phone lines. After natural disasters, such as tornadoes and hurricanes, tree technicians help clean up damaged trees.

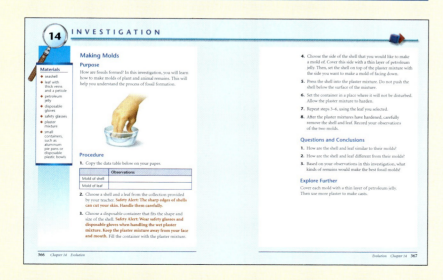

- Many features reinforce and extend student learning beyond the lesson content.

- Achievements in Science offers information about historic science-related events, achievements, or discoveries.

- Science in Your Life helps students relate chapter content to everyday life.

- Science at Work provides some examples of science careers.

- Investigation activities give students hands-on practice with chapter concepts. Students use critical thinking skills to complete each investigation.

Biology **T7**

Teacher's Edition Highlights

The comprehensive, wraparound Teacher's Edition provides instructional strategies at point of use. Everything from preparation guidelines to teaching tips and strategies are included in an easy-to-use format. Activities are featured at point of use for teacher convenience.

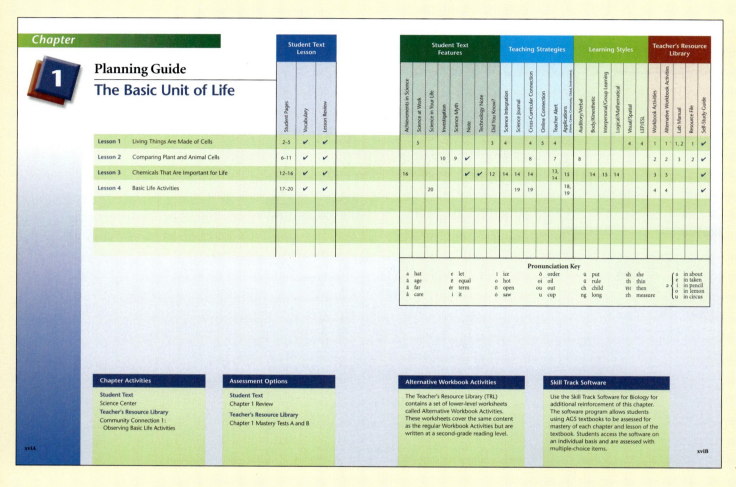

Chapter Planning Guides

- The Planning Guide saves valuable preparation time by organizing all materials for each chapter.
- A complete listing of lessons allows you to preview each chapter quickly.
- Assessment options are highlighted for easy reference. Options include:
  Lesson Reviews
  Chapter Reviews
  Chapter Mastery Tests, Forms A and B
  Midterm and Final Tests
- Page numbers of Student Text and Teacher's Edition features help customize lesson plans to your students.
- Many teaching strategies and learning styles are listed to support students with diverse needs.
- All activities in the Teacher's Resource Library are listed.
- A Pronunciation Key is provided to help you as you work with students to pronounce difficult words correctly.

Biology

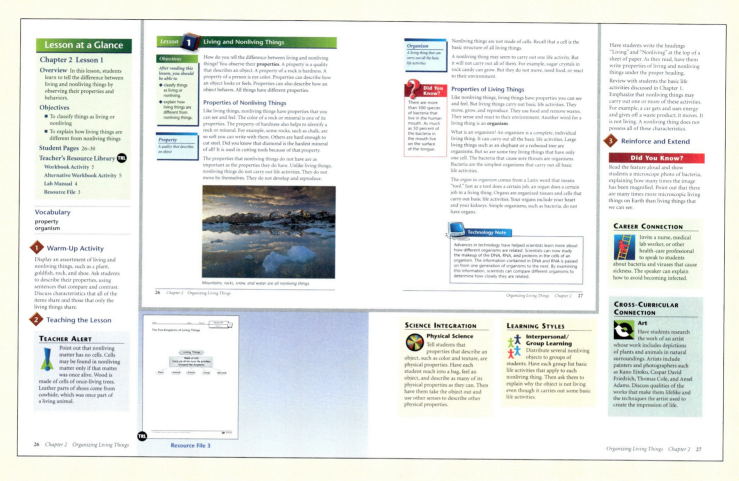

Lessons

- Quick overviews of chapters and lessons save planning time.
- Lesson objectives are listed for easy reference.
- Page references are provided for convenience.
- Easy-to-follow lesson plans in three steps save time: Warm-Up Activity, Teaching the Lesson, and Reinforce and Extend.
- Teacher Alerts highlight content that may need further explanation.
- Science Journal activities give students an opportunity to write about science.
- Cross-Curricular activities tie science to a variety of curriculum areas.
- Applications: Five areas of application—At Home, Career Connection, Global Connection, In the Community, and In the Environment—help students relate science to the world outside the classroom. Applications motivate students and make learning relevant.
- The Portfolio Assessment, which appears at the end of each lesson, lists items the student has completed for that lesson.
- Online Connections list relevant Web sites.
- Learning Styles provide teaching strategies to help meet the needs of students with diverse ways of learning. Modalities include Auditory/Verbal, Visual/Spatial, Body/Kinesthetic, Logical/Mathematical, and Interpersonal/Group Learning. Additional teaching activities are provided for LEP/ESL students.
- Answers for all activities in the Student Text appear in the Teacher's Edition. Answers for the Teacher's Resource Library, Student Workbook, and Lab Manual appear at the back of this Teacher's Edition and on the TRL CD-ROM.
- Worksheet, Workbook, Lab Manual, and Test pages from the Teacher's Resource Library are shown at point of use in reduced form.

Biology T9

Teacher's Resource Library Highlights

TRL All of the activities you'll need to reinforce and extend the text are conveniently located on the AGS Publishing Teacher's Resource Library (TRL) CD-ROM. All of the reproducible activities pictured in the Teacher's Edition are ready to select, view, and print. You can also preview other materials by linking directly to the AGS Publishing Web site.

Workbook Activities
Workbook Activities are available to reinforce and extend skills from each lesson of the textbook. A bound workbook format is also available.

Alternative Activities
These activities cover the same content as the Workbook Activities but are written at a second-grade reading level.

Lab Manual
These activities build critical thinking and teamwork skills. A bound format is also available.

Community Connection
Relevant activities help students extend their knowledge to the real world and reinforce concepts covered in class.

Resource File
These reference sheets on lesson content are tools for student study as well as teaching aids.

Self-Study Guide
An assignment guide provides the student with an outline for working through the text independently. The guide provides teachers with the flexibility for individualized instruction or independent study.

Mastery Tests
Chapter, Midterm, and Final Mastery Tests are convenient assessment options.

Answer Key
All answers to reproducible activities are included in the TRL and in the Teacher's Edition.

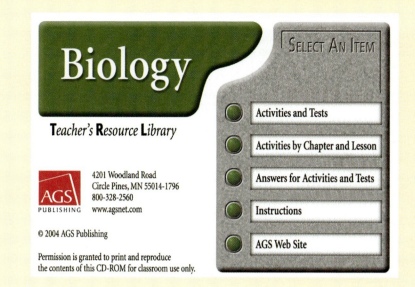

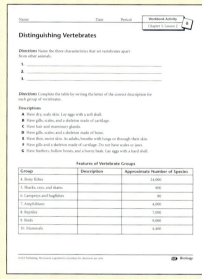
Workbook Activities

Lab Manual

Community Connections

Mastery Tests

T10 *Biology*

AGS Publishing Science Textbooks

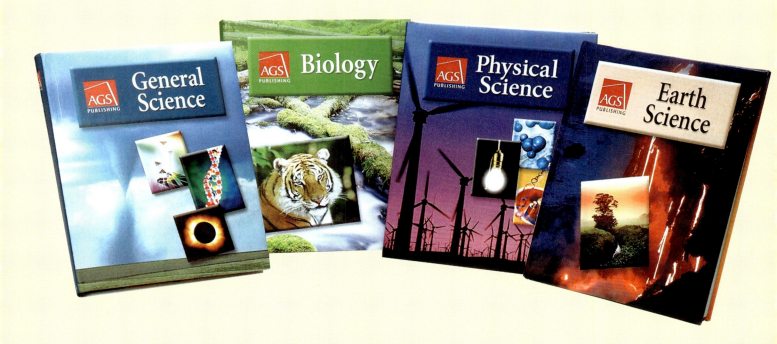

Enhance your science program with AGS Publishing textbooks—an easy, effective way to teach students the practical skills they need. Each AGS Publishing textbook meets your science curriculum needs. These exciting, full-color books use student-friendly text and real-world examples to show students the relevance of science in their daily lives. Each presents a comprehensive coverage of skills and concepts. The short, concise lessons will motivate even your most reluctant students. With readabilities of all the texts below fourth-grade reading level, your students can concentrate on learning the content. AGS Publishing is committed to making learning accessible to all students.

Correlation of Biology to the National Science Education Standards

STANDARD A Science as Inquiry

As a result of activities in grades 5-12, all students should develop:

◆ Abilities necessary to do scientific inquiry

Biology
Pages xiv–xv, 2–3, 10–11, 16, 28, 29–30, 49–50, 70, 75–76, 104, 105–106, 120, 121–122, 127, 161–162, 171, 179–180, 226–227, 254–255, 271–274, 286–287, 304–305, 339–340, 359, 366–367, 368, 392–393

◆ Understandings about scientific inquiry

Pages xiv–xv, 10–11, 14, 27, 29–30, 48, 49–50, 70, 75–76, 89, 98, 105–1–6, 121–122, 127, 160, 161–162, 171, 179–180, 186, 225, 226–227, 232, 245, 254–255, 271, 286–287, 304–305, 322, 333, 339–340, 344, 359, 366–367, 392–393

STANDARD C Life Science

As a result of activities in grades 5–12, all students should develop an understanding of:

◆ The cell

Biology
Pages 1–12, 14–17, 27, 31, 33, 56, 68–69, 77, 86–87, 90, 92, 95–97, 99, 101, 108, 123–124, 129, 131, 133, 135, 145, 149, 152–154, 156, 160, 168, 173, 176–177, 189, 211–217, 220, 225, 232, 242–243, 248, 271, 272, 279–282, 284, 353–354, 364

◆ Structure and function in living systems (5–8)

Pages 1–17, 27, 31, 33, 56, 68–69, 77, 86–111, 116–139, 145, 152–154, 156, 168–205, 211, 213–217, 220, 225, 232, 242–247, 279–282, 284, 312–323, 353–354

◆ Reproduction and heredity (5–8)

Pages 19, 95–98, 107–110, 156–163, 208–219, 264–292

◆ Molecular basis of heredity

Pages 6, 8, 15, 27, 60, 211–216, 220, 242, 264–293, 352–357, 364

◆ Biological evolution

Pages 24–39, 40–65, 66–83, 84–113, 350–357, 369–379

◆ Interdependence of organisms

Pages 294–325, 328

◆ Matter, energy, and organization in living systems

Pages 8–9, 12–17, 128, 130–131, 148–150, 185, 312–321, 331

◆ Diversity and adaptations of organisms (5–8)

Pages 208–234, 294–304, 314–315, 350–383

◆ Behavior of organisms

Pages 326–349, 369–375

STANDARD E Science and Technology

As a result of activities in grades 5-12, all students should develop:

◆ Abilities of technological design

Biology
The opportunity to address this objective is available on the following pages: 10–11, 29–30, 49–50, 75–76, 105–106, 121–122, 161–162, 179–180, 226–227, 254–255, 286–287, 304–305, 339–340, and 366–367.

◆ Understandings about science and technology

Pages 14, 16, 27, 28, 48, 62, 70, 74, 89, 94, 98, 110, 120, 127, 151, 160, 171, 172, 178, 186, 225, 232, 245, 247, 271, 290, 309, 322, 333, 344, 345, 359, 379

STANDARD F Science in Personal and Social Perspectives

As a result of activities in grades 5-12, all students should develop an understanding of:

Biology

- Personal and community health

    Pages 12–16, 242–246, 248–252, 256–258, 299–305, 313, 394–399

- Population growth

    Pages 297–310, 313–323, 352–353, 355–356

- Natural resources

    Pages 300–302, 306–310, 315, 317–321, 394–399

- Environmental quality

    Pages 294–325, 394–399

- Natural and human–induced hazards

    Pages 299–302

- Science and technology in local, national, and global challenges

    Pages 14, 16, 27, 28, 48, 62, 70, 74, 89, 94, 98, 110, 120, 127, 151, 160, 171, 172, 178, 186, 225, 232, 245, 247, 271, 290, 309, 322, 333, 344, 345, 359, 379

STANDARD G History and Nature of Science

As a result of activities in grades 5-12, all students should develop an understanding of:

Biology

- Science as a human endeavor

    Pages xiv–xv, 5, 16, 28, 35, 55, 62, 68, 74, 80, 94, 104, 110, 120, 127, 138, 147, 151, 172, 178, 204, 219, 235, 247, 253, 266–270, 276, 285, 290, 303, 309, 335, 337, 338, 345, 365, 369–370, 379

- Nature of Science (5–8)

    Pages xiv–xv, 10–11, 29–30, 49–50, 75–76, 105–106, 121–122, 161–162, 179–180, 226–227, 254–255, 286–287, 304–305, 339–340, 366–367

- Nature of scientific knowledge

    Pages xiv–xv, 10–11, 29–30, 49–50, 75–76, 105–106, 121–122, 161–162, 179–180, 226–227, 254–255, 286–287, 304–305, 339–340, 366–367

- History of Science (5–8)

    Pages 5, 14, 16, 27, 28, 35, 48, 55, 62, 68, 70, 74, 80, 89, 94, 98, 104, 110, 120, 127, 138, 147, 151, 160, 171, 172, 178, 186, 204, 219, 225, 232, 235, 245, 247, 253, 266–270, 271, 276, 285, 290, 303, 309, 322, 333, 335, 337, 338, 344, 345, 359, 365, 369–370, 379

- Historical perspectives

    Pages 14, 16, 27, 28, 48, 62, 70, 74, 89, 94, 98, 110, 120, 127, 151, 160, 171, 172, 178, 186, 225, 232, 245, 247, 271, 290, 309, 322, 333, 344, 345, 359, 379

Skills Chart

Biology

CHAPTER

Science Content

Behavior of Organisms					5	6	7						13	
Biological Evolution														14
The Cell	1									9				
Diversity and Adaptations of Organisms			2		4					9		11		14
Environmental Quality													12	
History and Nature of Science	1	2	3	4	5	6	7	8	9	10	11	12	13	14
Human Biology								8	9	10	11			
Inquiry and Investigation	1	2	3	4	5	6	7	8	9	10	11	12	13	14
Interdependence of Organisms					5	6	7					12		
Matter, Energy, Organization in Living Systems						6	7					12		
Natural Resources												12		
Populations and Ecosystems												12		
Reproduction and Heredity									9		11			
Science and Technology	1	2	3	4	5	6	7	8	9	10	11	12	13	14
Science in Personal and Social Perspectives	1	2	3	4	5	6	7	8	9	10	11	12	13	14

Process Skills

Communicating	1	2	3	4	5	6	7	8	9	10	11	12	13	14
Collecting Information	1	2	3	4	5	6	7	8	9	10	11	12	13	14
Describing		2		4	5	6		8				12		
Performing Experiments	1	2	3	4	5	6	7	8	9	10	11	12	13	14
Following Written Directions	1	2	3	4	5	6	7	8	9	10	11	12	13	14
Formulating Hypotheses	1	2	3	4	5	6	7	8	9	10	11	12	13	14
Listing	1	2	3		5		7							
Making and/or Using Models					5	6	7				11	12	13	14
Measuring	1				5		7	8	9	10		12		
Naming	1	2	3	4	5	6	7	8	9	10	11	12	13	14
Observing	1	2	3	4	5	6	7	8	9	10	11	12	13	14
Recording	1	2	3	4	5	6	7	8	9	10	11	12	13	14
Sequencing Steps in a Process					5		7	8	9			12		
Using Science Lab Equipment	1		3	4	5	6		8	9	10	11	12		14
Writing Aboiut Science	1	2	3	4	5	6	7	8	9	10	11	12	13	14

Thinking Skills														
Applying Information	1	2	3	4	5	6	7	8	9	10	11	12	13	14
Classifying and Categorizing	1	2	3	4	5	6	7	8	9	10	11	12	13	14
Comparing and Contrasting	1	2	3	4	5	6	7	8	9	10	11	12	13	14
Drawing Conclusions	1	2	3	4	5	6	7	8	9	10	11	12	13	14
Explaining Ideas	1	2	3	4	5	6	7	8	9	10	11	12	13	14
Formulating Questions	1	2	3	4	5	6	7	8	9	10	11	12	13	14
Identifying Similarities and Differences	1	2	3	4	5	6	7	8	9	10	11	12	13	14
Interpreting Data	1	2	3	4	5	6	7	8	9	10	11	12	13	14
Interpreting Visuals	1	2	3	4	5	6	7	8	9	10	11	12	13	14
Learning Science Vocabulary	1	2	3	4	5	6	7	8	9	10	11	12	13	14
Making Analogies	1	2	3	4	5	6	7	8	9	10	11	12	13	14
Making Decisions	1	2	3	4	5	6	7	8	9	10	11	12	13	14
Making Generalizations	1	2	3	4	5	6	7	8	9	10	11	12	13	14
Making Inferences	1	2	3	4	5	6	7	8	9	10	11	12	13	14
Organizing Information	1	2	3	4	5	6	7	8	9	10	11	12	13	14
Predicting Outcomes	1	2	3	4	5	6	7	8	9	10	11	12	13	14
Recalling Facts	1	2	3	4	5	6	7	8	9	10	11	12	13	14
Recognizing Cause and Effect	1	2	3	4	5	6	7	8	9	10	11	12	13	14
Recognizing Main Idea and Supporting Concepts	1	2	3	4	5	6	7	8	9	10	11	12	13	14
Recognizing Patterns	1	2	3	4	5	6	7	8	9	10	11	12	13	14
Summarizing	1	2	3	4	5	6	7	8	9	10	11	12	13	14
Understanding Concepts	1	2	3	4	5	6	7	8	9	10	11	12	13	14

Learning Styles

The learning style activities in the *Biology* Teacher's Edition provide activities to help students with special needs understand the lesson. These activities focus on the following learning styles: Visual/Spatial, Auditory/Verbal, Body/Kinesthetic, Logical/Mathematical, Interpersonal/Group Learning, LEP/ESL. These styles reflect Howard Gardner's theory of multiple intelligences. The writing activities suggested in this student text are appropriate for students who fit Gardner's description of Verbal/Linguistic Intelligence. The activities are designed to help teachers capitalize on students' individual strengths and dominant learning styles. The activities reinforce the lesson by teaching or expanding upon the content in a different way.

Following are examples of activities featured in the *Biology* Teacher's Edition:

Auditory/Verbal
Students benefit from having someone read the text aloud or listening to the text on audiocassette. Musical activities appropriate for the lesson may help auditory learners.

LEARNING STYLES

Auditory/Verbal
Provide a tape recording of animal sounds, such as bird calls, or have students tape-record the sounds of animals. Ask students to research information about one of the animals on the tape and to prepare their own audiotape presentation about it. Encourage them to begin or end their presentation with a recording of the animal making sounds, such as the trumpeting of an elephant. Place the tapes in the science center where students can access them.

LEP/ESL
Students benefit from activities that promote English language acquisition and interaction with English-speaking peers.

LEARNING STYLES

LEP/ESL
Many nouns in science vocabulary come from Latin and Greek words. These nouns may form plurals in ways that differ from most English plurals that are formed by adding *-s* or *-es* to the base words. In this lesson, several nouns have plural forms that may be unfamiliar to students. For example, *algae* is the plural form of the singular *alga*, *paramecia* is the plural form of *paramecium*, *cilia* is the plural form of *cilium*, and *flagella* is the plural form of *flagellum*.

Logical/Mathematical
Students learn by using logical/mathematical thinking in relation to the lesson content.

LEARNING STYLES

Logical/Mathematical
Give groups of students pictures of animals with backbones. Explain why these animals are classified together. Ask them to make a "tree" diagram showing how they would break this large group into smaller groups. Ask students to explain the logic of their classifications. After groups share their classifications, show students how the vertebrates are organized into classes: mammals, amphibians, reptiles, birds, fish.

Interpersonal/Group Learning
Learners benefit from working with at least one other person on activities that involve a process and an end product.

LEARNING STYLES

Interpersonal/Group Learning
Give small groups of students pictures of protists or a list of the different protists. Ask the groups to develop a system for organizing the protists into subgroups. Have the groups share their organizations with the class. Close the activity by pointing out that the Protist kingdom has been accepted by most biologists as an independent kingdom only since the second half of the 1900s.

Visual/Spatial
Students benefit from seeing illustrations or demonstrations beyond what is in the text.

LEARNING STYLES

Visual/Spatial
Have students use microscopes to observe a protozoan, or have them find detailed illustrations of a protozoan in a reference book. Ask students to draw their own picture of the organism and label parts that the organism uses to move.

Body/Kinesthetic
Learners benefit from activities that include physical movement or tactile experiences.

LEARNING STYLES

Body/Kinesthetic
Provide pairs of students with two cups of water, a spoon, and small amounts of sugar and salt. Have the students make sugar and salt solutions. Remind them that the ability of water to dissolve substances such as these enables nutrients to move from cell to cell in our bodies.

by
Charles J. LaRue

AGS Publishing
Circle Pines, Minnesota 55014-1796
800-328-2560

About the Author

Charles J. LaRue, Ph.D. The late Charles J. LaRue held a Ph.D. in science education, zoology, and botany from the University of Maryland, and an M.A. in zoology and botany from the University of Texas. He taught biology, zoology, and botany.

Photo and illustration credits for this textbook can be found on page 416.

The publisher wishes to thank the following consultants and educators for their helpful comments during the review process for *Biology*. Their assistance has been invaluable.

Susan B. Board, ESE Department Chair, Terry Parker High School (Duval County), Jacksonville, FL; **Brenda Bowman-Price,** Biology Teacher, Milby High School, Houston, TX; **Trish Duncan,** Special Education Teacher, Freedom High School, Tampa, FL; **Brian P. Johnson,** Science Teacher, Centennial High School, Circle Pines, MN; **Johnny McCarty,** Resource Science Teacher, Flour Bluff I.S.D. High School, Corpus Christi, TX; **Daniel A. McFarland,** Science Department Chair, Durant High School, Plant City, FL; **Helen M. Parke,** Ph.D., Meriwether Educational Designs, Greenville, NC; **Mary Schroder,** Science Department Head, St. Aloysius Gonzaga Secondary School, Mississauga, ON, Canada

Editorial and production services provided by Navta Associates, Inc. Publishing Services.

Publisher's Project Staff

Vice President, Product Development: Kathleen T. Williams, Ph.D., NCSP; Associate Director, Product Development: Teri Mathews; Senior Editor: Julie Maas; Editor: Susan Weinlick; Development Assistant: Bev Johnson; Senior Designer: Diane McCarty; Creative Services Manager: Nancy Condon; Project Coordinator/Designer: Katie Sonmor; Purchasing Agent: Mary Kaye Kuzma; Senior Marketing Manager/Secondary Curriculum: Brian Holl

© 2004 AGS Publishing
4201 Woodland Road
Circle Pines, MN 55014-1796
800-328-2560 • www.agsnet.com

AGS Publishing is a trademark and trade name of American Guidance Service, Inc.

All rights reserved, including translation. No part of this publication may be reproduced or transmitted in any form or by any means without written permission from the publisher.

Printed in the United States of America
ISBN 0-7854-3613-8
Product Number 93900
A 0 9 8 7 6 5 4 3 2 1

Contents

How to Use This Book: A Study Guide viii
The Nature of Science xiv

Chapter 1 The Basic Unit of Life 1
Lesson 1 Living Things Are Made of Cells 2
Lesson 2 Comparing Plant and Animal Cells 6
◆ Investigation 1: Comparing Cells 10
Lesson 3 Chemicals That Are Important for Life 12
Lesson 4 Basic Life Activities 17
◆ Chapter Summary 21
◆ Chapter Review 22
◆ Test-Taking Tip 23

Chapter 2 Organizing Living Things 24
Lesson 1 Living and Nonliving Things 26
◆ Investigation 2: Living or Nonliving? 29
Lesson 2 How Organisms Are Classified 31
◆ Chapter Summary 37
◆ Chapter Review 38
◆ Test-Taking Tip 39

Chapter 3 Classifying Animals 40
Lesson 1 How Biologists Classify Animals 42
◆ Investigation 3: Classifying Objects 49
Lesson 2 Vertebrates 51
Lesson 3 Invertebrates 56
◆ Chapter Summary 63
◆ Chapter Review 64
◆ Test-Taking Tip 65

Biology Contents **iii**

Chapter 4 Classifying Plant Groups 66
Lesson 1   How Plants Are Classified 68
Lesson 2   Seed Plants 71
◆ Investigation 4: Identifying Angiosperms and
   Gymnosperms 75
Lesson 3   Seedless Plants 77
◆ Chapter Summary 81
◆ Chapter Review 82
◆ Test-Taking Tip 83

Chapter 5 Bacteria, Protists, and Fungi 84
Lesson 1   Bacteria 86
Lesson 2   Protists 90
Lesson 3   How Protists Survive 95
Lesson 4   Fungi 99
◆ Investigation 5: Growing Bread Mold 105
Lesson 5   How Fungi Survive 107
◆ Chapter Summary 111
◆ Chapter Review 112
◆ Test-Taking Tip 113

Chapter 6 How Animals Stay Alive 114
Lesson 1   Animals Get and Digest Food 116
◆ Investigation 6: Studying Feeding in Hydras 121
Lesson 2   Respiration and Circulation 123
Lesson 3   Water Balance and Wastes 128
Lesson 4   Coordinating Bodily Activities 133
◆ Chapter Summary 139
◆ Chapter Review 140
◆ Test-Taking Tip 141

Chapter 7   How Plants Live **142**

    Lesson 1   The Vascular System in Plants............... 144
    Lesson 2   How Plants Make Food 148
    Lesson 3   How Plants Give Off Oxygen 152
    Lesson 4   How Plants Reproduce...................... 156
    ◆ Investigation 7: Growing an African Violet
       from a Leaf.. 161
    ◆ Chapter Summary 163
    ◆ Chapter Review 164
    ◆ Test-Taking Tip.. 165

Chapter 8   Human Body Systems **166**

    Lesson 1   How the Body Digests Food 168
    Lesson 2   How Materials Move to and from Cells 173
    ◆ Investigation 8: How Does Exercise Change
       Heart Rate?.. 179
    Lesson 3   How We Breathe........................... 181
    Lesson 4   How the Body Gets Rid of Wastes........... 184
    Lesson 5   How the Nervous System Controls the Body... 187
    Lesson 6   The Sense Organs 192
    Lesson 7   How the Endocrine System Controls
       the Body.. 196
    Lesson 8   How the Body Moves...................... 199
    ◆ Chapter Summary 205
    ◆ Chapter Review 206
    ◆ Test-Taking Tip.. 207

Chapter 9   Reproduction, Growth, and Development.... **208**

    Lesson 1   Where Life Comes From 210
    Lesson 2   How Organisms Reproduce................. 214
    Lesson 3   How Animals Grow and Develop 220
    ◆ Investigation 9: Graphing Gestation Times 226
    Lesson 4   How Humans Grow and Develop............ 228
    ◆ Chapter Summary 237
    ◆ Chapter Review 238
    ◆ Test-Taking Tip.. 239

Chapter 10 — Staying Healthy 240
- Lesson 1 How the Body Fights Disease 242
- Lesson 2 Good Nutrition 248
- ◆ Investigation 10: Reading Food Labels 254
- Lesson 3 Healthy Habits 256
- ◆ Chapter Summary 261
- ◆ Chapter Review 262
- ◆ Test-Taking Tip 263

Chapter 11 — Genetics 264
- Lesson 1 Heredity 266
- Lesson 2 Chromosomes 272
- Lesson 3 How Heredity Is Studied in Humans 279
- ◆ Investigation 11: Tracing a Genetic Disease 286
- Lesson 4 Applied Genetics 288
- ◆ Chapter Summary 291
- ◆ Chapter Review 292
- ◆ Test-Taking Tip 293

Chapter 12 — Ecology 294
- Lesson 1 Living Things and Nonliving Things 296
- ◆ Investigation 12: Testing the pH of Rain 304
- Lesson 2 Food Chains and Food Webs 305
- Lesson 3 How Energy Flows Through Ecosystems 312
- Lesson 4 How Materials Cycle Through Ecosystems ... 317
- ◆ Chapter Summary 323
- ◆ Chapter Review 324
- ◆ Test-Taking Tip 325

Chapter 13 — The Behavior of Organisms 326
- Lesson 1 Innate Behavior 328
- Lesson 2 Learned Behavior 335
- ◆ Investigation 13: Observing Learning Patterns 339
- Lesson 3 How Animals Communicate 341
- ◆ Chapter Summary 347
- ◆ Chapter Review 348
- ◆ Test-Taking Tip 349

Chapter 14 **Evolution** **350**
  Lesson 1 Change over Time 352
  Lesson 2 What Fossils Show 358
  ◆ Investigation 14: Making Molds.................. 366
  Lesson 3 The Theory of Evolution 369
  Lesson 4 What Humanlike Fossils Show 376
  ◆ Chapter Summary............................. 381
  ◆ Chapter Review 382
  ◆ Test-Taking Tip 383

Appendix A: Animal Kingdom..................................**385**

Appendix B: Plant Kingdom...................................**387**

Appendix C: Body Systems
        The Skeletal System...................**388**
        The Muscular System **389**
        The Nervous System **390**
        The Circulatory System **391**

Appendix D: Measurement Conversion Factors **392**

Appendix E: Alternative Energy Sources **394**

Glossary **400**

Index **410**

Using This Section

How to Use This Book: A Study Guide

Overview This section may be used to introduce the study of physical science, to preview the book's features, and to review effective study skills.

Objectives
- To introduce the study of biology
- To preview the student textbook
- To review study skills

Student Pages viii–xiii

Teacher's Resource Library
How to Use This Book 1–7

Introduction to the Book

Have volunteers read aloud the three paragraphs of the introduction. Discuss with students why studying science and developing scientific skills are important.

How to Study

Read aloud each bulleted statement, pausing to discuss with students why the suggestion is a part of good study habits. Distribute copies of the How to Use This Book 1, "Study Habits Survey," to students. Read the directions together and then have students complete the survey. After they have scored their surveys, ask them to make a list of the study habits they plan to improve. After three or four weeks, have students complete the survey again to see if they have improved their study habits. Encourage them to keep and review the survey every month or so to see whether they are maintaining and improving their study habits.

To help students organize their time and work in an easy-to-read format, have them fill out How to Use This Book 2, "Weekly Schedule." Encourage them to keep the schedule in a notebook or folder where they can refer to it easily. Suggest that they review the schedule periodically and update it as necessary.

Give students an opportunity to become familiar with the textbook features and the chapter and lesson organization and

viii *How to Use This Book: A Study Guide*

How to Use This Book: A Study Guide

Welcome to *Biology*. Life science, or biology, is the study of living things. Science touches our lives every day, no matter where we are—at home, at school, or at work. This book covers the area of biology. It also focuses on science skills that scientists use. These skills include asking questions, making predictions, designing experiments or procedures, collecting and organizing information, calculating data, making decisions, drawing conclusions, and exploring more options. You probably already use these skills every day. You ask questions to find answers. You gather information and organize it. You use that information to make all sorts of decisions. In this book, you will have opportunities to use and practice all of these skills.

As you read this book, notice how each lesson is organized. Information is presented in a straightforward manner. Tables, diagrams, and photos help clarify concepts. Read the information carefully. If you have trouble with a lesson, try reading it again.

It is important that you understand how to use this book before you start to read it. It is also important to know how to be successful in this course. Information in this first section of the book can help you achieve these things.

How to Study

These tips can help you study more effectively.
◆ Plan a regular time to study.
◆ Choose a quiet desk or table where you will not be distracted. Find a spot that has good lighting.
◆ Gather all the books, pencils, paper, and other equipment you will need to complete your assignments.
◆ Decide on a goal. For example: "I will finish reading and taking notes on Chapter 1, Lesson 1, by 8:00."
◆ Take a five- to ten-minute break every hour to stay alert.
◆ If you start to feel sleepy, take a break and get some fresh air.

viii *How to Use This Book: A Study Guide*

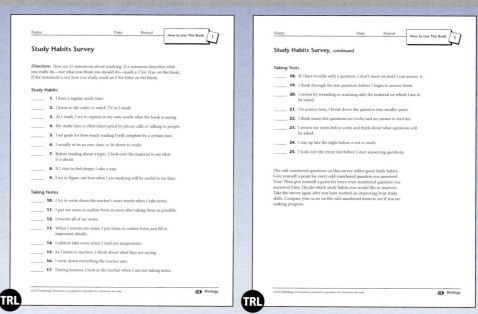

How to Use This Book 1, pages 1 and 2

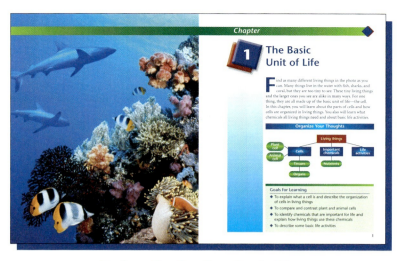

Before Beginning Each Chapter

◆ Read the chapter title and study the photo. What does the photo tell you about the chapter title?
◆ Read the opening paragraph.
◆ Study the Goals for Learning. The Chapter Review and tests will ask questions related to these goals.
◆ Look at the Chapter Review. The questions cover the most important information in the chapter.

Note These Features

Note
Points of interest or additional information that relates to the lesson

Did You Know?
Facts that add details to lesson content or present an interesting or unusual application of lesson content

Science Myth
Common science misconceptions followed by the correct information

How to Use This Book: A Study Guide   ix

Discuss with students why knowing these goals can help them when they are studying the chapter. Finally, have students skim the Chapter Summary to identify important information and vocabulary presented in the chapter. Have students turn to the Chapter Review and explain that it provides an opportunity to determine how well they have understood the chapter content.

Note These Features

Use the information on pages ix and x to identify features included in each chapter. As a class locate examples of these features in Chapter 1. Read the examples and discuss their purpose.

structure of *Biology*. List the following text features on the board: Table of Contents, Chapter Opener, Lesson, Lesson Review, Investigation, Chapter Summary, Chapter Review, Appendix A: Animal Kingdom, Appendix B: Plant Kingdom, Appendix C: Body Systems, Appendix D: Measurement Conversion Factors, Appendix E: Alternative Energy Sources, Glossary, Index.

Have students skim their textbooks to find these features. You may wish to remind students that they can use the Table of Contents to help identify and locate major features in the text. They can also use the Index to identify specific topics and the text pages on which they are discussed. Ask volunteers to call out a feature or topic and its page reference from the Table of Contents or Index. Have other students check to see that the specific features or topics do appear on the pages cited.

Before Beginning Each Chapter

When students begin their study of Chapter 1, you may wish to have them read aloud and follow each of the bulleted suggestions on page ix. Actually trying the suggestions will help students understand what they are supposed to do and recognize how useful the suggestions are when previewing a chapter. At the beginning of other chapters, refer students to page ix and encourage them to follow the suggestions. You may wish to continue to do this as a class each time or allow students to work independently.

In addition to the suggestions on page ix, the Teacher's Edition text for each Chapter Opener offers teaching suggestions for introducing the chapter. The text also includes a list of Teacher's Resource Library materials for the chapter.

Chapter Openers organize information in easy-to-read formats. Have a volunteer find and read the Chapter 1 title on page 1. Read aloud the second bulleted statement on page ix and have a volunteer read aloud the opening paragraphs of the chapter. Discuss the topics that students will study in the chapter. Have students examine the Organize Your Thoughts chart. It provides an overview of chapter content. Then have volunteers take turns reading aloud the Chapter 1 Goals for Learning.

Before Beginning Each Lesson

With students, read through the information in "Before Beginning Each Lesson" on page x. Then assign each of the four lessons in Chapter 1 to a small group of students. Have them restate the lesson title in the form of a statement or a question and make a list of features in their lesson. After their survey of the lesson, they should be prepared to report to the class on their findings.

Technology Note
Technology information that relates to the lesson or chapter

Science in Your Life
Examples of science in real life

Achievements in Science
Historical scientific discoveries, events, and achievements

Science at Work
Careers in science

Investigation
Experiments that give practice with chapter concepts

Before Beginning Each Lesson

Read the lesson title and restate it in the form of a question.

For example, write:
What chemicals are important for life?

Look over the entire lesson, noting the following:
- bold words
- text organization
- notes in the margins
- photos and diagrams
- lesson review questions

As You Read the Lesson

- Read the lesson title.
- Read the subheads and paragraphs that follow.
- Before moving on to the next lesson, see if you understand the concepts you read. If you do not understand the concepts, reread the lesson. If you are still unsure, ask for help.
- Practice what you have learned by completing the Lesson Review.

Using the Bold Words

Bold type
Words seen for the first time will appear in bold type

Glossary
Words listed in this column are also found in the Glossary

Knowing the meaning of all the boxed vocabulary words in the left column will help you understand what you read.

These words are in **bold type** the first time they appear in the text. They are often defined in the paragraph.

> All living things are made of **cells**. A cell is the basic unit of life.

All of the words in the left column are also defined in the **Glossary.**

> **Cell** (sel) the basic unit of life (p. 2)

Word Study Tips

- Start a vocabulary file with index cards to use for review.
- Write one term on the front of each card. Write the chapter number, lesson number, and definition on the back.
- You can use these cards as flash cards by yourself or with a study partner to test your knowledge.

Cell

the basic unit of life
Chapter 1, Lesson 1

How to Use This Book: A Study Guide **xi**

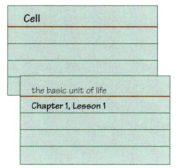

How to Use This Book 3

As You Read the Lesson

Read aloud the statements in the section "As You Read the Lesson" on page xi. Have students preview lessons in Chapter 1 and note lesson titles and subheads. Remind students as they study each lesson to follow this study approach.

Using the Bold Words

Read aloud the information on page xi. Make sure students understand what the term *bold* means. Explain to students that the words in bold are important vocabulary terms. Then ask them to look at the boxed words on page 2. Have a volunteer read the boxed term *cell* and then find and read the sentence in the text in which that word appears in bold type. Have another volunteer read the definition of the word in the box.

Point out that boxed words may appear on other pages in a lesson besides the first page. Have students turn to page 3 and look at the boxed words on that page. Explain that these words appear in a box here because they are used in the text on this page. Have volunteers find and read the sentences in the text in which the vocabulary words are used. Have students turn to the Glossary at the back of the book and read the definitions of the vocabulary words on page 3.

Word Study Tips

Have a volunteer read aloud the word study tips on page xi. You may wish to demonstrate how to make a vocabulary card by filling out an index card for the term *cell* and its definition (page 2).

Distribute copies of How to Use This Book 3, "Word Study," to students. Suggest that as they read, students write unfamiliar words, their page numbers, and their definitions on the sheet. Point out that having such a list will be very useful for reviewing vocabulary before taking a test. Point out that students can use words they listed on How to Use This Book 3 to make their vocabulary card file.

How to Use This Book: A Study Guide **xi**

Using the Summaries

Have students turn to page 21 and examine the Chapter 1 Summary. Emphasize that Chapter Summaries identify the main ideas of the chapter. Suggest that students can use the summary to focus their study of the chapter content. They might write each main idea in a notebook and add a few details that reinforce it. These notes will make a useful study tool.

Using the Reviews

Have students turn to page 5 and examine the Lesson 1 Review for Chapter 1. Emphasize that Lesson Reviews provide opportunities for students to focus on important content and skills developed in the lesson. Then have students turn to pages 22 and 23. Point out that the Chapter Review is intended to help them focus on and review the key terms, content information, and skills presented in the chapter before they are tested on the material. Suggest that they complete the review after they have studied their notes, vocabulary lists, and worksheets.

Preparing for Tests

Encourage students to offer their opinions about tests and their ideas on test-taking strategies. What do they do to study for a test? List their comments on the board. Then read the set of bulleted statements on page xii. Add these suggestions to the list on the board if they are not already there.

Discuss why each suggestion can help students when they are taking a test. Lead students to recognize that these suggestions, along with the Test-Taking Tips in their textbooks, can help them improve their test-taking skills.

Have students turn to the Chapter Review at the end of any chapter in the textbook and find the Test-Taking Tip. Ask several volunteers to read aloud the tips they find in the Chapter Reviews. Discuss how using the tips can help students study and take tests more effectively.

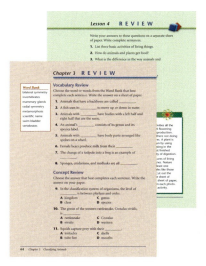

Using the Summaries

◆ Read each Chapter Summary to be sure you understand the chapter's main ideas.
◆ Make up a sample test of items you think may be on the test. You may want to do this with a classmate and share your questions.
◆ Read the vocabulary words in the Science Words box.
◆ Review your notes and test yourself on vocabulary words and key ideas.
◆ Practice writing about some of the main ideas from the chapter.

Using the Reviews

◆ Answer the questions in the Lesson Reviews.
◆ In the Chapter Reviews, answer the questions about vocabulary under the Vocabulary Review. Study the words and definitions. Say them aloud to help you remember them.
◆ Answer the questions under the Concept Review and Critical Thinking sections of the Chapter Reviews.
◆ Review the Test-Taking Tips.

Preparing for Tests

◆ Complete the Lesson Reviews and Chapter Reviews.
◆ Complete the Investigations.
◆ Review your answers to Lesson Reviews, Investigations, and Chapter Reviews.
◆ Test yourself on vocabulary words and key ideas.
◆ Use graphic organizers as study tools.

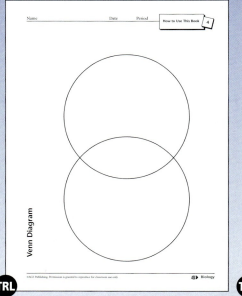

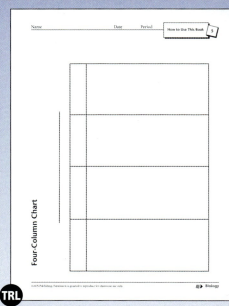

How to Use This Book 4 and 5

Using Graphic Organizers

A graphic organizer is a visual representation of information. It can help you see how ideas are related to one another. A graphic organizer can help you study for a test or organize information before you write. Here are some examples.

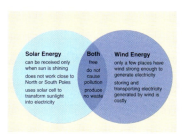

Venn Diagram
A Venn diagram can help you compare and contrast two things. For example, this diagram compares and contrasts solar energy and wind energy. The characteristics of solar energy are listed in the left circle. The characteristics of wind energy are listed in the right circle. The characteristics that both have are listed in the intersection of the circles.

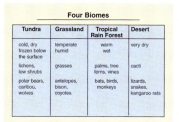

Column Chart
Column charts can help you organize information into groups, or categories. Grouping things in this format helps make the information easier to understand and remember. For example, this four-column chart groups information about each of the four biomes. A column chart can be divided into any number of columns or rows. The chart can be as simple as a two-column list of words or as complex as a multiple-column, multiple-row table of data.

Network Tree
A network tree organizer shows how ideas are connected to one another. Network trees can help you identify main ideas or concepts linked to related ideas. For example, this network tree identifies concepts linked to the concept of conservation. You can also use network trees to rank ideas from most important to least important.

How to Use This Book: A Study Guide **xiii**

Using Graphic Organizers

Explain to students that graphic organizers provide ways of visually organizing information to make it easier to understand and remember. Emphasize that there are many different kinds of graphic organizers including column charts, Venn diagrams, network trees, and word webs. Encourage students to look at the Organize Your Thoughts chart at the beginning of each chapter. Discuss how these graphic organizers provide a preview of the chapter content.

Tell students that they can use a variety of organizers to record information for a variety of purposes. For example, a Venn diagram is useful for comparing and contrasting information. Draw a Venn diagram on the board. Show students how to use the diagram to compare and contrast two items, such as a ball and a globe. Discuss how the diagram clearly shows the similarities and differences between the two items.

Display other organizers, such as a cause and effect chart, spider map, and two-column chart. Ask volunteers to suggest ways that these organizers can be used to record information. Then encourage students to record information on graphic organizers and use them as study tools.

Have students refer back to the pages in this section, "How to Use This Book," as often as they wish while using this textbook.

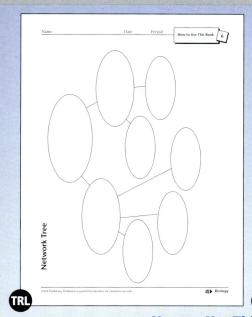
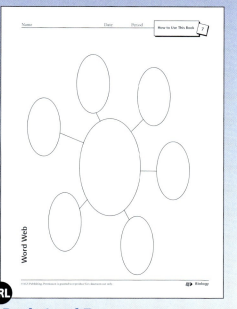

How to Use This Book 6 and 7

The Nature of Science

Write the word *science* on the board. Ask students what the word *science* brings to mind when they hear it. Record their ideas on the board. Guide the discussion to focus on the major areas of science, such as physical, earth, and life sciences; the achievements students credit to science; scientific investigations; and the kinds of information scientists need to do their work.

Following the discussion have students read the first paragraph on page xiv. Help them conclude that science has played an important role in the development of their way of life. From the time they wake up until the time they fall asleep, they benefit from scientific knowledge and discoveries.

Read and discuss the remaining paragraphs that focus on the scientific method. Emphasize the flexibility of the method and how the steps can be reordered or eliminated depending on the circumstances of the investigation. Discuss how the method can be cyclical. Results can lead to new questions and revised hypotheses and new or refined investigations. Explain that keeping accurate records of an investigation is an extremely important step because the data is used to support conclusions and verify results.

The Nature of Science

Science is an organized body of knowledge about the natural world. It encompasses everything from atoms to rocks to human health. Scientific knowledge is important because it solves problems, improves everyday life, and uncovers new opportunities. For example, scientists develop vaccines and antibiotics to prevent and cure diseases. Scientific knowledge helps farmers grow better and more crops. Science is behind the electricity we depend on every day. And science has launched space exploration, which continues to offer new opportunities.

Scientists use a logical process to explore the world and collect information. It is called the scientific method, and it includes specific steps. Scientists follow these steps or variations of these steps to test whether a possible answer to their question is correct.

1. Ask a question.
2. State a hypothesis, or make a prediction, about the answer.
3. Design an experiment, or procedure, to test the hypothesis.
4. Perform the experiment and gather information.
5. Analyze the data and organize the results.
6. State a conclusion based on the results, existing knowledge, and logic. Determine whether the results support the hypothesis.
7. Communicate the results and the conclusion.

As a scientist researches a question, he or she may do these steps in a different order or may even skip some steps. The scientific method requires many skills: predicting, observing, organizing, classifying, modeling, measuring, inferring, analyzing, and communicating.

Communication is an important part of the scientific method. Scientists all over the world share their findings with other scientists. They publish information about their experiments in journals and discuss them at meetings. A scientist may try another scientist's experiment or change it in some way. If many scientists get the same results from an experiment, then the results are repeatable and considered reliable.

Sometimes the results of an experiment do not support its hypothesis. Unexpected observations can lead to new, more interesting questions. For example, penicillin was discovered accidentally in 1928. Alexander Fleming observed that mold had contaminated one of his bacteria cultures. He noticed that the mold had stopped the growth of the bacterium. Since the mold was from the penicillium family, he named it penicillin. A decade later, researchers found a way to isolate the active ingredient. Since then, penicillin has been used to fight bacteria and save people's lives.

Once in a while, scientists discover something that dramatically changes our world, like penicillin. But, more often, scientific knowledge grows and changes a little at a time.

What scientists learn is applied to problems and challenges that affect people's lives. This leads to the development of practical tools and techniques. Tools help scientists accurately observe and measure things in the natural world. A new tool often provides data that an older tool could not. For example, computers help scientists analyze data more quickly and accurately than ever before. Our science knowledge grows as more advanced tools and technology make new discoveries possible.

Scientists use theories to explain their observations and data. A theory is a possible explanation for a set of data. A theory is not a fact. It is an idea. Theories are tested by more experiments. Theories may be confirmed, changed, or sometimes tossed out. For example, in 1808, John Dalton published a book describing his theory of atoms. His theory stated that atoms are solid spheres without internal structures. By the early 1900s, however, new tools allowed Ernest Rutherford to show that atoms are mostly empty space. He said that an atom consists of a tightly packed nucleus with electrons whizzing around it. This theory of the atom is still accepted today.

Theories that have stood many years of testing often become scientific laws. The law of gravity is one example. Scientists assume many basic laws of nature.

In this book, you will learn about biology. You will use scientific skills to solve problems and answer questions. You will follow some of the steps in the scientific method. And you will discover how important biology is to your life.

Display a variety of tools that students will use in their own scientific investigations. Identify each tool and its purpose. Discuss the proper care of the tools and the need to follow safety precautions when performing investigations. You may wish to help students develop a list of safety rules to follow. They can make a poster outlining the rules for display in the classroom.

Tell students that through the investigations in *Biology*, they have the opportunity to perform as scientists. Remind them to use the opportunity to follow the scientific method and develop the skills that will enhance their study and understanding of science.

Chapter 1

Planning Guide
The Basic Unit of Life

	Student Pages	Vocabulary	Lesson Review
Lesson 1 Living Things Are Made of Cells	2–5	✔	✔
Lesson 2 Comparing Plant and Animal Cells	6–11	✔	✔
Lesson 3 Chemicals That Are Important for Life	12–16	✔	✔
Lesson 4 Basic Life Activities	17–20	✔	✔

Student Text Lesson

Chapter Activities

Student Text
Science Center

Teacher's Resource Library
Community Connection 1:
　Observing Basic Life Activities

Assessment Options

Student Text
Chapter 1 Review

Teacher's Resource Library
Chapter 1 Mastery Tests A and B

Student Text Features								Teaching Strategies						Learning Styles						Teacher's Resource Library				
Achievements in Science	Science at Work	Science in Your Life	Investigation	Science Myth	Note	Technology Note	Did You Know?	Science Integration	Science Journal	Cross-Curricular Connection	Online Connection	Teacher Alert	Applications (Home, Career, Community, Global, Environment)	Auditory/Verbal	Body/Kinesthetic	Interpersonal/Group Learning	Logical/Mathematical	Visual/Spatial	LEP/ESL	Workbook Activities	Alternative Workbook Activities	Lab Manual	Resource File	Self-Study Guide
	5						3	4		4	5	4						4	4	1	1	1, 2	1	✔
			10	9	✔				8		7		8					2	2	3	2	✔
16					✔	✔	12	14	14	14		13, 14	15		14	15	14			3	3			✔
		20							19	19			18, 19							4	4			✔

Pronunciation Key

a	hat	e	let	ī	ice	ô	order	u̇	put	sh	she	ə {	a in about
ā	age	ē	equal	o	hot	oi	oil	ü	rule	th	thin		e in taken
ä	far	ėr	term	ō	open	ou	out	ch	child	₮H	then		i in pencil
â	care	i	it	ȯ	saw	u	cup	ng	long	zh	measure		o in lemon
													u in circus

Alternative Workbook Activities

The Teacher's Resource Library (TRL) contains a set of lower-level worksheets called Alternative Workbook Activities. These worksheets cover the same content as the regular Workbook Activities but are written at a second-grade reading level.

Skill Track Software

Use the Skill Track Software for Biology for additional reinforcement of this chapter. The software program allows students using AGS textbooks to be assessed for mastery of each chapter and lesson of the textbook. Students access the software on an individual basis and are assessed with multiple-choice items.

Chapter at a Glance

Chapter 1: The Basic Unit of Life
pages 1–23

Lessons
1. Living Things Are Made of Cells pages 2–5
2. Comparing Plant and Animal Cells pages 6–11

   Investigation 1 pages 10–11
3. Chemicals That Are Important for Life pages 12–16
4. Basic Life Activities pages 17–20

Chapter 1 Summary page 21

Chapter 1 Review pages 22–23

Skill Track Software for Biology

Teacher's Resource Library TRL

- Workbook Activities 1–4
- Alternative Workbook Activities 1–4
- Lab Manual 1–3
- Community Connection 1
- Resource File 1–2
- Chapter 1 Self-Study Guide
- Chapter 1 Mastery Tests A and B

(Answer Keys for the Teacher's Resource Library begin on page 420 of this Teacher's Edition. A list of supplies required for Lab Manual Activities in this chapter begins on page 449.)

Science Center

Set up one or two microscopes in an area of the classroom. Prepare a variety of slides for students to view as a way of introducing them to biology and some of the life activities of organisms. You might include pond water, onion cells, or prepared slides of tissues obtained from a biological supply company. This may be the first time some students have had an opportunity to look through a microscope. Demonstrate the use of this instrument.

xvi Chapter 1 The Basic Unit of Life

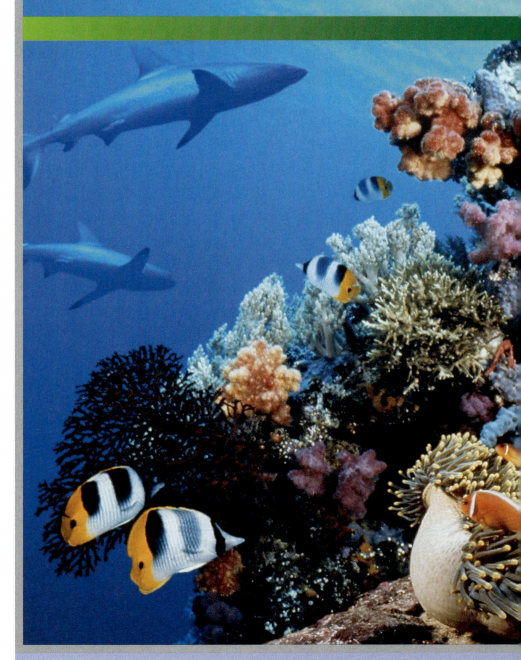

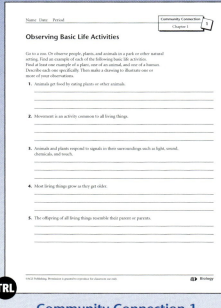

Community Connection 1

Chapter 1: The Basic Unit of Life

Find as many different living things in the photo as you can. Many things live in the water with fish, sharks, and coral, but they are too tiny to see. These tiny living things and the larger ones you see are alike in many ways. For one thing, they are all made up of the basic unit of life—the cell. In this chapter, you will learn about the parts of cells and how cells are organized in living things. You also will learn what chemicals all living things need and about basic life activities.

Organize Your Thoughts

- Living things
  - Cells
    - Plant cell
    - Animal cell
    - Tissues
    - Organs
  - Important chemicals
    - Nutrients
  - Life activities

Goals for Learning

- To explain what a cell is and describe the organization of cells in living things
- To compare and contrast plant and animal cells
- To identify chemicals that are important for life and explain how living things use these chemicals
- To describe some basic life activities

Introducing the Chapter

Ask students to think of some animals and plants. Discuss these organisms through a series of questions. What do students know about the animals or plants? Are these organisms made of any parts? What is the smallest part students know? Why do living things drink water? What do plants and animals have to do to stay alive? How do plants get their food?

Have students make lists of some of the things they know and some further questions they have. Then discuss what they have written.

After studying this chapter, have students evaluate their answers and discuss how their answers may have changed. Are they now able to at least partially answer any of their new questions?

Notes and Technology Notes

Ask volunteers to read the notes that appear in the margins throughout the chapter. Then discuss them with the class.

TEACHER'S RESOURCE

The AGS Teaching Strategies in Science Transparencies may be used with this chapter. The transparencies add an interactive dimension to expand and enhance the *Biology* program content.

CAREER INTEREST INVENTORY

The AGS Harrington-O'Shea Career Decision-Making System-Revised (CDM) may be used with this chapter. Students can use the CDM to explore their interests and identify careers. The CDM defines career areas that are indicated by students' responses on the inventory.

Chapter 1 Self-Study Guide

Lesson at a Glance

Chapter 1 Lesson 1

Overview In this lesson, students learn how living things are made up of cells that are organized into tissues and organs.

Objectives
- To describe a cell and explain some of its functions
- To explain what tissues are
- To explain what organs are

Student Pages 2–5

Teacher's Resource Library (TRL)
Workbook Activity 1
Alternative Workbook Activity 1
Lab Manual 1–2
Resource File 1

Vocabulary
cell	electron microscope
bacteria	tissue
microscope	organ
organelle	

Science Background
Biology

Biology, or life science, is the study of life and living things. Biologists, like all scientists, study the natural world not just by reading about it but by observing it and testing their observations with experiments.

Biology is a broad topic of study, and most biologists specialize by studying one group of organisms or life processes. For example, zoologists study animals, and botanists study plants. Further specialization includes ornithology (the study of birds) and entomology (the study of insects).

Basic Unit of Life

Most large things are not a solid mass. They are made up of units. For example, a very large wall may look solid from far away. But if you get closer, you may be able to see smaller units that make up the wall, such as bricks. Large living things, such as the plants and animals that students are familiar with, are made up of units called cells. All of the cells together make up a living organism.

2   Chapter 1   The Basic Unit of Life

Lesson 1 Living Things Are Made of Cells

Objectives
After reading this lesson, you should be able to
- describe a cell and explain some of its functions.
- explain what tissues are.
- explain what organs are.

Cell
The basic unit of life

Bacteria
The simplest single cells that carry out all basic life activities

Microscope
An instrument used to magnify things

All living things are made of **cells**. A cell is the basic unit of life. It is the smallest thing that can be called "alive." Most living things that you have seen are made of many cells. Depending on their size, plants and animals are made of thousands, millions, billions, or even trillions of cells. Cells are found in all parts of an animal: they make up blood, bone, skin, nerves, and muscle. Cells also are found in all parts of a plant: they make up roots, stems, leaves, and flowers.

Cells carry out many functions, or jobs. Some cells are specialized to do specific functions. Some of the functions of the specialized cells in your body are listed below.

- Skin cells: cover and protect
- Muscle cells: allow for movement
- Bone cells: support and protect
- Nerve cells: send and receive messages
- Blood cells: transport materials and fight diseases

Some living things, such as **bacteria**, are made of only one cell. Bacteria are the simplest single cells that carry out all basic life activities.

Observing Cells

Cells come in different sizes. However, most cells are so small that they are invisible to the unaided eye. They can be seen only with a **microscope**. A microscope is an instrument that scientists use to magnify small things in order to make them appear larger. Some microscopes are similar to a magnifying glass. You may have used a magnifying glass to look at tiny insects. Without the magnifying glass, an insect might look like just a black dot. With the magnifying glass, you can see the insect's tiny structures, such as legs. The same thing happens when scientists use a microscope. Tiny structures that were not visible before can be seen through the microscope.

2   Chapter 1   The Basic Unit of Life

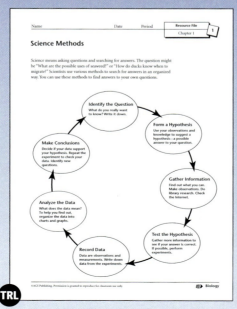

Resource File 1

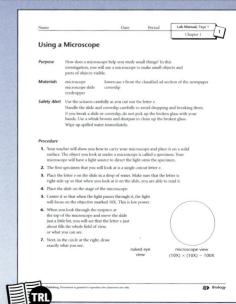

Lab Manual 1, pages 1–2

Organelle
A tiny structure inside a cell

Electron microscope
An instrument that uses a beam of tiny particles called electrons to magnify things

Tissue
Groups of cells that are similar and work together

Did You Know?
There are about 200 different types of cells in the human body.

When you look at cells through a microscope, the different shapes of the cells become visible. Cells may also be long, short, wide, or narrow.

Magnifying glasses and most microscopes use beams of light to magnify objects. When a cell is viewed under a light microscope, some of the tiny structures inside the cell can be seen. These tiny structures are called **organelles**. They perform specific functions of the cell. However, microscopes can magnify an object only so much. Then the image of the object gets blurry.

Scientists use special microscopes, called **electron microscopes**, to magnify objects even more. Electron microscopes use beams of tiny particles instead of light. When a cell is viewed under an electron microscope, its organelles can be seen very clearly.

Tissues

Groups of cells that are similar and act together to do a certain job are called **tissues**. For example, muscle cells in animals are joined together to make muscle tissues. These tissues include leg muscles, arm muscles, stomach muscles, and heart muscles. The cells in muscle tissues work together to make the body move. Other examples of tissues in animals are nerve tissue, bone tissue, and skin tissue.

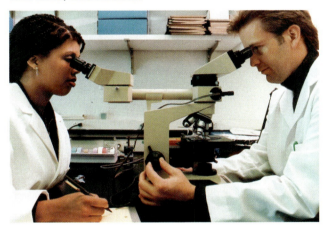
These technicians use a microscope to enlarge the cells they are studying.

Cells themselves are made up of still smaller structures called organelles. Organelles are essentially tiny organs inside cells.

 ### Warm-Up Activity

Have students look at a piece of granite that you hold up. You can do this demonstration with a brick or piece of concrete instead of granite. Looked at from far away, it looks like one solid piece. Now have students examine the rock closely, perhaps with a hand lens. They can now see that it is made up of tiny crystals. Not all of the crystals are alike. They are different colors, shapes, and sizes. All of the crystals together make up the granite. Now ask students to think about themselves. What parts come together to make up a person? (*A person is made of arms, legs, head, torso, and so on.*) What are those parts made of? (*Students may answer skin, muscles, bone, and other organs.*) Explain that organs in turn are made up of still smaller parts (tissues and cells).

Teaching the Lesson

As students read the lesson, have them find two topic sentences for the material under each subhead ("Observing Cells," "Tissues," and "Organs"). The topic sentences should represent main ideas. Alternatively, students can create their own topic sentences from the ideas presented. Ask students to read their sentences aloud.

Compare the specialization of cells to the "specialization" of rooms in a home. Different rooms have different functions and are equipped to perform those functions. For example, a kitchen may have appliances, a sink, and an exhaust fan to allow for cooking and cleaning of dishes.

Did You Know?
Explain that while the human body has 200 different types of cells, it has over 10 trillion total cells. Many cells are shaped according to the job they do. For example, muscle cells are long and thin while nerve cells have spreading branches in order to send and receive messages.

3 Reinforce and Extend

TEACHER ALERT

Students may think that all cells are so tiny that they cannot be seen without a microscope. There are many examples of cells that are large enough to be seen with the naked eye. These include fern spores, which are single cells about 100 micrometers across, and some algae cells, which are many centimeters long.

LEARNING STYLES

Visual/Spatial

Have students look at a piece of onion. Then use forceps to peel off a transparent piece of skin and mount it on a microscope slide. Have students examine it closely with the naked eye, then with a hand lens, and then under a microscope. You may wish to stain the tissue with a drop of iodine solution. Use caution when working with iodine.

LEARNING STYLES

LEP/ESL

To give students a better idea of what magnification means and how small most cells really are, have them draw a circle or other object at different magnifications (1x, 2x, 5x, 10x, 20x, etc.).

CROSS-CURRICULAR CONNECTION

Language Arts

There are many meanings of the word *cell*. Have students look up the definitions in a dictionary and discuss how the different definitions are similar to and different from the biological definition used in this chapter.

Organ
A group of different tissues that work together

Plants also have different kinds of cells, such as root cells, stem cells, and leaf cells. Similar cells are organized together into tissues. Different tissues carry out different functions necessary for plant growth. These functions include covering the plant and moving water and other substances in the plant.

Organs

Different kinds of tissues join together to form an **organ**. Organs are the main working parts of animals and plants. Organs do special jobs. Your heart is an organ. It pumps blood through your body. Your lungs are organs. They allow you to breathe. Other organs in your body include your stomach, liver, kidneys, and eyes.

You can see in the picture that the main organs of plants are roots, leaves, and stems. Roots take in water from the soil. Leaves make food for the plant. Stems support the plant and carry water and food to different parts of the plant.

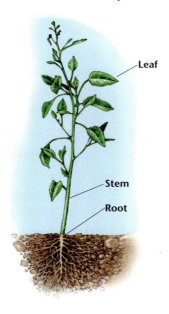

SCIENCE INTEGRATION

Technology

Explain that special kinds of electron microscopes give researchers even more power to study cells. Transmission electron microscopes enable researchers to see even smaller specimens than regular electron microscopes. Scanning electron microscopes project pictures of specimens on a screen and give a good three-dimensional view of a specimen. Have students research one of these microscopes and make a diagram of how it works.

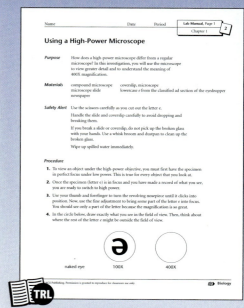

Lab Manual 2, pages 1–2

Lesson 1 REVIEW

Write your answers to these questions on a separate sheet of paper. Write complete sentences.

1. What are cells?
2. What are three functions of cells?
3. What are tissues made of?
4. What are organs made of?
5. Name four organs.

Science at Work

Microbiologist

A microbiologist needs good observation and communication skills. Microbiologists use microscopes and other equipment such as computers. They need a minimum of a two-year technical training degree. Most have at least a four-year college degree, and others have a master's degree or Ph.D.

A microbiologist is a scientist who studies living things that are too small to be seen without a microscope. Because of the variety of cells and tiny living things, microbiologists often specialize in a certain area. For example, some become cell biologists and study how cells function. Some study bacteria or other living things. Microbiologists often work as a part of a team. They might work in a laboratory at a hospital or in other industries. They might travel and study living things that cause disease. Microbiologists sometimes discover living things that have not been seen before.

The Basic Unit of Life Chapter 1 5

Lesson 1 Review Answers

1. Cells are the basic units of life.
2. Answers include covering and protection, movement, support, sending and receiving messages, transporting materials, and fighting diseases.
3. Tissues are made of groups of cells that are similar and act together to do a certain job.
4. Organs are made of different kinds of tissues.
5. Answers include heart, lungs, stomach, liver, kidneys, eyes, roots, stems, and leaves.

Science at Work

Have volunteers read the feature aloud. Point out that microbiologists work in many different fields. For example, medical microbiologists study microorganisms that cause disease; agricultural microbiologists study microorganisms that affect the health of soil and plants. Microbiologists in food industries work to make new foods and to preserve foods. Have students discuss some recent events in which microbiologists have probably played a part, such as the study of new diseases that have struck people and plants.

Portfolio Assessment

Sample items include:
- Topic sentences from Teaching the Lesson
- Lesson 1 Review answers

ONLINE CONNECTION

Students can find additional facts, activities, and review on cells at www.biology.arizona. edu/cell_bio/activities/cell_cycle/.

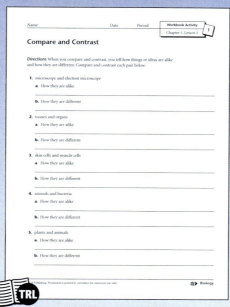

Workbook Activity 1

The Basic Unit of Life Chapter 1 5

Lesson at a Glance

Chapter 1 Lesson 2

Overview In this lesson, students learn the similarities and differences between plant cells and animal cells.

Objectives
- To identify ways that plant and animal cells are alike and different
- To describe the functions of structures in cells

Student Pages 6–11

Teacher's Resource Library
Workbook Activity 2
Alternative Workbook Activity 2
Lab Manual 3
Resource File 2

Vocabulary

cell membrane	Golgi body
cytoplasm	cell wall
nucleus	chloroplast
DNA	lysosome
ribosome	atom
mitochondrion	molecule
vacuole	ATP
endoplasmic reticulum	RNA

Science Background
Plant and Animal Cells

Because of the great variety in cells' sizes, shapes, and functions, there is really no typical cell. Still, it is useful for students to learn the things that all cells have in common.

Cells of multicellular plants are usually more stiff and angular than animal cells. The cells of humans and other multicellular animals vary greatly in shape but are usually more rounded and flexible.

Plants' cell walls contain cellulose. This makes a plant's stem stand up straight. A large part of many plants consists of vacuoles, which are cavities filled with fluid. Both plant and animal cells contain DNA, which carries the genetic program of the organism. DNA looks much the same in both plant and animal cells, but an organism's DNA is responsible for making it different from all other things.

6 Chapter 1 The Basic Unit of Life

Lesson 2 Comparing Plant and Animal Cells

Objectives

After reading this lesson, you should be able to

- identify ways that plant and animal cells are alike and different.
- describe the functions of structures in cells.

Cell membrane
A thin layer that surrounds and holds a cell together

Cytoplasm
A gel-like substance containing chemicals needed by the cell

Nucleus
Information and control center of the cell

DNA
The chemical inside cells that stores information about an organism

Ribosome
A protein builder of the cell

Mitochondrion
An organelle that uses oxygen to break down food and release energy in chemical bonds (plural is mitochondria)

You learned in Lesson 1 that plants have different kinds of cells than animals have. For example, plants have stem cells and you have muscle cells. If you look at a single plant cell, you can see that some parts are the same as the parts of an animal cell. You can also see that a plant cell has certain parts that make it different from an animal cell.

Plant and Animal Cells

All living things are made of cells. Some living things, such as plants and animals, are made up of many cells. The cells of plants and animals have many similarities and differences. Some of these similarities and differences are listed below.

Similarities

- Both animal and plant cells have **cell membranes** that enclose the cell.
- Both animal and plant cells are filled with **cytoplasm**, a gel-like substance containing chemicals needed by the cell.
- Both animal and plant cells have a **nucleus**, where **DNA** is stored. DNA controls many of the characteristics of living things. Inside the nucleus is the nucleolus.
- Both animal and plant cells have **ribosomes**, protein builders of the cell.
- Both animal and plant cells have **mitochondria** that use oxygen to break down food and release energy.
- Both kinds of cells have **vacuoles** that contain food, water, or waste products. Animal cells usually have many more vacuoles than plant cells do.

6   Chapter 1   The Basic Unit of Life

Vacuole
Stores substances such as food, water, and waste products

Endoplasmic reticulum
A system of tubes that process and transport proteins within the cell

Golgi body
Packages and distributes proteins outside the cell

Cell wall
The outer part of a plant cell that provides structure to the cell

Chloroplast
Captures the light energy from the sun to make food

Lysosome
Breaks down substances

- The cells of both plants and animals have **endoplasmic reticulum**, where a system of tubes transports proteins.
- The **Golgi bodies** in both plant and animal cells package and distribute proteins outside the cell.

Differences

- Plant cells have **cell walls** that provide structure, but animal cells do not.
- A few large animal cells have more than one nucleus, but plant cells always have just one.
- Plant cells have **chloroplasts** for photosynthesis. Animal cells do not.
- Animal cells use mitochondria for energy production. Plants primarily use chloroplasts to produce energy.
- Animal cells tend to have many small vacuoles. Mature plant cells may have only one large vacuole.
- Animals cells have **lysosomes**, but plant cells do not.

Examine the animal cell and the plant cell in the diagram below. Identify the parts you just read about and review the function of each part.

Animal and Plant Cells

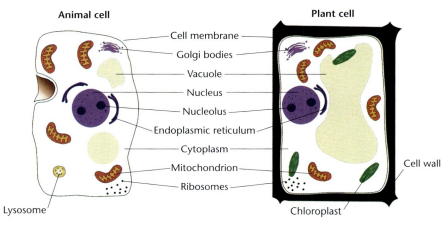

Recent DNA research has made progress in genetic testing, genetic engineering, and identification possible.

 Warm-Up Activity

Hold up a living plant and an inanimate object such as a stone. Ask students how the objects differ. (The plant is alive and the stone is not.) Then hold up a piece of wood. Ask students whether it is more like the plant or the stone. Encourage students to discuss the fact that, like the plant, the wood was once part of a living thing. Therefore, it had cells. Finally, ask how the plant is similar to their own body. Once again, students should discuss the fact that both are living and both are made of cells.

 Teaching the Lesson

Have students make headings for a 3-column chart. The headings should read "Animals," "Plants," and "Animals and Plants." Tell students that as they read the lesson they should write each cell part under the correct heading.

Copy the diagram (without labels) from page 7 onto the board. As you discuss each cell part, have a student come to the board and label the part on the diagram.

 Reinforce and Extend

 TEACHER ALERT

Make sure students realize that cells are three-dimensional. They may come to think of them as flat when they view them in diagrams and photographs and under microscopes. Also, tell students that when they view cells under a microscope, they see the colors that the cells have been dyed for easy viewing, not the true colors of the cells.

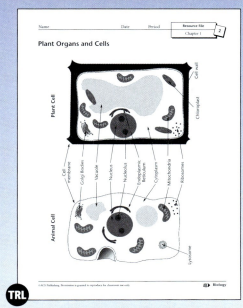

Resource File 2

LEARNING STYLES

Auditory/Verbal

Have pairs of students write on small slips of paper the 12 terms that describe plant and animal cells listed on pages 6–7. Then have them take turns choosing a term at random and defining it orally. Tell the pairs to continue until each student has defined each term correctly.

CROSS-CURRICULAR CONNECTION

Art

Have students look up pictures of various kinds of cells. Then have them draw and color one or more cells. Tell students to label the kind of cell and its parts, if applicable. Display the cell illustrations in the classroom.

Atom
The smallest particle of an element that still has the properties of that element

Molecule
Two or more atoms joined together

ATP
High-energy molecules that store energy in a form the cells can easily use

RNA
A molecule that works together with DNA to make proteins

When the bonds of larger molecules are broken down, smaller molecules are formed. Energy is released when the bonds are broken. These new molecules have chemical bonds that have less energy.

Cells Store and Use Energy

The energy that cells need comes originally from the sun. Chloroplasts in plant cells trap light energy from the sun and change it into chemical energy. This chemical energy is stored in chemical bonds between **atoms** of carbon, hydrogen, and oxygen. An atom is the smallest particle of an element that still has the properties of that element. Chemical bonds hold the atoms of carbon, hydrogen, and oxygen together to form **molecules** of sugar. A molecule is two or more atoms joined together. Cells can combine these molecules with other atoms to make larger molecules.

Both plant and animal cells break down these molecules by breaking their chemical bonds. Remember that energy is stored in these bonds. Energy is released when the bonds are broken. Cells can either use the energy or store it.

The mitochondria in plant and animal cells use oxygen to release the energy in chemical bonds. Cells store the chemical energy from food in high-energy **ATP** molecules. ATP stores energy in a form the cells can easily use. When the cell needs energy, the ATP is broken down to release energy.

Cells Store and Use Information

The nucleus of plant and animal cells is the control center of the cell. The nucleus contains DNA, a molecule that has instructions for all of a cell's activities. One of these activities is putting together protein molecules. Cells require thousands of proteins in order to work well. DNA and **RNA** molecules work together in the cell to make the proteins.

DNA in a cell's nucleus determines what kind of cell it is. When a cell divides and becomes two cells, the information needs to be passed on. To make certain that each cell has the information, the DNA doubles when a cell divides. Each of the two new cells contains the needed information to carry out all the cell's activities.

Lesson 2 REVIEW

Write your answers to these questions on a separate sheet of paper. Write complete sentences.

1. What are five things that plant cells and animal cells have in common?
2. What are three ways that a plant cell is different from an animal cell?
3. What is the function of mitochondria?
4. What is the function of a cell's nucleus?
5. What is the function of chloroplasts?

Science Myth

Animals grow as their cells get larger.

Fact: Most cells are about the same size. New cells that form when cells divide are about the same size as the original cells. Animals grow as their cells increase in number, not in size.

The Basic Unit of Life Chapter 1

Lesson 2 Review Answers

1. Both have cell membranes, cytoplasm, a nucleus containing DNA, ribosomes, mitochondria, vacuoles, endoplasmic reticulum, and Golgi bodies. 2. Plant cells have cell walls, and animal cells do not; all plant cells have only one nucleus; plant cells have chloroplasts, and animal cells do not; plant cells use chloroplasts to produce energy while animal cells use mitochondria; plant cells may have only one large vacuole while animal cells tend to have many small ones; plant cells do not have lysosomes as animal cells do. 3. Mitochondria use oxygen to break down food and release energy. 4. A cell's nucleus stores DNA, which controls many of the characteristics of living things. 5. Chloroplasts capture light from the sun to make food.

Portfolio Assessment

Sample items include:
- Chart from Teaching the Lesson
- Lesson 2 Review answers

Science Myth

Have one student read the myth portion and one the fact portion of the feature. Explain that if individual cells grew, they soon could no longer carry out the necessary cell functions. The larger cell would need more nourishment than could pass through the cell membrane in a timely manner.

Workbook Activity 2

The Basic Unit of Life Chapter 1

Investigation 1

This investigation will take approximately 45 minutes to complete. Students will use these process and thinking skills: observing, recording, comparing, making generalizations.

Preparation

- Make sure all of the microscopes are working properly and have working bulbs if the microscopes require them.
- Organize slides into animal, plant, and bacterial groups so that students can easily choose one in each category.
- Students may use Lab Manual 3 to record their data and answer the questions.

Procedure

- This investigation is best done with students working in pairs. Students should take turns viewing the slide under each level of magnification.
- Review with students the correct way to carry, use, and store the microscope.
- Be sure to check each group and make sure students know how the image looks in focus.
- Focus a slide for each group. Then move the slide slightly out of focus. Have each student look through the microscope and turn the knob slowly, bringing the slide into focus, through focus, and then back into focus again.
- Have students who successfully focus a slide help those who are having trouble.
- If your microscope has just one eyepiece, students can experiment with keeping the other eye open, closing it, or covering it with their hand to see how they can view the specimen better.
- Show students how the objective with the smallest magnification is useful for locating the specimen, and then the magnification can be increased for better viewing.
- Have students look at a single slide with different size objectives and compare what they can see.

INVESTIGATION 1

Comparing Cells

Purpose

Are there any differences among different types of cells? In this investigation, you will observe differences and similarities among different types of cells.

Materials
- prepared slides of animal, plant, and bacterial cells
- light microscope

Procedure

1. Choose one of the prepared slides. On a sheet of paper, record the type of cells you selected. Look at the slide without using the microscope. What can you see? **Safety Alert: Handle glass microscope slides with care. Dispose of broken glass properly.**

2. Place the slide on the stage of the microscope. Refer to the instructions for the microscope you are using. Focus and adjust the microscope so that you can see the cells on the slide clearly. Look at the cells under different levels of magnification. How does what you see now differ from what you saw without the microscope?

3. Observe one of the cells on the slide. What is its shape? Do you see any organelles inside the cell? If so, what do they look like? Make a drawing of the cell on your paper.

4. Repeat the procedure, observing at least one type of animal cell, one type of plant cell, and one type of bacterial cell.

SAFETY ALERT

- Be sure students handle glass slides carefully so that they do not get cut or break the slides.
- Have one student watch while the other is focusing the microscope to make sure the objective lens does not get forced into the slide accidentally.

Questions and Conclusions

1. What were some similarities between the plant cells and the animal cells? What were some differences?
2. How did the plant cells and animal cells differ from the bacterial cells?

Explore Further

Observe prepared slides of other cells or find out how to make your own slides of cells. Then draw the individual cells you observe and compare them.

Results

Students should observe that the sizes and shapes of the plant, animal, and bacterial cells differ.

Questions and Conclusions Answers

1. Similarities include size and shape and presence of organelles. Differences include shape and the presence of a cell wall, large vacuole, and green chloroplasts in plant cells but not in animal cells.
2. Differences include relative size, shape, and the absence of organelles in bacterial cells.

Explore Further Answers

Point out that if students observe an onion cell, they will not see chloroplasts. This is because the onion is part of the root of the plant and is therefore underground.

Assessment

Check students' drawings to make sure they include organelles that they can see. You might include the following items from this investigation in student portfolios:

- Drawings of cells
- Answers to Questions and Conclusions and Explore Further

Lab Manual 3

Lesson at a Glance

Chapter 1 Lesson 3

Overview In this lesson, students learn why chemicals are important for life and what some of the important chemicals are.

Objectives

- To explain why water is important to life
- To describe how living things use carbohydrates, fats, and proteins
- To discuss the importance of eating a variety of foods

Student Pages 12–16

Teacher's Resource Library TRL

Workbook Activity 3
Alternative Workbook Activity 3

Vocabulary

solution	amino acid
carbohydrate	mineral
fat	vitamin
protein	nutrient

Science Background
Life Chemicals

All matter is made up of combinations of chemicals. Some chemicals are important for life. Without them, life would not exist.

Water is one of the most important life chemicals. Most of the chemistry of life takes place in water solutions. Other important life chemicals are proteins, carbohydrates, fats, vitamins, and minerals. These chemicals are the nutrients our bodies need.

Plants make their food using the energy from the Sun. They get the chemicals to make their food from gases in the air and by absorbing chemicals out of the ground with their roots. Animals get these chemicals by eating food and drinking water.

Lesson 3 — Chemicals That Are Important for Life

Objectives

After reading this lesson, you should be able to
- explain why water is important to life.
- describe how living things use carbohydrates, fats, and proteins.
- discuss the importance of eating a variety of foods.

Solution
A mixture in which the particles are evenly mixed

Did You Know?
Humans can survive up to several weeks without food. However, they can survive only a few days without water.

Besides being composed of cells, living things are alike because they have similar chemicals. Living things use these chemicals to stay alive. Water is one of the chemicals all living things use.

Importance of Water

Life cannot exist without water. It is the most plentiful chemical in living things. Water is found in each of the approximately 100 trillion cells in the adult human body. It makes up about two-thirds of the weight of the cell.

Water is a useful chemical. Have you ever put sugar in a cup of tea? When you put sugar in tea, you stir the liquid until the sugar dissolves in the liquid. As the sugar dissolves, it breaks apart into tiny pieces that you can no longer see. The water in the tea is the chemical that does the dissolving. Special properties of water allow it to break things apart into tiny particles. When the water, sugar, and tea particles become equally mixed, they form a **solution**.

The ability to dissolve other chemicals is one of the most important properties of water for life. Cells are so small that the materials that go in and out of them must be very tiny. When a material dissolves into tiny pieces, it can move more easily from cell to cell.

Have you ever accidentally bitten your tongue so hard that it bled? Blood tastes salty. Your body fluids and the liquid in your cells are not pure water. They are a solution of many things, including salts. One example of a salt is sodium chloride. Another name for sodium chloride is table salt. The liquid found in living things is a solution of salts, water, and other chemicals.

Other important chemicals found in living things are carbohydrates, fats, and proteins. Each of these common chemicals has a job to do in the body of a living thing.

12 Chapter 1 The Basic Unit of Life

Did You Know?

Have a student read the Did You Know? feature on page 12 aloud. Explain that a person needs about 2.5 liters of water daily. This is about the same amount that is lost through respiration, perspiration, and excretion. People need to drink about 2 liters of liquids a day and maintain a healthy diet in order to remain properly hydrated.

Carbohydrate
A sugar or starch, which living things use for energy

Fat
A chemical that stores large amounts of energy

Carbohydrates are divided into two groups: simple and complex. Simple carbohydrates include several types of sugars. Complex carbohydrates are also made up of sugars but include fiber and starches.

Carbohydrates

Carbohydrates are sugars and starches. Sugar is a carbohydrate that is used to sweeten drinks and many foods. Fruits and vegetables, such as oranges and potatoes, contain sugar too. Starches are found in foods such as bread, cereal, pasta, rice, and potatoes. Plants use the energy from sunlight to make carbohydrates from carbon dioxide and water. Carbon dioxide is a gas found in air. Animals get energy from the carbohydrates that plants make.

Energy is needed to carry on various life activities. Energy comes from fuel. You can think of carbohydrates as fuel chemicals. Carbohydrates in your body work like gasoline in a car. Gasoline from the fuel tank gets to the engine, where it is broken down and energy is released. This released energy runs the engine. When carbohydrates are broken down in your body, energy is released. This energy powers your body. The same thing happens in other animals. Plants and other living things use carbohydrates for energy too.

Fats

Fats also can be thought of as fuel chemicals. Fats store large amounts of energy that are released when they are broken down. Of all the chemicals important for life, fats contain the most energy. They are found in foods such as meat, butter, cheese, and peanut butter. Fats are related to oils. Fats are solid at room temperature. Oils, such as corn oil used for frying foods, are liquid at room temperature.

Foods contain water, carbohydrates, fats, and other chemicals important for life.

1 Warm-Up Activity

Show students a clear jar and a pitcher of water. Tell them that the jar represents their body. Invite students to pour water into the jar to a certain level to show what percentage of the human body they think is made of water. Let several students pour the water. Make a drawing on the board of each student's guess. Take a class vote on which drawing students think is closest to the amount of water in the human body. Inform students that the body is about 70 percent water. Let a student show this amount with the water in the jar.

2 Teaching the Lesson

Have students write down the main types of chemicals covered in this lesson: water, carbohydrates, fats, proteins, minerals, and vitamins. As students read each section, have them write the main purpose of each type of chemical.

Have students think again about how their body is made up of parts that are made up of cells and organelles. The next level is chemicals, which make up the structures and fluids of cells.

3 Reinforce and Extend

TEACHER ALERT

Students may think of all fats as nutritionally bad. In fact, we must consume a certain amount of fat to be healthy. You may have students research some of the reasons our diets should include at least some fat.

TEACHER ALERT

Students may think the more vitamins they consume, the better. In fact, too much of some vitamins can be harmful. Vitamin supplements should be taken on the advice of a health-care professional.

LEARNING STYLES

Logical/Mathematical
Have students make a chart that compares and contrasts water, carbohydrates, fats, proteins, minerals, and vitamins. The charts should show how each nutrient is needed for life and the sources of each nutrient for people.

LEARNING STYLES

Body/Kinesthetic
Provide pairs of students with two cups of water, a spoon, and small amounts of sugar and salt. Have the students make sugar and salt solutions. Remind them that the ability of water to dissolve substances such as these enables nutrients to move from cell to cell in our bodies.

SCIENCE INTEGRATION

Physical Science
Tell students that the study of chemicals is called chemistry, and that the study of chemicals important to life is biochemistry. Have students check newspapers, magazines, or the Internet for articles on current biochemical research. Invite them to bring the articles to class for discussion.

Protein
A chemical used by living things to build and repair body parts and regulate body activities

Amino acid
Molecules that make up proteins

Proteins that have incorrect structures are found in the brains of people who have Alzheimer's and mad cow disease.

Proteins

Proteins are another kind of chemical important for life. Meats, such as beef, chicken, and fish, contain large amounts of proteins. Beans, nuts, eggs, and cheese also contain large amounts of proteins.

Like carbohydrates and fats, proteins provide energy for living things. But they have other important functions too. Proteins help to repair damaged cells and build new ones. Hair, muscles, and skin are made mostly of proteins. Proteins also help control body activities such as heart rate and the breaking down of food in the body.

Proteins are basic parts of living cells. The cell assembles molecules—mainly proteins—to carry out cellular functions. Proteins are made up of long chains of smaller molecules called **amino acids**. There are 20 different amino acids that can be arranged in different ways to make different proteins. This means that there can be many different kinds of proteins. For this reason, proteins have a wide range of jobs in the body. These include digesting food, fighting infections, controlling body chemistry, and keeping the body working smoothly.

Protein chains fold into a particular shape to carry out a particular function. Proteins with different shapes carry out different functions. If a protein twists into the wrong shape or has a missing part, it may not be able to do its job.

Technology Note

Scientists use computers to help them predict the structure of a protein. Computer models identify a protein's structure or shape based on its sequence of amino acids. A computer image shows the physical and chemical properties of the protein. It also provides clues about its role in the body. This information helps scientists understand a protein's role in health and disease. Scientists can then work to develop ways to treat disease.

14 Chapter 1 The Basic Unit of Life

CROSS-CURRICULAR CONNECTION

Physical Education
Point out that athletes have somewhat different nutritional needs than others. For example, training athletes often eat large amounts of lean protein. Some athletes, such as runners or swimmers, may eat carbohydrates before an event for extra energy. Have students find out more about the diets of athletes in a particular sport or event. Have them write an outline or paragraph describing their findings.

SCIENCE JOURNAL

Have students write a paragraph explaining why it is a good idea to eat a variety of foods every day. Encourage them to use examples from the lesson. Students should infer that a variety of foods provides the variety of nutrients that the body needs.

Mineral
A chemical found in foods that is needed by living things in small amounts

Vitamin
A chemical found in foods that is needed by living things in small amounts

Nutrient
Any chemical found in foods that is needed by living things

The DNA in a cell nucleus contains the code that's needed to produce a protein. When a cell receives a signal that a certain protein is needed, the DNA inside the nucleus reproduces the code. The code is then carried by RNA out into the cell. There, ribosomes read the RNA. The ribosomes join together the long chains of amino acids using the RNA code. A change in even one atom in the DNA molecule can change the protein that is produced.

Importance of Nutrients

Keeping your body working properly is not a simple job. You must get a regular supply of carbohydrates, proteins, and fats from the foods you eat. Each kind of food provides different chemicals your body needs. Therefore, it is important to eat a variety of foods every day.

In addition to water, carbohydrates, proteins, and fats, your body also needs **minerals** and **vitamins**. Your body needs these chemicals in small amounts only. Different foods contain different minerals, such as zinc, and vitamins, such as vitamin B_{12}. The chemicals that are needed for life and that come from foods are called **nutrients**. To be healthy, living things need to take in the right amounts of nutrients every day.

LEARNING STYLES

Interpersonal/Group Learning

Have students work in 4 or 6 groups. Have half the groups research the vitamins the human body needs, the foods each is found in, and why each is useful to the body. Have students put their information on a chart. Have the other groups research minerals and make a chart of their findings.

CAREER CONNECTION

The manufacture, development, and distribution of foods offer many career opportunities from food safety, to preparation, development, and delivery of food. For example, scientists have developed enriched food products that provide additional nutritional value and home economists have developed new food products in test kitchens. Have students contact a food manufacturer, a publisher of cookbooks or food magazines, or the restaurant and foodservice industry to find out about the kinds of career opportunities available and the educational requirements for each. Suggest that they compile the information for use in a classroom career center.

AT HOME

Have students keep a record of each food they eat for one day. Then have them make a chart and classify each food as a carbohydrate, fat, or protein. Suggest that they find a food pyramid or nutritional chart and discuss with family members whether the proportion of each kind of food is consistent with a healthy diet.

Lesson 3 Review Answers

1. One of the most important properties of water is its ability to dissolve other chemicals. **2.** The body uses carbohydrates and fats for energy. **3.** Proteins provide energy, help build body parts, and control body activities, such as heart rate and the breaking down of food in the body. **4.** Vitamins and minerals are nutrients that the body needs in small amounts only. **5.** Eating a variety of foods every day can provide you with all the nutrients you need.

Portfolio Assessment

Sample items include:
- Chemicals and their purposes from Teaching the Lesson
- Lesson 3 Review answers

Achievements in Science

Have students read aloud the Achievements in Science feature on page 16. Explain that Robert Hooke first observed and drew cells in 1665, but it was not until the early 1800s that scientists generally recognized the cell as the basic unit of life. Emphasize the importance of the microscope and its development to the understanding of the cell. Help students understand that achievements in one area of science often lead to achievements in others.

Lesson 3 REVIEW

Write your answers to these questions on a separate sheet of paper. Write complete sentences.

1. What is one of the most important properties of water for life?
2. How does your body use carbohydrates and fats?
3. What do proteins do in your body?
4. What are vitamins and minerals?
5. How can you get all the nutrients you need?

Achievements in Science

Plant Cells Observed

Before the microscope was invented, scientists could observe only what they could see with the unaided eye. Using a microscope, the world of tiny living organisms was revealed. Imagine how exciting it would be to see these organisms for the first time!

It is believed that the earliest microscope used drops of water to make things look larger. A later microscope was a tube that had a place for the object to be observed on one end. On the other end there was a lens that made something look about 10 times its actual size.

One English scientist, Robert Hooke, used a microscope to observe a slice of a cork. He saw that the cork was made up of tiny units that he called "cells." He thought that these cells existed only in plants. He believed that they were just containers, not the basic unit of life that we now know they are.

Workbook Activity 3

Lesson 4: Basic Life Activities

Objectives

After reading this lesson, you should be able to

- identify some basic life activities.
- compare how plants and animals get food, move, and respond.
- explain the difference between growth and development.

Most living things carry on the same kinds of activities. These activities allow living things to stay alive. Some examples of basic life activities are described below.

Getting Food

A familiar example of a life activity is getting food. Animals get food by eating plants or other animals. Plants make their own food. They use the energy from sunlight to make carbohydrates from carbon dioxide and water.

Using Food and Removing Wastes

Digestion is a life activity that breaks down food into chemicals that cells can use. **Respiration** is another basic life activity. During respiration, cells release the energy that is stored in the chemicals. Oxygen is used to release the stored energy. Cells use the energy to do work. Respiration also produces wastes. **Excretion** is the process that removes wastes from living things.

Movement

Movement is another activity that is common to living things. Plants do not move from one place to another, but they still move. Plants have roots that hold them in place, but their parts bend and move. For example, leaves may move to face sunlight. Animal movement is easier to see. Most animals move freely from place to place.

Digestion
The process by which living things break down food

Respiration
The process by which living things release energy from food

Excretion
The process by which living things get rid of wastes

The movements of most animals are obvious.

The Basic Unit of Life   Chapter 1   17

For each basic life activity, discuss why that activity by itself cannot be used to determine if something is alive. Have students give examples. (*Some nonliving things carry out one of the activities but not the others. For example, cars move and crystals grow.*)

3 Reinforce and Extend

IN THE COMMUNITY

Have students find out if any food is produced, manufactured, or packaged in their community. Have them discuss the information in class.

GLOBAL CONNECTION

Point out that many common foods and dishes in this country have their origins elsewhere. Have students research the origin of some of their favorite foods. Encourage them to report to the class and locate the country of origin on a map.

Homeostasis
The ability of organisms to maintain their internal conditions

Development
The changes that occur as a living thing grows

Besides outward movement, there is constant movement inside living things. The material inside plants and animals is always changing. Liquids are flowing, food is being digested, and materials are moving into and out of cells.

Sensing and Responding

Living things sense and respond. Animals and plants have tissues and organs that pick up, or sense, signals from their surroundings. These signals include light, sound, chemicals, and touch. Plants and animals change something, or respond, based on the kinds of signals they pick up. For example, some moths fly around lights at night. Fish swim to the top of a tank for food. Dogs respond to the sound of a human voice. Many flowers open in the morning light and close with night's darkness.

Homeostasis

Organisms have the ability to maintain their internal conditions. This ability is called **homeostasis**. An example of this is your body's ability to keep your temperature within a normal range.

Growth

Growing is part of being alive. You were once a baby, but you have grown into a larger person. You are still growing. You will continue to grow until you reach your adult size. Most living things go through a similar pattern of growth.

Development

Many living things develop as they grow. **Development** means becoming different, or changing, over time. Tadpoles hatch from eggs and develop by stages into frogs or toads. Notice in the photos on the next page that tadpoles look more like fish than like frogs. Unlike frogs, tadpoles have a tail and no legs. Tadpoles also have no mouth when they first hatch. As a tadpole develops, a mouth forms and changes in shape. The legs form, and the tail is absorbed into the body.

Reproduction
The process by which living things produce offspring

Reproduction

Living things produce offspring, or children, through the basic life activity of **reproduction**. Some living things reproduce by themselves. For example, bacteria reproduce by dividing in two. For other living things, such as humans, reproduction involves two parents. The offspring of all living things resemble their parent or parents.

CROSS-CURRICULAR CONNECTION

Health

Have students find information about their height and weight at birth, at present, and at two or more times in between. Have them make a graph that charts their growth. Encourage students to include artwork, such as silhouettes of themselves at different ages, on the chart. Display or discuss the charts, pointing out that although rates of growth differ for each individual, the general process is the same.

SCIENCE JOURNAL

Have students write a description of the basic life activities they are carrying out even as they sit at their desks.

IN THE ENVIRONMENT

Humans have impacted plant and animal life in many ways. For example, the breeding places of some sea turtles are beaches that people use. People, beach erosion, and artificial lighting can affect the nesting of females and survival rate of hatchlings. Have students research information about sea turtles and their nesting environments. They can find information in reference books and on the Internet. Ask students to find out what people can do to help sea turtles and to protect nesting environments. You might suggest that they go to www.co.broward.fl.us/agriculture/english/wildlife/ for recommendations about how to reduce lighting during nesting season in Florida.

Lesson 4 Review Answers

1. The basic life activities include getting food, using food, removing wastes, movement, sensing and responding, growth, development, and reproduction.
2. Animals must eat plants or other animals to get food. Plants use the energy from sunlight to make carbohydrates from carbon dioxide and water.
3. Animals can move from place to place. Plants cannot move from place to place but are able to move their parts, such as their stems and leaves. **4.** Growth is an increase in size. Development means changing over time. **5.** Sensing and responding means picking up signals from the surroundings and then changing because of these signals.

Portfolio Assessment

Sample items include:
- Sentences from Teaching the Lesson
- Lesson 4 Review answers

Science in Your Life

Remind students that many basic life activities cannot be observed directly. They take place over time or within an organism. For example, you usually cannot watch a plant and animal grow and develop, but you can easily see the result of the growth and development. Remind students to choose pictures that represent a basic life process. They could use a picture of a pet eating to represent digestion or a sequence of plant pictures to represent plant development.

Lesson 4   R E V I E W

Write your answers to these questions on a separate sheet of paper. Write complete sentences.

1. List three basic activities of living things.
2. How do animals and plants get food?
3. What is the difference in the way animals and plants move?
4. Contrast growth and development.
5. What does "sensing and responding" mean?

Science in Your Life

Can you identify life activities?

Living things around you carry out basic life activities all the time. Animals move around much of the time. A flowering plant being pollinated by a bee is involved in reproduction. Even when living things seem to be just sitting there not doing anything, they are carrying out basic life activities. A plant is constantly making food with energy from the sun by using carbon dioxide and water. The cells of a dog resting in the shade are carrying out respiration. If the dog just finished eating, it is also carrying out the basic life activity of digestion.

Can you recognize basic life activities? Take pictures of living things around you or look for photos in magazines. Nature magazines may be easiest to use. Try to find at least one example of each basic life activity. Some examples like those above may not be seen directly in the photos. Cut out the photos and arrange them as a collage on a large sheet of paper. Then number the images. On a separate sheet of paper, list the basic life activities that you can identify in each photo. Most photos will show more than one basic life activity.

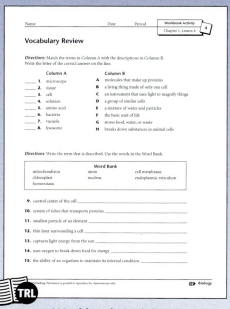

Workbook Activity 4

Chapter 1 SUMMARY

- Living things are made of cells. Cells carry out many different functions.
- Cells come in different shapes and sizes. Most cells are so small they can be seen only with a microscope.
- Cells are organized into tissues, which are organized into organs.
- Plant cells and animal cells have many of the same parts. Plant cells have cell walls and chloroplasts. Animal cells do not.
- Each organelle structure has a different function in the cell.
- An important property of water is its ability to dissolve things.
- Cells use carbohydrates, fats, and proteins for energy. Sugars and starches are carbohydrates.
- Plants use energy from sunlight to make carbohydrates from carbon dioxide and water.
- Fats store large amounts of energy. Fats are solid at room temperature. Oils are liquid at room temperature.
- Proteins provide energy, help to build and repair body parts, and control body activities.
- Minerals and vitamins are chemicals in foods. Your body needs them only in small amounts. Water, carbohydrates, proteins, fats, minerals, and vitamins are all called nutrients.
- All living things have basic life activities. These include getting and using food, removing wastes, moving, sensing and responding, growing, developing, and reproducing.

Science Words

amino acid, 14	development, 18	homeostasis, 18	protein, 14
atom, 8	digestion, 17	lysosome, 7	reproduction, 19
ATP, 8	DNA, 6	microscope, 2	respiration, 17
bacteria, 2	electron microscope, 3	mineral, 15	ribosome, 6
carbohydrate, 13	endoplasmic reticulum, 7	mitochondrion, 6	RNA, 8
cell, 2	excretion, 17	molecule, 8	solution, 12
cell membrane, 6	fat, 13	nucleus, 6	tissue, 3
cell wall, 7	Golgi body, 7	nutrient, 15	vacuole, 7
chloroplast, 7		organ, 4	vitamin, 15
cytoplasm, 6		organelle, 3	

Chapter 1 Summary

Have students read the Chapter Summary on page 21 to review the main ideas presented in Chapter 1.

Science Words

Direct students' attention to the Science Words box at the bottom of page 21. Ask them to read and review each term and its definition.

Chapter 1 Review

Use the Chapter Review to prepare students for tests and to reteach content from the chapter.

Chapter 1 Mastery Test

The Teacher's Resource Library includes two parallel forms of the Chapter 1 Mastery Test. The difficulty level of the two forms is equivalent. You may wish to use one form as a pretest and the other form as a posttest.

Review Answers

Vocabulary Review

1. carbohydrates 2. excretion 3. fats 4. proteins 5. microscopes 6. reproduction 7. respiration 8. vitamins 9. digestion 10. organs 11. development 12. organelles 13. tissues 14. nutrients 15. molecules

TEACHER ALERT

In the Chapter Review, the Vocabulary Review activity includes a sample of the chapter's vocabulary terms. The activity will help determine students' understanding of key vocabulary terms and concepts presented in the chapter. Other vocabulary terms used in the chapter are listed below:

amino acid	endoplasmic reticulum
atom	
ATP	Golgi body
bacteria	homeostasis
cell	lysosome
cell membrane	mineral
cell wall	mitochondrion
chloroplast	nucleus
cytoplasm	ribosome
DNA	RNA
electron microscope	solution
	vacuole

Chapter 1 REVIEW

Vocabulary Review

Choose the word from the Word Bank that best completes each sentence. Write the answer on a sheet of paper.

Word Bank
carbohydrates
development
digestion
excretion
fats
microscopes
molecules
nutrients
organelles
organs
proteins
reproduction
respiration
tissues
vitamins

1. Sugars and starches are _____.
2. Living things get rid of wastes by the process of _____.
3. Large amounts of energy are stored in _____.
4. _____ help control body activities.
5. Scientists observe the shapes and sizes of cells by using _____.
6. _____ produces offspring.
7. During _____, cells release energy stored in food.
8. Nutrients that living things need in small amounts are _____.
9. During _____, food is broken down into chemicals that cells can use.
10. Groups of different kinds of tissues form _____.
11. _____ is the change that occurs as a living thing grows.
12. Tiny structures found inside cells are called _____.
13. _____ are made of groups of similar cells that work together to do a certain job.
14. Water, carbohydrates, proteins, fats, vitamins, and minerals are _____.
15. Two or more atoms joined together form _____.

22 Chapter 1 The Basic Unit of Life

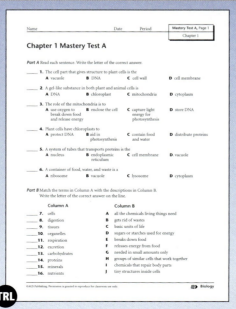
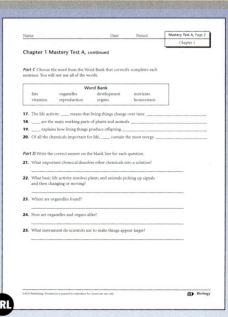

Chapter 1 Mastery Test A

Concept Review

Choose the answer that best completes each sentence. Write the letter of the answer on your paper.

16. All living things are made of _____.
    A only one cell   C many cells
    B one or more cells   D tissues

17. The most plentiful chemical in living things is _____.
    A water   C minerals
    B fat   D vitamins

18. Plants make _____ using carbon dioxide, water, and energy from the sun.
    A minerals   C carbohydrates
    B vitamins   D fats

19. Reproduction, digestion, movement, and growth are examples of _____.
    A nutrition   C development
    B homeostasis   D basic life activities

Critical Thinking

Write the answer to the following question.

20. What basic life activities are described in the following paragraph?

A kitten sees a ball of yarn and pounces on it. Just a few weeks ago, the kitten was not even able to walk. When it was born, the kitten was tiny. Its eyes were continually closed. Now, suddenly, the kitten stops and sniffs the air. Its mother has returned. The kitten walks over to its mother and begins to nurse, drinking milk.

Test-Taking Tip When answering multiple-choice questions, read the sentence completely using each choice. Then choose the choice that makes the most sense when the entire sentence is read.

Concept Review
16. B 17. A 18. C 19. D

Critical Thinking

20. Movement—pouncing; walking; crawling; drinking. Sensing and responding—sees, then pounces; sniffs, then goes to mother. Getting food—hunting; nursing. Reproduction—kitten was born. Growth—kitten was smaller when it was born. Development—learning to crawl, learning to walk; eyes open after being closed since birth.

ALTERNATIVE ASSESSMENT

Alternative Assessment items correlate to the student Goals for Learning at the beginning of this chapter.

- Give students pictures of organisms, organs, tissues, and cells. Have students group the pictures according to type.
- Give students diagrams of plant and animal cells that have the parts labeled but are not labeled "Plant" or "Animal." Have them analyze the cells and correctly title each diagram.
- Have each student bring in a nutrition label from a food package. Have students analyze the labels and decide which foods have the highest amount of protein, carbohydrates, and fat per serving.
- Display a picture of a baby. Have students describe how the baby exhibits the basic life activities.

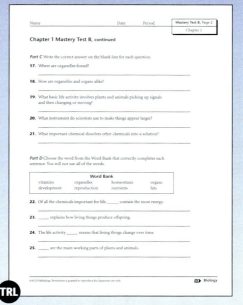

Chapter 1 Mastery Test B

Chapter 2

Planning Guide
Organizing Living Things

	Student Text Lesson		
	Student Pages	Vocabulary	Lesson Review
Lesson 1 Living and Nonliving Things	26–30	✔	✔
Lesson 2 How Organisms Are Classified	31–36	✔	✔

Chapter Activities

Student Text
Science Center

Teacher's Resource Library
Community Connection 2:
 Signs of Life: Observing Organisms

Assessment Options

Student Text
Chapter 2 Review

Teacher's Resource Library
Chapter 2 Mastery Tests A and B

24A

Student Text Features								Teaching Strategies						Learning Styles						Teacher's Resource Library				
Achievements in Science	Science at Work	Science in Your Life	Investigation	Science Myth	Note	Technology Note	Did You Know?	Science Integration	Science Journal	Cross-Curricular Connection	Online Connection	Teacher Alert	Applications (Home, Career, Community, Global, Environment)	Auditory/Verbal	Body/Kinesthetic	Interpersonal/Group Learning	Logical/Mathematical	Visual/Spatial	LEP/ESL	Workbook Activities	Alternative Workbook Activities	Lab Manual	Resource File	Self-Study Guide
28			29			✔	27	27	28	27		26	27, 28			27				5	5	4	3	✔
	35	36		34	✔		33	32	33	33, 36	34	32	32, 33, 36	34	34		34	33	36	6	6	5, 6	4	✔

Pronunciation Key

a	hat	e	let	ī	ice	ô	order	ù	put	sh she
ā	age	ē	equal	o	hot	oi	oil	ü	rule	th thin
ä	far	ėr	term	ō	open	ou	out	ch	child	ŦH then
â	care	i	it	ȯ	saw	u	cup	ng	long	zh measure

ə { a in about, e in taken, i in pencil, o in lemon, u in circus }

Alternative Workbook Activities

The Teacher's Resource Library (TRL) contains a set of lower-level worksheets called Alternative Workbook Activities. These worksheets cover the same content as the regular Workbook Activities but are written at a second-grade reading level.

Skill Track Software

Use the Skill Track Software for Biology for additional reinforcement of this chapter. The software program allows students using AGS textbooks to be assessed for mastery of each chapter and lesson of the textbook. Students access the software on an individual basis and are assessed with multiple-choice items.

Chapter at a Glance

Chapter 2: Organizing Living Things
pages 24–39

Lessons

1. **Living and Nonliving Things**
   pages 26–30

   Investigation 2 pages 29–30

2. **How Organisms Are Classified**
   pages 31–36

Chapter 2 Summary page 37

Chapter 2 Review pages 38–39

Skill Track Software for Biology

Teacher's Resource Library

   Workbook Activities 5–6

   Alternative Workbook Activities 5–6

   Lab Manual 4–6

   Community Connection 2

   Resource File 3–4

   Chapter 2 Self-Study Guide

   Chapter 2 Mastery Tests A and B

(Answer Keys for the Teacher's Resource Library begin on page 420 of this Teacher's Edition. A list of supplies required for Lab Manual Activities in this chapter begins on page 449.)

Science Center

Have students divide a bulletin board in half and label one side "Nonliving" and the other "Living." Have them put the categories "Plant," "Animal," "Protist," Fungi," and "Monerans" under "Living." While they read the chapter, ask students to bring in pictures of different organisms and place them on the bulletin board under the correct headings. Encourage variety in the types of organisms. Science magazines are a good source of pictures of protists, fungi, and monerans.

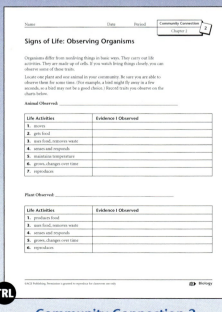

Community Connection 2

Chapter 2
Organizing Living Things

The world around you is made up of living and nonliving things. Notice the nonliving things in the photo. You can see rocks, water, and ice. Although you can't see it, air is another nonliving thing. You know that the penguins in the picture are living things. If you were there, you might see fish in the water. Many other living things are so small you would not be able to see them. In this chapter, you will learn what makes living things different from nonliving things. You also will learn how living things are divided into groups based on their similarities and differences.

Organize Your Thoughts

Living things
- Made of cells
- Carry out all the basic life activities
- Grouped into kingdoms

Plant · Animal · Protist · Fungi · Monera

Goals for Learning

◆ To identify the differences between living and nonliving things

◆ To name the five kingdoms and learn what kinds of organisms belong to each

◆ To describe the similarities and differences between living things in different kingdoms

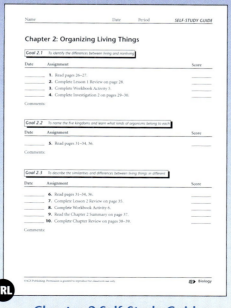

Chapter 2 Self-Study Guide

Introducing the Chapter

Direct students' attention to the photograph on page 24. Have them name the things they see in the photo on a sheet of paper and state whether each is living or nonliving. Ask them how they decided which things were living.

Together brainstorm a list of living things on the board. (Be sure some plants are listed.) Challenge students to organize the different living things on the list into categories. They should be able to organize the organisms as plants and animals. Point out that on the graphic organizer, plants and animals are two of the five groups, or kingdoms, into which scientists organize living things. In this chapter, they will learn what kinds of organisms belong to each of the five kingdoms.

Have students read the Goals for Learning on page 25 and have them rewrite each goal as a question that they should answer as they read the chapter.

Notes and Technology Notes

Ask volunteers to read the notes that appear in the margins throughout the chapter. Then discuss them with the class.

TEACHER'S RESOURCE

The AGS Teaching Strategies in Science Transparencies may be used with this chapter. The transparencies add an interactive dimension to expand and enhance the *Biology* program content.

CAREER INTEREST INVENTORY

The AGS Harrington-O'Shea Career Decision-Making System-Revised (CDM) may be used with this chapter. Students can use the CDM to explore their interests and identify careers. The CDM defines career areas that are indicated by students' responses on the inventory.

Lesson at a Glance

Chapter 2 Lesson 1

Overview In this lesson, students learn to tell the difference between living and nonliving things by observing their properties and behaviors.

Objectives

- To classify things as living or nonliving
- To explain how living things are different from nonliving things

Student Pages 26–30

Teacher's Resource Library

Workbook Activity 5
Alternative Workbook Activity 5
Lab Manual 4
Resource File 3

Vocabulary

property
organism

Warm-Up Activity

Display an assortment of living and nonliving things, such as a plant, goldfish, rock, and shoe. Ask students to describe their properties, using sentences that compare and contrast. Discuss characteristics that all of the items share and those that only the living things share.

Teaching the Lesson

Teacher Alert

Point out that nonliving matter has no cells. Cells may be found in nonliving matter only if that matter was once alive. Wood is made of cells of once-living trees. Leather parts of shoes come from cowhide, which was once part of a living animal.

Lesson 1 Living and Nonliving Things

Objectives

After reading this lesson, you should be able to

- classify things as living or nonliving.
- explain how living things are different from nonliving things.

Property
A quality that describes an object

How do you tell the difference between living and nonliving things? You observe their **properties**. A property is a quality that describes an object. A property of a rock is hardness. A property of a person is eye color. Properties can describe how an object looks or feels. Properties can also describe how an object behaves. All things have different properties.

Properties of Nonliving Things

Like living things, nonliving things have properties that you can see and feel. The color of a rock or mineral is one of its properties. The property of hardness also helps to identify a rock or mineral. For example, some rocks, such as chalk, are so soft you can write with them. Others are hard enough to cut steel. Did you know that diamond is the hardest mineral of all? It is used in cutting tools because of that property.

The properties that nonliving things do not have are as important as the properties they do have. Unlike living things, nonliving things do not carry out life activities. They do not move by themselves. They do not develop and reproduce.

Mountains, rocks, snow, and water are all nonliving things.

26 Chapter 2 Organizing Living Things

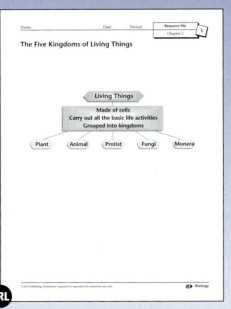

Resource File 3

Organism
A living thing that can carry out all the basic life activities

Did You Know?
There are more than 500 species of bacteria that live in the human mouth. As much as 50 percent of the bacteria in the mouth live on the surface of the tongue.

Nonliving things are not made of cells. Recall that a cell is the basic structure of all living things.

A nonliving thing may seem to carry out one life activity. But it will not carry out all of them. For example, sugar crystals in rock candy can grow. But they do not move, need food, or react to their environment.

Properties of Living Things

Like nonliving things, living things have properties you can see and feel. But living things carry out basic life activities. They move, grow, and reproduce. They use food and remove wastes. They sense and react to their environment. Another word for a living thing is an **organism**.

What is an organism? An organism is a complete, individual living thing. It can carry out all the basic life activities. Large living things such as an elephant or a redwood tree are organisms. But so are some tiny living things that have only one cell. The bacteria that cause sore throats are organisms. Bacteria are the simplest organisms that carry out all basic life activities.

The *organ* in *organism* comes from a Latin word that means "tool." Just as a tool does a certain job, an organ does a certain job in a living thing. Organs are organized tissues and cells that carry out basic life activities. Your organs include your heart and your kidneys. Simple organisms, such as bacteria, do not have organs.

Technology Note

Advances in technology have helped scientists learn more about how different organisms are related. Scientists can now study the makeup of the DNA, RNA, and proteins in the cells of an organism. The information contained in DNA and RNA is passed on from one generation of organisms to the next. By examining this information, scientists can compare different organisms to determine how closely they are related.

Organizing Living Things Chapter 2 27

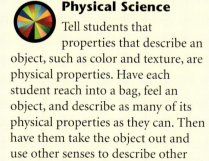

SCIENCE INTEGRATION

Physical Science
Tell students that properties that describe an object, such as color and texture, are physical properties. Have each student reach into a bag, feel an object, and describe as many of its physical properties as they can. Then have them take the object out and use other senses to describe other physical properties.

LEARNING STYLES

Interpersonal/Group Learning
Distribute several nonliving objects to groups of students. Have each group list basic life activities that apply to each nonliving thing. Then ask them to explain why the object is not living even though it carries out some basic life activities.

Have students write the headings "Living" and "Nonliving" at the top of a sheet of paper. As they read, have them write properties of living and nonliving things under the proper heading.

Review with students the basic life activities discussed in Chapter 1. Emphasize that nonliving things may carry out one or more of these activities. For example, a car gets and uses energy and gives off a waste product; it moves. It is not living. A nonliving thing does not possess all of these characteristics.

3 Reinforce and Extend

Did You Know?
Read the feature aloud and show students a microscope photo of bacteria, explaining how many times the image has been magnified. Point out that there are many times more microscopic living things on Earth than living things that we can see.

CAREER CONNECTION

Invite a nurse, medical lab worker, or other health-care professional to speak to students about bacteria and viruses that cause sickness. The speaker can explain how to avoid becoming infected.

CROSS-CURRICULAR CONNECTION

Art
Have students research the work of an artist whose work includes depictions of plants and animals in natural surroundings. Artists include painters and photographers such as Kano Eitoku, Caspar David Friedrich, Thomas Cole, and Ansel Adams. Discuss qualities of the works that make them lifelike and the techniques the artist used to create the impression of life.

Organizing Living Things Chapter 2 27

IN THE ENVIRONMENT

Ask students to define *natural habitat* (*the place where an organism naturally lives*). Explain that sometimes organisms are removed from their natural habitat and taken to a new one. Have students investigate what impact the change can have on the new environment in which the organism is introduced. You may want to suggest that they investigate the introduction of kudzu into the southeastern United States or the rabbit into Australia.

SCIENCE JOURNAL

Have students write a narrative about a newly discovered thing and how scientists discover whether it is living or nonliving.

Lesson 1 Review Answers

1. living and nonliving **2.** Answers will vary. Examples could include rocks, clouds, plastic, gasoline, cars, computers. **3.** Answers will vary. Examples could include plants, animals, and bacteria. **4.** Answers will vary. A representative list might include any of the basic life activities—movement, growth, reproduction, food use and removal of waste, sensing and reacting to the environment—and state that living things are made of cells. **5.** An organism is a complete, living thing that carries out all basic life activities.

Portfolio Assessment

Sample items include:
- Table of properties from Teaching the Lesson
- Lesson 1 Review answers

Lesson 1 REVIEW

Write your answers to these questions on a separate sheet of paper. Write complete sentences.

1. Into what two groups are most things in the world divided?
2. What are three examples of nonliving things?
3. What are three examples of living things?
4. List some properties of living things.
5. What is an organism?

Achievements in Science

Microscopic Bacteria Observed

Microscopes with more than one lens, which are called compound microscopes, were first used during the 1600s. But they often changed the shape and color of images. This problem became worse as lenses were made larger and stronger. To solve this problem, the Dutch inventor Antonie van Leeuwenhoek designed a new kind of lens. The lens was very small and made objects look very clear.

Using this invention, van Leeuwenhoek was able to see things that had never been seen before. He examined water from ponds and found tiny creatures swimming in the water. We now know that these creatures were protists, which are organisms having one cell. He was probably the first person to see bacteria and protozoans. His work was an enormous benefit to the scientific community because he carefully described the things he observed.

Achievements in Science

Ask volunteers to read the Achievements in Science feature on page 28. Explain that van Leeuwenhoek's microscope used only one lens, not the multiple lenses of the microscopes students may be familiar with. However, his lenses could magnify up to 270 times. He called the protists he saw in pond water "animalcules."

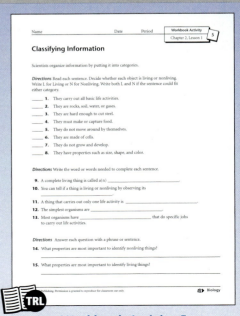

Workbook Activity 5

INVESTIGATION 2

Living or Nonliving?

Purpose
What are the differences between living and nonliving things? In this investigation, you will practice classifying things as living or nonliving.

Materials
- 5 pictures from a magazine or book (numbered 1–5)

Procedure

1. Copy the table below on a sheet of paper. Label it Picture 1. Leave blank columns beneath the headings. Different pictures will have different numbers of objects. You may need more than one line to list the properties of an object.

Object	Properties

2. Look at the first picture. In your table, list all of the objects that you see in the picture. Leave several blank lines between objects to list properties.

3. List the properties of each object. For example, is it a solid, liquid, or gas? What is its shape and color? Do you know if it moves, grows, reproduces, senses things, and reacts? Does it get food and remove wastes?

4. Make a new table for each of the other four pictures. Label them Picture 2 through Picture 5. List the objects in the pictures and their properties.

Organizing Living Things Chapter 2

Investigation 2

This investigation will take approximately 30 minutes to complete. Students will use these process and thinking skills: observing, describing, classifying, comparing and contrasting, and drawing conclusions.

Preparation

- Make sure some of the pictures include living things and some have nonliving things. Pictures might show people, different animals, plants, and minerals such as in jewelry, cars, computers, and furniture.
- Magazine pictures can be torn out of the magazine and numbers written on the pictures.
- Students may use Lab Manual 4 to record their data and answer the questions.

Procedure

- This investigation is best done with students working in pairs. Students may write down properties independently first. Then have partners compare their lists and decide together which objects are living and nonliving.
- Be sure students can distinguish a solid, liquid, and gas before completing step 3. Provide examples.
- If students bring in their own pictures, collect and mix them up. Then distribute a random group of pictures to each pair. Have students number the pictures before they begin.

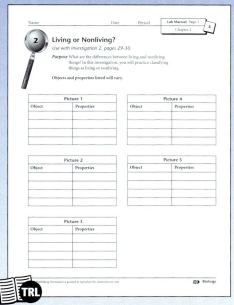

Lab Manual 4, pages 1–2

Organizing Living Things Chapter 2

- For the Table in step 5, you might suggest that students include only a few living and nonliving things from each picture.
- Challenge students to explain why certain nonliving things possess some basic life activities yet are not living.

Results

Students should find that the pictures include living and nonliving objects. Living objects have a number of properties, such as cell structure and ability to move on their own, that nonliving objects do not have.

Questions and Conclusions Answers

1. Answers will vary depending on the pictures chosen.
2. Answers will vary depending on the pictures chosen.
3. Answers will vary. Possible answers: A lawn mower gets energy from gasoline and it moves. But the lawn mower does not reproduce, develop, or grow, so it is not a living thing.
4. Answers may vary, but most pictures will show living things and nonliving things together.

Explore Further Answers

Living objects to list might include other students, the teacher, plants, insects, and class pets. Possible nonliving objects could include desks, floor, walls, chalkboard, pencil, pen, clothing, shoes, backpacks, and paper.

As an alternative, you might show students an aquarium and have them classify its contents as living and nonliving.

Assessment

Check to be sure that students make a new table for each picture and list the objects and their properties. You might include the following items from this investigation in student portfolios:
- Investigation 2 tables
- Questions and Conclusions and Explore Further answers

5. Copy the table below on a sheet of paper. Review the properties you recorded for each object. Use the properties to decide if the object is living or nonliving. Write the name of each object in the correct column.

Picture	Nonliving Things	Living Things
1		
2		
3		
4		
5		

Questions and Conclusions

1. List three living things that you observed in the pictures.
2. List three nonliving things that you observed in the pictures.
3. Did any of the nonliving things carry out a basic life activity, such as growing? Give an example. Then explain why you decided that it was a nonliving thing.
4. Did any of the pictures show *only* living things? List the living things in the picture.

Explore Further

Copy the table above again. Look around you and classify everything you see as living or nonliving. Write the name of each object in the correct column.

Lesson 2: How Organisms Are Classified

Living things are more like one another than they are like nonliving things. For example, living things all carry out the basic life activities. However, living things can be very different from one another. A cat is different from a dog. A bird and a tree are even more different from each other.

Scientists divide the world of living things into five groups, or **kingdoms**. These kingdoms are plant, animal, protist, fungi, and monera. Biological classifications are based on how organisms are related. The study of living things is called **biology**. The science of classifying organisms based on the features they share is called **taxonomy**. Most of the living things you know are either in the plant kingdom or the animal kingdom. There are three other kingdoms that you may not know very well.

Objectives

After reading this lesson, you should be able to
- explain how living things are divided into kingdoms.
- list and describe the five kingdoms of living things.

Kingdom
One of the five groups into which living things are classified

Biology
The study of living things

Taxonomy
The science of classifying organisms based on the features they share

The Plant Kingdom

Most plants are easy to recognize. Examples of plants are trees, grasses, ferns, and mosses. Plants don't move from place to place like animals. They don't need to do so. Plants make their own food, using sunlight and other substances around them. All plants have many cells. These cells are organized into tissues. Many plants also have organs.

The Animal Kingdom

Animals have many different sizes and shapes. You probably recognize dogs, turtles, and fish as animals. Corals, sponges, and insects are animals too.

Animals get their food by eating plants or by eating other animals that eat plants.

While most scientists follow the five kingdom classifications, some want to add a sixth kingdom for viruses. As new information becomes known, the five groupings may change.

In 1969, Robert Whittaker proposed the first 5-kingdom classification system. He placed fungi in their own kingdom. He also placed prokaryotic cells (cells that contain no membrane-bound organelles) in their own kingdom—the monerans.

Organizing Living Things Chapter 2 31

1 Warm-Up Activity

Display a number of small items such as washers, coins, pencils, pens, chalk, paper, cards, books, notebooks, and paper clips. Together with students discuss the properties of the objects. Invite students to suggest ways to organize the objects into groups and subgroups, for example, things that are round, things that are long, things used to write. Explain that scientists group living things into kingdoms using this procedure: finding their similarities and differences.

2 Teaching the Lesson

Before assigning the reading of this lesson, have students refer to the margins to read the vocabulary terms as you say them so that students become familiar with their pronunciations.

Write the five kingdoms on the board as column headings. As you discuss each kingdom with students, write down some properties of the organisms in each. List example organisms with which students may be familiar.

After they have finished reading, have students write original sentences using the new vocabulary words.

3 Reinforce and Extend

TEACHER ALERT

Students might think that all organisms that are not plants or animals are microscopic. Point out the larger size of some protists, such as seaweed, and fungi, such as mushrooms.

GLOBAL CONNECTION

Algae are important around the world as a source of food, fertilizer, and agar—a jelly-like material used for growing bacteria in labs. Have students report on ways algae affect the world economy.

Microorganism
An organism that is too small to be seen without a microscope

Protist
An organism that usually is one-celled and has plant-like or animal-like properties

Algae
Protists that make their own food and usually live in water

Protozoan
A protist that has animal-like qualities

Animals cannot make their own food. They get their food from other living things. They eat plants, or they eat other animals that eat plants. Most animals move around to capture or gather their food. Moving also helps them to find shelter, escape danger, and find mates. All animals have many cells. These cells form tissues in all animals except the sponge. In most animals, the tissues form organs.

The Protist Kingdom

At one time, biologists divided the living world into only two kingdoms, plant and animal. Then the microscope was invented. When biologists used the microscope, they discovered tiny organisms. They called them **microorganisms** because biologists could see these organisms only under a microscope. These organisms did not fit into either the plant or the animal kingdom. Biologists placed them in a separate kingdom. They called the organisms **protists**.

Most protists have only one cell. A few have many cells. Some protists make their own food. Others absorb food from other sources. **Algae** are plant-like protists. **Protozoans** are animal-like protists. Some protozoans have properties of plants and animals. All protists can carry out the basic life activities.

Algae live in lakes, streams, rivers, ponds, and oceans. You have probably seen the algae that grow as a green scum on a pond. The green scum is thousands of tiny algae. Like plants, algae can make their own food. Algae are food for the organisms that live in waters around the world. Many larger algae are called seaweeds. Some seaweeds can become as long as a football field. Algae also produce oxygen that other organisms use. At one time, biologists classified algae as plants. Algae, however, are simpler than plants and have more in common with protists.

SCIENCE INTEGRATION

Earth Science
Have students research how scientists use fossils to help them learn how kinds of animals developed over time. Imprints of early reptiles, for example, show strong similarities to amphibians. Students can report on how these data help biologists classify animals into the right kingdom.

Flagellum
A whip-like tail that helps some one-celled organisms move (plural is flagella)

Cilia
Hair-like structures that help some one-celled organisms move

Pseudopod
Part of some one-celled organisms that sticks out like a foot to move the cell along

Amoeba
A protozoan that moves by pushing out parts of its cell

Fungus
An organism that usually has many cells and decomposes material for its food (plural is fungi)

Did You Know?
Athlete's foot is a disease caused by a fungus.

Protozoans live in water, soil, and the bodies of animals. Most protozoans are harmless. But a few, such as *Giardia*, cause disease. This protozoan infects the small intestine of humans and other animals. It causes tiredness, weight loss, and stomach pain.

Protozoans behave like animals by getting food and moving. Different kinds of protozoans have different methods of moving. Protozoans can use **flagella**, **cilia**, or **pseudopods** to move. **Amoebas** push out a part of their cell. This part is called a pseudopod. It looks like a foot and pulls the amoeba along. Some protozoans have tails, or flagella, that move them back and forth. Others use cilia to move. Cilia are tiny hair-like structures that beat like boat paddles.

Euglenas are protozoans that behave like both plants and animals. Like plants, they make their own food when sunlight is present. Like animals, they can absorb food from the environment. They absorb food when sunlight is not present.

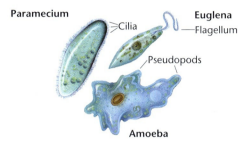

The Fungi Kingdom

You are probably more familiar with organisms in the fourth kingdom, **fungi**. Mushrooms and the mold that grows on bread are fungi. Most fungi have many cells. At one time, fungi were classified as plants. Like plants, fungi do not move around by themselves. But unlike plants, fungi do not make their own food. They absorb food from other organisms.

Organizing Living Things   Chapter 2   **33**

Did You Know?

Read aloud the Did You Know? feature on page 33. Inform students that there are over 100,000 species of fungi—some helpful and some harmful to humans. Yeast and penicillin are two helpful fungi. Discuss the basic life activities of a fungus that lives in skin by asking questions such as *How could the fungus get food? How could it remove wastes?*

LEARNING STYLES

 Visual/Spatial
Have students use microscopes to observe a protozoan, or have them find detailed illustrations of a protozoan in a reference book. Ask students to draw their own picture of the organism and label parts that the organism uses to move.

CROSS-CURRICULAR CONNECTION

 Home Economics
Have a small group of students describe yeast and then observe what happens when yeast is mixed with warm water and sugar. The group can follow recipe directions to make bread or rolls. Ask them to summarize the effects of the yeast's life activities on the dough.

SCIENCE JOURNAL

 Ask students to imagine that they are no bigger than a cell. Have them write a journal entry about an encounter with a paramecium, euglena, or amoeba, describing the organism's appearance and behavior.

AT HOME

 Students might conduct an at-home search to find products that are related to fungi. For example, they can list foods that are fungi or contain them, medicines to prevent or fight fungal infections, and poisons (fungicides) that kill fungi. Advise students to ask an adult to help them during their search.

Organizing Living Things   Chapter 2   **33**

Science Myth

Have volunteers read the Science Myth feature on page 34. If possible, display photographs of several poisonous varieties of mushrooms. Explain that mushrooms for sale in grocery stores are produced by professional growers.

LEARNING STYLES

Auditory/Verbal
Read aloud characteristics of one kingdom. (For example, this kingdom contains microorganisms. Most have one cell. Some are plant-like, and some are animal-like.) Students try to name the kingdom with the fewest clues possible. (protists) Repeat for all five kingdoms.

LEARNING STYLES

Body/Kinesthetic
Use tape and labels to make a large chart on the classroom floor. Label the following areas: Living, Nonliving, Plants, Animals, Monerans, Protists, and Fungi. Give a student a card labeled with the name of a living or nonliving thing. The student reads the name and works through the chart as he or she explains how to classify by characteristics.

LEARNING STYLES

Logical/Mathematical
Give groups of students pictures of animals with backbones, including several mammals, birds, and fish. Explain why these animals are classified together. Ask them to make a "tree" diagram showing how they would break this large group into smaller groups. Ask students to explain the logic of their classifications. Each subgroup could then be divided further—birds by type of beaks and claws, and so on.

Decompose
To break down or decay matter into simpler substances

Parasite
An organism that absorbs food from a living organism and harms it

Moneran
An organism, usually one-celled, that does not have organelles

Science Myth

Mushrooms that grow in nature are safe to eat if they look like mushrooms in grocery stores.

Fact: Some mushrooms are poisonous. Only experts can safely identify mushrooms that are safe to eat. The mushrooms in stores are grown especially for food or identified by experts.

ONLINE CONNECTION

The following Web address is useful for students to explore the principles of taxonomy in greater depth: anthro.palomar.edu/animal/default.htm. In addition to an introduction, the site offers an explanation of the principles of classification and specifics about various levels of classification. It also links the user to related sites, gives a table of kingdoms, and provides an in-depth glossary of terms.

Because of the way fungi get food, they are important to other organisms. Fungi release special chemicals on dead plant and animal matter. The chemicals break down, or **decompose**, the matter. The fungi then absorb the decomposed material. But some of the decomposed matter also gets into the soil. Other organisms, such as plants, can then use it.

Some fungi are **parasites**. They absorb food from a living organism. Some fungi harm plants. For example, Dutch elm disease kills elm trees. Other fungi harm animals. A fungus causes ringworm, a human skin disease.

The Monera Kingdom

The last of the kingdoms contains **monerans**. *Monera* means "alone." This kingdom has only one kind of organism, which is bacteria. Monerans are usually one-celled organisms. Like animals, some can move and get food. Like plants, some stay put and make their own food. You may wonder why bacteria are not placed in the protist kingdom. The cells of bacteria are different from the cells of all other organisms. Bacteria do not have organelles in their cells. Organelles are tiny structures in cells that do certain jobs. The cells of all other organisms have organelles.

Some bacteria cause disease. For example, bacteria cause strep throat. Most bacteria, though, are harmless. Many are even helpful. Like fungi, bacteria help to decompose the remains of plants and animals. People also use bacteria to make foods such as cheese and yogurt.

Some bacteria are green like plants and make their own food.

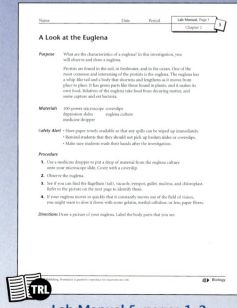

Lab Manual 5, pages 1–2

Lesson 2 REVIEW

Write your answers to these questions on a separate sheet of paper. Write complete sentences.

1. Name the five kingdoms of living things.
2. List two differences between plants and animals.
3. How are protists similar to plants and animals?
4. What is an important function of fungi?
5. Why are bacteria placed in a kingdom by themselves?

Science at Work

Taxonomist

Taxonomists should have good organization and research skills and be able to classify data. A college degree in botany, zoology, or biology is needed. Some taxonomists may need a master's degree or a Ph.D.

A taxonomist is a scientist who studies organisms and classifies them into groups. Taxonomists not only classify the organisms into groups, they study all the information about a specific group. They label thousands of species and collect various data. Because so many different kinds of organisms exist, taxonomists usually specialize by choosing which organisms to study. Taxonomists need to keep up with new technologies. New information could lead to changes in the way organisms are classified.

Some taxonomists travel around the world to study organisms. They may even discover a kind of organism no one else has identified before. Some taxonomists work in museums, zoos, or botanical gardens and study plants and animals that have been collected. Other taxonomists study organisms in laboratories or teach at universities.

Organizing Living Things Chapter 2 35

Lesson 2 Review Answers

1. animals, plants, protists, fungi, and monerans **2.** Plants make their own food while animals must eat plants or other animals for energy. Animals move from place to place while plants do not. **3.** Like plants, some protists make their own food. Others are like animals in that they absorb food from other sources and move about. **4.** Fungi decompose, or break down, dead plant and animal matter into a form that can be used by plants. Both plants and animals benefit. **5.** Bacteria lack organelles in their cells. All other organisms have organelles.

Portfolio Assessment

Sample items include:
- Sentences using vocabulary words from Teaching the Lesson
- Lesson 2 Review answers

Science at Work

Write *taxonomy* and *taxonomist* on the board. Pronounce the terms and discuss basics of the method of classifying organisms. (At seven different levels, organisms are grouped according to similarities. At each successive level, the group contains fewer organisms with more characteristics in common. For example, there are many more animals in the group mammals than in the group carnivores [mammals that eat meat].)

Read the Science at Work feature on page 35 together. To give students a taste of the logic of taxonomy, display pictures of a variety of arthropods (insects, spiders, crabs, lobsters, centipedes). Have students tell different ways they are similar. Then they can decide on ways to classify them into different groups, such as number of legs, types of appendages, or types of mouth parts.

Organizing Living Things Chapter 2 35

Science in Your Life

Have volunteers read aloud the paragraphs of Science in Your Life on page 36. Ask students to identify the sentence in each paragraph that states its main point. (*The first sentence in each paragraph gives the main idea.*)

You may want to assign small groups each one paragraph of the feature. They can do research on the Internet or by using encyclopedias to find out more about where different bacteria live, how they help or harm us, and how they are studied in laboratories. Invite groups to share what they learned with the class.

CROSS-CURRICULAR CONNECTION

Language Arts
Have students create an acrostic from a kingdom of their choice. They should write the kingdom name vertically on their papers. Each letter in the kingdom name becomes the first letter of a word or phrase that relates to the kingdom. Acrostics can serve as a mnemonic device for remembering characteristics of organisms in that kingdom.

LEARNING STYLES

LEP/ESL
Label five boxes with the names of kingdoms. Have students cut out or draw pictures of organisms and place them in the appropriate boxes. At the same time, they should write the name of each organism on a card. Students can practice vocabulary by drawing a card, reading the organism name, and finding the matching picture.

Science in Your Life

Are bacteria helpful or harmful?

Bacteria live everywhere. They are in the ocean, on top of mountains, in polar ice, on your hands, and even inside you. You cannot go anywhere without coming in contact with bacteria.

Most bacteria are helpful. Bacteria in soil break down plant and animal material and release nutrients. They also take in gases from the air, such as nitrogen. They change nitrogen into a form that plants and animals can use. Bacteria in your intestines make vitamin K. This vitamin helps your blood clot when you are cut.

Bacteria can also be harmful. Many bacteria cause diseases in people. Bacteria cause diseases such as tuberculosis, tetanus, and cholera. Food that is contaminated by certain kinds of bacteria can cause illness.

Microbiologists are scientists who work with bacteria and other microorganisms. Some microbiologists help to identify bacteria that cause disease. They grow the bacteria on special plates. Each cell multiplies until it forms millions of bacteria cells, called a colony. It is impossible to see a single bacterium without a microscope. But it is easy to see colonies of bacteria.

There are ways to get rid of most harmful bacteria. Antibiotics are drugs that kill bacteria in people and animals. Pasteurization, or rapid heating, kills harmful bacteria in milk. Drinking water is purified to remove bacteria and other microorganisms. Sewage is treated so that it will not pollute water supplies. Sometimes bacteria help to clean up sewage. Helpful bacteria break down material in the sewage so that it does not harm people.

IN THE COMMUNITY

Have students investigate methods used by the water treatment plant in your community to prepare the water for consumption. Their goal is to find out how harmful protists and monerans are removed before water is pumped into homes for cooking, drinking, and cleaning. Ask students to report on tests performed on water to be sure it is safe.

Chapter 2 SUMMARY

- Living and nonliving things have different properties.
- Living things are different from nonliving things in two main ways. Living things are made of cells. Nonliving things are not. Living things can carry out all the basic life activities. Nonliving things cannot.
- Biology is the study of living things, or organisms.
- Living things are divided into five kingdoms based on how they are related. The kingdoms are plant, animal, protist, fungi, and monera.
- Plants make their own food and do not move around from place to place. They have many cells that are organized into tissues and sometimes organs.
- Animals eat other organisms for food. They can move around to get their food. They have many cells that are organized into tissues and organs.
- Protists include algae, seaweeds, and protozoans. Most are one-celled. They have properties of both animals and plants. Some make their own food. Others absorb their food. Some do both.
- At one time, fungi were classified as plants. But fungi do not make their own food. They absorb their food from other organisms or the remains of other organisms. Some are parasites.
- Monerans are bacteria. They are usually one-celled organisms that do not have organelles. Some make their own food and others absorb it.

Science Words

algae, 32	flagellum, 33	organism, 27	pseudopod, 33
amoeba, 33	fungus, 33	parasite, 34	taxonomy, 31
biology, 31	kingdom, 31	property, 26	
cilia, 33	microorganism, 32	protist, 32	
decompose, 34	moneran, 34	protozoan, 32	

Chapter 2 Review

Use the Chapter Review to prepare students for tests and to reteach content from the chapter.

Chapter 2 Mastery Test

The Teacher's Resource Library includes two parallel forms of the Chapter 2 Mastery Test. The difficulty level of the two forms is equivalent. You may wish to use one form as a pretest and the other form as a posttest.

Review Answers

Vocabulary Review

1. biology 2. property 3. cilia
4. kingdom 5. organism 6. decompose
7. fungi 8. microorganism 9. algae
10. parasite 11. protozoan 12. moneran
13. protist

TEACHER ALERT

In the Chapter Review, the Vocabulary Review activity includes a sample of the chapter's vocabulary terms. The activity will help determine students' understanding of key vocabulary terms and concepts presented in the chapter. Other vocabulary terms used in the chapter are listed below:

amoeba
flagellum
pseudopod
taxonomy

Chapter 2 REVIEW

Word Bank
algae
biology
cilia
decompose
fungi
kingdom
microorganism
moneran
organism
parasite
property
protist
protozoan

Vocabulary Review

Choose the word from the Word Bank that best completes each sentence. Write the answer on a sheet of paper.

1. _____ is the study of living things.
2. The hardness of a rock is a(n) _____.
3. Some protozoans use hair-like structures called _____ to move.
4. Mushrooms are in the fungi _____.
5. Another word for a living thing is a(n) _____.
6. Fungi release chemicals that _____ matter.
7. _____ usually have many cells and break down dead plants and animals.
8. An organism that is too small to be seen without a microscope is a(n) _____.
9. Protists that make their own food and usually live in water are _____.
10. An organism that lives on a living organism and harms it is a(n) _____.
11. A(n) _____ is an animal-like protist.
12. An organism that usually has one cell and does not have organelles is a(n) _____.
13. One type of _____ is a protozoan.

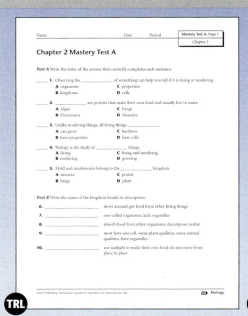
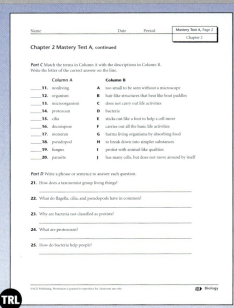

Chapter 2 Mastery Test A

Concept Review

Choose the answer that best completes each sentence. Write the letter of the answer on your paper.

14. Nonliving things do not have _____.
  A cells   C color
  B properties   D hardness

15. Most animals need to _____ to get food.
  A use sunlight   C reproduce
  B move   D breathe

16. Plants can make their own _____.
  A minerals   C flagella
  B water   D food

17. A new kingdom was discovered when _____ were seen under a microscope.
  A plants   C microorganisms
  B fungi   D animals

18. _____ were once classified as plants, but they do not make their own food.
  A Fungi   C Animals
  B Algae   D Mosses

19. Bacteria are organisms in the _____ kingdom.
  A animal   C protist
  B fungi   D monera

Critical Thinking

20. What are some of the properties that are used to divide living things into kingdoms? Give some examples.

 **Test-Taking Tip** Answer all questions you are sure of first. Then go back and answer the others.

Organizing Living Things Chapter 2 39

Concept Review
14. A **15.** B **16.** D **17.** C **18.** A **19.** D

Critical Thinking
20. Answers may include, but are not limited to, ability to make food versus ability to get food, means of moving about, type of food eaten, number of cells in the organism, whether cells contain organelles with membranes, body structure, and type of reproduction.

ALTERNATIVE ASSESSMENT
Alternative Assessment items correlate to the student Goals for Learning at the beginning of this chapter.

- Give each student one organism and one nonliving object to classify. First have the student list the properties of each thing. Then ask how he or she determined the properties and whether the object was living or nonliving.

- Divide the class into teams. Write the five kingdoms on the board as headings. Each team must give an example of an organism that belongs to each kingdom. Move in order across each kingdom as turns progress. Teams gain a point for an accurate identification and lose a point for an incorrect one.

- Give students the names of pairs of organisms that belong in different kingdoms (penguin, mushroom; algae, bacteria; cat, amoeba; paramecium, oak tree). Have them first identify the kingdom in which each organism belongs and then tell one way the organisms are alike and one way they are different.

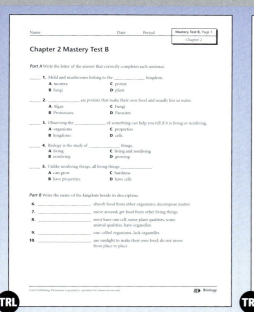

Chapter 2 Mastery Test B

Chapter 3

Planning Guide
Classifying Animals

	Student Text Lesson		
	Student Pages	Vocabulary	Lesson Review
Lesson 1 How Biologists Classify Animals	42–50	✔	✔
Lesson 2 Vertebrates	51–55	✔	✔
Lesson 3 Invertebrates	56–62	✔	✔

Chapter Activities

Student Text
Science Center

Teacher's Resource Library
Community Connection 3: Vertebrates and Invertebrates in Your Life

Assessment Options

Student Text
Chapter 3 Review

Teacher's Resource Library
Chapter 3 Mastery Tests A and B

40A

Student Text Features							Teaching Strategies						Learning Styles						Teacher's Resource Library				
Achievements in Science	Science at Work	Science in Your Life	Investigation	Science Myth	Note	Technology Note	Did You Know?	Science Integration	Science Journal	Cross-Curricular Connection	Online Connection	Teacher Alert	Applications (Home, Career, Community, Global, Environment)	Auditory/Verbal	Body/Kinesthetic	Interpersonal/Group Learning	Logical/Mathematical	Visual/Spatial	LEP/ESL	Workbook Activities	Alternative Workbook Activities	Lab Manual	Resource File	Self-Study Guide
		47	49			✓	45		45	44			47	44			45		46	7	7	7, 8	5	✓
	55			54	✓				54	53, 54	53	52	53, 54	54	53		53	54		8	8			✓
62							57	58, 61	58, 59, 61	59	57, 58	58, 60	59		58	60	61			9	9	9	6	✓

Pronunciation Key

a	hat	e	let	ī	ice	ô	order	u̇	put
ā	age	ē	equal	o	hot	oi	oil	ü	rule
ä	far	ėr	term	ō	open	ou	out	ch	child
â	care	i	it	ȯ	saw	u	cup	ng	long

sh	she		a	in about
th	thin		e	in taken
ᵺ	then	ə	i	in pencil
zh	measure		o	in lemon
			u	in circus

Alternative Workbook Activities

The Teacher's Resource Library (TRL) contains a set of lower-level worksheets called Alternative Workbook Activities. These worksheets cover the same content as the regular Workbook Activities but are written at a second-grade reading level.

Skill Track Software

Use the Skill Track Software for Biology for additional reinforcement of this chapter. The software program allows students using AGS textbooks to be assessed for mastery of each chapter and lesson of the textbook. Students access the software on an individual basis and are assessed with multiple-choice items.

Chapter at a Glance

Chapter 3: Classifying Animals
pages 40–65

Lessons

1. **How Biologists Classify Animals** pages 42–50

   Investigation 3 pages 49–50

2. **Vertebrates** pages 51–55

3. **Invertebrates** pages 56–62

Chapter 3 Summary page 63

Chapter 3 Review pages 64–65

Skill Track Software for Physical Science

Teacher's Resource Library

Workbook Activities 7–9

Alternative Workbook Activities 7–9

Lab Manual 7–9

Community Connection 3

Resource File 5–6

Chapter 3 Self-Study Guide

Chapter 3 Mastery Tests A and B

(Answer Keys for the Teacher's Resource Library begin on page 420 of the Teacher's Edition. A list of supplies required for Lab Manual Activities in this chapter begins on page 449.)

Science Center

As they read the chapter, ask students to bring in animal pictures. Challenge them to find at least one animal for each group of vertebrates and invertebrates discussed in Lessons 2 and 3. Provide a poster for each group. Have students determine which poster each picture belongs on and mount the picture.

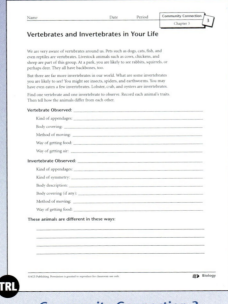

Community Connection 3

Chapter 3

Classifying Animals

The first animals you probably notice in the photo are the birds. If you study their features, you will notice that there are different kinds of birds. The two largest birds are the same kind of bird. They are more related to each other than they are to the other birds. Look closely at the animals at the edge of the water. It may surprise you to learn that these horseshoe crabs are more related to spiders than to birds. In this chapter, you will learn how biologists classify animals into groups. You also will learn what features the animals in each group have in common.

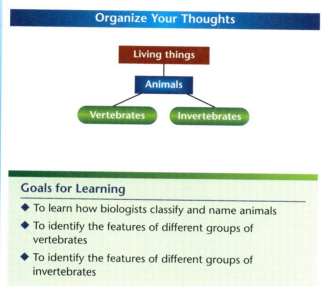

Organize Your Thoughts

Living things → Animals → Vertebrates, Invertebrates

Goals for Learning

- ◆ To learn how biologists classify and name animals
- ◆ To identify the features of different groups of vertebrates
- ◆ To identify the features of different groups of invertebrates

Introducing the Chapter

Direct students' attention to the photograph on page 40. Ask them to identify different kinds of animals they see and to describe ways they are alike and different.

Read the introductory paragraph together. Ask students to predict why the horseshoe crabs are more like spiders than birds. Have students look at the organizer. Write the word *vertebrate* on the board. Ask a volunteer to tell what *vertebrate* means. (*having backbones*) Add the letters *in* to make *invertebrate*. Have students predict what it means. (*without backbones*)

Have students read the Goals for Learning on page 41. Remind them to keep these goals in mind as they read the lessons.

Notes and Technology Notes

Ask volunteers to read the notes that appear in the margins throughout the chapter. Then discuss them with the class.

Teacher's Resource

The AGS Teaching Strategies in Science Transparencies may be used with this chapter. The transparencies add an interactive dimension to expand and enhance the *Biology* program content.

Career Interest Inventory

The AGS Harrington-O'Shea Career Decision-Making System-Revised (CDM) may be used with this chapter. Students can use the CDM to explore their interests and identify careers. The CDM defines career areas that are indicated by students' responses on the inventory.

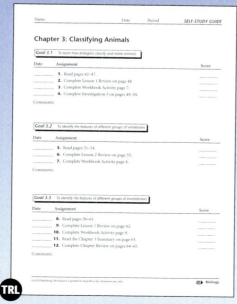

Chapter 3 Self-Study Guide

Lesson at a Glance

Chapter 3 Lesson 1

Overview In this lesson, students learn how animals are classified and why biologists give each species of animal a scientific name.

Objectives
- To explain how biologists classify animals
- To name the seven levels in the classification system of organisms
- To identify the two parts of a scientific name

Student Pages 42–50

Teacher's Resource Library TRL
- Workbook Activity 7
- Alternative Workbook Activity 7
- Lab Manual 7, 8
- Resource File 5

Vocabulary

classify	species
phylum	scientific name
genus	

Science Background
Early Classification Systems

Aristotle (384–322 B.C.) devised the first classification system, dividing living things as plants or animals and subdividing animals into land dwellers, water dwellers, and air dwellers. During the Middle Ages, scholars began assigning Latin names to individual species. Each name consisted of a string of descriptive terms. This naming system was difficult to use, however, since some names contained more than a dozen terms. In the mid-18th century, the Swedish biologist Carolus Linnaeus (1707–1778) devised the seven-level classification system that biologists use today. Linnaeus also began the practice of giving a two-word scientific name to each species.

Lesson 1 — How Biologists Classify Animals

Objectives

After reading this lesson, you should be able to
- explain how biologists classify animals.
- name the seven levels in the classification system of organisms.
- identify the two parts of a scientific name.

Classify
Group things based on the features they share

Biologists have identified more than one million different kinds of animals in the world. More kinds of animals are added to the list every day. To deal with such a large list, biologists need a way to divide it into smaller groups.

Classifying Based on Shared Features

Look at the vehicles in the photo. How could you **classify** the vehicles, or divide them into groups? One way would be to think of how some of the vehicles are similar. For example, you could put the passenger cars in one group and the SUVs in another group. You could then divide the passenger cars and the SUVs by their size. You might divide each of those groups into smaller groups based on color.

You can classify these vehicles based on the features they share.

42 Chapter 3 Classifying Animals

Resource File 5

Lab Manual 7, pages 1–2

42 Chapter 3 Classifying Animals

Biologists divide animals into groups based on their similarities too. For example, falcons, sparrows, and geese are classified as birds because they all have feathers. All birds have feathers, but no other type of animal does.

The bird group is divided into smaller groups based on other features. Falcons are birds of prey. All members of that group have feet with sharp claws that can grab prey. Sparrows are perching birds. Their feet have toes that are good for gripping branches. Geese are waterbirds. Like other waterbirds, geese use their webbed feet for swimming. Try to match these three bird groups with the bird feet shown in the photos.

Biologists use the similar features of organisms to determine how the organisms are related. Appearance is only one feature biologists use. They also use similarities in cell structure, hereditary material in cells, and the way the organisms get nutrients and reproduce. Organisms that have more features in common are more related to each other than to other organisms. Biologists classify all organisms based on how the organisms are related. The science of classifying organisms is called taxonomy.

Which groups of birds have feet like these?

1 Warm-Up Activity

List the following animals and show pictures of them: angelfish, silverfish, starfish, crayfish, and jellyfish. Have students group the animals as water dwellers or land dwellers (*all but silverfish live in water*). Then regroup them as hard-bodied or soft-bodied (*hard: crayfish; soft: angelfish, silverfish, jellyfish; starfish could be either*). Finally, regroup them as having legs or no legs (*silverfish and crayfish have legs*). Explain that biologists consider characteristics such as these when organizing animals into groups. Students should also recognize that animals with similar common names may belong in different groups.

2 Teaching the Lesson

Have students copy the lesson objectives on a sheet of paper. As they read the lesson, they can write one or two sentences that meet each objective.

Give students the seven classification names for orangutans and gorillas. (Both belong to kingdom Animalia, phylum Chordata, class Mammalia, order Primates, and family Pongidae. The orangutan belongs to the genus *Pongo* and species *pygmaeus*. The gorilla belongs to the genus *Gorilla* and species *gorilla*.) Have students use this information to prepare a diagram for these animals similar to the one on page 45.

Write the following scientific names on the board: *Thamnophis proximus, Thamnophis melanogaster, Drosophila melanogaster*. Ask students which two animals are more similar. Have them explain their reasoning. (*Animals with the same genus name are more similar to each other than animals in a different genus.*) Repeat the activity for a second group of names: *Canis latrans, Canis lupus, Anarhichas lupus*.

Have students select an animal from the diagram on page 45. Ask them to find the names of other classes in its phylum, other orders in its class, and so on.

3 Reinforce and Extend

CROSS-CURRICULAR CONNECTION

History
Have students investigate the life of Carolus Linnaeus, the person who devised the seven-level classification system for organisms. Ask them to find out what else Linnaeus tried to classify. (*minerals and diseases*)

LEARNING STYLES

Auditory/Verbal
Have students come up with their own mnemonic device for learning the sequence of levels in the biological classification system. Give them an example, such as **K**aren **p**laced **C**arl's **o**rder **f**or **g**reen **s**ocks—kingdom, phylum, class, order, family, genus, and species.

CROSS-CURRICULAR CONNECTION

Language Arts
Have students analyze the Library of Congress Classification System for books in schools and libraries. Ask them to model how the number on the spine of a book was generated. Have them tell which part of the identification number is like the genus and species name in taxonomy. Ask students to list ways the two systems are similar. (*Both provide many levels of classification; both are exact; both provide a way to include new items.*)

Phylum
Subdivision of a kingdom (plural is phyla)

Genus
A group of living things that includes separate species

Species
A group of organisms that can breed with each other to produce offspring like themselves

The Seven Levels of Classification

Recall the example of organizing vehicles into groups. In that example, there were three levels of organization. The highest level contained two groups, which were passenger cars and SUVs. The middle level contained groups based on the sizes of the vehicles. The lowest level contained groups based on color.

Biologists also use different levels to classify living things. The diagram shows that there are seven levels in the classification system of organisms: kingdom, **phylum**, class, order, family, **genus**, and **species**.

Kingdoms represent the highest level in the classification system. You learned in Chapter 2 that biologists classify all organisms into five kingdoms. The animal kingdom is one of the five kingdoms. Each kingdom is divided into groups called phyla. The phyla represent the second-highest level of classification. More organisms are included in a kingdom than in any one of its phyla. Each phylum is divided into classes, each class is divided into orders, and so on.

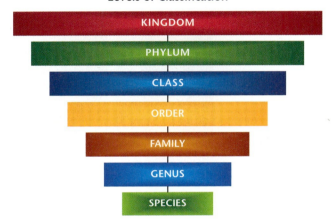

Levels of Classification
KINGDOM
PHYLUM
CLASS
ORDER
FAMILY
GENUS
SPECIES

Did You Know?

The Greek philosopher Aristotle was the first person to classify organisms. More than 2,000 years ago, he divided all living things into the two groups of plants and animals.

The lowest level in the classification system is the species. Each species represents a single type of organism. Members of the same species can breed and produce offspring like themselves. A group of separate but related species belongs to the same genus.

A Place for Every Organism

Every organism that has been identified has its own place in the classification system. The diagram below shows how biologists classify four species of animals. Notice that the African elephant, the red tree mouse, and the heather mouse belong to the same phylum. The boll weevil belongs to a different phylum. This means that these three animals are more similar to each other than they are to the boll weevil. Notice also that the two mice belong to the same order. The elephant belongs to a different order. Thus, the red tree mouse and the heather mouse are more similar to each other than they are to the elephant. Organisms that are very similar belong to the same genus. Which animals in the diagram belong to the same genus?

Classification of Four Animals

Some classification groups contain a large number of species. For example, the order Coleoptera contains over 360,000 species, including the boll weevil. Other orders may have just a few species. For example, the African elephant and the Asian elephant are the only two species in the order Proboscidea.

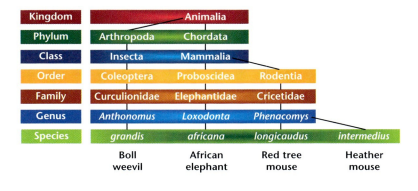

Learning Styles

Logical/Mathematical
Have students consider why the levels of classification appear as an inverted triangle on page 44. That is, why is it appropriate for the bars to become smaller as you descend through the levels of classification? (*Each successive subgroup becomes smaller, or has fewer kinds of animals in it, until you reach the species level, in which there is only one kind of animal.*)

Science Journal

Have students write an explanation for why a mouse and a rabbit are more closely related than a mouse and a dog.

Did You Know?

Read aloud the Did You Know? on page 45. Have students tell why they think Aristotle did not have difficulty classifying microorganisms and fungi as plants or animals. (*He would not have known about microorganisms because there were no microscopes. Fungi could definitely be excluded from the animals' group even though they differ from most plants.*)

Learning Styles

LEP/ESL
Have students who are learning English label pictures of familiar plants and animals with the common names in their native language. Provide one or more common names for the animals and plants in English and have students add these labels. Then have students use an encyclopedia to find the scientific name of each plant and animal. Invite volunteers to "introduce" the organisms to the class. They might compare organisms' common names to "nicknames" and their scientific names to "real names."

Scientific name
The name given to each species, consisting of its genus and its species label

Scientific Names

Most people call animals by their common names, such as mockingbird and mountain lion. However, using common names can be confusing. The mountain lion in the photo has at least four other common names: puma, cougar, catamount, and American panther. All five names refer to the same species. People who use one of these names may not know that the other names refer to the same species. The opposite problem occurs with the common name "June bug." At least a dozen different beetle species have that name. When someone says "June bug," you have no way of knowing which species that is. The same animal may have different names in different languages too. For example, an owl is called *gufo* in Italian, *hibou* in French, and *búho* in Spanish.

To overcome these problems, biologists give each species a **scientific name**. An organism's scientific name consists of two words. The first word is the organism's genus, and the second word is its species label. For example, the scientific name of the mountain lion is *Felis concolor*. Thus, the mountain lion belongs to the genus *Felis* and the species *concolor*. Look again at the diagram on page 45. What is the scientific name of the African elephant?

The mountain lion has several common names but only one scientific name: Felis concolor.

The scientific name given to each species is unique. This means that different species have different scientific names, even if they have the same common name. Scientific names are in Latin, so they are recognized by biologists around the world. For example, *Felis concolor* means the same thing in France, the United States, and Mexico. As you may have noticed, scientific names are always printed in *italics* or are underlined. The first word in the name is capitalized, but the second word is not.

Science in Your Life

Has everything been classified?

You may think that every kind of organism on Earth has already been studied, classified, and named. In fact, biologists continue to discover species that no one has identified before. Many of the newly found species are insects. Some biologists think there could be millions of insect species that still have not been identified.

Some new species may be useful in finding new medicines. Others may help control pests that damage crops. To learn how a new species might be useful, biologists must study the organism closely. They must learn how it carries out its life activities. Then they can classify the organism.

Science In Your Life

Read aloud the boldfaced question in the Science in Your Life feature on page 47. Ask students to predict the answer and give the reasons they think as they do. Have volunteers read the paragraphs aloud. Discuss regions where students think it is likely there are undiscovered species. (*deep sea habitats, tropical rainforests, deserts, and mountains—places where it is difficult for scientists to explore in any depth*)

IN THE COMMUNITY

Have students work in groups to brainstorm a list of dairy foods. Then have them check in a grocery store to see which items are stocked together. Finally, have groups set up a classification system for the products with categories based on common characteristics. Each category should then be broken down into subcategories, with two or more levels based on common characteristics. Have groups set up diagrams for their systems and then compare their diagrams to those of others.

Lesson 1 Review Answers

1. on the similarities of organisms' features, such as appearance, cell structure, hereditary material, and means of getting food and reproducing
2. kingdom, phylum, class, order, family, genus, and species
3. a single type of organism, capable of breeding with other members of its species to produce offspring like themselves
4. banana slug
5. *Tyto alba*

Portfolio Assessment

Sample items include:
- Sentences from Teaching the Lesson
- Lesson 1 Review answers

Lesson 1 REVIEW

Write your answers to these questions on a separate sheet of paper. Write complete sentences.

1. On what do biologists base their classification of organisms?
2. List the seven levels of classification of organisms, from highest to lowest.
3. What is a species?
4. The banana slug and the cuttlefish belong to the same phylum. The clownfish belongs to a different phylum. Is the banana slug or the clownfish more similar to the cuttlefish?
5. The barn owl belongs to the genus *Tyto* and the species *alba*. What is the barn owl's scientific name?

Technology Note

As scientists continue to research new and existing species of animals, they use computers to store and study their data. Computers also allow scientists to share information with other scientists. Large amounts of information about animal species are available to you on the Internet. You can get this information by using an Internet search engine. Search for words such as *vertebrates*, *invertebrates*, *mammals*, and *birds*. Also search for *zoos*, *aquariums*, and *universities*. These places often have Web sites that provide interesting information about animal groups.

Workbook Activity 7

INVESTIGATION 3

Classifying Objects

Purpose
How are objects classified? In this investigation, you will make a classification system for objects found in your classroom.

Materials
- assortment of objects found in a classroom

Procedure

1. Form a team with two or three other students. On a sheet of paper, make a list of objects in your classroom. Include objects that may be on shelves or in drawers and cabinets.

2. Divide the objects on your list into groups based on their similarities. Name each group.

3. Make up a classification system for the objects on your list. Your system should have several levels. Each level should include all of the groups in the next-lower level.

4. Write your classification system on a sheet of paper. List the objects that belong in each group. Show how the different levels are related to each other.

5. Compare your classification system with the system made up by other student teams.

Investigation 3

This investigation will take approximately 30 minutes to complete. Students will use these process and thinking skills: observing, listing, classifying, and identifying similarities and differences.

Preparation

- Rather than having students classify only classroom objects, you may wish to supplement by bringing in various household objects. Include objects made of different materials with a range of shapes, sizes, and uses.

- Objects could include desk items (pens, pencils, paper clips, staples, tape, erasers, rulers, rubber bands, index cards, sticky notes), eating utensils, lab instruments, shoes, combs, brushes, and small hardware (nuts, bolts, screws, washers).

- Students may use Lab Manual 8 to record their data and answer the questions.

Procedure

- This investigation is best done with students working in groups of four. Have all students participate in identifying and grouping objects, naming groups, and devising a classification system.

- You might have students list the objects on index cards and work together to establish groups. Assign one group member to write the names of classification groups and subgroups the students organize.

- Encourage students to think of as many features as possible to help them with the classification process. They might consider composition, shape, size, use, and color.

- If students discover that some objects can be placed into more than one group, tell them to choose the placement that makes the most sense to them.

SAFETY ALERT

♦ If you bring in objects, avoid items, such as pins and glassware, that are sharp or otherwise dangerous.

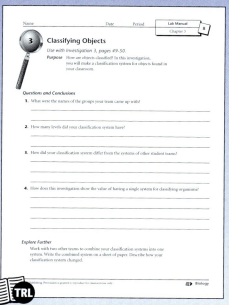

Lab Manual 8

Results

The results of this investigation will depend on the objects that are available for students to classify and on the students' creativity. Expect students to devise a variety of classification systems.

Questions and Conclusions Answers

1. The names of groups will vary. Possible groups include eating utensils, fasteners, toiletry supplies, things found in a desk.
2. The number of levels will vary, but students should be able to come up with at least three levels.
3. Groups' systems will probably vary in their levels and names for groups.
4. The investigation shows that objects can be classified in more than one way. Using a single system of classification eliminates confusion and duplication of naming.

Explore Further Answers

Combining classification systems will mean that the new chart looks like neither of the original two. Encourage groups to preserve the most logical parts of each system. Emphasize that they may choose to change names of some levels completely.

Assessment

Check that students' classification systems place each object in only one classification group. For example, if a classification system contains the groups "desk items" and "small hardware" at the same level, paper clips could fit in either group but not both. Include the following items from this investigation in student portfolios.

- Classification system
- Answers to Questions and Conclusions
- Explore Further revisions and descriptions

Questions and Conclusions

1. What were the names of the groups your team came up with?
2. How many levels did your classification system have?
3. How did your classification system differ from the systems of other student teams?
4. How does this investigation show the value of having a single system for classifying organisms?

Explore Further

Work with two other teams to combine your classification systems into one system. Write the combined system on a sheet of paper. Describe how your classification system changed.

Lesson 2: Vertebrates

Objectives

After reading this lesson, you should be able to
- list the main features of all vertebrates.
- describe the features of the different vertebrate classes.

Vertebrate
An animal with a backbone

Cartilage
A soft material found in vertebrate skeletons

Vertebra
One of the bones or blocks of cartilage that make up a backbone

The animals that are probably most familiar to you are animals with backbones. These animals are called **vertebrates**. Vertebrates include tiny hummingbirds and enormous blue whales. Humans also are vertebrates. Altogether, there are nearly 50,000 species of vertebrates in the world.

Features of Vertebrates

Vertebrates have three features that set them apart from other animals. First, all vertebrates have an internal skeleton, which is inside their body. The skeleton of vertebrates is made of bone or a softer material called **cartilage**. Some other animals also have an internal skeleton, but it is made of different materials.

The second feature of vertebrates is their backbone. A backbone is made up of many small bones or blocks of cartilage. For example, the human backbone contains 26 bones. Each bone or block of cartilage in the backbone is called a **vertebra**. That is why animals with backbones are known as vertebrates.

The third feature of vertebrates is the skull. The skull surrounds and protects the brain. Look for the backbone and skull in the skeleton of this cow.

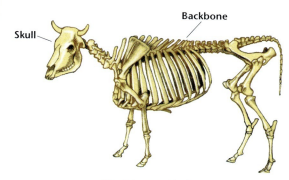

Vertebrate Skeleton

Classifying Animals Chapter 3   51

1 Warm-Up Activity

Display a model of a vertebrate skeleton and have students observe how bones are joined to allow movement. Ask students to point out the backbones and skull. Allow them to observe that these structures are composed of many individual bones. In the skull, some of the joints are fixed, or fused together by connecting material. Contrast the movable joints in the backbone with the fixed joints in the skull. Have students speculate about the advantages of the different kinds of joints in these structures. (*The backbone's movable joints allow for bending and twisting. The skull must protect the brain, so it needs fixed joints to make it rigid.*)

2 Teaching the Lesson

Have students make a table with the headings "Bony Fish," "Cartilage Fish," "Jawless Fish," "Amphibians," "Reptiles," "Birds," and "Mammals." As they read the lesson, they can list the characteristics of each type of vertebrate under the appropriate heading.

Fill a 1-L beaker with water. Ask students to observe what happens when you do the following: Place a small, inflated balloon on the water. (*It floats.*) Remove the balloon and place a drop of cooking oil on the water. (*It floats.*) Drop a chicken bone into the beaker. (*It sinks.*) Explain that bony fishes use a swim bladder filled with air to keep from sinking. Sharks avoid sinking by storing oil in their bodies. Otherwise, both kinds of fishes would sink like the bone.

Have students name ways amphibians are like fish and ways they are like reptiles. (*For part of their lives, amphibians breathe with gills and live in water; as adults, they breathe with lungs and live on land, although they must stay moist and lay eggs in water.*) Explain that scientists believe amphibians evolved from fishlike ancestors, and their life cycle mimics that evolution. They are considered a "link" between fish and reptiles.

Gill
A structure used by some animals to breathe in water

Swim bladder
A gas-filled organ that allows a bony fish to move up and down in water

Amphibian
A vertebrate that lives at first in water and then on land

Metamorphosis
A major change in form that occurs as some animals develop into adults

---

Vertebrates are divided into seven classes. Three of the classes consist of different types of fish. The other four classes are amphibians, reptiles, birds, and mammals.

Fish

Biologists have identified about 24,000 species of fish. There are more species of fish than of any other kind of vertebrate. All fish live in water and breathe with structures called **gills**.

A bony fish is covered with scales.

Most fish have a skeleton made of bone and are called bony fish. This first type includes bass, trout, salmon, and many others. You can see in the photo that the body of a bony fish is covered with scales that overlap like roof shingles. The scales protect the fish and give it a smooth surface. Many bony fish have an organ called a **swim bladder** that is filled with gas. By changing the amount of gas in its swim bladder, the fish can move up or down in water.

Sharks, rays, and skates make up the second type of vertebrate. They have a skeleton made of cartilage instead of bone. Many of these fish have powerful jaws and rows of sharp teeth. Their tiny, toothlike scales make their skin feel like sandpaper. Lampreys and hagfish, the third type, are jawless fish. They also have a skeleton made of cartilage, but they have no jaws or scales.

Amphibians

Amphibians include about 5,000 species of frogs, toads, and salamanders. The word *amphibian* comes from two Greek words meaning "double life." This refers to the fact that many amphibians spend part of their life in water and part on land. Recall from Chapter 1 that frogs begin their life as tadpoles that live in water. After a while, a tadpole grows legs, loses its gills and tail, and develops into an adult frog. This change is called **metamorphosis**. The frog may spend much of its life on land.

TEACHER ALERT

Students may feel that fish should be classified with other marine life such as lobsters and crabs. Explain that scales are not an external skeleton but a body covering, like hair or feathers. All fish have a backbone while invertebrates, such as lobsters and crabs, have an external skeleton.

Reptile
An egg-laying vertebrate that breathes with lungs

Cold-blooded
Having a body temperature that changes with temperature of surroundings

Adult amphibians breathe with lungs or through their skin. The skin is thin and moist. To keep from drying out, amphibians must stay near water or in damp places. Since amphibian eggs do not have shells, they must be laid in water or where the ground is wet.

Reptiles

Snakes, lizards, turtles, alligators, and crocodiles are **reptiles**. There are about 7,000 species of reptiles. Some reptiles, such as sea turtles, live mostly in water. Others, such as tortoises, live on land. The skin of reptiles is scaly and watertight, so reptiles can live in dry places without drying out. Some tortoises, for example, live in deserts where water is scarce. Most reptiles lay eggs on land. The eggs have a soft shell that keeps the young inside from drying out. All reptiles breathe with lungs. Reptiles that live in water must come to the surface to breathe.

Dinosaurs were reptiles. The first dinosaurs appeared about 235 million years ago. Some dinosaurs were taller than a four-story building and heavier than ten elephants. However, many dinosaurs were no bigger than a house cat. All dinosaurs became extinct about 65 million years ago.

Fish, amphibians, and reptiles are **cold-blooded** animals. Their body temperature changes with the temperature of their surroundings.

Amphibians have smooth, moist skin. The skin of reptiles is dry and scaly.

3 Reinforce and Extend

CROSS-CURRICULAR CONNECTION

Geography
Have students use a field guide to learn about a species of North American bird that migrates. They can find where it spends the winter and how far it migrates. Ask students to compare the climate at this time in both places where the bird lives.

ONLINE CONNECTION

Have students enjoy a virtual tour of an aquarium, where they can see many species of sea life. They can visit the Web site www.neaq.org/vtour/aqtour.html or do a search by typing in aquarium tour. They can then choose a site to visit. Ask students to identify the animal on the tour that most interested them and why.

LEARNING STYLES

Logical/Mathematical
Give groups of students pictures of animals with backbones. Explain why these animals are classified together. Ask them to make a "tree" diagram showing how they would break this large group into smaller groups. Ask students to explain the logic of their classifications. After groups share their classifications, show students how the vertebrates are organized into classes: mammals, amphibians, reptiles, birds, fish.

GLOBAL CONNECTION

Have students investigate the kinds of vertebrates that people in other cultures use as food. Challenge them to look for examples that we might consider "unusual." Examples include the axolotl, a salamander that people in parts of Mexico eat, and the pufferfish, which the Japanese consume as *fugu* though it contains a deadly poison in its internal organs. Students can share their information with the class.

LEARNING STYLES

Body/Kinesthetic
Have students prepare a skit to reinforce the meaning of the term *metamorphosis*. They might use props and costuming to show the stages that a frog goes through. Ask students to present their skit to the class.

Birds

There are more than 9,000 species of birds, and almost all of them can fly. Feathers make flight possible by providing lift and smoothing the lines of the body. Birds also have hollow bones, which keep their skeleton light. Flying requires a lot of energy, so birds cannot go long without eating. Feathers act like a warm coat that keeps heat inside the bird's body. All birds breathe with lungs and have a horny beak. Birds lay eggs that are covered by a hard shell. As **warm-blooded** animals, birds and mammals have a body temperature that stays the same.

Warm-blooded
Having a body temperature that stays the same

Mammary gland
A milk-producing structure on the chest or abdomen of a mammal

Penguins, ostriches, and emus are classified as birds, but they do not fly. Like other birds, they have lungs, feathers, beaks, and lay eggs covered by a hard shell. Ostriches and emus have strong legs and can run quickly. Penguins are good swimmers.

Mammals

Mammals are named for their **mammary glands**, which are milk-producing structures on the chest or abdomen. As shown in the photo, female mammals nurse their young with milk from these glands. Mammals also have hair covering most of their body. Hair helps keep in body heat. Most mammals live on land, but some, such as whales and porpoises, live in water. All mammals have lungs.

More than 4,000 species of mammals have young that develop inside the mother. These mammals include bears, elephants, mice, and humans. About 300 species of mammals, including opossums and kangaroos, have young that develop in a pouch on the mother. The duck-billed platypus and the spiny anteater are the only mammals that lay eggs.

Science Myth

Whales, dolphins, and porpoises are classified as fish.

Fact: Whales, dolphins, and porpoises are mammals. They do look similar to fish and live in the ocean. However, they have lungs instead of gills and must swim to the surface to breathe air. They also have hair and mammary glands.

Mammals feed their young with milk produced by mammary glands.

54 Chapter 3 Classifying Animals

CROSS-CURRICULAR CONNECTION

Mathematics
Have students list the number of species of fishes, amphibians, reptiles, birds, and mammals there are. Ask students to use their list as a source of information for a bar graph. After the bar graphs are made, use them to compare and contrast the number of species in the classes of vertebrates.

SCIENCE JOURNAL

Have students write a short story with themselves as a fledgling bird trying to learn to fly. What are their experiences? How do they feel when they succeed?

Science Myth

Show photographs of a whale, dolphin, and porpoise. Invite students to tell what they know about these animals. Read aloud the myth portion (first sentence) of the feature on page 54. Have students predict what is false about the statement. After reading the "Fact," ask students to check their prediction. Point out to students that because humans and whales are both mammals, humans are more closely related to whales than fish are.

LEARNING STYLES

Auditory/Verbal
Provide a tape recording of animal sounds, such as bird calls, or have students tape-record the sounds of animals. Ask students to research information about one of the animals on the tape and to prepare their own audiotape presentation about it. Encourage them to begin or end their presentation with a recording of the animal making sounds, such as the trumpeting of an elephant. Place the tapes in the science center where students can access them.

LEARNING STYLES

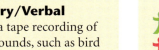
Visual/Spatial
Have students find out about the sizes, colors, and markings of eggs that different birds lay. Ask them to prepare a chart, drawing, or models of the eggs. Each illustration should identify the bird that laid the egg and note the egg's color, size, and markings. Encourage students to provide information about at least ten eggs, including the largest and the smallest bird eggs.

CAREER CONNECTION

Invite a zoo or pet store worker to speak to students about caring for a variety of vertebrates. The speaker might talk about how animals are kept healthy, what they need to eat, and what environment each needs.

Lesson 2 REVIEW

Write your answers to these questions on a separate sheet of paper. Write complete sentences.

1. What three features do all vertebrates have?
2. How does a trout's skeleton differ from a shark's skeleton?
3. What happens during metamorphosis in a frog?
4. Why are a reptile's eggs able to survive in dry places?
5. What two features do mammals have that other vertebrates do not have?

Science at Work

Zookeeper

Zookeepers should have good observation skills and strong communication skills. They should be able to deal with emergencies and solve problems. Usually a college degree in zoology, biology, or an animal-related field is needed.

Because of the variety of animals in a zoo, a zookeeper might take care of mammals, birds, fish, amphibians, reptiles, or even invertebrates. The job may include making sure the animals are clean and that they get enough food and exercise. The zookeeper may also have to keep the animals' living area clean. Zookeepers must also notice any changes in the appearance or behavior of the animals. Changes could mean the animal is sick or has another problem.

Zookeepers talk to people who visit the zoo. They answer questions and help teach visitors about the animals.

Classifying Animals Chapter 3 55

Lesson 2 Review Answers

1. All vertebrates have an internal skeleton made of bone or cartilage, a backbone, and a skull. 2. A trout's skeleton is made of bone whereas a shark's skeleton is made of cartilage. 3. A tadpole grows legs, loses its gills and tail, and develops into an adult frog. In the process, it changes from a water dweller to an organism that is capable of living on land. 4. The eggs have a soft shell that helps keep the young inside from drying out. 5. Mammals have mammary glands to feed young and hair to hold in body heat.

Portfolio Assessment

Sample items include:
- Diagram showing characteristics of vertebrate classes from Teaching the Lesson
- Lesson 2 Review answers

Science at Work

Read the Science at Work feature on page 55 together. Invite volunteers to describe several emergency situations at a zoo that a zookeeper might need to handle. Ask how students think the zookeeper could deal with the situation. Then have students explain why good powers of observation are essential to this worker.

Students might read further about this career by visiting zoo or career sites online. Invite interested students to write about which kind of animals they would most like to care for at a zoo, and why.

Workbook Activity 8

Lesson at a Glance

Chapter 3 Lesson 3

Overview In this lesson, students learn characteristics of eight phyla of invertebrates and explore the diversity of this large group.

Objectives

- To describe the features of sponges and cnidarians
- To distinguish between flatworms, roundworms, and segmented worms
- To describe the features of mollusks, echinoderms, and arthropods

Student Pages 56–62

Teacher's Resource Library

Workbook Activity 9
Alternative Activity 9
Lab Manual 9
Resource File 6

Vocabulary

invertebrate	mollusk
cnidarian	arthropod
radial symmetry	molting
tentacle	crustacean
flatworm	arachnid
bilateral symmetry	complete metamorphosis
roundworm	incomplete metamorphosis
segmented worm	pupa
	tube foot

Science Background
Invertebrates

Most of the world's animals are invertebrates, a diverse group with little in common except the absence of a backbone. Sponges are the simplest invertebrates, unique in their lack of tissues and type of development. All other invertebrate phyla fall into two main groups on the basis of body symmetry. Cnidarians (such as jellyfish) have radial symmetry. They have a top and bottom but no left and right sides or front and rear ends. Most radially symmetrical animals either live attached to one place or drift through the water.

56 Chapter 3 Classifying Animals

Lesson 3 — Invertebrates

Objectives

After reading this lesson, you should be able to
- describe the features of sponges and cnidarians.
- distinguish between flatworms, roundworms, and segmented worms.
- describe the features of mollusks, echinoderms, and arthropods.

Every animal that is not a vertebrate is called an **invertebrate**. An invertebrate is an animal that does not have a backbone. Invertebrates make up about 97 percent of all animal species and belong to more than 30 phyla. You will learn about eight of those phyla in this lesson.

Sponges

Sponges are the simplest animals. Their bodies consist of two layers of cells without any tissues or organs. All 10,000 species of sponges live in water. Sponges strain food particles out of the water as the water moves through their body. The water enters through pores in the body wall. If you use a natural bath sponge, you are using the skeleton of a dead sponge.

Cnidarians

Cnidarians include animals such as jellyfish, corals, and hydras. There are about 10,000 species of cnidarians. All live in water. Cnidarians have body parts that are arranged like spokes on a wheel. This type of arrangement of body parts is known as **radial symmetry**. You can see the radial symmetry of a sea anemone in the photo. Cnidarians have armlike **tentacles** with stinging cells. The tentacles capture small prey and push them into the body, where they are digested.

Invertebrate
An animal that does not have a backbone

Cnidarian
An invertebrate animal that includes jellyfish, corals, and hydras

Radial symmetry
An arrangement of body parts that resembles the arrangement of spokes on a wheel

Tentacle
An armlike body part in invertebrates that is used for capturing prey

The tentacles of this sea anemone show radial symmetry.

56 Chapter 3 Classifying Animals

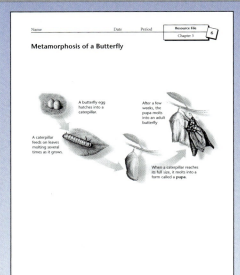

Resource File 6

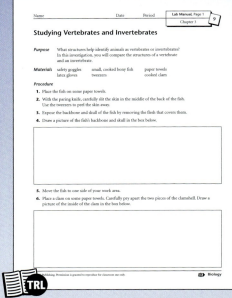

Lab Manual 9, pages 1–2

Flatworm
A simple worm that is flat and thin

Bilateral symmetry
A body plan that consists of left and right halves that are the same

Roundworm
A worm with a smooth, round body and pointed ends

Segmented worm
A worm whose body is divided into sections, such as earthworms or leeches

 Did You Know?
Leeches can be very helpful to doctors. A doctor might place a leech on a patient's bruise to help drain the blood from the injured tissue.

Flatworms

As their name suggests, **flatworms** are flat and thin. Their bodies have a left half and a right half that are the same. This type of body plan is known as **bilateral symmetry**. There are more than 20,000 species of flatworms. Most are parasites that live on or inside other animals. An example of a flatworm that is a parasite is the tapeworm. Tapeworms live in the intestines of vertebrates, including humans. In the intestine, tapeworms absorb nutrients through their skin. People can get tapeworms when they eat infected meat that has not been cooked completely.

Roundworms

Roundworms have long, round bodies that come to a point at the ends. Like flatworms, roundworms have bilateral symmetry. Most of the 80,000 species of roundworms are not parasitic. They may live in the soil or in water. Some soil-dwelling roundworms help plants by eating insect pests. About 150 species of roundworms are parasites, and many of them live in humans. For example, hookworms settle in the intestine and feed on blood. Hookworms enter the body by boring through the skin. That usually happens when people walk barefoot in places that are not clean.

Segmented Worms

Segmented worms have a body that is divided into many sections, or segments. These worms may live in the soil, in freshwater, or in the ocean. The earthworm is the most familiar of the 15,000 species of segmented worms. Earthworms tunnel through the soil, eating small food particles. Their tunnels loosen the soil and allow air to enter it, which helps plants grow. Leeches are another kind of segmented worm. Many leeches eat small invertebrates, but some leeches are parasites. Leeches that are parasites attach to the skin of a vertebrate and feed on its blood. While feeding, leeches release a chemical that keeps the blood flowing.

Classifying Animals Chapter 3 57

Did You Know?

After they read Did You Know? on page 57, tell students that leeches are also called bloodsuckers. Ask why this might be a particularly appropriate nickname. Then tell students that not all leeches are parasitic; some feed on dead plants and animals.

Teacher Alert

 Because they have less interaction with invertebrates, many students may be surprised to learn that there are many times more types of invertebrates than there are vertebrates. Point out that, not only are there many more species of invertebrates, but they tend to produce more offspring and therefore to be more numerous.

Their symmetry suits them to interact with their environment equally well in any direction. Flatworms, roundworms, segmented worms, mollusks, and arthropods have bilateral symmetry (as do vertebrates). Most bilaterally symmetrical animals move actively through their environment. They usually have sensory structures concentrated at the front end, which better equip them to detect food and danger as they move.

Echinoderms (such as sea stars and sand dollars) are radially symmetrical as adults but bilaterally symmetrical during development. Because their development resembles that of vertebrates in several ways, biologists consider echinoderms to be the vertebrates' closest relatives among the invertebrates.

 ### Warm-Up Activity

Display photographs of invertebrates that represent a variety of phyla, such as earthworms, jellyfish, snails, crabs, spiders, beetles, mosquitoes, starfish, and pill bugs. Have students observe each animal and describe its body structure, body covering, and habitat if shown. Discuss characteristics the invertebrates have in common (e.g., no backbone) and differences among them (e.g., kind and number of appendages).

 ### Teaching the Lesson

As they read the lesson, have students list each group that is described and organize them in a diagram showing how they are classified. (*Crustaceans, arachnids, insects should be listed under arthropods while entries for mollusks, sponges, and others align with the arthropod entry.*)

Display photos of a sponge, a cnidarian, and an arthropod. Ask students how many ways they could slice straight through the body of each animal to produce symmetrical halves. (*The sponge, with no body symmetry, cannot be sliced this way. There are many ways to divide the cnidarian, which has radial symmetry. In the arthropod, which has bilateral symmetry, there is one way—along the middle of the body.*)

Classifying Animals Chapter 3 57

TEACHER ALERT

Point out to students that spiders are very beneficial to humans. They kill more harmful insects than all the birds in the world do. Spiders also kill more insects than commercial insecticides. Without spiders, the insect population could not be kept under control.

3 Reinforce and Extend

LEARNING STYLES

Interpersonal/Group Learning

Have small groups of students design and carry out an experiment to observe how earthworms change soil. Have them combine layers of various types of soil in a jar and add some humus and apple peelings on top. They can carefully dig up earthworms and add them to the jar along with enough water to keep the soil damp. The jar should be kept in a cool place and covered with dark paper. Groups can remove the paper once a day to observe the jar and record their observations. After a week, have them conclude how earthworms live in the soil and how they change it.

CROSS-CURRICULAR CONNECTION

Health

Have students investigate the temperatures to which meats must be cooked to ensure that parasites such as tapeworms and trichinella worms are killed. They can report on these worms' life cycles and the way they infect humans.

Mollusk
An invertebrate divided into three parts

Arthropod
A member of the largest group of invertebrates, which includes insects

Molting
The process by which an arthropod sheds its external skeleton

Mollusks

There are more than 112,000 species of **mollusks**. These are invertebrates that are divided into three parts: head, body, and foot. Some live on land, while others live in freshwater or in the ocean. Snails and slugs make up the largest group of mollusks. Snails have a coiled shell, but slugs have no shell at all. Another group of mollusks includes clams, scallops, and oysters. Their shell is made of two hinged pieces that can open and close. Squids and octopuses have no outer shells. These mollusks can swim quickly as they hunt for fish and other animals.

A squid uses its tentacles to capture prey.

Arthropods

Arthropods are the largest group of invertebrates. They make up more than three-fourths of all animal species. The major groups of arthropods are crustaceans, arachnids, centipedes, millipedes, and insects. Arthropods are segmented animals with jointed legs. Most arthropods also have antennae, which they use to feel, taste, or smell.

Arthropods shed their external skeleton as they grow.

All arthropods have an external skeleton that supports the body and protects the tissues inside. If you ever cracked open the claw of a crab, you know how hard this skeleton can be. Arthropods can bend their bodies because they have joints in their legs and between their body segments. However, an external skeleton is not able to grow as an internal skeleton does. For that reason, an arthropod must shed its skeleton to grow in size. The shedding process is called **molting**, which is shown in the photo. An arthropod begins to produce a new skeleton before it molts. After the animal molts, the skeleton takes a few days to harden completely. The soft-shelled crabs served in restaurants are crabs that have just molted.

58 Chapter 3 Classifying Animals

AT HOME

Have students search at home for foods that contain mollusks or crustaceans and interview adults at home to find out different forms of mollusks they have eaten. Examples of mollusks include abalone, oysters, snails, clams, and squid. Examples of crustaceans might include crab, lobster, and crayfish.

SCIENCE INTEGRATION

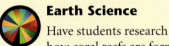

Earth Science

Have students research how coral reefs are formed, their locations, and their effects on the surrounding geographic area. After students report, discuss possible results of destruction of coral reefs due to pollution and rough use. (*Coral reefs provide homes for diverse populations of organisms, which would be endangered if coral reefs die; they also protect shores from erosion, a process which could not be stopped if the reefs were no longer present.*)

Crustacean
A class of arthropods that includes crabs, lobsters, crayfish, and sow bugs

Arachnid
A class of arthropods that includes spiders, scorpions, mites, and ticks

Crabs, lobsters, and crayfish are **crustaceans**. Most of the 40,000 species of crustaceans live in rivers, lakes, and oceans. Crustaceans have five pairs of legs. Some of the legs have small claws that help the animal handle food. The two legs closest to the head usually have powerful claws used for protection. Sow bugs and pill bugs are crustaceans that live on land. You can often find them under rocks and in other moist places.

Spiders, scorpions, mites, and ticks are **arachnids**. There are about 70,000 species of arachnids. Almost all arachnids live on land. They have four pairs of legs. Spiders produce threads of silk to spin webs and build nests. Most spiders eat insects, but some also catch small fish or frogs. Spiders capture their prey by injecting it with a poison. Scorpions also use poison to capture prey. They use a stinger to inject the poison. Mites and ticks include species that live on the human body. Mites feed on hair and dead skin. Ticks pierce the skin and feed on blood.

All of the 2,500 species of centipedes and the 10,000 species of millipedes live on land. Their bodies have up to 175 segments. Notice in the photo that centipedes have one pair of legs on each body segment. They can run quickly because their legs are long. Centipedes use their poison claws to kill insects and other prey. Millipedes have two pairs of legs on each body segment. Their legs are short, so millipedes move slowly. Most millipedes eat dead plant matter in the soil.

Centipedes have a pair of legs on each body segment.

LEARNING STYLES

Auditory/Verbal
Display a live centipede and millipede in separate aerated jars. Ask students to observe these arthropods and point out similarities and differences. Discuss which is more likely to be a predator and which probably feeds on vegetable matter. (*Centipedes move quickly to catch prey; millipedes move slowly and eat mostly decaying vegetable matter.*) Handle the animals carefully and release them when the activity is complete.

CROSS-CURRICULAR CONNECTION

Art
Have students observe spider webs and the spiders that make them and then create drawings or wall hangings illustrating them. Invite artists to discuss the shapes used in creating webs and the textures, colors, and types of lines necessary to render the spider web.

ONLINE CONNECTION

Students will find interesting resources about invertebrates by visiting the following site:can-do.com/uci/lessons98/Invertebrates.html.

The site has good lessons and visuals to reinforce text learning about invertebrates.

LEARNING STYLES

Logical/Mathematical Have students add the figures in the lesson to determine how many more species of arthropods there are than in all other invertebrate phyla combined. Then have them calculate the percentage of invertebrates that are insects.

IN THE ENVIRONMENT

Have students research alternatives to chemical insecticides, such as importing organisms to control harmful insect pests on farms. Suggest that students name some of the organisms used for this purpose and tell what insects they eat. (*For example, humans might try to establish a bat nesting area in an area that needs the mosquito population reduced.*)

Complete metamorphosis
Changes in form during development in which earlier stages do not look like the adult

Incomplete metamorphosis
Changes in form during development in which earlier stages look like the adult

Pupa
A stage in the development of some insects that leads to the adult stage

Insects are also arthropods. The nearly one million species of insects live almost everywhere except in the deep ocean. Insects include mosquitoes, flies, ants, and beetles. Insects have three pairs of legs. Most have one or two pairs of wings. Insects are the only invertebrates that can fly. Like frogs, most insects go through metamorphosis. Study the metamorphosis of a butterfly in the diagram. Notice that the butterfly metamorphosis has four stages. The first three stages look nothing like the adult butterfly in stage four. This kind of metamorphosis is known as **complete metamorphosis**. Frogs also develop by complete metamorphosis. A tadpole looks very different from an adult frog.

Some insects such as grasshoppers develop by **incomplete metamorphosis**. Grasshopper eggs hatch into a stage that looks similar to an adult grasshopper. Crickets and cockroaches also develop by incomplete metamorphosis.

Many insects are pests. Grasshoppers and caterpillars destroy crops. Fleas and mosquitoes may carry microorganisms that cause diseases. However, many insects are helpful to humans. Bees and other insects spread pollen from flower to flower. Without pollen, the flowers could not produce fruits. Insects also make useful products, such as honey, wax, and silk.

Stage 1: A butterfly egg hatches into a caterpillar.

Stage 2: A caterpillar feeds on leaves, molting several times as it grows.

Stage 3: When a caterpillar reaches its full size, it molts into a form called a pupa.

Stage 4: After a few weeks, the pupa molts into an adult butterfly.

Complete Metamorphosis

Tube foot
A small structure used by echinoderms for movement

Echinoderms

Echinoderms include sea stars, sea urchins, sand dollars, and sea cucumbers. All 7,000 species of echinoderms live in the ocean. Like cnidarians, echinoderms have radial symmetry. Find the echinoderm's **tube feet** in the photo. The tube feet attach firmly to surfaces. Echinoderms use their tube feet to move.

Echinoderms, such as this sea star, are invertebrates with tube feet and radial symmetry.

SCIENCE INTEGRATION

Earth Science
Ask students to read articles that debate the wisdom of insecticide use. Why do critics believe that the harm pesticides cause outweighs the benefits? Why do supporters insist that pesticides are essential if farmers are to prosper? Ask students to choose a position and debate this issue.

CROSS-CURRICULAR CONNECTION

Literature
Have students read Rachel Carson's *Silent Spring* and prepare a report on the book. Encourage them to be creative in their report, which might be a video, poster, or written presentation.

LEARNING STYLES

Visual/Spatial
Have students draw a picture of the stages of metamorphosis for a butterfly on four separate sheets. Mix up the order and ask students to identify the stages and place them in the correct order.

SCIENCE INTEGRATION

Physical Science
Have students study the appendages of an arthropod and observe the animal using its appendages in differing ways. Encourage students to compare the appendages to the three types of levers.

Lesson 3 Review Answers

1. Sponges strain food particles out of the water that moves through their bodies.
2. In radial symmetry, body parts are arranged like spokes on a wheel. In bilateral symmetry, the body has identical left and right halves.
3. Tapeworms are flatworms, hookworms are roundworms, and earthworms and leeches are segmented worms.
4. Their external skeleton does not grow, so they must shed their skeleton to grow in size.
5. They attach their tube feet to surfaces and pull themselves along.

Portfolio Assessment

Sample items include:
- Diagram of invertebrates from Teaching the Lesson
- Lesson 3 Review answers

Achievements in Science

Read the Achievements in Science feature on page 62 together. Then ask students to summarize ways that invertebrates are important to ecosystems.

Assign an interested student the task of reading about Lamarck's life in more detail and reporting additional facts about his life.

Lesson 3 REVIEW

Write your answers to these questions on a separate sheet of paper. Write complete sentences.

1. How do sponges feed?
2. Contrast radial symmetry and bilateral symmetry.
3. Give an example of a flatworm, roundworm, and segmented worm.
4. Explain why arthropods molt.
5. How do echinoderms move?

Achievements in Science

Study of Invertebrates Begins

In the late 1700s, few scientists thought that insects and worms were important enough to study. The word *invertebrate* did not even exist. The classification of this group of organisms was not made until a French professor began to study insects and worms.

In 1793 Jean Lamarck became a professor of insects and worms at France's National Museum of Natural History. He was not familiar with these organisms, and the collection at the museum was poorly organized. Lamarck studied, researched, and classified these organisms. He was the first person to use the word *invertebrate*. His work resulted in the study of invertebrates becoming a new field of biology.

By studying invertebrates, we know how important they are to our ecosystems. Invertebrates help build a healthy environment, and they serve as food for countless other animals. Today scientists continue to study and identify all the different invertebrates.

Workbook Activity 9

Chapter 3 SUMMARY

- Biologists classify animals based on the animals' similar features.
- The classification system used by biologists has seven levels.
- Every species has a two-word scientific name consisting of its genus and its species label.
- Vertebrates are animals that have a backbone.
- Fish live in water and breathe with gills.
- Amphibians live in water or in damp places on land.
- Reptiles can live in dry places because their skin is watertight.
- Birds have feathers and hollow bones. These features make flight possible.
- Mammals have hair and feed their young with milk from mammary glands.
- Invertebrates are animals without a backbone.
- Sponges have no tissues or organs.
- Cnidarians have radial symmetry and use tentacles to capture prey.
- Most flatworms and some roundworms are parasites.
- Segmented worms have a body that is divided into many segments.
- Many mollusks have a hard shell. Squids and octopuses capture prey with their tentacles.
- Arthropods have a segmented body, external skeleton, and jointed legs.
- Echinoderms have radial symmetry and use tube feet to move.

Science Words

amphibian, 52
arachnid, 59
arthropod, 58
bilateral symmetry, 57
cartilage, 51
classify, 42
cnidarian, 56
cold-blooded, 53
complete metamorphosis, 60
crustacean, 59
flatworm, 57
genus, 44
gill, 52
incomplete metamorphosis, 60
invertebrate, 56
mammary gland, 54
metamorphosis, 52
mollusk, 58
molting, 58
phylum, 44
pupa, 60
radial symmetry, 56
reptile, 53
roundworm, 57
scientific name, 46
segmented worm, 57
species, 44
swim bladder, 52
tentacle, 56
tube foot, 61
vertebra, 51
vertebrate, 51
warm-blooded, 54

Chapter 3 Review

Use the Chapter Review to prepare students for tests and to reteach content from the chapter.

Chapter 3 Mastery Test

The Teacher's Resource Library includes two parallel forms of the Chapter 3 Mastery Test. The difficulty level of the two forms is equivalent. You may wish to use one form as a pretest and the other form as a posttest.

Review Answers

Vocabulary Review

1. vertebrates 2. swim bladder
3. bilateral symmetry 4. scientific name
5. radial symmetry 6. mammary glands
7. metamorphosis 8. invertebrates

Concept Review

9. B 10. C 11. A 12. C 13. B 14. A
15. D 16. A 17. D 18. A

TEACHER ALERT

In the Chapter Review, the Vocabulary Review activity includes a sample of the chapter's vocabulary terms. The activity will help determine students understanding of key vocabulary terms and concepts presented in the chapter. Other vocabulary terms used in the chapter are listed below:

amphibian	mollusk
arachnid	molting
arthropod	phylum
cartilage	pupa
classify	reptile
cnidarian	roundworm
cold-blooded	segmented
complete	worm
metamorphosis	species
crustacean	tentacle
flatworm	tube foot
genus	vertebra
gill	warm-blooded
incomplete	
metamorphosis	

Chapter 3 REVIEW

Word Bank
bilateral symmetry
invertebrates
mammary glands
radial symmetry
metamorphosis
scientific name
swim bladder
vertebrates

Vocabulary Review

Choose the word or words from the Word Bank that best complete each sentence. Write the answer on a sheet of paper.

1. Animals that have a backbone are called _____.
2. A fish uses its _____ to move up or down in water.
3. Animals with _____ have bodies with a left half and right half that are the same.
4. An animal's _____ consists of its genus and its species label.
5. Animals with _____ have body parts arranged like spokes on a wheel.
6. Female bears produce milk from their _____.
7. The change of a tadpole into a frog is an example of _____.
8. Sponges, cnidarians, and mollusks are all _____.

Concept Review

Choose the answer that best completes each sentence. Write the answer on your paper.

9. In the classification system of organisms, the level of _____ is between phylum and order.
   A kingdom C genus
   B class D species

10. The genus of the western rattlesnake, *Crotalus viridis*, is _____.
    A rattlesnake C *Crotalus*
    B *viridis* D western

11. Squids capture prey with their _____.
    A tentacles C shells
    B tube feet D mouths

64 Chapter 3 Classifying Animals

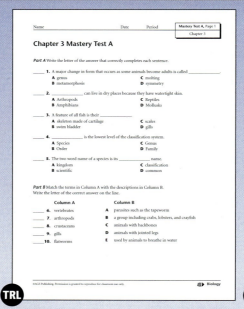
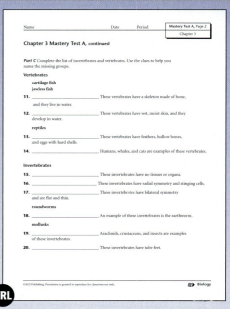

Chapter 3 Mastery Test A

12. Sharks have _____ scales.
    A smooth   C toothlike
    B large   D colorful

13. _____ allows crustaceans to grow larger in size.
    A Bilateral symmetry   C Metamorphosis
    B Molting   D Bone growth

14. The bodies of _____ are made of segments.
    A insects   C cnidarians
    B flatworms   D sponges

15. A bird's feathers and _____ help it to fly.
    A beak   B feet   C gills   D hollow bones

16. All vertebrates have a(n) _____ skeleton.
    A internal   B external   C bony   D soft

17. _____ allow animals to live on land.
    A Gills   C Scales
    B Swim bladders   D Lungs

18. Fish, sponges, and mollusks belong to the same _____.
    A kingdom   C class
    B phylum   D genus

Critical Thinking

Write the answer to each of the following questions.

19. A biologist is studying vertebrates in the desert. Would she be more likely to find amphibians or reptiles during her studies? Explain your answer.

20. Suppose you found a small arthropod under a rock. Using only a hand lens, how could you tell whether the animal is an arachnid or an insect?

Test-Taking Tip When answering multiple-choice questions, first identify the choices you know are untrue.

Classifying Animals Chapter 3 65

Critical Thinking

19. The biologist would be more likely to find reptiles, which have watertight skin and are able to live in dry places without drying out. Amphibians, with thin skin, must stay near water or in damp places, or they will dry out. **20.** You could count its legs; arachnids have four pairs while insects have three pairs. The presence of wings would also be a good clue, since most insects have wings, while arachnids do not.

ALTERNATIVE ASSESSMENT

Alternative Assessment items correlate to the student Goals for Learning at the beginning of this chapter.

- Provide students with the following classification. Have them explain the relationship of each level to the next level, name other animals that can be found in each level with this organism, and tell the scientific name of the animal. [Kingdom: Animalia; Phylum: Chordata; Class: Mammalia; Order: Carnivora; Family: Ursidae; Genus: *Ursus*; Species: *maritimus*]

- Write the classes of vertebrates on the board as headings. Name specific characteristics and have students write them in the column where they belong. Include some characteristics that are common to all vertebrates.

- Pin names of invertebrate phyla on a bulletin board. Provide pictures of various invertebrates. Ask students to place them under the correct name on the bulletin board and to name some characteristics of this type of invertebrate, such as its type of body symmetry, type of appendages, or body covering.

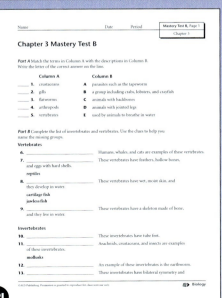
Chapter 3 Mastery Test B

Classifying Animals Chapter 3 65

Chapter 4

Planning Guide
Classifying Plant Groups

	Student Text Lesson		
	Student Pages	Vocabulary	Lesson Review
Lesson 1 How Plants Are Classified	68–70	✔	✔
Lesson 2 Seed Plants	71–76	✔	✔
Lesson 3 Seedless Plants	77–80	✔	✔

Chapter Activities

Student Text
Science Center

Teacher's Resource Library
Community Connection 4:
 Garden Plants in Your Community

Assessment Options

Student Text
Chapter 4 Review

Teacher's Resource Library
Chapter 4 Mastery Tests A and B

66A

Student Text Features							Teaching Strategies							Learning Styles						Teacher's Resource Library				
Achievements in Science	Science at Work	Science in Your Life	Investigation	Science Myth	Note	Technology Note	Did You Know?	Science Integration	Science Journal	Cross-Curricular Connection	Online Connection	Teacher Alert	Applications (Home, Career, Community, Global, Environment)	Auditory/Verbal	Body/Kinesthetic	Interpersonal/Group Learning	Logical/Mathematical	Visual/Spatial	LEP/ESL	Workbook Activities	Alternative Workbook Activities	Lab Manual	Resource File	Self-Study Guide
					✔	✔	69	70		69		69	70						69	10	10			✔
74		75	73						72	73	72		72, 73, 74	72	73		73			11	11	10, 11, 12	7	✔
	80	79					79	79	79	78, 79						80		78		12	12		8	✔

Pronunciation Key

a	hat	e	let	ī	ice	ô	order	ù	put	sh	she
ā	age	ē	equal	o	hot	oi	oil	ü	rule	th	thin
ä	far	ėr	term	ō	open	ou	out	ch	child	ᴛʜ	then
â	care	i	it	ȯ	saw	u	cup	ng	long	zh	measure

ə { a in about, e in taken, i in pencil, o in lemon, u in circus }

Alternative Workbook Activities

The Teacher's Resource Library (TRL) contains a set of lower-level worksheets called Alternative Workbook Activities. These worksheets cover the same content as the regular Workbook Activities but are written at a second-grade reading level.

Skill Track Software

Use the Skill Track Software for Biology for additional reinforcement of this chapter. The software program allows students using AGS textbooks to be assessed for mastery of each chapter and lesson of the textbook. Students access the software on an individual basis and are assessed with multiple-choice items.

Chapter at a Glance

Chapter 4: Classifying Plant Groups
pages 66–83

Lessons
1. **How Plants Are Classified** pages 68–70
2. **Seed Plants** pages 71–76
   Investigation 4 pages 75–76
3. **Seedless Plants** pages 77–80

Chapter 4 Summary page 81

Chapter 4 Review pages 82–83

Skill Track Software for Biology

Teacher's Resource Library
- Workbook Activities 10–12
- Alternative Workbook Activities 10–12
- Lab Manual 10–12
- Community Connection 4
- Resource File 7–8
- Chapter 4 Self-Study Guide
- Chapter 4 Mastery Tests A and B

(Answer Keys for the Teacher's Resource Library begin on page 420 of the Teacher's Edition. A list of supplies required for Lab Manual Activities in this chapter begins on page 449.)

Science Center

Make a bulletin board with a large version of the concept map shown on page 67. Have students collect leaves of different plants, dry them, mount them on pieces of paper, and identify them with their scientific name. Then have them attach the papers to the bulletin board under the appropriate classification. As an alternative, have students attach their mounted leaves from the Investigation 4 Explore Further to the bulletin board under the appropriate classification.

Community Connection 4

Chapter 4: Classifying Plant Groups

Notice the different kinds of plants in the photo. Some have flowers. Some have needles. Some have green leaves. Plants come in different shapes and sizes. If you live near a desert, you might be familiar with these plants. All the plants in the photo are classified as vascular plants and seed plants. In this chapter, you will learn how plants are classified into different groups. You will learn what structures the plants in each group have in common.

Organize Your Thoughts

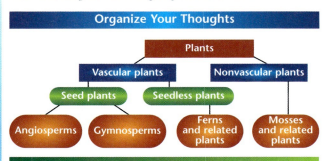

Goals for Learning

- To understand that plants are classified according to their similar structures
- To explain the difference between vascular and nonvascular plants
- To explain the differences and similarities between plants with seeds and seedless plants
- To describe angiosperms, gymnosperms, ferns, and mosses

Introducing the Chapter

Ask students to describe the plants they see in the picture on page 66. Ask them to describe some similarities and differences with other plants with which they are familiar. Then have students name criteria that could be used to organize plants into different groups. (*examples: size, the kinds of leaves they have, form and shape of plant body parts, how the plants reproduce*) Have students read the titles of Lessons 2 and 3 in this chapter. Explain that one important characteristic by which plants are classified is whether they have seeds.

Notes and Technology Notes

Ask volunteers to read the notes that appear in the margins throughout the chapter. Then discuss them with the class.

TEACHER'S RESOURCE

The AGS Teaching Strategies in Science Transparencies may be used with this chapter. The transparencies add an interactive dimension to expand and enhance the *Biology* program content.

CAREER INTEREST INVENTORY

The AGS Harrington-O'Shea Career Decision-Making System-Revised (CDM) may be used with this chapter. Students can use the CDM to explore their interests and identify careers. The CDM defines career areas that are indicated by students' responses on the inventory.

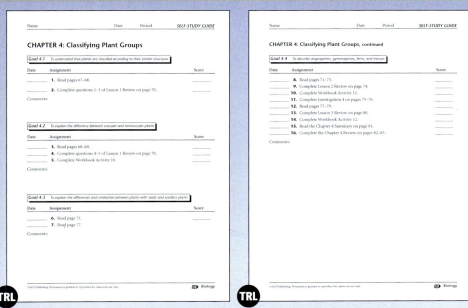

Chapter 4 Self-Study Guide

Lesson at a Glance

Chapter 4 Lesson 1

Overview This lesson gives students a brief history of the classification and scientific naming of organisms. Students learn that plants can be classified as vascular and nonvascular plants.

Objectives
- To explain how plants are classified
- To describe the history of the classification of plants
- To tell the difference between vascular and nonvascular plants

Student Pages 68–70

Teacher's Resource Library TRL
Workbook Activity 10
Alternative Workbook Activity 10

Vocabulary

seed	vascular plant
fern	vascular tissue
moss	nonvascular plant

Science Background
Scientific Classification

Early systems of classification of organisms were based on their functions and their relations to humans. For example, plants were classified according to their use by humans—as medicines, herbs for cooking, and so on. Our modern system of taxonomy is based on the classification system of Linnaeus, who gave all living things a two-part name. The first part of the name is the genus. The second part of the name is the species. The names are usually Latin, Greek, or a Latinized version of a name in another language.

Warm-Up Activity

Ask students to imagine that they have a collection of 1 million stamps. Prompt them to think about how they might organize the collection in order to enjoy it, show it, or add to it. Ask: Why couldn't you just keep all the stamps in a huge bag?

Lesson 1 How Plants Are Classified

Objectives

After reading this lesson, you should be able to
- explain how plants are classified.
- describe the history of the classification of plants.
- tell the difference between vascular and nonvascular plants.

Seed
A plant part that contains a beginning plant and stored food

Fern
A seedless vascular plant

Moss
A nonvascular plant that has simple parts

Vascular plant
A plant that has tubelike cells

Vascular tissue
A group of plant cells that form tubes through which food and water move

Scientists have identified more than 260,000 kinds of plants. That sounds like a lot. However, scientists think even more kinds have yet to be discovered. About 1,000,000 kinds of plants may exist that have not been found and named. Most of these plants live in the tropical rain forests.

Scientists divide this huge number of plants into groups to make them easier to study. They classify plants according to whether they have body parts such as **seeds**, tubes, roots, stems, and leaves. The three main groups of plants are seed plants, **ferns**, and **mosses**. The groups that contain ferns and mosses also contain related plants. However, ferns and mosses form the greatest number in each of these groups.

History of Classification

The classification of plants started more than 2,000 years ago. The Greek philosopher Aristotle first classified plants and animals. His student Theophrastus listed the names of over 500 plants. In 1753, Carolus Linnaeus, a Swede, developed a new method to classify plants and animals. Today, organisms are classified based on his system.

Under this system, organisms have a two-word name. The first word is the genus. For example, maple trees belong to the genus *Acer*. The scientific name of all maple trees begins with the word *Acer*. The second word is the species. Each kind of maple tree has its own species name. The scientific name of the sugar maple tree is *Acer saccharum*. The scientific name of the red maple is *Acer rubrum*.

Vascular and Nonvascular Plants

Seed plants and ferns are **vascular plants**. Vascular plants have tubelike cells. *Vascular* means "vessel" or "tube." These cells form tissue called **vascular tissue**. The tissue forms tubes that transport food and water through the plant. Vascular plants have well-developed leaves, stems, and roots.

68 *Chapter 4 Classifying Plant Groups*

Nonvascular plant
A plant that does not have tubelike cells

Did You Know?
The tallest living thing in the world is a redwood tree in California. It is 112 meters tall. The oldest living thing in the world is a bristlecone pine tree in California, which is over 4,700 years old.

Water in a nonvascular plant moves from one cell to another. Since water can't travel far this way, each cell must be near water. This limits the number of cells a nonvascular plant can have.

Vascular tissue is important in two ways. First, it allows food and water to be transported through the plant. The plant can grow larger because its leaves and stems do not need to be near water. Second, vascular tissue is thick and provides support for a plant. This also allows plants to grow tall.

Mosses are **nonvascular plants**. Nonvascular plants do not have tubelike cells. These plants are short and must have constant contact with moisture. They do not have tubes to transport water or to support them. These small plants usually grow in damp, shady places on the ground and on the sides of trees and rocks. Unlike vascular plants, nonvascular plants do not have true leaves, stems, or roots.

The veins of a leaf are vascular tissue.

Lead students to the conclusion that you would want to put the stamps in albums by category, such as country, date, and so on. Remark that scientists have worked to classify plants so they can easily study and compare them.

 Teaching the Lesson

Bring several leaves to class or have each student bring a leaf. Have students examine the photograph on page 69. Point out that the vein pattern is part of the vascular system. Have students find the vein patterns in their leaves.

Point out that scientific names in print, such as those on page 68, are italicized. The genus is capitalized but the species is not. Write the following words on the board: *cucurbita pepo*. Say that *cucurbita* is the genus name for a familiar plant and *pepo* is the species name. Have students write the scientific name for the plant correctly on a piece of paper. Students will want to know that *Cucurbita pepo* is the scientific name for the pumpkin plant. Have them label their papers with this information.

 Reinforce and Extend

LEARNING STYLES

LEP/ESL
Make sure students understand the meanings of the words *tube* and *vessel*. Write the words on the board. Demonstrate the concept with a drinking straw. Point out that the straw draws up liquid from a cup. In the same way, a tube or vessel in a plant draws water and sap up into plant parts.

Did You Know?

Have students read the feature on page 69. Compare the height of the 112-meter redwood to the height of a 40-story skyscraper. Draw a timeline to illustrate 4,700 years. The bristlecone pine began living in about 2700 B.C.

TEACHER ALERT
Students may get the impression that mosses do not transport water. However, water does move very slowly from cell to cell through tiny openings. In vascular plants, by contrast, water moves in hollow cells, like water moving up a straw, so it can move much faster.

CROSS-CURRICULAR CONNECTION

 Math
Draw attention to the first paragraph on page 68. Ask: If all the plant species in the world were discovered and classified, how many might there be in all? (*1,260,000; the text states that 260,000 kinds of plants have been identified and that 1,000,000 more may yet remain to be discovered.*)

SCIENCE INTEGRATION

Physical Science

A force called *capillarity* causes liquid to rise in the vascular tissue of a plant—even of a very tall tree. *Capillary* is another word for tube. The walls of the plant's vascular tubes attract water molecules more powerfully than the water molecules attract each other. This force of attraction between molecules draws the water upward.

GLOBAL CONNECTION

Emphasize that scientists all over the world use the same scientific names for plants, regardless of the common name in their own languages. For example, the common weed known to English speakers as the dandelion is called *pissenlit* in French, *löwvenzahn* in German, and *diente de león* in Spanish. The scientific name of this pesky weed is *Taraxacum officinale*.

Lesson 1 Review Answers

1. The three main groups of plants are seed plants, ferns, and mosses. **2.** Carolus Linnaeus **3.** genus and species **4.** Vascular plants have tubelike cells. Nonvascular plants do not. **5.** Vascular tissue transports food and water and supports the plant.

Portfolio Assessment

Sample items include:
- The scientific name of the pumpkin plant from Teaching the Lesson
- Lesson 1 Review answers

70 Chapter 4 Classifying Plant Groups

Lesson 1 REVIEW

Write your answers to these questions on a separate sheet of paper. Write complete sentences.

1. What are the three main groups of plants?
2. Who developed the classification system of organisms that is used today?
3. What two-word name makes up the scientific name of each kind of organism?
4. How do vascular and nonvascular plants differ?
5. What are two ways that vascular tissue is important?

Technology Note

To study some plant tissues, scientists use equipment to make very thin slices of a plant's stem. These thin slices are used to prepare microscope slides. When scientists first used microscopes, they had to draw what they saw. Today, they can use special cameras to take pictures of what they see. They can also store the pictures on a computer. This allows scientists to share their work with others.

70 Chapter 4 Classifying Plant Groups

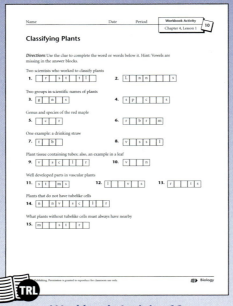

Workbook Activity 10

Lesson 2 — Seed Plants

Objectives

After reading this lesson, you should be able to
- explain how seed plants are different from the other plant groups.
- explain the differences between angiosperms and gymnosperms.
- explain the differences between dicots and monocots.

Embryo
A beginning plant

Angiosperm
A flowering plant

Recall that scientists classify plants into three main groups. Seed plants are different from the other plant groups because they use seeds to reproduce. A seed is a plant part that contains a beginning plant and stored food. The beginning plant is called an **embryo**. A seed has a seed coat that holds in moisture. When conditions are right, the embryo grows into a full-sized plant.

Seed plants have the most advanced vascular tissue of all plants. They have well-developed leaves, stems, and roots.

Seed plants come in many sizes and shapes. The duckweed plant that floats on water may be just one millimeter long. Giant redwood trees are the largest plants in the world. A pine tree has long, thin needles. A rose has soft petals. The different sizes and shapes of seed plants help them to live in many different places. Grass, trees, garden flowers, bushes, vines, and cacti are all seed plants.

Seed plants are the largest group of plants. They are divided into two subgroups. One group is flowering plants and the other group is nonflowering plants.

Angiosperms

Most species of plants are **angiosperms**, or flowering plants. The word *angiosperm* is made from the Greek words *angeion*, "capsule," and *sperma*, "seed." A capsule, or fruit, protects the seeds of angiosperms. The fruit forms from part of the flower. Flowers come in many shapes and colors.

The flowers of some plants are colorful and showy.

Classifying Plant Groups   Chapter 4   71

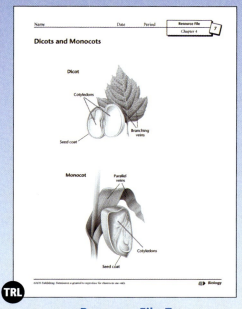
Resource File 7

Lesson at a Glance

Chapter 4 Lesson 2

Overview Students learn the similarities and differences between the two types of seed plants—angiosperms and gymnosperms—and the two different types of angiosperms—monocots and dicots.

Objectives
- To explain how seed plants are different from the other plant groups
- To explain the differences between angiosperms and gymnosperms
- To explain the differences between dicots and monocots

Student Pages 71–76

Teacher's Resource Library
Workbook Activity 11
Alternative Activity 11
Lab Manual 10–12
Resource File 7

Vocabulary

embryo   cotyledon
angiosperm   gymnosperm
monocot   conifer
dicot

Science Background
Angiosperms and Gymnosperms

Angiosperms are called flowering plants because they reproduce by seeds formed after pollination of flowers. Angiosperms can be divided into dicots and monocots.

Dicots have
- two cotyledons
- netlike leaf veins
- vascular bundles in rings
- a taproot (usually)
- flower parts in multiples of 4 or 5

Monocots have
- one cotyledon
- parallel leaf veins
- vascular bundles not in rings
- fibrous roots
- flower parts in multiples of 3

Classifying Plant Groups   Chapter 4   71

Warm-Up Activity

Bring to class seeds from a flowering plant and a pinecone. Ask: "Do all plants have flowers?" Ask students if pine or spruce trees have flowers. (*No*) Point out that all the plants discussed make seeds but not all have flowers. Gymnosperms have seeds but no flowers. Students are probably familiar with cone-bearing gymnosperms such as pine, spruce, cedar, and juniper. On the pinecone you display, students may be able to observe the "naked" seeds attached to its scales.

 Teaching the Lesson

Tell students that some seeds we eat are beans, peas, corn, and nuts. We also eat seeds in some fruits, such as tomatoes and strawberries.

Draw students' attention to the pictures of the dicot and monocot. Emphasize the prefixes *di-* (meaning "two") and *mono-* ("one"). Make sure students can distinguish between the two cotyledons of the dicot and the one cotyledon of the monocot.

Have students write the headings in the lesson on paper. For each heading, they should write one or two main ideas discussed in the lesson.

 Reinforce and Extend

LEARNING STYLES

 Auditory/Verbal
Have students write riddles for each of the vocabulary words in the lesson. Volunteers read their riddles and class members identify the words.

AT HOME

 Students can observe the moisture that is stored inside a seed by popping popcorn with the help of an adult. Steam pouring out of a bag of microwave-popped corn is evidence of moisture that was stored in the kernels.

Monocot
An angiosperm that has one seed leaf

Dicot
An angiosperm that has two seed leaves

Cotyledon
A structure in the seeds of angiosperms that contains food for the plant

Dicots and Monocots

Angiosperms are divided into two kinds of plants, **monocots** and **dicots**. Most angiosperms are dicots. Dicots have two **cotyledons** inside the seed that contain food for the developing plant. You can see this by looking at a bean, which is a large dicot seed. If you split a bean apart, you may be able to see the two leaves on the tiny embryo. When a bean is planted, the plant appears with two leaves. If you look closely at the leaves, you will see that the veins are branched, or netlike. This pattern of veins is another property of dicots.

The number of species of dicots is more than 175,000. Some dicots, such as oak and ash, are trees that provide shade or produce wood for furniture. Animals eat dicots in the form of fruits and vegetables. Most flowering plants are dicots. Examples are roses and sunflowers.

Monocots have only one cotyledon. When a monocot starts growing from a seed, a single leaf appears. The veins in the leaf of a monocot are parallel. This means they all go in the same direction. You can see parallel veins in a blade of grass, for example. There are more than 50,000 species of monocots. They include corn, wheat, and rice. Grasses that cattle eat are monocots. Some flowers, such as lilies and orchids, are monocots.

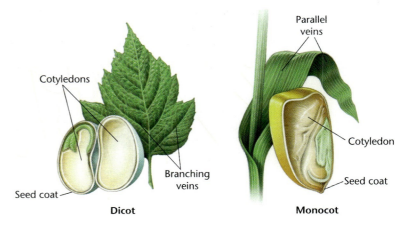

Dicot / Monocot

SCIENCE JOURNAL

 Have students record the name of plants and identify the plant parts that they eat in a week. Ask them to use the information as a springboard to a paragraph discussing the importance of plants in their diet.

ONLINE CONNECTION

 Students can find many pictures of plants at Web sites of botanical gardens. Most of these sites have a click-on selection for plants that are currently in bloom. The following are Web sites of two major U.S. botanical gardens:

- Missouri Botanical Garden (St. Louis): http://www.mobot.org
- New York Botanical Garden (Bronx): http://www.nybg.org

Gymnosperm
A nonflowering seed plant

Conifer
A cone-bearing gymnosperm

Gymnosperms

Nonflowering seed plants are called **gymnosperms**. They do not produce flowers. The word *gymnosperm* means "naked seed." The seeds of gymnosperms are not surrounded by a fruit. The seeds are produced inside cones. For example, the seeds of pine trees form on the scales of cones.

Conifers and Other Gymnosperms

There are over 700 species of gymnosperms. The major group of gymnosperms is **conifers**. Conifers are cone-bearing gymnosperms. There are about 600 species of conifers. All conifers are woody shrubs or trees. They make up 30 percent of the forests around the world. Pines, spruces, and firs are conifers. Plants such as junipers, yews, and spruces decorate the landscape of many homes.

Most conifers have green leaves all year. Therefore, they are called evergreens. They lose only some of their leaves at any time. The leaves of conifers are shaped like needles. They do not lose water as easily as the broad leaves on other trees do. This makes it easier for conifers to live in dry places where trees must store water for a long time.

Besides conifers, there are other gymnosperms. The ginkgo tree is one of the most familiar. Ginkgo trees have peculiar fan-shaped leaves. These trees are planted along many city streets because they are able to survive pollution better than other trees.

Science Myth

All evergreens are gymnosperms, and all gymnosperms are evergreens.

Fact: Some angiosperms are also evergreens. Angiosperms, such as holly, magnolia, live oak, and rhododendron, do not lose their leaves in the winter. Some gymnosperms, such as the ginkgo, are not evergreens and do lose their leaves.

Conifers have cones and needle-shaped leaves.

The leaves of the ginkgo tree are shaped like fans.

In the Environment

Some people worry that farm crops are not genetically diverse enough. In a worst-case scenario, a disease or pest might wipe out identical or similar crops, threatening people with famine. For this reason, some farmers and gardeners are growing heirloom plants. These are plants that were grown as crops or garden plants in the past but have been set aside for hybridized plants. Today, more and more garden centers are offering seeds of heirloom plants, and more small farmers are raising and selling heirloom vegetables.

Career Connection

Botanic gardens offer a variety of career opportunities to plant lovers. The gardens often employ botanists, landscape architects, and maintenance workers. Have students contact a botanic garden to find out about the opportunities there.

Lesson 2 Review Answers

1. angiosperms and gymnosperms
2. Dicots have two seed leaves and a crisscross pattern of veins. Monocots have one seed leaf and a parallel pattern of veins.
3. Answers will vary. Roses and beans are dicots. Grasses and corn are monocots.
4. Seeds in gymnosperms are not enclosed in a fruit.
5. Their leaves are shaped like needles, so they do not lose water as easily as do broad leaves.

Portfolio Assessment

Sample items include:
- Main Ideas listed from Teaching the Lesson
- Science Journal paragraph on plant foods in students' diets
- Lesson 2 Review answers

74 Chapter 4 Classifying Plant Groups

Lesson 2 REVIEW

Write your answers to these questions on a separate sheet of paper. Write complete sentences.

1. What are other names for flowering plants and nonflowering seed plants?
2. What are the differences between dicots and monocots?
3. Name two plants that are dicots and two that are monocots.
4. Why are the seeds of gymnosperms called "naked seeds"?
5. Why are conifers able to live where other plants cannot?

Achievements in Science

World's Oldest Flowering Plant Discovered

A fossil of the oldest flowering plant ever found was discovered recently in China. It was in rock that used to be at the bottom of a lake. The plant did not have a flower with petals, but it did have characteristics of a flowering plant. It had seeds in an undeveloped fruit. Only flowering plants have seeds in fruits. This discovery has challenged existing ideas about the ancestors of flowering plants. Their ancestors were thought to be shrubs similar to small trees. This fossil plant is at least 125 million years old and was only about 20 inches high. Scientists believe it lived underwater with its thin stems extending upward above the water. Now scientists wonder if the ancestors of today's flowering plants were aquatic plants.

74 Chapter 4 Classifying Plant Groups

Achievements in Science

Have students read Achievements in Science on page 74. Encourage students to identify how the oldest flowering plant was alike and different from most of today's flowering plants.

Workbook Activity 11

INVESTIGATION 4

Identifying Angiosperms and Gymnosperms

Materials
- 5 different kinds of leaves

Purpose
How can you tell whether a plant is an angiosperm or a gymnosperm? In this investigation, you will learn to identify angiosperms and gymnosperms by looking at their leaves.

Procedure

1. Copy the data table below on a sheet of paper.

Leaf	Leaf Shape	Angiosperm or Gymnosperm	Vein Pattern	Monocot or Dicot
1				
2				
3				
4				
5				

2. Look at the leaves that your teacher shows you. On another sheet of paper, draw each leaf. Be sure to include some of the veins in your sketch. Number the leaves 1 through 5.

3. Examine the shape of each leaf. Is it needle-shaped? Is it round and flat? Is the edge smooth or ragged? You might think of other ways to describe the leaf. Record what you see in the first column on your table.

Classifying Plant Groups Chapter 4 75

Investigation 4

This Investigation will take approximately 50 minutes to complete.

Students will use these process and thinking skills: observing, classifying, describing, comparing, and recalling facts.

Preparation

- Collect leaves from outside or from a florist or garden store. Include the leaves of at least one conifer (examples: pine, spruce, juniper), one dicot (examples: roses, African violets, ivy, oak), and one monocot (examples: grasses, orchids, irises). If possible, include ginkgo leaves.
- Divide the leaves into groups of five. Be sure each group has at least one gymnosperm, one dicot, and one monocot.
- Make and distribute copies of the "Dicots and Monocots" diagram from the Resource File for background information.
- Students may use Lab Manual 10 to record their data and answer the questions.

Procedure

- Verify that students have copied the data table before the investigation begins.
- Form teams of two or three students. Give each team five leaves to examine.
- Encourage students to examine the leaves carefully and make accurate observations. Point out the veins in leaves if students need assistance. Suggest that students pay attention to other leaf features, such as thickness and texture.
- Make sure each student examines all five leaves.
- Check that students are filling in data tables as they proceed.

SAFETY ALERT

- Be careful to avoid poisonous plants such as poison ivy, poison oak, and poison sumac. Use a field guide to determine what these plants look like.
- Have students wash their hands thoroughly with soap and water after examining the leaves.
- Caution students not to touch their eyes or to put their hands in their mouths during the activity.

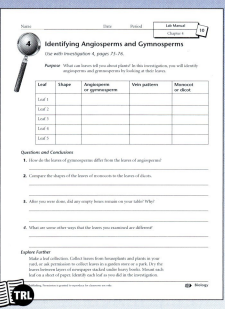

Lab Manual 10

Classifying Plant Groups Chapter 4 75

Results

Students should note that gymnosperms have needle-shaped leaves while angiosperms have broad, flat leaves. (Ginkgo leaves represent a variation on needle form; students should observe that the texture of a ginkgo leaf is similar to that of a pine needle.) Students should note that veins in a dicot leaf are netlike while the veins in a monocot leaf are long and parallel.

Questions and Conclusions Answers

1. The shape of leaves from gymnosperms such as conifers is needlelike. Angiosperms have flat, broader leaves.

2. Monocots have long, narrow leaves with smooth edges. Dicots have broader leaves that can have lobes (rounded, protruding parts) and wavy or serrated (zigzag) edges.

3. Some empty boxes might remain because the veins in the leaves of many gymnosperms cannot be seen, and they are not classified as monocots or dicots.

4. Answers will vary. Students may mention size, thickness, or texture. If you picked clusters of leaves from a rose bush or a pecan or walnut tree, students may recognize that leaves may grow singly (simple) or in groups (compound).

Explore Further Answers

Display students' collections in the classroom. Names of plants will vary depending on the collection.

Assessment

Check drawings to make sure students are including some veins and that the shapes are reasonably accurate. Check the data tables for detailed descriptions and appropriately classified leaves. Only angiosperms should be classified as monocots or dicots. You might include the following items from this investigation in student portfolios:

- Drawings of leaves
- Data table
- Questions and Conclusions and Explore Further answers

4. Classify each leaf as belonging to an angiosperm or a gymnosperm. Use the shape of the leaf to decide. Record your answer on your table.

5. Observe the vein pattern in each leaf. Are the veins netlike or parallel? Record your findings on your table. If you cannot see veins in a leaf, leave its box blank.

6. Classify each leaf that belongs to an angiosperm as a dicot or a monocot. Use the vein pattern of the leaf to decide. Record your answer on your table.

7. Draw a line through any empty boxes that remain on your table.

Questions and Conclusions

1. How do the leaves of gymnosperms differ from the leaves of angiosperms?

2. Compare the shapes of the leaves of monocots to the leaves of dicots.

3. After you were done, did any empty boxes remain on your table? Why?

4. What are some other ways that the leaves you examined are different?

Explore Further

Make a leaf collection. Collect leaves from houseplants and plants in your yard, or ask permission to collect leaves in a garden store or a park. Dry the leaves between layers of newspaper stacked under heavy books. Mount each leaf on a sheet of paper. Identify each leaf as you did in the investigation. Use reference books to find the two-word scientific names of the plants.

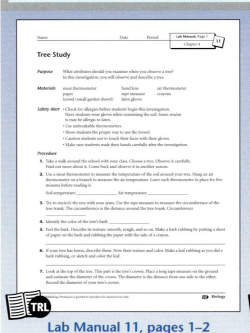

Lab Manual 11, pages 1–2

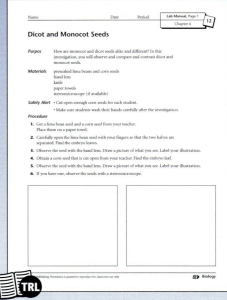

Lab Manual 12, pages 1–2

Lesson 3: Seedless Plants

Objectives

After reading this lesson, you should be able to
- list vascular and nonvascular plants that are seedless.
- describe similarities and differences between ferns and seed plants.
- describe mosses and how they grow.

Frond
A large feathery leaf of a fern

Sori
Clusters of reproductive cells on the underside of a frond

Spore
The reproductive cell of some organisms

Rhizome
A plant part that has shoots aboveground and roots belowground

There are two main groups of seedless plants. The largest group includes ferns and related plants. Like seed plants, ferns are vascular plants. Unlike seed plants, they do not have seeds. The second group of seedless plants includes mosses and related plants. They are different from ferns because they are nonvascular plants.

Ferns

The largest group of seedless vascular plants is ferns. There are over 10,000 species of ferns in the world. Many of them are tropical plants. They range in size from tiny plants to large treelike plants. Like other vascular plants, ferns have well-developed leaves, stems, and roots.

The leaves, or **fronds**, usually are large and flat. They are divided into small sections, or leaflets, that spread out from a center rib. If you look closely, you can see new fronds that are curled up. They uncurl as they grow.

On the underside of fronds, you can see small dots called **sori**. Sori are clusters that contain the reproductive cells of ferns. These cells are called **spores**. When the spores are ripe, the sori burst open and release the spores into the air.

The **rhizome** is a plant part that has shoots aboveground and roots belowground.

Fern

Classifying Plant Groups Chapter 4 77

Lesson at a Glance

Chapter 4 Lesson 3

Overview In this lesson, students learn about the two main groups of seedless plants—ferns and mosses—and how they reproduce by spores rather than seeds.

Objectives
- To list vascular and nonvascular plants that are seedless
- To describe similarities and differences between ferns and seed plants
- To describe mosses and how they grow

Student Pages 77–80

Teacher's Resource Library TRL
  Workbook Activity 12
  Alternative Activity 12
  Resource File 8

Vocabulary

frond	rhizome
sori	rhizoid
spore	humus

Science Background
Ferns and Mosses

Both ferns and mosses are seedless plants, but ferns are vascular plants. Ferns are structurally much more developed and complex than mosses although both produce spores and have two distinct generations: the spore generation, which we usually do not see; and the spore-producing generation, which is the form of the plant we typically see.

1 Warm-Up Activity

Remind students that they learned about angiosperms and gymnosperms in the previous lesson. Ask: What do angiosperms and gymnosperms have in common? (*Both produce seeds.*) Remark that, in this lesson, students will learn about some plants that do not have seeds.

Resource File 8

Classifying Plant Groups Chapter 4 77

If possible, show a fern frond to the class and point out the spore clusters on the underside of the leaflets. Ask students if they would expect to see similar structures on oak or maple leaves or on pine needles. (*no*) Point out that the spores are the way ferns reproduce, but they are different from seeds.

 Teaching the Lesson

Display a fern frond. Draw attention to the sori on the underside of the leaflets. Let students examine the sori with a magnifying glass. Have a volunteer explain what role sori play in the life cycle of a fern.

Demonstrate a characteristic of moss by bringing to class a patch of sphagnum moss, available from florists or garden supply stores. Place the moss in a pan and slowly pour water over it. Let students observe how the moss acts as a sponge to absorb much of the water. Point out that sphagnum moss absorbs over 20 times its mass in water.

Ask students to write the lesson objectives on paper, leaving space between the objectives. As they read the lesson, they can record information that will help them meet the objective.

 Reinforce and Extend

Cross-Curricular Connection

 Art
Have students make a silhouette of a fern frond as a piece of art. Instruct them to tape the stem at several places (using clear tape) to dark-colored construction paper. Have them place the paper, frond side up, under bright light for several days. The light causes the paper to fade. When they remove the frond, they will have its outline on paper. Encourage students to frame their artwork with construction paper of contrasting color.

Rhizoid
A tiny rootlike thread of a moss plant

After they are released, spores must land in a moist place. Spores that drop in a moist place produce a tiny plant. The plant must have constant moisture to grow. Seeds, on the other hand, have food stored inside and seed coats. The seed coat protects a seed until it has the right conditions to grow. Seeds usually survive longer than spores when conditions are dry. This explains why there are more seed plants than seedless plants.

Mosses

Scientists have found more than 9,000 species of mosses. A moss is a nonvascular plant that has simple leaflike and stemlike parts. It does not have well-developed leaves, stems, and roots. Mosses do not have vascular tissue to transport water. They must live in moist, shady places.

Mosses grow best where the air is full of moisture and the soil is wet. They get water through rootlike threads called **rhizoids**. Woodlands and the edges of streams are common homes for mosses. Mosses look like little trees and often form carpetlike mats on the forest floor.

Like ferns, mosses reproduce by means of spores. Millions of tiny spores form inside spore cases on special stalks. The spore case breaks open when it is ripe. It shoots the spores into the air. The spores make new plants when they fall on moist soil. One reason for moss survival is that mosses produce great numbers of spores.

Mosses cover the roots of this tree.

Learning Styles

 Visual/Spatial
Encourage students to make a comparison-and-contrast chart. Show the chart format on the board. Under the title "Ferns and Mosses," make two columns. Label one column "Alike" and the other "Different." Have students copy the format on paper and fill in the chart. (*Alike: both are plants; have no seeds; reproduce by spores. Different: fern is vascular, moss is nonvascular; fern has true roots, stem, and leaves while moss has poorly developed parts.*)

Humus
Decayed plant and animal matter that is part of the topsoil

Did You Know?
The bodies of people who died 2,000 years ago have been found in bogs in Denmark. The lack of oxygen in bogs keeps the bodies well preserved.

Soil and Bog Builders

Because mosses are so tiny, they can grow in places where other plants cannot take root. Mosses grow within the bark of fallen trees, within cracks in rocks, and in thin soils. In this way, they help to form soil. When moss plants die, they form **humus**. Humus is the part of the soil made by dead plant and animal matter. It is very rich and helps plants to grow.

Mosses of the genus *Sphagnum* are known as bog builders. A bog is wet, spongy ground that is formed from rotted moss and other plant matter. Air does not reach the dead plants. The lack of oxygen keeps the plants from breaking down, or decaying, quickly. Over time, this plant matter becomes tightly pressed together and forms peat. Most of the plant matter in peat is sphagnum moss. Peat moss is mixed with soil to improve gardens.

Science in Your Life

How do ferns and mosses provide energy?

Ferns that lived millions of years ago are important to your life today. About 300 million years ago, forests and swamps contained many ferns, some the size of trees. Over time, these ferns died, and layers of dead ferns and other plants built up. Pressure and heat on the deep layers of plant material caused coal to form. Coal is burned to produce steam in power plants, which produce electricity.

Peat is also used as a source of energy, especially in parts of Europe. In Ireland, for example, peat is cut into loaf-sized chunks and burned in stoves and fireplaces. The peat burns slowly, like charcoal.

Classifying Plant Groups Chapter 4 **79**

Science Integration

Earth Science

Fertile soil—soil in which strong, healthy plants can grow—must contain the right mix of chemical elements and compounds. Three chemical elements are especially important: nitrogen, phosphorous, and potassium. Plants need these substances to carry out photosynthesis and to build structures such as roots, leaves, and stems. Humus from rotting mosses and other dead plants adds nitrogen, potassium, phosphorous, and traces of other chemicals to the soil.

Cross-Curricular Connection

History

The ways people lived their daily lives thousands of years ago is mostly lost to us. The European bog mummies give valuable information about the daily lives of such people. The bodies are so well preserved that scientists can determine what the people ate by examining the contents of their stomachs and intestines. For example, one bog person ate a thick soup made of vegetables, seeds, and herbs. Some of these materials came from farms while others were from the wild. Scientists have also found leather caps and belts, knitted woolen scarves, jewelry, and tools on bog people. They have found bog bodies with hair perfectly preserved.

Science Journal

Prompt students to think about the plants they have studied in this chapter and choose a favorite. Suggest they make their choice by considering both the plant's beauty (aesthetics) and its usefulness (utility). Have students write a paragraph describing the plant and explaining why they have selected it as their favorite.

Science in Your Life

Have students read Science in Your Life on page 79. Point out that although coal and peat were both formed from dead plant material and provide energy, they are otherwise quite different. Coal is much older than peat and was put under so much pressure and heat deep underground that it formed a hard, dense black mineral. Coal has formed over millions or hundreds of millions of years. Peat, on the other hand, can form in thousands or even hundreds of years.

Did You Know?

Have students read Did You Know? on page 79. Explain that the remains of dead organisms usually decompose over time. Point out that being able to examine body tissue of people who died thousands of years ago helps scientists learn about the lives of people from the distant past.

Learning Styles

Interpersonal/Group Learning

Divide students into teams of two or three. Challenge each team to make a list of people's first names that are based on plant names. (*Possible examples: Daisy, Fern, Holly, Hyacinth, Iris, Ivy, Lily, Rose, Rosemary, Violet*) Extend the activity by having students identify each plant as an angiosperm or gymnosperm (seed plant) or as a nonseed plant.

Science at Work

Have students read the feature on page 80. Ask what *technician* means. (*a person who has special skills to do a particular activity*) If appropriate, expand the feature into a discussion of careers available in horticulture, or cultivation of gardens. The list might include garden designer, working for a gardening service or a garden center, and so on.

Lesson 3 Review Answers

1. Ferns, mosses, and related plants do not have seeds. **2.** *Nonvascular* means that the plants do not have highly developed vascular tissue for transporting water. **3.** Seeds usually survive longer than spores when conditions are dry. **4.** Unlike seeds, spores do not have a seed coat to hold in moisture. **5.** Mosses need to live in moist, humid places because they do not have vascular tissue to transport water.

Portfolio Assessment

Sample items include:
- Main ideas supporting lesson objectives from Teaching the Lesson
- Lesson 3 Review answers

Lesson 3 REVIEW

Write your answers to these questions on a separate sheet of paper. Write complete sentences.

1. What groups of plants do not have seeds?
2. What does *nonvascular* mean?
3. Why are there more seed plants than seedless plants?
4. Why do spores need a moist place to land?
5. Where do mosses need to live and why?

Science at Work

Tree Technician

Tree technicians need strength and balance. They need the ability to climb and to use equipment such as chainsaws and wood chippers. Tree technicians need knowledge of trees and their growth. Tree technicians need a high school diploma. A two-year degree is encouraged but not required.

Tree technicians trim trees, get rid of dead limbs, and remove trees. To do this, they often climb up into a tree and use ropes and pulleys to keep from falling. Other times they use trucks with buckets that lift them into the air. The work of tree technicians helps trees grow correctly and helps prevent problems.

Tree technicians might be hired by homeowners to trim trees on their property. They also might be hired by companies to trim trees that could damage electrical wires or phone lines. After natural disasters, such as tornadoes and hurricanes, tree technicians help clean up damaged trees.

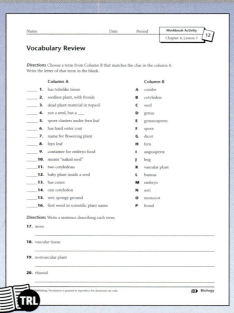

Workbook Activity 12

Chapter 4 SUMMARY

- Plants are classified according to whether they have body parts such as seeds, tubes, roots, stems, and leaves.

- The three main groups of plants are seed plants, ferns and related plants, and mosses and related plants.

- Vascular plants have vascular tissue that forms tubes for transporting food and water.

- Nonvascular plants do not have vascular tissue, so they cannot transport water far.

- Flowering plants are called angiosperms. The seeds of angiosperms are surrounded by a fruit.

- Angiosperms are divided into two groups. Dicots have two cotyledons. Their leaves have a netlike pattern. Monocots have one cotyledon. The veins in their leaves are parallel.

- Gymnosperms are nonflowering plants that have seeds. Their seeds are not surrounded by a fruit.

- Conifers are the major group of gymnosperms.

- Conifers bear cones and have needlelike leaves. Most are evergreens.

- Gingko trees are another kind of gymnosperm. They have fan-shaped leaves.

- Ferns and mosses have no seeds. Ferns are vascular plants. Mosses are nonvascular plants.

- Ferns and mosses reproduce by spores.

Science Words

angiosperm, 71	fern, 68	moss, 68	seed, 68
conifer, 73	frond, 77	nonvascular plant, 69	sori, 77
cotyledon, 72	gymnosperm, 73		spore, 77
dicot, 72	humus, 79	rhizoid, 78	vascular plant, 68
embryo, 71	monocot, 72	rhizome, 77	vascular tissue, 68

Classifying Plant Groups Chapter 4

Chapter 4 Summary

Have students read the Chapter Summary on page 81 to review the main ideas presented in Chapter 4.

Science Words

Direct students' attention to the Science Words at the bottom of page 81. Ask students to review each term and its definition and to use the term in a sentence.

Chapter 4 Review

Use the Chapter Review to prepare students for tests and to reteach content from the chapter.

Chapter 4 Mastery Test

The Teacher's Resource Library includes two parallel forms of the Chapter 4 Mastery Test. The difficulty level of the two forms is equivalent. You may wish to use one form as a pretest and the other form as a posttest.

Review Answers

Vocabulary Review

1. rhizoids 2. humus 3. frond 4. seed
5. nonvascular plants 6. monocots
7. gymnosperm 8. angiosperm 9. conifer
10. fern 11. spores 12. dicot 13. vascular plant 14. embryo

TEACHER ALERT

In the Chapter Review, the Vocabulary Review activity includes a sample of the chapter's vocabulary terms. The activity will help determine students' understanding of key vocabulary terms and concepts presented in the chapter. Other vocabulary terms used in the chapter are listed below:

cotyledon
moss
rhizome
sori
vascular tissue

Chapter 4 REVIEW

Word Bank
angiosperm
conifer
dicot
embryo
fern
frond
gymnosperm
humus
monocots
nonvascular plants
rhizoids
seed
spores
vascular plant

Vocabulary Review

Choose the word or words from the Word Bank that best complete each sentence. Write the answer on a sheet of paper.

1. Mosses get water through rootlike threads called _____.
2. _____ is the part of soil made by dead plants and animals.
3. The large feathery leaf of a fern is called a(n) _____.
4. A(n) _____ has a protective coat around a plant embryo.
5. Mosses are examples of _____.
6. The leaves of _____ have parallel veins.
7. Any nonflowering seed plant is a(n) _____.
8. The seeds of a(n) _____ are usually surrounded by fruit.
9. A(n) _____ is a cone-bearing gymnosperm.
10. A(n) _____ is a seedless, vascular plant.
11. _____ are reproductive cells of ferns and mosses.
12. An angiosperm that has cotyledons is called a(n) _____.
13. A plant that has tissue that forms tubes is called a(n) _____.
14. A seed contains stored food and a(n) _____.

82 Chapter 4 Classifying Plant Groups

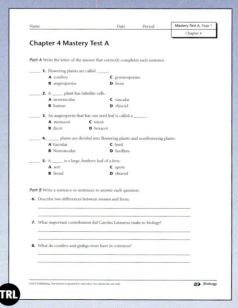
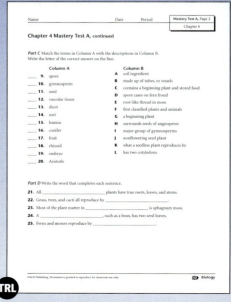

Chapter 4 Mastery Test A

Concept Review

Choose the answer that best completes each sentence. Write the letter of the answer on your paper.

15. Vascular tissue forms _____ that transport food and water.
  A tubes
  B hollows
  C leaves
  D embryos

16. All angiosperms have _____.
  A one cotyledon
  B two cotyledons
  C cones
  D flowers

17. Most conifers have _____ that are green all year round.
  A broad leaves
  B flowers
  C needle-shaped leaves
  D seeds

18. Plants that must live in moist, shady places are _____.
  A angiosperms
  B mosses
  C conifers
  D gymnosperms

Critical Thinking

Write the answer to each of the following questions.

19. Explain how ferns that lived millions of years ago provide energy for us today.

20. Suppose a plant does not have seeds. Can you tell if the plant is vascular or nonvascular? Explain your answer.

Test-Taking Tip When taking a test where you must write your answer, read the question twice to make sure you understand what is being asked.

Concept Review
15. A 16. D 17. C 18. B

Critical Thinking
19. Ferns from millions of years ago became coal, a fossil fuel, over time.
20. You cannot tell if the plant is vascular or nonvascular. Ferns do not have seeds, and they are vascular. Mosses do not have seeds, and they are nonvascular.

ALTERNATIVE ASSESSMENT

Alternative Assessment items correlate to the student Goals for Learning at the beginning of this chapter.

- Show students a corn leaf or a large blade of a grass and a conifer needle, such as pine or spruce. Prompt students to find three differences between these plant parts. Have them state how these characteristics could be used to classify the plants.

- Give students a drinking straw and a shallow bowl of clean water. Ask: What is the easiest way to drink the water? (*Take up the straw and drink from the bowl using the straw.*) Have students write a paragraph explaining how this demonstration illustrates the advantage that vascular plants have over nonvascular plants.

- Have students make a Venn diagram that compares and contrasts plants with seeds and seedless plants.

- Have students prepare a scrapbook of plants with sections for angiosperms, gymnosperms, ferns, and mosses. Have them illustrate their scrapbooks with photos, drawings, dried leaf or flower specimens, and any other relevant information they wish to include.

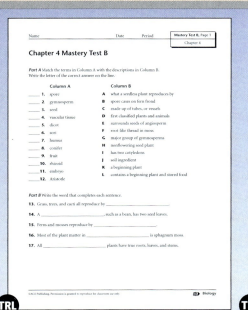
Chapter 4 Mastery Test B

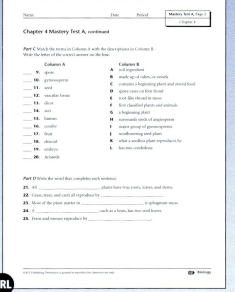

Chapter 5

Planning Guide
Bacteria, Protists, and Fungi

			Student Text Lesson		
			Student Pages	Vocabulary	Lesson Review
Lesson 1	Bacteria		86–89	✔	✔
Lesson 2	Protists		90–94	✔	✔
Lesson 3	How Protists Survive		95–98	✔	✔
Lesson 4	Fungi		99–106	✔	✔
Lesson 5	How Fungi Survive		107–110	✔	✔

Chapter Activities

Student Text
Science Center

Teacher's Resource Library
Community Connection 5: Helpful Bacteria, Protists, and Fungi in Your Community

Assessment Options

Student Text
Chapter 5 Review

Teacher's Resource Library
Chapter 5 Mastery Tests A and B

Student Text Features								Teaching Strategies						Learning Styles						Teacher's Resource Library				
Achievements in Science	Science at Work	Science in Your Life	Investigation	Science Myth	Note	Technology Note	Did You Know?	Science Integration	Science Journal	Cross-Curricular Connection	Online Connection	Teacher Alert	Applications (Home, Career, Community, Global, Environment)	Auditory/Verbal	Body/Kinesthetic	Interpersonal/Group Learning	Logical/Mathematical	Visual/Spatial	LEP/ESL	Workbook Activities	Alternative Workbook Activities	Lab Manual	Resource File	Self-Study Guide
					✓	✓	86	87		87, 88	89	88	89						87	13	13		9	✓
94							91			92	93	91	92, 93			93	92		92	14	14			✓
						✓		97	97			96			96					15	15	13	10	✓
	104	103	105	102			100		101	102, 103			101	102		103			101	100	16	16	14, 15	✓
110								109	109	108, 109	109	108							108	17	17			✓

Pronunciation Key

a	hat	e	let	ī	ice	ô	order	ù	put	sh she
ā	age	ē	equal	o	hot	oi	oil	ü	rule	th thin
ä	far	ėr	term	ō	open	ou	out	ch	child	ŦH then
â	care	i	it	ȯ	saw	u	cup	ng	long	zh measure

ə { a in about / e in taken / i in pencil / o in lemon / u in circus }

Alternative Workbook Activities

The Teacher's Resource Library (TRL) contains a set of lower-level worksheets called Alternative Workbook Activities. These worksheets cover the same content as the regular Workbook Activities but are written at a second-grade reading level.

Skill Track Software

Use the Skill Track Software for Biology for additional reinforcement of this chapter. The software program allows students using AGS textbooks to be assessed for mastery of each chapter and lesson of the textbook. Students access the software on an individual basis and are assessed with multiple-choice items.

Chapter at a Glance

Chapter 5: Bacteria, Protists, and Fungi
pages 84–113

Lessons
1. Bacteria pages 86–89
2. Protists pages 90–94
3. How Protists Survive pages 95–98
4. Fungi pages 99–106

Investigation 5 pages 105–106

5. How Fungi Survive pages 107–110

Chapter 5 Summary page 111

Chapter 5 Review pages 112–113

Skill Track Software for Biology

Teacher's Resource Library TRL
  Workbook Activities 13–17
  Alternative Workbook Activities 13–17
  Lab Manual 13–15
  Community Connection 5
  Resource File 9–10
  Chapter 5 Self-Study Guide
  Chapter 5 Mastery Tests A and B

(Answer Keys for the Teacher's Resource Library begin on page 420 of the Teacher's Edition. A list of supplies required for Lab Manual Activities in this chapter begins on page 449.)

Science Center

Ask students to prepare index cards with information about how bacteria, protists, and fungi may harm or help people. Each card should describe a specific organism and its relationship with humans. Encourage students to include a picture of the organism if possible. Have students assemble the cards in a display titled "Bacteria, Protists, and Fungi in Our Lives."

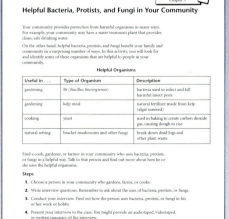

Community Connection 5

Chapter 5
Bacteria, Protists, and Fungi

If you study the photo, you can see Spanish moss growing on a cypress tree. You also can see the leaves of the green plants floating on the surface of the water. There are other organisms too small for you to see without a microscope. These tiny organisms, algae, and fungi are not plants or animals. They are bacteria, protists, and fungi. In this chapter, you will learn about the properties and life activities of these organisms.

Organize Your Thoughts

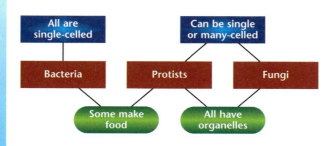

Goals for Learning

- ◆ To identify the properties of bacteria
- ◆ To explore the different kinds of protists and their features.
- ◆ To describe the properties and life activities of fungi

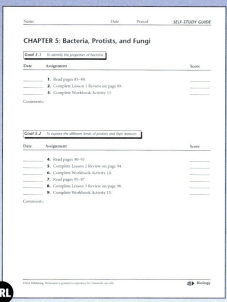

Chapter 5 Self-Study Guide

Introducing the Chapter

Have students work in small groups to answer the following questions: What kinds of organisms exist as single cells? What kinds have bodies made of many cells? Why does food spoil if it isn't refrigerated? What causes tetanus and strep throat? What makes bread dough rise? Is a mushroom a plant, an animal, or something else? Ask each group to share its answers with the class. Refer to the answers as you proceed through the chapter.

Notes and Technology Notes

Ask volunteers to read the notes that appear in the margins throughout the chapter. Then discuss them with the class.

TEACHER'S RESOURCE

The AGS Teaching Strategies in Science Transparencies may be used with this chapter. The transparencies add an interactive dimension to expand and enhance the *Biology* program content.

CAREER INTEREST INVENTORY

The AGS Harrington-O'Shea Career Decision-Making System-Revised (CDM) may be used with this chapter. Students can use the CDM to explore their interests and identify careers. The CDM defines career areas that are indicated by students' responses on the inventory.

Lesson at a Glance

Chapter 5 Lesson 1

Overview In this lesson, students learn how bacteria differ from other organisms and how bacteria can either harm or help other organisms, including humans.

Objectives

- To describe the properties of bacteria
- To explain how bacteria get energy and reproduce
- To list ways in which bacteria are harmful and helpful

Student Pages 86–89

Teacher's Resource Library (TRL)
- Workbook Activity 13
- Alternative Workbook Activity 13
- Resource File 9

Vocabulary

methane	commensalism
binary fission	toxin
saprophyte	endospores
mutualism	

Science Background
Bacteria

Bacteria lack a nucleus, the organelle that stores DNA. For this reason, bacteria are called *prokaryotes*, which means "before nucleus." All other organisms have organelles, including a nucleus, and are therefore known as *eukaryotes*, meaning "true nucleus." Some species of bacteria respond to harsh conditions by forming resistant structures called endospores. An endospore consists of a bacterial cell's DNA surrounded by a durable wall. Endospores can survive in a dormant state without food or water for centuries. Even boiling does not usually destroy an endospore.

1 Warm-Up Activity

Display the nodules on the roots of a leguminous plant, such as a pea, soybean, or clover plant. Explain to students that the nodules contain bacteria, and the bacteria and the plant live in a mutually beneficial relationship.

Lesson 1 Bacteria

Objectives

After reading this lesson, you should be able to

- describe the properties of bacteria.
- explain how bacteria get energy and reproduce.
- list ways in which bacteria are harmful and helpful.

The greatest number of organisms in the world are bacteria. They are members of the kingdom Monera. More than 10,000 known species of bacteria live almost everywhere on Earth.

Properties of Bacteria

Bacteria are single-celled organisms. Most bacterial cells are very small. They can be seen only by using a microscope. The cells of other organisms are usually at least 50 times larger.

Bacteria are the simplest organisms. The inside of a bacterial cell lacks organelles. Recall that organelles are tiny structures inside cells that do particular jobs. All other organisms have organelles. This characteristic sets bacteria apart from all other organisms.

Fossil evidence suggests that bacteria appeared on Earth at least 3.5 billion years ago. For about 1 billion years, they were the only form of living thing on Earth. Some modern-day bacteria may be like those first bacteria.

One way that biologists classify bacteria is by their shape. Bacteria can look like rods, spirals, or spheres. Spherical, or round, bacteria often form long chains. Some form clusters that look like a bunch of grapes.

Did You Know?

Some bacteria grow in extremely salty lakes, such as the Great Salt Lake and the Dead Sea. Other bacteria live in hot springs and around volcanic vents deep in the ocean. They can survive in water that may reach 110°C.

  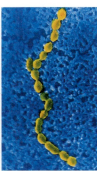

Most bacteria are shaped like rods, spirals, or spheres.

Did You Know?

Have students read the Did You Know? feature on page 86. Ask: How hot is 110°C? (*Very hot; it is above the boiling point of water.*) Remind students that most living things cannot survive in water that hot.

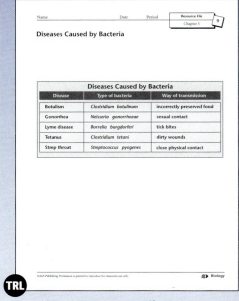

Resource File 9

Methane
A gas produced by bacteria from hydrogen and carbon dioxide

Binary fission
Reproduction in which a bacterial cell divides into two cells that look the same as the original cell

Saprophyte
An organism that decomposes dead organisms or waste matter

Mutualism
A closeness in which two organisms live together and help each other

Commensalism
A relationship in which one organism benefits and the other is not affected

Life Activities of Bacteria

Although bacteria are single-celled and simple, they are complete organisms. Each bacterial cell carries out all of the basic functions of life. For example, bacteria use many different ways to obtain energy. Some bacteria break down dead organisms or waste matter. Others use sunlight to make food. Still other bacteria get energy from minerals, such as iron. In fact, bacteria have more ways to get energy than any other kind of organism.

Many kinds of bacteria need oxygen to carry out respiration. However, some bacteria are poisoned by oxygen. They live only where oxygen is not present, such as at the bottom of swamps. These bacteria produce **methane** out of hydrogen and carbon dioxide. The methane bubbles out of the water as marsh gas.

Bacteria reproduce by a process called **binary fission**. The bacterial cell divides into two cells that look the same as the original cell. Some bacteria can reproduce every 20 minutes.

Helpful Bacteria

Many bacteria are helpful to other organisms in two ways. First, some bacteria recycle nutrients, such as carbon and nitrogen. They break down, or decompose, dead organisms or waste matter. Organisms that do this are called **saprophytes**. Some of the broken-down nutrients are returned to the soil. Plants use them. Animals get these nutrients by eating the plants or by eating other animals that eat plants.

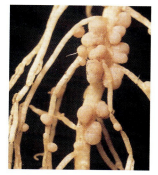

The swellings on the roots of this pea plant contain bacteria that change nitrogen into ammonia.

Second, bacteria help plants to get nitrogen. Nitrogen is plentiful in the atmosphere. However, plants cannot absorb, or take in, nitrogen, from the air. The nitrogen must be changed into ammonia before the plants can use it. Some kinds of bacteria change the nitrogen into ammonia. They live inside the roots of particular plants. In return, the plants provide the bacteria with food. The plants and the bacteria benefit, or help, each other. A closeness in which two organisms live together and help each other is called **mutualism**. A relationship in which one organism benefits and the other is not affected is called **commensalism**.

Bacteria, Protists, and Fungi Chapter 5 **87**

2 Teaching the Lesson

Before students read the lesson, have them scan for unfamiliar words. Ask students to make a list of the unfamiliar words and then to write definitions for each word on their lists.

Teach the vocabulary word *binary fission* by pointing out that *binary* means "two" and *fission* means "splitting"; therefore, *binary fission* means "splitting in two."

Review the fundamental life processes. See Chapter 1 pages 17–19. Students will refer to such processes repeatedly through Chapter 5.

Remind students of the fundamental distinction between food-producing organisms (plants, for example) and organisms that must obtain food from outside sources (animals, for example). Encourage them to think about this distinction as they read the chapter.

The day before class, inoculate two sterile Petri dishes with the bacterium *Bacillus subtilis*, obtained from a biological supply company. **Safety Alert:** Seal each dish with tape and label it as a biohazard. Do not open the dishes once they have been sealed. Place one dish in a refrigerator and leave the other at room temperature. The next day, ask students to observe both dishes and describe their observations. (*Bacterial colonies should be visible on the dish stored at room temperature, but not on the refrigerated dish.*)

3 Reinforce and Extend

LEARNING STYLES

LEP/ESL
The helpful/harmful dichotomy is used throughout Chapter 5, so students must completely understand the terms *helpful* and *harmful*. Explain that in the context of this lesson, *helpful* and *harmful* have opposite meanings. An organism that is helpful does something for another organism. A harmful organism does something that works against or hurts another organism.

SCIENCE INTEGRATION

Physical Science
Write the formulas for nitrogen gas and ammonia on the board: N_2 (nitrogen) and NH_3 (ammonia). Ask students to explain how these formulas show that ammonia can be a source of nitrogen. (*N is the chemical formula for nitrogen. NH_3 is the formula for ammonia, a chemical compound made up of nitrogen and hydrogen [H]. The presence of nitrogen in the formula for ammonia shows that ammonia can be a source of nitrogen.*)

CROSS-CURRICULAR CONNECTION

Math
Bacteria have lived on Earth for 3.5 billion years. Help students grasp the magnitude of a billion by prompting them to think about stacks of pennies. About 6 pennies make a stack 1 cm high, and about 16 pennies make a 1-inch stack. Ask: How high do you think a stack of 1 billion pennies would be? (*A stack of 1 billion pennies would be almost 1,000 miles high—or slightly higher than 1,500 km.*)

Bacteria, Protists, and Fungi Chapter 5 **87**

TEACHER ALERT

Read the information in the bottom box on page 88. It presents the term *endospores*, a lesson vocabulary word. Write the word on the board and underline the part that spells *spores*. Ask students what they remember about the spores of plants, such as ferns and mosses. (*The spores are a means of reproduction.*) Tell students that bacterial spores are similar to plant spores in that they help the organism spread to new places. However, bacteria reproduce by binary fission, a completely distinct and separate activity from spore formation.

CROSS-CURRICULAR CONNECTION

Health

Botulism is one of several bacterial food-borne illnesses that may appear when food is prepared or handled improperly. Other bacteria that cause illness include *campylobacter*, *salmonella*, and *E. coli*. Write the names of these bacterial species and *clostridium botulinum* on the board. Have students research information about one of the bacteria and report their findings to the class. Ask them to include information about how people can avoid and treat an infection by the bacteria.

Toxin
A poison produced by bacteria or other organisms

Endospores
Bacteria that dry up and form into cells with thick walls

Some bacteria can dry up and form cells with thick walls. These **endospores** can survive for years. Tetanus, botulism, and anthrax are diseases caused by bacteria that form endospores.

Many bacteria are useful to humans in two ways. First, bacteria help to produce some of the food we eat. Do you ever eat cheese, sour cream, or yogurt? Bacteria break down milk to produce these foods. Bacteria also change cucumbers into pickles and cabbage into sauerkraut.

Second, bacteria can produce many different materials that are helpful to humans. For example, chemical companies use bacteria to make vitamins. Some bacteria make antibiotics, materials that kill other kinds of bacteria.

Harmful Bacteria

Some bacteria cause food to spoil. For example, pasteurized milk still contains some bacteria. They are harmless but eventually will sour the milk. Refrigerating food helps to prevent spoiling. Bacteria grow slowly at low temperatures. Food that looks or tastes bad must be thrown away instead of eaten.

Other bacteria are more than just a nuisance. Some bacteria cause disease. The table below lists several diseases that bacteria cause. Some bacteria harm the body by producing poisons called **toxins**. Bacterial toxins are among the most powerful poisons known. A single gram of botulism toxin can kill a million people.

Diseases Caused by Bacteria		
Disease	**Type of bacteria**	**Way of transmission**
Botulism	*Clostridium botulinum*	incorrectly preserved food
Gonorrhea	*Neisseria gonorrhoeae*	sexual contact
Lyme disease	*Borrelia burgdorferi*	tick bites
Tetanus	*Clostridium tetani*	dirty wounds
Strep throat	*Streptococcus pyogenes*	close physical contact

Doctors prescribe, or give, antibiotics, such as penicillin, to fight bacterial diseases. However, bacteria often develop ways to resist the antibiotic. Household products, such as dish soaps, often contain antibacterial agents. They also may cause bacteria to become resistant. As more bacteria become resistant, scientists must search for new ways to kill the bacteria.

Lesson 1 REVIEW

Write your answers to these questions on a separate sheet of paper. Write complete sentences.

1. Explain how bacteria are different from all other organisms.
2. What are the three shapes that most bacteria have?
3. How do bacteria obtain energy?
4. How do bacteria reproduce?
5. Name two ways in which bacteria are useful to humans.

Technology Note

Before the use of technology, scientists used other ways to classify bacteria. One of these was a staining method developed by Hans Christian Gram in 1884. Known as the Gram Stain, it is still used today to identify bacteria. Gram-positive bacteria stain a blue-green color. Gram-negative bacteria stain red. Staining helps doctors know what kind of antibiotic can help fight a disease. Gram-negative bacteria are more resistant to antibiotics, especially penicillin.

Bacteria, Protists, and Fungi Chapter 5 89

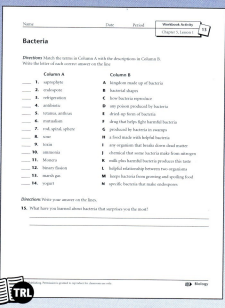
Workbook Activity 13

ONLINE CONNECTION

 Encourage students to learn more about food safety by visiting a Web page maintained by the U.S. Food and Drug Administration/Center for Food Safety and Applied Nutrition. The site, which features activities and games that promote food safety education, is at www.cfsan.fda.gov/~dms/educate.html.

IN THE ENVIRONMENT

 Tell students that bacteria are used to solve certain environmental problems, such as cleaning up oil and chemical spills and breaking down plastics in garbage. Have students do research on one of these topics and write a short report on their findings.

Lesson 1 Review Answers

1. Bacteria are single-celled organisms that are usually much smaller than other organisms. Bacteria cells lack organelles that are found in the cells of other organisms. **2.** Most bacteria are shaped like rods, spirals, or spheres. **3.** Bacteria may break down the chemicals in dead organisms, capture the energy in sunlight, or obtain energy from minerals. **4.** Bacteria reproduce by dividing into two cells that look the same. **5.** Answers may include that bacteria help to produce foods, vitamins, antibiotics, and special chemicals.

Portfolio Assessment

Sample items include:

- Definitions of terms from Teaching the Lesson
- Lesson 1 Review answers

Bacteria, Protists, and Fungi Chapter 5 89

Lesson at a Glance

Chapter 5 Lesson 2

Overview In this lesson, students are introduced to several types of plant-like protists, or algae, and animal-like protists, or protozoans.

Objectives

- To describe the features of several types of algae
- To describe the features of several types of protozoans

Student Pages 90–94

Teacher's Resource Library TRL

Workbook Activity 14
Alternative Activity 14

Vocabulary

diatom   trypanosomes
paramecium   sporozoan

Science Background
Protists

Protists are eukaryotes (organisms with a true nucleus and other organelles) and are therefore much more complex than bacteria. Protista is the most diverse kingdom in biology. It includes organisms that make their own food, like plants; that consume food, like animals; and that absorb food, like fungi. It also includes single-celled as well as a few multicelled organisms, such as algal seaweeds. However, these multicelled organisms do not have the high degree of tissue differentiation that plants and animals do. Biologists are still debating the classification of protist organisms and the scientific consensus for this classification scheme may change over time.

Lesson 2 Protists

Objectives

After reading this lesson, you should be able to
- describe the features of several types of algae.
- describe the features of several types of protozoans.

The Protista kingdom contains more than 60,000 species of organisms. Protists may have features like those of plants, animals, or fungi. However, all plants and animals and most fungi have many cells. Most protists are single-celled, like bacteria. What makes protists different from bacteria? Unlike bacteria, the cells of protists have organelles inside their cells. The cells of plants, animals, and fungi also have organelles.

Algae

Plant-like protists are known as algae. Like plants, algae use the energy in sunlight to make food from carbon dioxide and water.

Most algae are aquatic, which means they live in freshwater or salt water. Most are single-celled. Tiny algae float or swim near the surface of the water in vast numbers. Although each is only a single cell, together they produce nearly half the world's carbohydrates. Algae provide food for all the other organisms that live in lakes or the sea. Algae also release oxygen as they make food. About half the oxygen that enters the atmosphere comes from algae.

One kind of single-celled alga is *Euglena gracilis*. Inside its cell are many chloroplasts. These structures capture the light energy from the Sun that is used to make food. All other algae also have chloroplasts. *Euglena gracilis* lives in freshwater and moves with a long, whiplike structure called a flagellum.

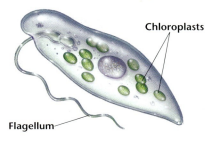

90   Chapter 5   *Bacteria, Protists, and Fungi*

Diatom
Microscopic alga that has a hard shell

Did You Know?

When you brush your teeth, you may feel the crunch of shells from diatoms. Diatoms are used in making toothpaste.

Other kinds of single-celled algae are **diatoms**. They are found in great numbers in freshwater and in the ocean. They have hard shells that contain silica. This is the same material found in glass. As diatoms die, their shells build up in deposits at the bottom of lakes and the sea. These deposits are mined to make metal polish, soaps, and scouring powders.

Although most algae are single-celled, some are many-celled. One familiar example is seaweed. One kind of seaweed is commonly called sea lettuce. Its blades look like lettuce leaves and can grow up to 60 centimeters long. Giant seaweeds known as kelps are even larger. Their bodies may stretch 100 meters from a shallow ocean floor to the surface. Kelps are harvested and used to thicken food.

Not all algae live in water. Some live in moist places on land, such as in soil or on stones. Algae in the genus *Protococcus* often grow on trees and fence posts. The next time you walk in the woods, look for these algae on the shaded side of tree trunks.

Kelps are large, many-celled algae.

Bacteria, Protists, and Fungi   Chapter 5   **91**

1 Warm-Up Activity

Ask students if they have ever observed a stagnant pond or lake. The water surface may have been covered with bright green, slimy matter, and it probably smelled bad. Ask what students think the green material is. (*Students should recognize it as a form of life; they may mistake it for a plant because it is green.*) Ask students if they think the pond slime has leaves, roots, and stems like most plants. (*no*) Point out that green pond slime is not a plant or an animal. It is a protist, one of many organisms that are not bacteria, plants, or animals.

TEACHER ALERT

Remind students that bacteria do not have a nucleus or any other organelles. Make sure students understand that this characteristic distinguishes bacteria from all other organisms, including the single-celled protists featured in this lesson.

2 Teaching the Lesson

Have students copy two sentences for each vocabulary term in the lesson. Then have them write a new sentence using that term.

Obtain a culture of live amoebas from a biological supply company. Place a drop of the culture on a microscope slide and add a coverslip. Have students examine the slide under a microscope to observe how the amoebas move by extending their pseudopods. Emphasize that amoebas are protozoans, animal-like protists. Prompt students to contrast amoebas with algae. (*Algae are plant-like; they have chlorophyll and make their own food. Amoebas are animal-like; they have to find and eat food.*)

Did You Know?

Read aloud Did You Know? on page 91. The shells of diatoms sink to the ocean floor when the diatoms die. Land that was under the sea in ancient times is a source of diatomaceous earth that is processed for use in the toothpaste.

TEACHER ALERT

Students may wonder why many-celled algae, such as kelp, are classified as protists rather than plants. Even though some algae have many cells, their cells are actually more closely related to cells of other protists than to plant cells. Also, in plants, tissues are much more differentiated than in many-celled algae.

3 Reinforce and Extend

CROSS-CURRICULAR CONNECTION

Art
Diatoms display an amazing variety of intricate, beautiful forms. Prompt students to look in reference books or online for photomicrographs of diatoms. Have them select a specimen they particularly like and produce a work of art from it. Students might make interesting patterns from tracings of many diatom pictures or even produce a colorful three-dimensional model of a diatom.

AT HOME

Explain that carageenan is a chemical that is extracted from red algae and used as a thickener in foods. Have students read package labels to find foods at home that contain carageenan. Examples may include puddings, salad dressings, ice cream, chocolate milk, whipped toppings, and processed cheeses.

LEARNING STYLES

Logical/Mathematical
Have students develop a diagram showing the two main subgroups (algae and protozoans) in the Protist kingdom and listing representative organisms in each subgroup. Prompt students to consider what type of graphic organizer would be most appropriate to represent logical relationships between the subgroups and among organisms in the subgroups. The organizer they choose should enable them to depict the two subgroups as equivalent within the Protist kingdom and species subordinate to each subgroup.

Protozoans

Paramecium
A protozoan that moves by using its hairlike cilia (plural is paramecia)

Animal-like protists are called protozoans, which means "first animals." All protozoans are single-celled and can be seen only under a microscope. They live in water, on land, and inside other organisms. Like animals, protozoans cannot make their own food. Instead, they eat bacteria, other protists, or dead organisms. Protozoans are divided into four groups by the way they move.

Amoebas are protozoans that move by pushing out parts of their cell. The pushed-out part is called a pseudopod. *Pseudopod* means "false foot." It looks like a foot and pulls the amoeba along. In this way, amoebas change their shape. Amoebas also use their pseudopods to surround and trap other protists for food. Amoebas live on rocks and on plants in ponds.

Another group of protozoans is covered with short, hairlike structures known as cilia. A **paramecium** is a common example of a protozoan that has cilia. Like amoebas, paramecia often are found in ponds. The cilia on a paramecium move back and forth like tiny oars. They all move in the same direction at the same time. For example, when the cilia push backward, the paramecium moves forward. You can see a diagram of a paramecium on page 95.

An amoeba (blue-green) pushes out pseudopods, parts of its cell, to move and to capture food.

92 Chapter 5 Bacteria, Protists, and Fungi

LEARNING STYLES

LEP/ESL
Many nouns in science vocabulary come from Latin and Greek words. These nouns may form plurals in ways that differ from most English plurals that are formed by adding *-s* or *-es* to the base words. In this lesson, several nouns have plural forms that may be unfamiliar to students. For example, *algae* is the plural form of the singular *alga*, *paramecia* is the plural form of *paramecium*, *cilia* is the plural form of *cilium*, and *flagella* is the plural form of *flagellum*.

Trypanosome
A protozoan that is a parasite and lives in blood; may cause sleeping sickness

Sporozoan
A protozoan that is a parasite and lives in blood; may cause malaria

A third group of protozoans uses flagella to push or pull themselves around. These protozoans include *Giardia lamblia*, which is a parasite in the intestines of animals. Humans can get this parasite from water that contains wastes from infected humans or animals. The disease causes tiredness, weight loss, and diarrhea. It is usually not fatal. A much more serious disease is sleeping sickness. This disease is caused by **trypanosomes**. These protozoans live as parasites in the blood of humans and other mammals. They have a flagellum that moves them through the blood. Trypanosomes are spread by the bite of tsetse flies, which live only in Africa. Sleeping sickness is fatal unless treated.

The fourth group of protozoans has no means of moving as adults. They reproduce by forming spores, so they are called **sporozoans**. All sporozoans are parasites. They live in the blood of their hosts. The sporozoan *Plasmodium* causes malaria. Mosquitoes in the genus *Anopheles* spread malaria when they draw blood from an infected person. The sporozoans enter the mosquito and reproduce. The mosquito transfers the sporozoans when it bites another person. Malaria can be deadly. Every year, it affects about half a billion people and kills one to three million.

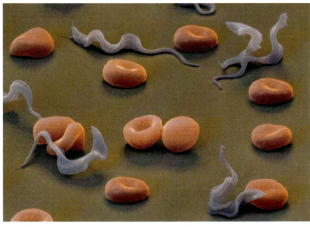

Trypanosomes are parasitic protozoans that cause sleeping sickness. They live among the red blood cells.

IN THE COMMUNITY

Point out that good health isn't just an individual concern. Communities need to use good health practices, too. A focus on the health of an entire community is called public health. Students may be aware of public health departments in their communities. Suggest that they think about what health issues might be especially important to public health. (*Examples: infectious diseases such as influenza, the quality of available drinking water, and even accident rates.*)

CAREER CONNECTION

The arena of public health offers a wide variety of careers, many of them in government agencies. Some of these careers are directly tied to the medical profession, such as those of medical doctors and nurses working for a federal agency. Other related careers may include health-related data collection; disaster relief; medical research to track and develop treatments for diseases; inspection of hospitals, pharmacies, clinics, and nursing homes; and environmental safety regulation. Have students investigate a public health career and identify educational requirements.

LEARNING STYLES

Interpersonal/Group Learning
Give small groups of students pictures of protists or a list of the different protists. Ask the groups to develop a system for organizing the protists into subgroups. Have the groups share their organizations with the class. Close the activity by pointing out that the Protist kingdom has been accepted by most biologists as an independent kingdom only since the second half of the 1900s.

ONLINE CONNECTION

The Centers for Disease Control (CDC) maintains "BAM!," a Web page on public health at www.bam.gov. Students can learn more about diseases and other public health topics there.

Lesson 2 Review Answers

1. Unlike bacteria, the cells of protists have organelles. 2. Aquatic algae produce nearly half of the world's carbohydrates and oxygen. 3. Diatoms are single-celled algae with hard shells. Kelps are large, many-celled algae. 4. Protozoans move by thrusting out pseudopods, bending their cilia back and forth, or using flagella to push or pull themselves around. 5. Malaria is caused by a sporozoan. Mosquitoes spread malaria when they draw blood from an infected person and then bite other people.

Portfolio Assessment

Sample items include:
- Sentences from Teaching the Lesson
- Protist kingdom diagram from Learning Styles
- Lesson 2 Review answers

Achievements in Science

Read Achievements in Science on page 94. Lake Vida is in a dry area of Antarctica that receives less than 10 cm of precipitation per year. Through the use of scientific data and tools, scientists learned that Lake Vida has water in the form of liquid under its ice. This liquid zone is extremely salty and cold.

Lesson 2 REVIEW

Write your answers to these questions on a separate sheet of paper. Write complete sentences.

1. How are protists different from bacteria?
2. List two reasons why aquatic algae are important.
3. How are diatoms and kelps different from each other?
4. Describe the three ways that protozoans move.
5. Explain what causes malaria and how the disease is spread.

Achievements in Science

Ancient Bacteria Discovered in Antarctic Ice

In October 1996, a team of scientists took two ice core samples from Lake Vida in the Antarctic. The drilling took two weeks in temperatures below −35°C. The scientists discovered bacteria and algae that were at least 2,800 years old in the ice. Amazingly, the microorganisms began carrying out life activities when exposed to liquid water.

Scientists will need to study the microorganisms to find out what allows them to survive at such cold temperatures. Scientists also can learn about the history of microorganisms by studying their ancient DNA. Although it sounds strange, this discovery may also help scientists know where to look for signs of ancient life on Mars. Scientists believe that the surface of Mars had water in the past. It's possible that organisms could have lived in areas similar to Lake Vida before all the water on Mars froze.

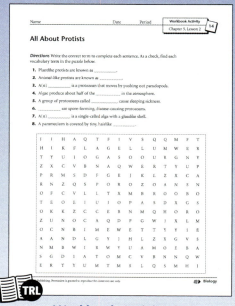

Workbook Activity 14

Lesson 3: How Protists Survive

Objectives

After reading this lesson, you should be able to
- describe how protists digest food.
- explain how protists maintain water balance.
- discuss how protists react to their environment and reproduce.

Gullet
The opening through which a paramecium takes in food

Food vacuole
A bubblelike structure where food is digested inside a protozoan

Anal pore
The opening through which undigested food leaves a paramecium

Like all other organisms, protists must carry out all of the basic life activities. Though most protists are single-celled, their cells are not simple. Each cell must perform the duties of tissues and organs in a plant or an animal.

Getting and Digesting Food

You have already learned that algae make their own food but protozoans do not. However, some algae can change the way they feed themselves. For example, *Euglena gracilis* uses its chloroplasts to make food when sunlight is present. In the dark, it acts like a protozoan and eats its food.

Different kinds of protozoans have different ways of getting food. As you read earlier, amoebas trap other protists with their pseudopods. Paramecia use their cilia to sweep food particles over their surface. The food moves into an opening called the **gullet** on the paramecium's side. The gullet is like the mouth of an animal.

The diagram below shows what happens to the food that enters a paramecium's gullet. The gullet encloses the food within a bubblelike structure called a **food vacuole**. These small packets of food travel all through the paramecium. Chemicals in each food vacuole break down the food. The food then leaves the food vacuole to be used by the paramecium. Food that is not digested leaves the paramecium through an opening called the **anal pore**. Amoebas and other protozoans also digest food inside food vacuoles.

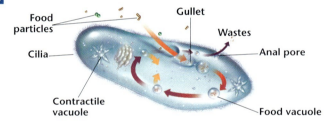

Bacteria, Protists, and Fungi Chapter 5 95

Lesson at a Glance

Chapter 5 Lesson 3

Overview In this lesson, students explore some of the life activities of protists, including how protists obtain food and reproduce.

Objectives
- To describe how protists digest food
- To explain how protists maintain water balance
- To discuss how protists react to their environment and reproduce

Student Pages 95–98

Teacher's Resource Library
- Workbook Activity 15
- Alternative Activity 15
- Lab Manual 13
- Resource File 10

Vocabulary
gullet
food vacuole
anal pore
osmosis
contractile vacuole
eyespot
asexual reproduction
sexual reproduction

Science Background
Life Processes of Protozoans

Protozoans are single-celled protists that consume food in the same way as animals do. Animals have tissue groups and organs to perform specialized tasks such as digestion. Protists such as the paramecium and the amoeba perform digestion and other life processes at the cellular level.

1 Warm-Up Activity

Prompt students to think about what happens inside their bodies after they eat a meal. Emphasize that humans have many different organs, including the mouth, esophagus, stomach, large and small intestines, liver, and gall bladder, that play a role in the digestive process. Tell students that protists do not have digestive organs; instead, their one-celled bodies must do the work.

Bacteria, Protists, and Fungi Chapter 5 95

Teaching the Lesson

Have students rephrase the lesson objectives as questions. Students should answer each question as you teach the lesson.

Remind students that cilia on a paramecium are hairlike structures that enable the creature to move and, to a limited degree, sense conditions in the outside world. Make an analogy with the tiny, sensitive hairs on a person's arm. If something brushes against your arm enough to ruffle these hairs, you probably react immediately by moving your arm.

Have students study the illustration of the paramecium on page 95. Then have them make a free-hand drawing of the paramecium without looking at the picture in the textbook. Remind them to include and label the gullet, anal pore, food vacuole, and cilia.

Distribute copies of Resource File 10. Have students construct a compare-and-contrast chart identifying the similarities and differences between the *Euglena* and paramecium shown.

 Reinforce and Extend

> **TEACHER ALERT**
>
> As students read about how a paramecium senses and responds to an obstacle, remind them that a protist such as the paramecium does not have a brain to direct its activities. It cannot evaluate a situation that arises in its environment. The paramecium's cilia sense the barrier, and the paramecium starts spinning. It will probably be pointing in a different direction when it stops spinning. However, it has not thought out in advance what to do.

Osmosis
The movement of water through a cell membrane

Contractile vacuole
A structure in a protist that removes water that is not needed

Maintaining Water Balance

Most protists live in a watery environment. Water moves easily through the cell membrane. The cell membrane is a thin layer that surrounds and holds a cell together. This movement of water creates a problem. To understand how, consider a single-celled protist that lives in freshwater. The water inside the protist is mixed with many other chemicals. The water outside the protist contains few chemicals but more water molecules. The water molecules are more concentrated, or crowded together. This difference in concentration causes water to move into the protist from the outside. **Osmosis** is the movement of water through a cell membrane. Water molecules move from an area of high concentration to an area of low concentration. This is a type of passive transport, or movement, that doesn't use cellular energy.

Protists need some water to survive. However, too much water causes the cell to burst. The diagram shows what would happen if the cell could not get rid of some of the water. As more water entered, the protist would grow larger and larger. At some point, the cell membrane would stretch to its breaking point. The protist would burst and die.

To avoid bursting, protists release water that is not needed. Structures called **contractile vacuoles** collect the water. They contract, or pull together, to squeeze the water out of the protist. Find a contractile vacuole in the diagram of the paramecium on page 95. Contractile vacuoles carry out the same function that your kidneys perform when you drink too much water. Maintaining water balance is an important function for all organisms.

96   Chapter 5   *Bacteria, Protists, and Fungi*

> **LEARNING STYLES**
>
> **Body/Kinesthetic**
> Have students act out how a paramecium senses a barrier with its cilia, spins around, and then takes off in a new direction. As several students role-play the process of sensing, other students can describe the process.

Lab Manual 13, pages 1–2

Eyespot
A structure on many protists that senses changes in the brightness of light

Asexual reproduction
Reproduction that involves one parent and no egg or sperm

Sexual reproduction
Reproduction that involves two parents, and usually an egg and sperm

Sensing and Reacting

All organisms must be able to sense and react to signals in their environment. Many protists have an **eyespot**, which can sense changes in the brightness of light. The eyespot allows algae to move to areas where the light is brighter. With brighter light, they can make food more quickly.

Protozoans can sense food in their environment and move toward it. Protozoans also can detect and move away from harmful chemicals. They can move away from objects that are in their way. For example, when a paramecium bumps into an object, it reverses the movement of its cilia. This causes the paramecium to back up. The paramecium then spins briefly and heads off in a new direction.

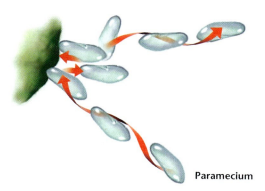
Paramecium

Reproduction

Many single-celled protists reproduce by dividing into two cells. The result is two protists that look the same as the original cell. This process is an example of **asexual reproduction**.

Some protists also can reproduce in pairs. Each member of the pair gives some hereditary material to the offspring. The offspring are different from either parent. *Hereditary* refers to properties that are passed from parent to offspring. This type of reproduction is known as **sexual reproduction**.

SCIENCE INTEGRATION

Physical Science

Osmosis is an important concept in the science of solutions, a branch of physics. It is also important to many living things, including plants and protists. For osmosis to occur, a semipermeable membrane must be between two liquids. One liquid could be a solution such as sugar dissolved in water; the other liquid could be pure water. If the semipermeable membrane has holes that are too small for dissolved sugar, the solute, to pass through, then only the water molecules, the solvent, will pass through the tiny holes. The water will move from the pure water into the sugar solution. The tendency of the solvent to move across the semipermeable membrane into the solution is called osmotic pressure.

The description of the way in which a protist maintains its water balance illustrates the process of osmosis. The protist's cell membrane is the semipermeable membrane. The osmotic pressure is from the water outside the cell to the solution of water and other chemicals inside the cell.

SCIENCE JOURNAL

Ask students to write a description of a typical day in the life of a single-celled protist living in a pond. Students should identify the kind of protist, its ways of moving and obtaining food, and its way of sensing and reacting to its environment.

Lesson 3 Review Answers

1. Food enters the gullet, is enclosed in a food vacuole, and travels all through the paramecium; food waste is expelled through the anal pore. **2.** Water continually enters a protist by osmosis. Without a means of removing the water, the protist would swell and burst. **3.** Water is entering the protist by osmosis and is not being expelled; eventually the protist bursts and dies. **4.** By sensing changes in the brightness of light, eyespots allow algae to move to areas where the light is brighter. With brighter light, they can make food more quickly. **5.** A single-celled protist divides into two cells.

Portfolio Assessment

Sample items include:
- Compare-and-contrast chart from Teaching the Lesson
- Lesson 3 Review answers

Lesson 3 REVIEW

Write your answers to these questions on a separate sheet of paper. Write complete sentences.

1. Describe the path of food in a paramecium.
2. Why is living in freshwater a problem for single-celled protists?
3. Describe what is happening in the diagram below.

4. How are eyespots useful to algae?
5. Describe one way that protists reproduce asexually.

Technology Note

Scientists use satellites to detect and study what is known as "red tide." Certain factors cause harmful, single-celled sea organisms to increase. Sometimes this algae causes the water to become a reddish-brown. This is called a red tide. Shellfish such as clams and oysters eat the algae. Toxins from the red-tide algae can then enter the food chain. There is no way to control red-tide algae.

Workbook Activity 15

Lesson 4 Fungi

Objectives

After reading this lesson, you should be able to
- identify the properties of fungi.
- describe the features of several types of fungi.
- list ways in which fungi are helpful and harmful.

Hyphae
Thin, tubelike threads produced by a fungus

Mycelium
A mass of hyphae (plural is mycelia)

More than 50,000 species belong to the kingdom Fungi. These organisms grow in the coldest and the hottest places on Earth. They inhabit water and soil. Fungi even live on and inside other organisms, including humans. Whenever you eat mushrooms or bread made with yeast, you are eating fungi.

Properties of Fungi

What features set fungi apart from organisms in other kingdoms? Unlike animals, fungi do not move from place to place. They live in one spot the way plants do. But unlike plants, fungi cannot make their own food. They have no chloroplasts as algae and plants do. Fungi reproduce by releasing spores. You will learn more about reproduction in fungi later in this chapter. Most fungi are many-celled. A few are single-celled.

Most fungi grow in moist, dark, warm places. Because they cannot make food, they must absorb it. Most are saprophytes because they decompose waste or dead matter. Fungi do not grow tall or make any woody tissue. They grow by producing fine tubelike threads called **hyphae**. These hyphae extend into the soil or the material they feed on. A mass of hyphae is called a **mycelium**. You can see a diagram of hyphae and a mycelium on page 107. Mycelia often look white and fuzzy. The hyphae in a mycelium transport nutrients through the fungus.

Bacteria, Protists, and Fungi Chapter 5 **99**

Lesson at a Glance

Chapter 5 Lesson 4

Overview In this lesson, students identify different kinds of fungi, including club fungi, rusts, smuts, molds, and yeasts. They learn how some fungi are helpful and others are harmful to humans.

Objectives
- To identify the properties of fungi
- To describe the features of several types of fungi
- To list ways in which fungi are helpful and harmful

Student Pages 99–106

Teacher's Resource Library

  Workbook Activity 16
  Alternative Activity 16
  Lab Manual 14, 15

Vocabulary

hyphae	cyclosporine
mycelium	aflatoxin

Science Background
Fungi

Fungi grow by extending their tubelike threads called hyphae into their environment. A branching chain of cells separated by walls make up most hyphae. The walls have pores that allow nutrients and other substances to pass easily from cell to cell throughout a fungus. With many thin hyphae increasing its surface area, a fungus can absorb nutrients effectively. For example, 16 cubic centimeters (1 cubic inch) of soil may contain 1.6 km (1 mile) of hyphae, presenting an enormous amount of surface area.

1 Warm-Up Activity

Ask students to identify fungi that they have seen. If necessary, remind them that they may have seen mushrooms growing on the ground, mold growing on food, or breads made of yeasts. Point out that the mushrooms, mold, and yeast are fungi.

Lab Manual 14, pages 1–2

2 Teaching the Lesson

Read the list of vocabulary words so students can hear their pronunciations. As students read the lesson, have them copy down a sentence from the lesson that uses each term.

Have students examine prepared slides of mushroom hyphae with a microscope. Ask them to draw pictures of the hyphae and to write a verbal description.

Inform students that scientists used to classify fungi as part of the plant kingdom. Ask students why they think the classification of fungi was changed. (*Organisms in the plant kingdom have chlorophyll so that they can make their own food. No fungus, however, has chlorophyll.*)

LEARNING STYLES

LEP/ESL
This lesson presents additional scientific vocabulary derived from Greek. Point out the following singular nouns and their plurals.

Singular	Plural
fungus	fungi
hypha	hyphae
mycelium	mycelia

Did You Know?

Have students read Did You Know? on page 100. Ask: What is distinctive about the honey mushroom? (*It is the largest organism in the world.*)

Did You Know?

The largest known organism in the world is a fungus called the honey mushroom, *Armillaria bulbosa*. One was discovered in Michigan that had a mycelium covering about 150,000 square meters. It has been estimated to weigh more than 90,000 kilograms.

Club Fungi

Club fungi are the most familiar fungi because they are often the easiest to see. They come in many colors. White and brown are the most common. They also can be black, purple, red, and yellow. Club fungi also are called mushrooms. However, a mushroom is just the aboveground part. Most of the fungus is a mycelium that grows underground. A single mycelium may be quite large and produce dozens of mushrooms. A mushroom is made of a stalk that supports a cap. On the underside of the cap are rows of gills. The gills are not used for breathing. Their job is to release millions of spores into the air.

Mushrooms are the aboveground part of club fungi.

Rusts and Smuts

Rusts and smuts are fungi that live as parasites on plants. They often grow on cereal plants, such as wheat, corn, barley, and rye. Although they are related to club fungi, they do not produce mushrooms. Rusts get their name because they are reddish brown, the color of rust. Smuts are noted because they stink. They change plant organs into dark masses of spores. Both rusts and smuts can kill the plants that serve as their hosts.

Corn smut, a parasitic fungus, produced the gray masses on this ear of corn.

The mold that grows on bread is a fungus.

Molds

What happens if you leave a slice of bread uncovered for several days? If the bread doesn't become too dry, it will probably develop a fuzzy coating. That coating is a fungus called bread mold. Other kinds of mold may form on old books or clothes stored in a damp basement. Molds will grow on just about any food or object made from something living.

Bread mold looks fuzzy because it produces thousands of hyphae. At the top of each hypha is a small knob that releases spores. You can see these knobs in the hyphae shown on page 107. The spores scatter in the air. They can survive freezing or drying for months. If they land on moist food, such as a slice of bread, they produce mycelia. The bread you buy in stores often contains preservatives that slow the growth of mold.

Yeasts

Yeasts live in liquid or moist environments, such as plant sap and animal tissues. Yeasts are normally single-celled. Under some conditions, however, some yeasts can develop mycelia.

Bakers have used the yeast *Saccharomyces cerevisiae* for centuries. This yeast digests carbohydrates to produce carbon dioxide and other substances. The carbon dioxide released by the yeast in bread dough makes bread rise.

3 Reinforce and Extend

LEARNING STYLES

Visual/Spatial
Ask students to make a two-column chart—one column labeled "Fungi" and one labeled "Plants." Read the following characteristics to students and ask them to place each characteristic in the appropriate column of their charts: make food, eat food, hyphae, roots, mycelium, stems, leaves, spores, chloroplasts, single-celled, many-celled. (*fungi: eat food, hyphae, mycelium, spores, single-celled, and many-celled; plants: make food, roots, stems, leaves, chloroplasts, many-celled*) Students can use their charts as a study guide.

GLOBAL CONNECTION

Have students read about how Native Americans have traditionally used club fungi known as puffballs, which produce spores inside round sacs. (*Puffballs were used as food, medicine, incense, to stop bleeding, to make certain ceremonial necklaces, and in games.*)

SCIENCE JOURNAL

Prompt students to recall an experience of finding food with mold on it. Was the mold fuzzy? Was it an unusual color? After students have stirred their recollections, have them write a journal entry describing the mold in detail.

Science Myth

Ask volunteers to read aloud the Science Myth on page 102. Emphasize the importance of discarding any food with mold growing on it. Discuss ways to protect foods from mold, such as wrapping the food in airtight containers.

CROSS-CURRICULAR CONNECTION

History
Have students read reference books about the discovery of penicillin by Alexander Fleming in 1928 and about penicillin's application as a disease-fighting drug during World War II. Ask students to find out why scientists asked people to send them their garbage. (*Mold growing on the garbage was a source of penicillin.*)

LEARNING STYLES

Auditory/Verbal
Have volunteers debate whether people should regard fungus as harmful or helpful. Tell students to find details to support their positions in the lesson text. Make an audio recording of the debate and let the participants listen to the playback. Solicit constructive criticisms about the debate performances.

Cyclosporine
A drug that is produced from mold and that helps prevent the rejection of transplanted organs

Science Myth

Moldy foods are safe to eat if the mold is scraped off.

Fact: Some molds that grow on foods produce poisonous substances. Even if the visible part of the mold is scraped off or scooped out, the mold's mycelium grows into the food and may not be visible. There could be poisons around the mycelium and throughout the food.

Helpful Fungi

Many fungi are useful to humans and other organisms. You learned that yeasts are used to make bread. Many mushrooms can be eaten and are quite tasty, although many are poisonous. Fungi known as truffles and morels are considered delicacies. Some cheeses owe their special flavors to certain species of molds.

These fungi make this dead tree's nutrients ready for use by other organisms.

Fungi also are used to make products besides food. For example, molds called *Penicillium* produce the antibiotic penicillin. Another species of mold produces **cyclosporine**. This drug helps prevent the rejection of transplanted organs. Yeasts make alcohol, which is added to gasoline to create gasohol.

Like bacteria, fungi play an important role in the recycling of nutrients. They break down dead leaves, dead animals, and other waste matter for food. As the material is broken down, some of it is returned to the soil. Plants can absorb the nutrients from the soil and use them to grow. When a plant dies, fungi begin immediately to decompose it, continuing the cycle.

Harmful Fungi

Fungi can grow on nearly anything made from something living. You already know that molds spoil bread and other foods. Rusts and smuts cause millions of dollars of damage to crop plants every year. Fungi break down the wood in buildings and the leather in boots. They can even digest the images on photographs and negatives.

Aflatoxin

A chemical that causes liver cancer and is produced by molds growing on stored crops

Some fungi cause disease in humans. The skin disease known as ringworm is caused by a fungus, not a worm. Fungi also cause the itching and burning of athlete's foot. Many people are allergic to fungal spores. Fungi can cause deadly diseases, especially in people who are not healthy.

Some fungi are extremely poisonous. Some molds produce chemicals called **aflatoxins**, which cause liver cancer. These molds can grow on poorly stored peanuts or grains, such as corn. Some mushrooms in the genus *Amanita* contain a toxin that destroys the liver. Their common names are "death cap" and "destroying angel." Eating one of these mushrooms is enough to kill you.

The mushroom *Amanita muscaria* is poisonous.

Science in Your Life

Can a fungus save your life?

You have already read that scientists can cause bacteria to make specific chemicals. Scientists have had similar successes with yeasts. Like bacteria, yeasts are single-celled and easy to grow in the laboratory. They also reproduce rapidly.

Scientists used the yeast *Saccharomyces cerevisiae* to develop a vaccine for hepatitis B. Hepatitis B is a life-threatening disease caused by a virus. The scientists inserted genetic material from the virus into the yeast. The yeast cells then produced proteins that are normally part of the virus. People who are injected with these proteins will not become sick with hepatitis B.

Lesson 4 Review Answers

1. Fungi do not move from place to place as most animals do. Fungi cannot make their own food as plants do. 2. The gills release millions of spores into the air. 3. The hyphae make bread mold look fuzzy. 4. Fungi break down waste matter and return nutrients to the soil. 5. Fungi cause ringworm and athlete's foot; they also cause allergies in some people.

Portfolio Assessment

Sample items include:
- Drawings and description of hyphae from Teaching the Lesson
- Lesson 4 Review answers

Science at Work

Have student volunteers read aloud Science at Work on page 104. Ask: What life forms does a mycologist study? (*all types of fungi*) Point out that mycology is the science of fungi, just as botany is the science of plants.

Lesson 4 REVIEW

Write your answers to these questions on a separate sheet of paper. Write complete sentences.

1. How are fungi different from animals and plants?
2. What do the gills of a mushroom do?
3. Why does bread mold look fuzzy?
4. How do fungi help to recycle nutrients?
5. What are two diseases that fungi cause?

Science at Work

Mycologist

Mycologists use microscopes and other equipment, including computers, to study fungi. They work with other scientists, experimenting, observing, and communicating. Most jobs require a four-year college degree. A master's degree or Ph.D. is needed for some jobs.

Most mycologists are biologists or microbiologists that specialize in fungi. A mycologist may study mushrooms, molds, yeasts, rusts, or smuts. Some mycologists perform experiments to discover fungi that might be harmful. Other mycologists might discover fungi that are useful as medicines or in industry. While some mycologists work in laboratories, others are out in the field collecting data. A biologist with a background in mycology might work in areas as different as agriculture, the environment, veterinary medicine, human health and disease, forestry, marine biology, or the food industry.

Workbook Activity 16

INVESTIGATION 5

Growing Bread Mold

Purpose
How do environmental conditions affect the growth of mold? In this investigation, you will study the effects of light and moisture on the growth of mold.

Procedure

1. Copy the data table below on a sheet of paper.

Storage Condition	After 1 Day	After 2 Days	After 3 Days
Dry bread in light			
Moist bread in light			
Dry bread in dark			
Moist bread in dark			

2. Use the knife to cut the slice of bread into four equal-sized pieces. Place one piece in the bottom of each petri dish.

3. Add a small amount of water to the bread in two of the dishes. Use enough water to soak the bread. Pour off any excess water.

4. Let all four dishes sit uncovered for 30 minutes. Then cover all four dishes with their lids and seal them with masking tape.

5. Place one dish containing a dry piece of bread and one dish containing moist bread in a dark place, such as a closet or cabinet.

Materials
- slice of dried white bread
- table knife
- 4 petri dishes with lids
- water
- masking tape
- stereo-microscope

Bacteria, Protists, and Fungi Chapter 5 105

SAFETY ALERT

◆ Several days before conducting the investigation, ask if any students have allergies and if they are allergic to molds.

◆ During the activity, make sure students leave the petri dishes sealed.

◆ To avoid having students cut the bread, you may wish to cut it yourself.

Lab Manual 15, pages 1–2

Investigation 5

This investigation will take approximately 45 minutes to complete.

Students will use these process and thinking skills: observing, describing, comparing, making inferences, drawing conclusions.

Preparation

- Set up the investigation on a Monday or Tuesday in a week without days off so that students can make observations for three consecutive days.
- Take several slices of white bread out of the bread wrapper and leave them out to dry for several hours ahead of time.
- Wash and sterilize the petri dishes with hot soapy water. Rinse thoroughly.
- Wash, rinse, and dry the knife thoroughly.
- Students may use Lab Manual 15 to record their data and answer the questions.

Procedure

- Have students work in pairs. One student can prepare the dry samples and the other the moist samples. Be sure students label their dishes with their names.
- Verify that students have copied the data table before the investigation begins.
- Supervise students carefully as they use the knife, or cut the bread yourself before the activity.
- Carefully time the 30 minutes during which the bread in the petri dishes is to be exposed to air.
- Guide students to choose a lighted place that is out of direct sunlight to avoid heating up the dishes.
- Arrange for students to check the petri dishes at the same time each day.
- Encourage students to make careful observations. To observe the dishes that have been positioned in a dark place, students will have to move them to a lighted area. Caution students to handle these dishes gently; make sure they return them to the same place.

Bacteria, Protists, and Fungi Chapter 5 105

Results

Mold may appear in all the dishes. The growth will most likely be heaviest on the moistened bread. It will probably be heavier in the petri dish kept in the dark than in the one kept in the light.

Questions and Conclusions Answers

1. Students' answers should accurately reflect their observations. The dish with the moist bread kept in the dark should have the most mold. The dish with moist bread in the light may also have significant mold growth.
2. Students' answers should accurately reflect their observations. Both dishes with the dry bread should have little or no mold growth.
3. The results indicate that the mold grows best in the presence of moisture and in the dark.
4. It was important to let the dishes sit uncovered for a period of time so that mold spores from the air could fall into them. The spores could then produce mycelia that grew into the bread if it was moist.
5. If the dark and light places were not at the same temperature, you could not conclude that differences in mold growth were due to differences in light or moisture. The difference in temperature could also be a factor.

Explore Further Answers

If students need help choosing another environmental condition to test, suggest any of the following:

- Temperature
- Growth medium (that is, using food other than bread)
- Variable amounts of water

Suggest that students use the Investigation text on pages 105–106 as their model.

6. Place the other two dishes in a place where they will be exposed to light during the day. The lighted place should not be warmer or colder than the dark place.
7. Over the next three days, record the appearance of each piece of bread in your data table. Safety Alert: Do not unseal the dishes. Look through the lids instead.
8. If you see signs of mold growing on the bread in any dish, examine the bread closely with a stereomicroscope. Draw what you see.

Questions and Conclusions

1. Which dish had the most mold growth after three days?
2. Which dish had the least mold growth after three days?
3. What do your results indicate about the effects of light and moisture on the growth of mold?
4. Why was it important to let the dishes sit uncovered for 30 minutes before sealing them?
5. Why was it important that the lighted place should not be warmer or colder than the dark place?

Explore Further

Plan an investigation to study the effects of another environmental condition on mold growth. Write a purpose, materials list, and procedure. Then perform your investigation.

Assessment

Check drawings to make sure students are including an appropriate level of detail. Check the data tables to ensure that students are making reasonably detailed and accurate descriptions. You might include the following items from this investigation in student portfolios:

- Drawings of magnified mold
- Data table
- Answers to Questions and Conclusions
- Written investigation from Explore Further

Lesson 5: How Fungi Survive

Objectives

After reading this lesson, you should be able to
- describe how fungi digest food.
- explain asexual and sexual reproduction in fungi.
- discuss the features of mycorrhizae and lichens.

Digestive enzyme
A chemical that helps break down food

You might think that fungi are placed in their own kingdom because they do not fit anywhere else. However, fungi have special ways of feeding, reproducing, and living with other organisms. The organisms in the fungi kingdom are alike in many ways.

Getting and Digesting Food

As you know, fungi get food by breaking down waste and dead matter. Some animals and protists also feed on dead matter. But there is a difference. Animals and protists digest their food after it enters their body. Fungi digest their food outside their body. Fungi release **digestive enzymes** onto their food. Digestive enzymes are chemicals that help break down food. The food is then small enough to be absorbed by the fungi. Fungi use the nutrients in the food for energy and growth.

Although fungi cannot hunt food, they can grow their hyphae quickly into a food source. The hyphae of a mycelium branch all through a food source, such as the orange in the diagram. The hyphae provide a large surface for absorbing nutrients.

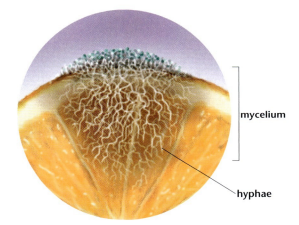

The Hyphae of a Mycelium

1 Warm-Up Activity

Tape half a sheet of black construction paper to a sheet of white paper so that the combined sheet is half black and half white. Two days before class, place the caps from several different kinds of mushrooms on the paper, gill side down, straddling the black-white line. Cover the mushrooms with a shoe box to keep them from drying out.

In class, remove the box and caps and allow students to look at the "spore prints" on the paper. Explain that the spores are reproductive structures that can produce new fungi if they land in a moist place with food. **Safety Alert:** Unless you know the source of the mushrooms and are sure they are harmless, caution students not to touch or handle them. Have students wash their hands immediately and thoroughly with soap and water if they touch the mushrooms.

2 Teaching the Lesson

Have students outline the lesson as they read. Suggest they use the text subheads as the three main outline heads.

Draw an analogy between the digestive enzymes that fungi release on their food and the meat tenderizers that some cooks sprinkle on meat before cooking. Explain that meat tenderizers contain a digestive enzyme that begins to break down some of the protein in meat.

Be sure students understand that humans and other animals complete the digestion of food within their bodies, whereas fungi digest food entirely outside their bodies.

Teacher Alert

 Make sure students understand that many-celled fungi reproduce by spores, whether sexually or asexually.

Budding
Reproduction in which part of an organism pinches off to form a new organism

Mycorrhiza
A mutualism between a fungus and the roots of a plant (plural is mycorrhizae)

Reproduction

When conditions are good for growth, fungi usually reproduce asexually. The most common way of asexual reproduction is the release of spores. Each spore usually contains a single cell. If it lands in a moist place containing food, the spore produces a mycelium. The mycelium looks the same as the fungus that released the spore.

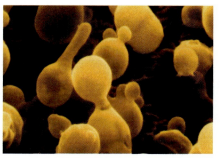

Yeasts bud to produce new cells.

Yeasts use another way of asexual reproduction, called **budding**. During budding, part of a yeast cell pinches off to make a smaller cell. The smaller cell grows rapidly to the same size as the first cell. The cells look the same.

Many fungi also reproduce sexually, especially when growing conditions are poor. Fungi do not have male and female sexes. They have opposite mating types called "plus" and "minus." When a plus hypha touches a minus hypha, the two can join. The joined hyphae then form a structure that releases spores. These spores will produce new mycelia that are different from either parent.

Living with Other Organisms

Some fungi live close to the roots of plants. The fungus absorbs minerals from the soil and shares them with the plant. In exchange, the plant provides the fungus with some of the food it makes. The plant and the fungus both benefit. This example of mutualism is called a **mycorrhiza**.

Learning Styles

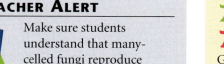

LEP/ESL
This lesson presents additional scientific vocabulary derived from Greek. Point out the Greek-derived vocabulary word *mycorrhiza* and its plural *mycorrhizae*. Help students decode the words correctly when they encounter them in the lesson text.

Cross-Curricular Connection

 Music
Do fungi respond positively to music by growing faster? Pose the question to students. Suggest that students could modify the chapter investigation to check for the effect of music on fungal growth. Challenge them to write a proposal to check for this environmental variable. Once approved by you, they can perform the investigation and report their findings to the class.

Lichen
An organism that is made up of a fungus and an alga or a bacterium

More than 90 percent of plants have mycorrhizae on their roots. The next time you see an oak or pine tree, look for mushrooms around its base. They probably sprouted from fungi that have formed mycorrhizae with the tree.

Another type of mutualism is two organisms that grow together. Algae or certain bacteria grow in a tangle among the hyphae of fungi. The two organisms together are called a **lichen**. The algae or bacteria use sunlight to make food and provide it to the fungi. In return, the fungi provide moisture and shelter for the other organism. In lichens that grow on rocks, the fungi produce acids that release minerals from the rock. Both organisms use the minerals for growth.

Lichens can withstand great extremes of temperature. They can grow where other organisms cannot survive, such as on rocks. For this reason, lichens are often the first organisms to inhabit a new area. They break down rock and begin to form soil for other organisms to use.

Lichens are sensitive to air pollution. They absorb water from moist air. They can be killed by air that contains pollutants. The presence of healthy lichens in an area usually means that the air is clean.

The colorful patches on this rock are lichens.

3 Reinforce and Extend

SCIENCE INTEGRATION

Physical Science
A lichen is used to make litmus paper. Chemists use litmus paper to test whether a substance is an acid or a base. If the substance is acid, it turns blue litmus paper red. If it is a base, it turns red litmus paper blue. A familiar example of an acid is vinegar; an example of a base is baking soda.

ONLINE CONNECTION

For a stunning visual presentation of lichens of varied form and color, visit "Lichens of North America" at www.lichen.com. The site is maintained by the authors of the book *Lichens of North America* published by Yale University Press in 2001. The authors are Irwin M. Brodo and nature photographers Sylvia and Stephen Sharnoff.

SCIENCE JOURNAL

Challenge students to write an imaginary dialogue between an alga and a fungus that decide to join up and live together as a lichen. Tell students to use the science from this lesson in their dialogue. Alternatively, artistically inclined students could make a cartoon strip to develop the same idea.

CROSS-CURRICULAR CONNECTION

Home Economics
Bread mold is extremely common. Ask students if their families take any special steps to keep bread from getting moldy. Several possibilities might include refrigerating bread during warm, humid weather or freezing part of a large loaf when the package is opened. Ask students to share what adults at home have told them about keeping other kinds of food fresh and safe.

Achievements in Science

Read aloud Achievements in Science on page 110. *Geobacter metallireducens* are rod-shaped bacteria that live far underground where there is no sunlight or oxygen. They feed on metals such as iron, manganese, uranium, and plutonium.

Lesson 5 Review Answers

1. The function of digestive enzymes is to help break down food. **2.** Part of a yeast cell pinches off to make a smaller cell, which rapidly grows to the size of the first cell. **3.** When a plus hypha touches a minus hypha, the two can join and form a structure that releases spores. The spores produce mycelia that are different. **4.** Mycorrhiza is a mutualism between a fungus and the roots of a plant. **5.** Lichens can be killed by air pollution. The presence of healthy lichens usually means that the air is clean.

Portfolio Assessment

Sample items include:
- Lesson outline from Teaching the Lesson
- Science Journal dialogue or cartoon strip
- Lesson 5 Review answers

Lesson 5 REVIEW

Write your answers to these questions on a separate sheet of paper. Write complete sentences.

1. What is the function of digestive enzymes?
2. Describe asexual reproduction in yeasts.
3. How do fungi produce offspring that are not identical?
4. What is a mycorrhiza?
5. How do lichens show air quality?

Achievements in Science

Bacteria's Energy Source Discovered

A species of bacteria called *Geobacter metallireducens*, or *Geobacter*, uses metals as its source of energy. Scientists thought that *Geobacter* could not move. They had wondered how the bacteria find metals and get to them. In April 2002, a group of scientists published the answers.

They observed *Geobacter* that were growing on iron oxide and found that they had flagella and were swimming. They also set up microscope slides that required the bacteria to move in order to get to metals. They discovered that *Geobacter* can sense the presence of metals. The *Geobacter* then grow flagella to swim to the metals they sense. They also can grow flagella to swim in search of metals. After reaching their food source, *Geobacter* grow short, hairlike structures and attach to the metal. These traits could make these bacteria useful in helping to clean up areas that are contaminated with metals.

110 Chapter 5 *Bacteria, Protists, and Fungi*

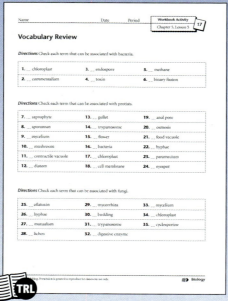

Workbook Activity 17

Chapter 5 Summary

- Bacteria are very small, single-celled organisms. They do not have organelles inside their cells.
- Bacteria reproduce by splitting into two cells that look the same.
- Some bacteria are used to produce food and different kinds of chemicals. Others spoil food and cause diseases.
- Bacteria play an important role in recycling nutrients.
- Protists include plantlike algae and animal-like protozoans.
- Algae make their own food by using the energy in sunlight. Most algae are single-celled and live in water.
- Protozoans are single-celled and must absorb their food. They move by using pseudopods, flagella, or cilia.
- Protists use contractile vacuoles to remove water that is not needed. They can sense and respond to their environment. They can reproduce asexually and sexually.
- Most fungi are many-celled. They absorb nutrients by growing hyphae into food.
- Like bacteria, some fungi spoil food and cause diseases. Others are highly poisonous. Some fungi are used to produce food and medicines.
- Fungi are important nutrient recyclers.
- Fungi reproduce by releasing spores. The spores may be produced by asexual or sexual reproduction.
- Mycorrhizae and lichens are examples of mutualism between fungi and other organisms.

Science Words

aflatoxin, 103
anal pore, 95
asexual reproduction, 97
binary fission, 87
budding, 108
commensalism, 87
contractile vacuole, 96
cyclosporine, 102
diatom, 91
digestive enzyme, 107
endospores, 88
eyespot, 97
food vacuole, 95
gullet, 95
hyphae, 99
lichen, 109
methane, 87
mutualism, 87
mycelium, 99
mycorrhiza, 108
osmosis, 96
paramecium, 92
saprophyte, 87
sporozoan, 93
sexual reproduction, 97
toxin, 88
trypanosome, 93

Chapter 5 Review

Use the Chapter Review to prepare students for tests and to reteach content from the chapter.

Chapter 5 Mastery Test

The Teacher's Resource Library includes two parallel forms of the Chapter 5 Mastery Test. The difficulty level of the two forms is equivalent. You may wish to use one form as a pretest and the other form as a posttest.

Review Answers

Vocabulary Review

1. D 2. A 3. H 4. F 5. G 6. E 7. B 8. C

Concept Review

9. A 10. D 11. B 12. B 13. C 14. A 15. D 16. C 17. A 18. B

TEACHER ALERT

In the Chapter Review, the Vocabulary Review activity includes a sample of the chapter's vocabulary terms. The activity will help determine students' understanding of key vocabulary terms and concepts presented in the chapter. Other vocabulary terms used in the chapter are listed below:

- aflatoxin
- anal pore
- asexual reproduction
- binary fission
- budding
- commensalism
- cyclosporine
- diatom
- digestive enzyme
- endospore
- eyespot
- food vacuole
- gullet
- hyphae
- mutualism
- osmosis
- paramecium
- saprophyte
- sexual reproduction
- sporozoan
- toxin
- trypanosome

Chapter 5 REVIEW

Vocabulary Review

Match each word in Column A with the correct phrase in Column B. Write the letter of the correct phrase on a sheet of paper.

Column A	Column B
___ 1. alga	A short, hairlike structures
___ 2. cilia	B mass of hyphae
___ 3. contractile vacuole	C fungus around plant roots
___ 4. flagellum	D a plantlike protist
___ 5. lichen	E a gas produced by bacteria
___ 6. methane	F long, whiplike structure
___ 7. mycelium	G fungus with alga or bacterium
___ 8. mycorrhiza	H removes water from a cell

Concept Review

Choose the answer that best completes each sentence. Write the letter of the answer on your paper.

9. Bacteria are _____.
   A single-celled
   B many-celled
   C not made of cells
   D made of one or more cells

10. Most disease-causing bacteria harm the body by producing _____.
    A antibiotics
    B nitrogen
    C carbon dioxide
    D toxins

11. Single-celled algae with hard shells that contain silica are called _____.
    A kelps
    B diatoms
    C amoebas
    D protozoans

112 Chapter 5 Bacteria, Protists, and Fungi

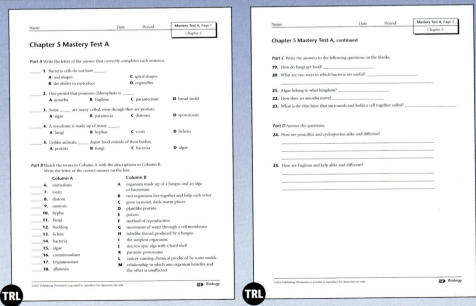

Chapter 5 Mastery Test A

12. The disease sleeping sickness is caused by _____.
    A bacteria   C fungi
    B protozoans   D amoebas

13. A paramecium moves away from an object by _____.
    A pushing out a pseudopod
    B pushing with its flagellum
    C reversing the movement of its cilia
    D using its contractile vacuole

14. The gills on the underside of a mushroom are used for _____.
    A releasing spores   C digestion
    B breathing   D making food

15. Yeasts reproduce by a process called _____.
    A binary fission   C mutualism
    B osmosis   D budding

16. The fungus in a lichen provides _____.
    A food   C moisture
    B light   D toxins

A

B

Use the photo and diagram to answer questions 17 and 18.

17. Which organism makes its own food?

18. Which organism reproduces by releasing spores?

Critical Thinking

Write the answer to each of the following questions.

19. Biologists used to classify *Euglena gracilis* as a protozoan. Explain why they may have classified it that way.

20. Why might producing an antibiotic such as penicillin be useful to a fungus?

Test-Taking Tip Don't get stuck on a hard question. Keep moving and go back to it later.

Bacteria, Protists, and Fungi   Chapter 5   113

Critical Thinking

19. Like some protozoans, *Euglena gracilis* uses a flagellum to move through the water. Also, *Euglena gracilis* can act like a protozoan and eat its food if it is kept in the dark. **20.** Fungi and bacteria grow in the same places and get food by breaking down dead organisms. They often try to get the same food. A fungus that produces an antibiotic can stop bacteria from growing nearby. The fungus then gets the food.

ALTERNATIVE ASSESSMENT

Alternative Assessment items correlate to the student Goals for Learning at the beginning of this chapter.

- Have students draw a concept web and use it to identify the characteristics of bacteria and to show ways that bacteria are helpful or harmful.

- On the board, draw a two-column chart with the column headings "Plant-like Protists" and "Animal-like Protists." Have students complete the chart by naming examples and characteristics of each type of protist.

- Have students write one or more paragraphs that describe fungi, their properties, and their life activities. Tell students to use and underline *all* of the vocabulary words from Lessons 4 and 5 in their paragraphs.

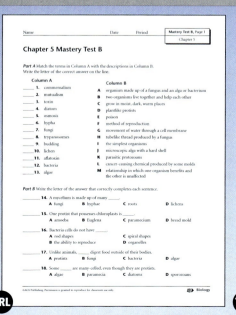

Chapter 5 Mastery Test B

Chapter 6

Planning Guide
How Animals Stay Alive

	Student Text Lesson		
	Student Pages	Vocabulary	Lesson Review
Lesson 1 How Animals Get and Digest Food	116–122	✔	✔
Lesson 2 Respiration and Circulation	123–127	✔	✔
Lesson 3 Water Balance and Wastes	128–132	✔	✔
Lesson 4 Coordinating Bodily Activities	133–138	✔	✔

Chapter Activities

Student Text
Science Center

Teacher's Resource Library
Community Connection 6:
    Vertebrate Body Systems at Work

Assessment Options

Student Text
Chapter 6 Review

Teacher's Resource Library
Chapter 6 Mastery Tests A and B

Student Text Features								Teaching Strategies						Learning Styles						Teacher's Resource Library				
Achievements in Science	Science at Work	Science in Your Life	Investigation	Science Myth	Note	Technology Note	Did You Know?	Science Integration	Science Journal	Cross-Curricular Connection	Online Connection	Teacher Alert	Applications (Home, Career, Community, Global, Environment)	Auditory/Verbal	Body/Kinesthetic	Interpersonal/Group Learning	Logical/Mathematical	Visual/Spatial	LEP/ESL	Workbook Activities	Alternative Workbook Activities	Lab Manual	Resource File	Self-Study Guide
120			121		✓		117		119	118, 119		117	118, 119	118	118		118	119		18	18	16, 17	11	✓
						✓	125	125		126	125	124				126			125	19	19	18		✓
		132	131				129	130	130		129	130				131	131			20	20			✓
	138						135	135, 137	137	136	136	134	135		135			136	137	21	21		12	✓

Pronunciation Key

a	hat	e	let	ī	ice	ô	order	ù	put
ā	age	ē	equal	o	hot	oi	oil	ü	rule
ä	far	ėr	term	ō	open	ou	out	ch	child
â	care	i	it	ȯ	saw	u	cup	ng	long

sh	she		a in about
th	thin		e in taken
ŦH	then	ə {	i in pencil
zh	measure		o in lemon
			u in circus

Alternative Workbook Activities

The Teacher's Resource Library (TRL) contains a set of lower-level worksheets called Alternative Workbook Activities. These worksheets cover the same content as the regular Workbook Activities but are written at a second-grade reading level.

Skill Track Software

Use the Skill Track Software for Biology for additional reinforcement of this chapter. The software program allows students using AGS textbooks to be assessed for mastery of each chapter and lesson of the textbook. Students access the software on an individual basis and are assessed with multiple-choice items.

Chapter at a Glance

Chapter 6:
How Animals Stay Alive
pages 114–141

Lessons

1. **How Animals Get and Digest Food** pages 116–122

   Investigation 6 pages 121–122

2. **Respiration and Circulation** pages 123–127

3. **Water Balance and Wastes** pages 128–132

4. **Coordinating Bodily Activities** pages 133–138

Chapter 6 Summary page 139

Chapter 6 Review pages 140–141

Skill Track Software for Biology

Teacher's Resource Library

   Workbook Activities 18–21

   Alternative Workbook Activities 18–21

   Lab Manual 16–18

   Community Connection 6

   Resource File 11–12

   Chapter 6 Self-Study Guide

   Chapter 6 Mastery Tests A and B

(Answer Keys for the Teacher's Resource Library begin on page 420 of the Teacher's Edition. A list of supplies required for Lab Manual Activities in this chapter begins on page 449.)

Science Center

Display large diagrams of the internal structures of a flatworm, an insect, and a fish. Provide each student a smaller copy of the diagrams. As you progress through the chapter, write the life activity or activities described in each lesson on the diagram. Have students draw lines between each life activity and the appropriate internal structures.

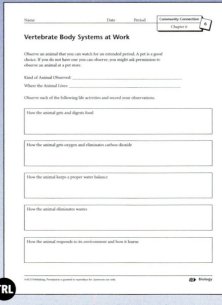

Community Connection 6

Chapter

6 How Animals Stay Alive

The bald eagle in the photo has just snatched a fish out of the water with its talons. The bird's keen eyesight and coordination made catching this meal possible. Imagine the eagle's heart pumping and its lungs taking in oxygen as it flies away with its catch. In this chapter, you will learn about the systems that help animals stay alive. You will learn about getting and digesting food. You also will learn about respiration and circulation, the nervous system, other systems, and how these systems vary in different animals.

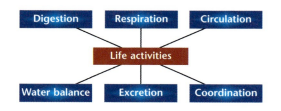

Organize Your Thoughts

Digestion · Respiration · Circulation → Life activities ← Water balance · Excretion · Coordination

Goals for Learning

- To understand how animals obtain and digest food
- To explore respiration and circulation in animals
- To explain how animals preserve water balance and excrete wastes
- To describe the functions of the endocrine system and the nervous system

Introducing the Chapter

Direct students' attention to the photo on page 114. Ask them to identify the animal and describe what it is doing. Invite volunteers to predict what is going on inside the eagle as it carries out these life activities.

Read the introductory paragraph together. Then write each life activity from the concept map on page 115 on an overhead transparency. Ask students to define or describe each term to find out what they know about the life activities of animals. List all of the body structures students think are involved in each activity. As students work through the chapter, encourage them to change or expand upon the answers and to use new vocabulary terms.

Notes and Technology Notes

Ask volunteers to read the notes that appear in the margins throughout the chapter. Then discuss them with the class.

TEACHER'S RESOURCE

The AGS Teaching Strategies in Science Transparencies may be used with this chapter. The transparencies add an interactive dimension to expand and enhance the *Biology* program content.

CAREER INTEREST INVENTORY

The AGS Harrington-O'Shea Career Decision-Making System-Revised (CDM) may be used with this chapter. Students can use the CDM to explore their interests and identify careers. The CDM defines career areas that are indicated by students' responses on the inventory.

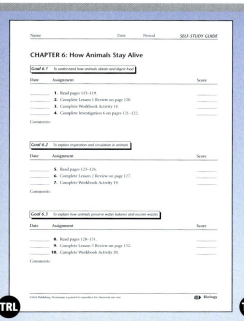

Chapter 6 Self-Study Guide

Lesson at a Glance

Chapter 6 Lesson 1

Overview In this lesson, students explore the ways that different animals ingest food. They learn how digestion differs in a gastrovascular cavity and a digestive tract.

Objectives

- To describe three main ways animals get food
- To explain the importance of digestion
- To tell the difference between a gastrovascular cavity and a digestive tract

Student Pages 116–122

Teacher's Resource Library (TRL)

- Workbook Activity 18
- Alternative Workbook Activity 18
- Lab Manual 16, 17
- Resource File 11

Vocabulary

filter feeding	digestive tract
herbivore	crop
carnivore	gizzard
secrete	anus
enzyme	
gastrovascular cavity	

Science Background
Digestion

A gastrovascular cavity digests and distributes nutrients throughout the body. An animal with such a system has no circulatory system. (It is not needed to move nutrients to the cells.) A digestive tract with two openings is more efficient at digestion than a gastrovascular cavity is. Because food moves through a digestive tract in one direction, storage, digestion, and absorption can occur in different places at the same time. A digestive tract has separate regions where different digestive enzymes break down food. In vertebrates, the stomach provides an acidic environment in which the enzyme pepsin works best.

116 *Chapter 6 How Animals Stay Alive*

Lesson 1 How Animals Get and Digest Food

Objectives

After reading this lesson, you should be able to

- describe three main ways animals get food.
- explain the importance of digestion.
- tell the difference between a gastrovascular cavity and a digestive tract.

Unlike plants, algae, and some bacteria, animals cannot make their own food. Animals must get food from other organisms. Different animals have different ways of getting food.

Filter Feeding

Many animals that live in water get food by filtering, or straining, it. This way of getting food is called **filter feeding**. Sponges strain bacteria and protists from the water that passes through their body. Sponges cannot move around as adults, and filter feeding allows them to gather food without chasing it. Barnacles also remain in place. They collect food particles with their legs. The legs act as screens. Mollusks, such as clams and oysters, tend to remain in one spot. They use their gills to strain food out of the water. Some filter-feeding animals do move. Many whales harvest millions of tiny animals by swimming with their mouths open.

Filter feeding
A way of getting food by straining it out of the water

Feeding on Fluids

Some animals get food from the fluids of plants or other animals. The fluids are rich in nutrients. Aphids and cicadas are insects that have piercing mouthparts. They draw sap from roots, leaves, and stems. Bees, butterflies, and hummingbirds draw nectar from flowers. Spiders and assassin bugs capture insects and suck the fluid from their bodies. Leeches, mosquitoes, and horseflies feed on the blood of vertebrates, including humans.

116 *Chapter 6 How Animals Stay Alive*

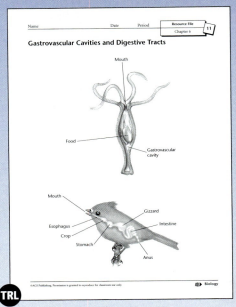

Resource File 11

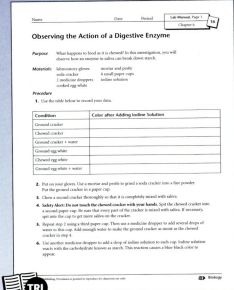

Lab Manual 16, pages 1–2

Herbivore
An animal that eats plants

Carnivore
An animal that eats other animals

  Did You Know?

The Komodo monitor lizard can weigh 75 kilograms and can hunt deer and water buffalo.

Consuming Large Pieces of Food

Most animals consume, or eat, large pieces of solid food. Sometimes they eat entire organisms. Such animals use different kinds of body structures to capture and consume their food. For example, hydras, jellyfish, and other cnidarians have tentacles armed with stinging cells. The tentacles catch small animals in the water and bring them to the mouth. Cnidarians consume their food whole.

Many insects have mouthparts that are suited for cutting and chewing. The mouthparts turn the food into pieces that are small enough to swallow. Grasshoppers, termites, and beetles use their chewing mouthparts to feed on plants. Animals that eat plants are known as **herbivores**. Dragonflies and praying mantises also have chewing mouthparts but eat other insects. Animals that eat other animals are called **carnivores**.

Vertebrates are the only animals that have teeth. Mammals have teeth of different shapes and sizes. Each kind of tooth does a certain job. Chisel-like teeth at the front of the mouth cut food into pieces. Long, pointed teeth grip and pierce food. Teeth that have a flat surface grind and crush food. A mammal's teeth indicate what kind of food it eats. Carnivores have sharp, pointed teeth that tear flesh. Herbivores have large teeth that have a flat surface. These teeth are suited for grinding plants. Look at the teeth of these skulls. Which is the skull of a carnivore? of an herbivore?

 Warm-Up Activity

Set up a small aquarium with a pump and a clean filter. Fill the aquarium with water and thoroughly stir in enough yeast to make the water cloudy. Let students see how the pump and filter gradually clear the water. Turn off the pump, remove the filter, and observe the yeast accumulated on the filter. Tell students that some animals that live in water use filter-like structures to collect food from the water.

2 **Teaching the Lesson**

As they read the lesson, have students write each vocabulary term and its definition.

Ask students to draw an example of a gastrovascular cavity and a digestive tract and label their parts. Have students write a paragraph below the illustrations telling how these systems are different and alike.

Let students infer what animal skulls are shown on page 117 (*left: lion; right: antelope*).

Set up two clear glasses of water. Into one, drop a piece of hard candy. Into the second, place smashed bits of the same candy. Swirl the water in both glasses. After several minutes, have students examine the glasses and describe what they see. (*The whole piece of candy will have dissolved little; the bits should be mostly dissolved.*) Explain that the smaller pieces represent food broken into small pieces with mouthparts, teeth, stomach, or gizzard. This action increases the surface area of food exposed to digestive enzymes. The smaller pieces are digested more quickly.

Did You Know?

Ask students to describe a lizard. Then display a picture of the Komodo "dragon" or monitor lizard. Read aloud Did You Know? on page 117. Ask students to point out characteristics of the lizard that suggest it is a carnivore.

TEACHER ALERT

 Digestion can be further broken down into mechanical and chemical digestion. Mechanical digestion occurs as teeth cut and grind food into smaller pieces. Chemical digestion, which requires the help of enzymes, is responsible for breaking down nutrients in food into simpler molecules that can dissolve in water. Both kinds of digestion begin in the mouth, where an enzyme changes starch into sugar.

3 Reinforce and Extend

LEARNING STYLES

Body/Kinesthetic
Provide assorted materials such as pipe cleaners, foam shapes, aluminum foil, buttons, cardboard, cotton swabs, paper clips, tape, and glue. Have students work in pairs to make a model of the head of an imaginary animal, including mouthparts. They can then use the animal like a puppet to demonstrate how the mouthparts enable it to ingest food.

IN THE COMMUNITY

Have students observe a variety of insects such as bees, caterpillars, flies, mosquitoes, ants, and beetles. They can draw mouthparts and classify each insect into a group: Feeds on Fluids, Feeds on Plant Parts, or Feeds on Other Animals.

LEARNING STYLES

Logical/Mathematical
Have students locate and bring in pictures of various animals. Ask them to work in groups to analyze each animal and decide whether it is a herbivore or a carnivore. Discuss how students reached these conclusions.

Secrete
Form and release, or give off

Enzyme
A substance that speeds up chemical changes

Gastrovascular cavity
A digestive space with a single opening

Digesting Food

Foods usually contain fats, proteins, and carbohydrates. These chemicals provide the energy an animal needs. However, they are too large for most animal cells to absorb. These large chemicals must be broken down into smaller chemicals before cells can absorb them. The process of breaking down food into small chemicals is digestion. Animals digest food by **secreting** digestive **enzymes**. An enzyme is a substance that speeds up chemical changes. *Secrete* means to "form and release."

In sponges, the digestive enzymes work inside cells. These cells line the inside of the sponge. The cells trap food that enters the sponge. They package the food in food vacuoles like those in paramecia. Digestive enzymes break down the food into small chemicals. The cells then absorb the chemicals.

Digesting food inside cells has one drawback. The food must be small enough to fit inside food vacuoles. This means that sponges can eat only tiny food particles. Most other animals digest their food outside of cells. They have a space where digestion begins. These animals can eat much larger foods.

Gastrovascular Cavities

Cnidarians, such as the hydra in the diagram, and flatworms digest food in a hollow space called a **gastrovascular cavity**. This space has only one opening, the mouth. Food enters through the mouth. Special cells line the gastrovascular cavity. These cells secrete digestive enzymes. The enzymes break down the food into small particles. The cells can then absorb the particles. Material that is not digested leaves through the mouth.

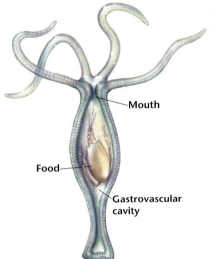

CROSS-CURRICULAR CONNECTION

Drama
Ask interested students to write a skit about digestion. The "characters" can be enzymes and molecules of fat, protein, and carbohydrate. Assign a group of students to act out the roles to show how food is broken down in the digestive tract. For example, several students might join hands and play the role of a protein molecule as it is broken into smaller molecules.

LEARNING STYLES

Auditory/Verbal
Ask students who learn best by speaking and listening to make an audiotape explaining three ways animals get food and how they digest it. Suggest that recorders insert comprehension questions into the tape for listeners to answer. Make the tape available as students review the chapter.

Digestive tract
A tubelike digestive space with an opening at each end

Crop
The part of the digestive tract of some animals where food is stored

Gizzard
The part of the digestive tract of some animals that grinds food

Anus
The opening through which material that is not digested leaves the digestive tract

Digestive Tracts

Animals that are more developed have a **digestive tract**. This is a tubelike digestive space with an opening at each end. Food moves through a digestive tract in one direction. Different parts of the tract carry out different functions. The main functions of digestive tracts are storing food, digesting food, and absorbing nutrients.

Most digestive tracts are organized the same way. The digestive tract of a bird provides a good example. Food enters the digestive tract through the mouth. It passes down the esophagus to the **crop**, where it is stored. In the stomach, the food mixes with acid and digestive enzymes. The mixture moves to the **gizzard**. The gizzard grinds it into a watery paste. More digestive enzymes are added in the intestine. This is where digestion is completed. The walls of the intestine absorb the small chemicals. Material that is not digested leaves the digestive tract through an opening called the **anus**.

The digestive tracts of animals have some differences. For example, humans do not have a crop or a gizzard. The human stomach carries out the functions of those organs. You will learn more about the human digestive tract in Chapter 8.

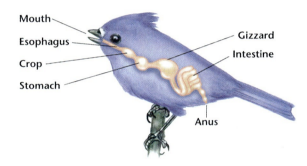

How Animals Stay Alive Chapter 6 **119**

SCIENCE JOURNAL
Ask students to imagine what it would be like to be a filter feeder such as a clam or sponge. Have them write a journal entry describing how the experience of eating would be different.

LEARNING STYLES

Visual/Spatial
Have students copy the bird digestive system on page 119 onto a sheet of paper; then find a picture of and draw a human digestive system. Ask students to label each part of each system and highlight the parts that are similar. Invite volunteers to describe differences in the two systems.

CROSS-CURRICULAR CONNECTION

Language Arts
Write the words *filter*, *cavity*, and *tract* on the board and ask volunteers to tell what meanings each word can have. Add *feeding* after *filter* and have students explain the meaning of the compound word. Discuss how the meaning of the word *filter* fits into the compound's meaning. Repeat the process for *gastrovascular cavity* and *digestive tract*.

IN THE ENVIRONMENT

Tell students that people in developing countries eat far less meat than people in developed countries, such as the United States and Canada. Ask students to investigate the potential impact on world food supplies if more of the world's human population were to switch to a diet high in meat. Students should discover that grain supplies go much farther when eaten by people than when fed to livestock that people eat.

How Animals Stay Alive Chapter 6 **119**

Lesson 1 Review Answers

1. Students' answers may include, but are not limited to, sponges, barnacles, mollusks such as clams and oysters, and whales. 2. Cicadas have mouthparts that pierce and draw up liquid. Dragonflies have mouthparts that cut and chew. 3. Molecules of fat, protein, and carbohydrate must be broken down into smaller chemicals that cells can use for energy. 4. Cnidarians have a gastrovascular cavity, which is a hollow space with one opening, the mouth. 5. stores ingested food, digests food, and absorbs nutrients

Portfolio Assessment

Sample items include:
- Definitions and illustrations with paragraph from Teaching the Lesson
- Lesson 1 Review answers

Achievements in Science

Ask students whether they have a fish tank at home. Encourage those who do to talk about the responsibilities of caring for the fish and the enjoyment of watching them. Then have students read Achievements in Science on page 120. Explain that Villepreux-Power was a self-taught naturalist. Ask how her achievement helped advance scientific understanding of sea animals.

Lesson 1 REVIEW

Write your answers to these questions on a separate sheet of paper. Write complete sentences.

1. Name three kinds of animals that use filter feeding to get food.
2. How are cicadas different from dragonflies in the way they feed?
3. Why must animals digest their food?
4. Name and describe the type of digestive space found in a cnidarian.
5. What functions does the digestive tract perform?

Achievements in Science

Aquariums Invented

Before the 1800s, scientists could study animals such as fish only in their natural habitat. Studying animals underwater was difficult. The scientists could not always gather complete information over a long period of time. The animals could not survive outside their aquatic environments, so it didn't help to take them out of the water. To study underwater animals well, the scientists would need to confine them in some way.

Jeanne Villepreux-Power was a self-taught naturalist and marine biologist. She became interested in the living organisms and the environment around her when she lived on an island in the Mediterranean Sea. In 1832, she invented aquariums. She was the first scientist to use aquariums to do experiments in an aquatic environment. Aquariums allowed scientists to confine aquatic organisms while they studied them.

Workbook Activity 18

INVESTIGATION 6

Studying Feeding in Hydras

Purpose
What stimuli trigger feeding responses in a simple animal? In this investigation, you will observe feeding behavior in a hydra and record its responses to different stimuli.

Materials
- microscope well slide
- hydra culture
- 2 medicine droppers
- stereo-microscope
- water flea culture
- filter paper
- scissors
- forceps
- beef broth
- dish labeled "used hydras"

Procedure

1. Copy the data table below on a sheet of paper.

Stimulus	What Happened?
Water fleas	
Moving filter paper	
Touched by filter paper	
Beef broth on filter paper	

2. Use a medicine dropper to place a hydra on a well slide. Make sure the hydra is in water.

3. Use another medicine dropper to add some water fleas to the slide. Water fleas are small crustaceans on which hydras often feed.

4. Place the slide on the microscope stage. Using high power, observe the feeding responses of the hydra. Record your observations. If the hydra does not respond after several minutes, get another hydra and try again.

5. Place any hydras you have used in the dish labeled "used hydras." Rinse the slide with water.

6. Place a new hydra on your slide. Move the slide to the microscope stage.

How Animals Stay Alive   Chapter 6   **121**

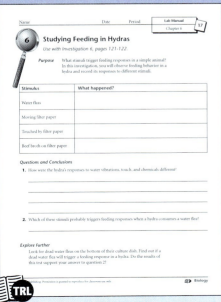

Lab Manual 17

Investigation 6

This investigation will take approximately 45 minutes to complete.

Students will use these process and thinking skills: observing, describing, comparing, making inferences, drawing conclusions.

Preparation

- Obtain live hydras and water fleas from a pet store or a biological supply company. Feed some of the water fleas to the hydras daily or every other day. Stop feeding the hydras one or two days before the lab.
- Students may use Lab Manual 17 to record their data and answer the questions.

Procedure

- This investigation is best done with students working in groups of three. They should work cooperatively to set up the investigation, clean up, and discuss and record their observations. Assign one student to test for responses to vibrations from moving filter paper, another to test for responses to being touched by filter paper, and the third to test for responses to beef broth.
- Instruct students to handle hydras gently and follow directions carefully. They will need to transfer hydras quickly from the culture to the slide to prevent them from attaching to the medicine dropper
- Caution students not to let their slide dry out while it is on the microscope. If the slide is drying, have them add a drop of water to it.
- After students have observed the hydras feed on water fleas, tell them they will find out which stimuli trigger the hydra's feeding response. They will test the hydra's response to water vibrations, touch, and chemicals released by prey.
- Encourage students to be patient, as the hydras may not respond immediately to certain stimuli.

SAFETY ALERT

- Remind students to be careful when handling microscope slides and to dispose of broken slides in a container marked for glass only.
- Suggest that students wear protective gloves during the investigation.
- Have students wash their hands thoroughly after the investigation.

Results

When a hydra is exposed to water fleas, students should see the hydra wrap its tentacles around a water flea and bring the water flea toward its mouth. It may take 30 minutes or more for a hydra to completely ingest a water flea. Expect variable results for the other three stimuli. The filter paper soaked in beef broth should come closest to mimicking a natural prey stimulus and trigger the initial steps in feeding behavior.

Questions and Conclusions Answers

1. Answers may vary. Typically, hydras do not respond to water vibrations; shorten their tentacles and body when touched; and bend toward the filter paper soaked in beef broth.
2. Chemical stimuli probably trigger feeding responses.

Explore Further Answers

If a hydra shows a feeding response when a dead, motionless water flea is nearby but not touching the hydra, then the answer to question 2 would be supported.

Assessment

Check that students have correctly recorded responses in their data table. Include the following items from this investigation in student portfolios:
- Data table
- Answers to Questions and Conclusions and Explore Further

7. Use scissors to cut a small piece of filter paper. Using forceps, move the filter paper in the water near the hydra. Be careful not to touch the hydra. You are testing the hydra's response to water vibrations. Record your observations.
8. Gently touch different parts of the hydra with the filter paper. You are testing the hydra's response to touch. Record your observations.
9. Dip the filter paper in beef broth. Hold the filter paper in the water near the hydra. Again, do not touch the hydra. You are testing the hydra's response to chemicals. Record your observations.
10. Place the hydra in the dish labeled "used hydras." Clean up your work space.

Questions and Conclusions

1. How were the hydra's responses to water vibrations, touch, and chemicals different?
2. Which of these stimuli probably triggers feeding responses when a hydra consumes a water flea?

Explore Further

Look for dead water fleas on the bottom of their culture dish. Find out if a dead water flea will trigger a feeding response in a hydra. Do the results of this test support your answer to question 2?

Lesson 2: Respiration and Circulation

Objectives

After reading this lesson, you should be able to
- explain gas exchange in simple and complex animals.
- tell the difference between open and closed circulatory systems.
- trace the flow of blood through a bird or mammal.

Respire
Take in oxygen and give off carbon dioxide

Diffusion
The movement of materials from an area of high concentration to an area of low concentration

To obtain the energy in food, all animals must carry out chemical reactions. In these reactions, food molecules join with oxygen. Energy is released. Carbon dioxide forms as a waste product. Thus, animals must bring oxygen into their body. They must eliminate carbon dioxide. This process of gas exchange is called **respiration**. Animals respire, or take in oxygen and give off carbon dioxide, in different ways.

Gas Exchange in Simple Animals

The body wall of sponges and cnidarians is made of just two cell layers. Water outside the animal touches the cells in one layer. Water inside the animal touches cells in the other layer. Both layers of cells get oxygen and get rid of carbon dioxide by **diffusion**. Diffusion is movement from an area of high concentration to an area of low concentration. The concentration of oxygen is higher in the water than in the cells. Therefore, oxygen diffuses from the water into the cells. The concentration of carbon dioxide is higher in the cells than in the water. Therefore, carbon dioxide diffuses from the cells into the water.

Flatworms have very thin body walls too. Most cells touch water, either outside or in the gastrovascular cavity. Gas exchange in flatworms also happens by diffusion. All cells in a hydra can exchange oxygen and carbon dioxide with the surrounding water.

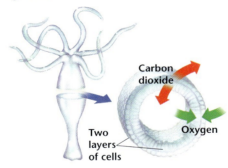

How Animals Stay Alive Chapter 6

Lesson at a Glance

Chapter 6 Lesson 2

Overview In this lesson, students learn about the functioning of different animal respiratory and circulatory systems.

Objectives
- To explain gas exchange in simple and complex animals
- To tell the difference between open and closed circulatory systems
- To trace the flow of blood through a bird or mammal

Student Pages 123–127

Teacher's Resource Library
Workbook Activity 19
Alternative Activity 19
Lab Manual 18

Vocabulary
respire   atrium
diffusion   ventricle
circulatory
open circulatory system
annelid
closed circulatory system

Science Background
Respiration and Circulation

Animals must constantly respire to support the energy-releasing reactions that occur in their cells. For very small animals, oxygen and carbon dioxide diffuse between cells and the environment. However, diffusion is far too slow for most animals. Insects and some spiders overcome this problem by pumping outside air to cells through a system of fine tubes. Most other animals use a circulatory system to carry gases in blood between a respiratory surface (gills, lungs, or skin) and the rest of the body. In addition to transporting gases, blood carries nutrients, wastes, and hormones.

Lab Manual 18

1 Warm-Up Activity

Have students observe the respiration of a live animal with gills. If possible, collect larval salamanders or aquatic insects and have students examine them with a hand lens. (Gills are located on each side of a salamander's head or in various locations behind an insect's head.) Ask students to describe the gills and the movements by which the animal makes water pass over them. Have volunteers predict how the structures are like lungs.

2 Teaching the Lesson

Have students write three sentences beginning "I want to remember . . ." Then ask them to use the lesson objectives to finish each sentence.

Give each student a drinking straw and a test tube half full of limewater. Note the appearance of the limewater. (clear) Tell students that limewater turns cloudy when it reacts with carbon dioxide. Have students exhale gently through the straw into the limewater. Caution them not to inhale through the straw. Note the appearance of the limewater. (slightly cloudy) Have students inhale, hold their breath for ten seconds, and exhale into the limewater. Ask them how the limewater has changed and why. (*It should be much cloudier. Holding the breath causes carbon dioxide to build up in the lungs. The exhaled air contains more carbon dioxide than normal.*)

As they read about vertebrate circulation patterns, ask students to use the diagram on page 126 to trace the movement of blood to and from the heart.

TEACHER ALERT

Oceans contain nearly 3.5 percent salt. Freshwater contains less than 0.006 percent salt. Salinity is not the only factor in determining which animals may live in the sea. Water pressure, which increases greatly with depth, is another factor. In addition, light and temperature limit marine life. Only organisms adapted to extreme cold and darkness may live in the deepest reaches of the oceans.

Amphibians, such as frogs, have lungs for gas exchange. They also rely on gas exchange through their skin. Oxygen from the air diffuses through the moist skin that covers their bodies.

Gas Exchange in Other Animals

Most animals are not just two cell layers thick. They contain many cells deep inside the body. These cells cannot exchange gases directly with the outside environment. Animals like these must have a special organ for gas exchange. Such organs come in many different forms.

Animals that live in water usually have gills. Fish have one type of gills. Tadpoles, lobsters, and clams also have gills. Gills often have a feathery structure. This structure provides a large surface area. This allows diffusion to happen quickly. Oxygen diffuses from the water into the gills. Carbon dioxide diffuses in the opposite direction.

The gills on this brown trout provide a large surface area for exchanging gases quickly.

Land animals exchange oxygen and carbon dioxide with the air. Insects use a system of tubes to carry air into the body. The tubes have very fine branches that reach almost all of the animal's cells. The entrances to the tubes are scattered over the insect's body. Watch a bee that has landed on a flower. You will see its abdomen move in and out. It is pumping air into and out of the tubes.

Most other land animals use lungs for gas exchange. Lungs are like balloons inside the body. When you inhale, or breathe in, you draw air into your lungs. Exhaling, or breathing out, forces the air back out. Like gills, lungs provide a large surface area for gas exchange.

Circulatory
Flowing in a circle

Open circulatory system
A system in which blood makes direct contact with cells

Annelid
A segmented worm or a worm whose body is divided into sections

Closed circulatory system
A system in which blood stays inside vessels at all times

Circulatory Systems

Animals must transport oxygen from their gills or lungs to the rest of their body. They must transport carbon dioxide from the rest of their body to their gills or lungs. A **circulatory** system performs these jobs. *Circulatory* means "flowing in a circle." This system moves blood through the body. In the gills or lungs, oxygen enters the blood. Carbon dioxide leaves. As the blood circulates, it delivers oxygen and picks up carbon dioxide. Blood also carries nutrients from the digestive tract to cells.

All circulatory systems have a set of tubes and one or more pumps. The tubes are called blood vessels. The pumps are called hearts. When a heart contracts, or pulls together, it squeezes blood through the blood vessels.

Arthropods and most mollusks have an **open circulatory system**. The grasshopper in the diagram below has an open circulatory system. In this system, blood leaves the vessels and enters spaces around the organs. The blood flows slowly through the spaces and makes direct contact with cells.

Annelids are segmented worms or worms whose body is divided into sections. Annelids, such as earthworms, vertebrates, and some mollusks, have a **closed circulatory system**. The blood stays inside vessels at all times. The smallest vessels have very thin walls. Oxygen and carbon dioxide diffuse into or out of the blood across these walls.

Did You Know?
A shrew's heart can beat as fast as 1,300 times per minute. An elephant's heart beats only about 25 times per minute.

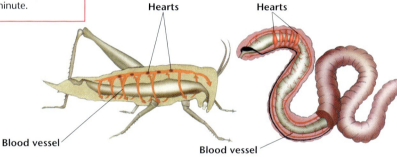

Open Circulatory System   Closed Circulatory System

Online Connection
Students can explore the human circulation system using Pathfinders for Kids: The Circulatory System—The Life Pump at infozone.imcpl.org/kids_circ.htm. It provides basic facts and lists books and Web sites where students can find specific information.

3 Reinforce and Extend.

SCIENCE INTEGRATION

Physical Science
Provide students with two sheets of tracing paper or tissue the same size. Tell them they are to use one sheet to make a body covering for an imaginary animal. The other paper they are to fold to create "gills" for a second imaginary animal. Have students explain how one animal could use its skin and the other its gills to exchange gases with the environment. Discuss their conclusions about how gills increase an animal's ability to obtain oxygen and get rid of carbon dioxide.

Did You Know?

Show students pressure points in the wrist and neck where they can feel their pulse. Have them count how many beats they can hear in 15 seconds and calculate their heart rate per minute. Read aloud Did You Know? on page 125. Ask students to compare the size of the shrew, elephant, and a person. Then compare their heart rates. What conclusion can they draw? (*The smaller the animal, the faster its heart beats to circulate blood; the larger the animal, the slower its heart beats.*)

LEARNING STYLES

LEP/ESL
To help students who are learning English build vocabulary, write the words *respire, respiration, respiratory system; circulate, circulation,* and *circulatory system* on the board in two columns. Have students underline the root of each word, then look up its meaning. Point out the endings *-tion* and *-atory*. Explain that they are added to some words to change the part of speech from verb to noun or adjective. Ask students to use each term in sentence to show its part of speech and meaning.

CROSS-CURRICULAR CONNECTION

Music
Ask students to make a list of all the songs they can find with the word *heart* in the title or lyrics. Then have them indicate whether each song refers to the heart as an organ that pumps blood or as a center of emotion. Ask them to research why people use the heart as a symbol of emotion instead of other organs, such as the liver or kidneys.

LEARNING STYLES

Interpersonal/Group Learning
Have a group of students make a chalk or tape diagram of a vertebrate circulatory system on the floor. They should include the four-chamber heart. Then ask group members to role play. One can act as a red blood cell that circulates. One person can stand in each chamber of the heart and explain its role in circulation and then send the red blood cell on its way.

Atrium
A heart chamber that receives blood returning to the heart (plural is atria)

Ventricle
A heart chamber that pumps blood out of the heart

Vertebrate Circulatory Systems

The circulatory system of a vertebrate includes a single heart. The heart is divided into enclosed spaces called chambers. The **atria** are chambers that receive blood that returns to the heart. The **ventricles** are chambers that pump blood out of the heart. Fish have one atrium and one ventricle. Amphibians and most reptiles have two atria and one ventricle. Birds, mammals, and some reptiles have two atria and two ventricles.

The diagram below shows how blood circulates through the body of a mammal or bird. The left atrium receives blood from the lungs. This blood has a lot of oxygen that was picked up in the lungs. The blood has little carbon dioxide. The left atrium sends the blood to the left ventricle. The left ventricle pumps it to the rest of the body. The blood delivers oxygen to body tissues. It picks up carbon dioxide that has formed as waste. The blood returns to the right atrium. The blood has little oxygen and a lot of carbon dioxide. The blood moves from the right atrium to the right ventricle. The right ventricle pumps the blood to the lungs. In the lungs, oxygen enters the blood. Carbon dioxide leaves the blood. The carbon dioxide is exhaled as waste. The blood returns to the left atrium, completing the cycle.

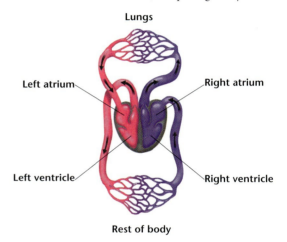

Lesson 2 REVIEW

Write your answers to these questions on a separate sheet of paper. Write complete sentences.

1. How are gases exchanged in a sponge?
2. Describe the system that insects use to respire.
3. What is the difference between an open and a closed circulatory system?
4. What is the function of the atria in a vertebrate heart?
5. In a bird's heart, where does blood go from the right ventricle?

Technology Note

Studying animals that live in water became easier after the invention of the Aqua-Lung. This equipment allows people to breathe underwater. Jacques Cousteau and Emile Gagnan invented the Aqua-Lung in 1943. Today, it is more commonly known as scuba (**s**elf-**c**ontained **u**nderwater **b**reathing **a**pparatus). The equipment consists of a tank of compressed air, regulators, hoses, pressure chambers, valves, and a mouthpiece. All these things make exploring animals underwater possible and safer.

Lesson 2 Review Answers

1. Gases are exchanged through the body wall of a sponge, which is only two cell layers thick. **2.** A system of tubes carries air into the insect's body; it branches very fine, carrying gases into every part. **3.** In an open circulatory system, blood leaves the vessels and makes direct contact with cells. In a closed circulatory system, blood stays inside vessels at all times. **4.** Atria receive blood that is returning to the heart. **5.** to the lungs

Portfolio Assessment

Sample items include:
- Sentences and circulatory diagrams from Teaching the Lesson
- Lesson 2 Review answers

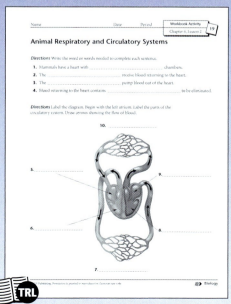

Workbook Activity 19

Lesson at a Glance

Chapter 6 Lesson 3

Overview In this lesson, students learn how animals regulate movement of water into and out of their bodies and how they get rid of wastes.

Objectives
- To explain how animals maintain water balance in the sea, in freshwater, and on land
- To explain how animals excrete wastes

Student Pages 128–132

Teacher's Resource Library

Workbook Activity 20
Alternative Activity 20

Vocabulary
excrete
flame cell
kidney

Science Background
Water Balance

All animals must maintain the proper balance of water between their cells and the environment. Most animals can limit water exchange with the environment because they have a body covering that is virtually watertight. However, an animal's respiratory surface must exchange oxygen and carbon dioxide, so it cannot be watertight. This is why much of the water that an animal gains or loses moves across its gills, lungs, or skin.

Excretion

Ammonia, a highly toxic substance, forms in the body as a result of the breakdown of proteins. In aquatic animals, ammonia diffuses across the gills or entire body surface into the surrounding water. Land animals convert ammonia into urea or uric acid, chemicals that are much less toxic than ammonia. (If they had to excrete ammonia in their urine, they would have to urinate frequently to eliminate the poison. This would make water balance difficult to maintain.)

Lesson 3: Water Balance and Wastes

Objectives

After reading this lesson, you should be able to
- explain how animals maintain water balance in the sea, in freshwater, and on land.
- explain how animals excrete wastes.

What happens if you exercise hard on a hot day? You sweat a lot. You drink fluids to replace the water you lost. What happens if you drink too much water? You get rid of it by producing more urine. Your body works to keep a normal balance of water. Other animals also have ways to maintain, or keep, water balance.

Water Balance in the Sea

Seawater is water and salt. The fluids of animals also contain water and salt. The more salt a fluid has, the lower its water concentration is. Recall from Chapter 5 that water will move into an area where its concentration is lower. This is called osmosis. This movement can cause problems if too much water gets into an animal.

Most sea invertebrates avoid getting too much water in their bodies. The water concentration in their fluids equals the water concentration in seawater. Therefore, water does not move between their bodies and the surrounding seawater. The same is true for sharks, rays, and skates.

In bony fish, their body fluids have a higher water concentration than seawater. As a result, water moves from their bodies into the sea. Like all animals, bony fish need some water. If all the water left their bodies, they would shrink and die. These fish

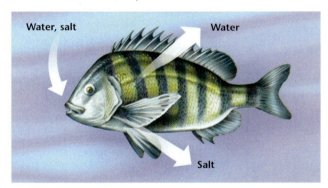

Bony Fish in the Sea

128 Chapter 6 How Animals Stay Alive

Excrete
Get rid of wastes or substances that are not needed

Flame cell
A cell that collects excess water in a flatworm

Did You Know?
The sea turtle has organs of excretion near the corners of its eyes. The sea turtle sheds salty "tears" to get rid of excess salt.

drink seawater to replace the water they lose by osmosis. Drinking seawater brings a lot of salt into their bodies. They **excrete** the extra salt through their gills. To excrete means to get rid of wastes or substances that are not needed.

Water Balance in Freshwater

Animals that live in freshwater have too much water coming into their bodies. You may recall that protists have this same problem. The concentration of water outside their bodies is higher than the concentration inside. Therefore, water constantly moves into them.

Freshwater animals use special organs to remove excess, or too much, water. Flatworms have a system of tiny tubes all through their bodies. The tubes are connected to cells called **flame cells**. These cells got their name because they have a tuft of cilia. When the cilia move, they look like a flickering flame. Flame cells collect excess water inside the flatworm. Their cilia push the water along the tubes. Water leaves the tubes through pores, or openings, in the body wall.

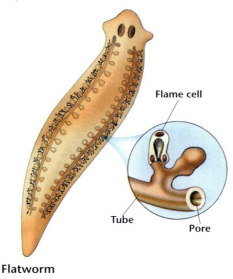

Flatworm

1 Warm-Up Activity

Have students imagine people adrift on the ocean in a raft with no fresh water. Explain that seawater has more chemicals (or a lower concentration of water) than blood has. Ask students to explain why the people cannot drink seawater. (*It would add more chemicals to blood, lowering the concentration of water. The people would be dehydrated by the seawater and could die.*) Have students predict why animals that live in the sea do not have this problem. (*They have ways to get rid of the extra salt while retaining the water.*)

2 Teaching the Lesson

After they read each section, have students summarize its main ideas in a sentence or two.

Review the process of osmosis with students. Explain that an environment with a high water concentration is like water in a high place. It will flow into lower places (environments with a low water concentration).

Draw a large rectangle on the board and divide it in half with a dashed line. Draw 20 circles on each side of the line. Fill in 3 circles on one side and 12 circles on the other side. Explain that the filled circles represent chemicals such as salt, and the open circles represent water molecules. The dashed line represents a cell membrane. Have students tell which side has a lower concentration of water. (*the one with 12 filled circles*) Ask them which direction water would move by osmosis. (*toward the side with 12 filled circles— the "salty" side—to balance water concentration on both sides of the membrane*)

Compare the kidney's removal of wastes to the clearing of a table at a restaurant. Everything is removed from the table and taken into the kitchen. There plates, glasses, and utensils are kept to be reused while wastes are discarded. The kidney removes a lot of material from the blood. Some of it (sugar, salt, water) is returned to the blood while wastes are excreted as urine.

Did You Know?
Read aloud Did You Know? on page 129. Tell students that most sea turtles live in warm ocean waters, but the leatherback will visit cold waters. Suggest that students find out in what other ways the leatherback differs from other sea turtles.

TEACHER ALERT

Many Americans do not drink enough water to maintain appropriate water balance. Point out to students that half to three-fourths of their body weight is water; humans can live much longer without food than without water. Six to eight 8-oz glasses of liquid are recommended per day. While the body takes water it needs from all liquids—and even from foods with high water content—drinking water is preferable, since it puts far less strain on the kidneys.

Freshwater bony fish and bony fish in the sea have opposite problems. In a lake or stream, water enters through the gills of a fish at all times. To get rid of the excess water, the fish use their **kidneys** to excrete urine. Kidneys are organs of excretion found in vertebrates. *Excretion* means "getting rid of wastes from the body." The urine is mostly water, but it also contains some salt. The fish must replace the salt. The fish absorbs salt with its gills.

Kidney
An organ of excretion found in vertebrates

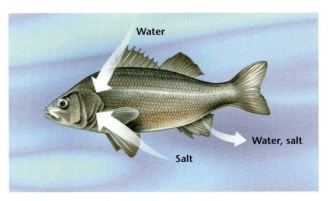

Freshwater Bony Fish

Water Balance on Land

In land animals, the biggest problem with water balance is drying out. Animals have ways to limit water loss. Land snails withdraw into their shell. Insects have a waxy layer that covers their outside skeleton. The wax slows evaporation of water from their body. Reptiles, birds, and mammals have a watertight skin.

The entire body of a land animal cannot be watertight. Animals have to respire. They give off water when oxygen and carbon dioxide are exchanged. Land animals also lose water in their urine, feces, and sweat.

---

3 Reinforce and Extend

SCIENCE INTEGRATION

Earth Science
Global warming has the potential to melt glaciers relatively quickly. Discuss with students what could happen to the concentration of water in oceans if this occurs. What adjustments would marine animals have to make to maintain their water balance?

AT HOME

Encourage students to observe fish in an aquarium at home or at a friend's home. They can talk with the fish caregivers (or read in a fish care manual) about how to maintain the water so that fish's needs are met. Then ask them to compare how water balance and wastes of a land-dwelling mammal such as a cat or dog are controlled.

CROSS-CURRICULAR CONNECTION

Health
Have students research how much water they should drink each day. (*six to eight 8-oz glasses per day*) Then have them record their water consumption for several days. Discuss what happens when they are especially active. Why does the need for water increase?

GLOBAL CONNECTION

In many developing countries, lack of a supply of clean water is a leading cause of illness and death. Have interested students read articles about what is being done to help people in these countries obtain a healthful, reliable source of water. They can form a panel to lead discussions about technology that can help solve this problem.

Most land animals can maintain water balance by drinking water. If open water is scarce, animals may get water by eating leaves, fruits, or roots. Some animals are especially suited for getting water from food. For example, the kangaroo rat of the American Southwest never drinks water. It gets its water from seeds and other plant matter.

Kidneys are the main organs in keeping water balance in mammals and birds. What happens when too much water gets into the body? The kidneys excrete urine with a high water concentration. In this way, they remove excess water. Suppose the water concentration in the body is too low. The kidneys excrete urine with a low water concentration. This keeps more water in the body.

Excreting Wastes

If an animal needs to save water, why does it produce urine? Producing urine does more than maintain water balance. It also removes dangerous wastes from the body. One type of waste is ammonia. It is formed when proteins break down. Ammonia is poisonous to cells. Animals need a way to get rid of it.

In most animals that live in water, the ammonia moves out into the water. The ammonia does not build up inside the body. Animals that live on land must get rid of ammonia in another way. They change ammonia into chemicals that are less poisonous. Then they excrete these chemicals in their urine.

Science Myth

All animals that live on land excrete urine.

Fact: Mammals change ammonia to urea. Urea dissolved in water is urine. Birds change ammonia to uric acid. Uric acid is not dissolved in water and is excreted as a thick liquid or semisolid.

Science Myth

After students read "Excreting Wastes," have them read the Science Myth feature on page 131. After reading the myth (first sentence) aloud, have students predict what land animals do not excrete urine. Have students recall places they have seen bird droppings. Explain that the white solid is uric acid.

LEARNING STYLES

Logical/Mathematical
Remind students that water and the salts dissolved in it seek an equal concentration on both sides of a membrane. Have students use math symbols $>$, $<$, and $=$ to create formulas showing water and salt relationships between saltwater fish and their environment and then between freshwater fish and their environment. (*E.g., saltwater fish: seawater concentration of water $<$ fish's body fluids concentration of water; therefore, water leaves fish and it must drink seawater. Seawater intake $-$ salt excreted $=$ water loss to sea.*)

LEARNING STYLES

Interpersonal/Group Learning
Have students work in groups to determine the effects of salt water versus freshwater on tissues. Each group will need 6 jars of tap water, salt, a spoon, and two small equal sized pieces of apple, two raisins, and two grapes. Let groups design a way to set up saltwater and freshwater "environments" and place one set of foods in each. After an hour, have them observe the samples, record the results, and draw conclusions about what happened. (*Water is drawn out of foods in salt water; it diffuses into foods in freshwater.*)

Lesson 3 Review Answers

1. They keep an internal water concentration equal to that of seawater so they do not have to pump water out of their bodies. 2. to collect excess water inside flatworms and push it via cilia into tubes for excretion 3. They get rid of excess water by excreting urine using their kidneys. 4. Land animals rely on a covering that stops or slows evaporation of water: land snails have a shell; insects a waxy layer; reptiles, birds, and mammals a watertight skin. 5. Ammonia is a by-product of protein digestion; this poisonous waste material must be eliminated.

Portfolio Assessment

Sample items include:
- Summary sentences from Teaching the Lesson
- Lesson 3 Review answers

Science in Your Life

Read the Science in Your Life feature on page 132 together. Display a $1\frac{1}{2}$ liter container of water and point out that this is the volume of blood the kidneys filter every minute. Have students calculate how many liters they would filter in an hour and a day (*90 l/hr; 2,160 l/day*).

Invite a nurse or medical technician to class to explain how a dialysis machine works and the process of dialysis. You might ask your guest to also talk about peritoneal dialysis, in which blood is purified by the membrane lining the abdominal wall.

Lesson 3 REVIEW

Write your answers to these questions on a separate sheet of paper. Write complete sentences.

1. How do most sea invertebrates maintain water balance?
2. What is the function of flame cells?
3. How do freshwater fish maintain water balance?
4. List three ways some land animals limit water loss.
5. Why is the breakdown of proteins a problem for animals?

Science in Your Life

Can you live without kidneys?

Kidneys are important organs. They constantly filter wastes from the blood. Although they weigh only 300 grams, they receive about $1\frac{1}{2}$ liters of blood every minute. You can survive just fine with only one kidney. A person who loses both kidneys, though, cannot survive without help.

A dialysis machine, or artificial kidney, can remove wastes from the blood of a person whose kidneys no longer function. In this machine, the patient's blood circulates on one side of a very thin membrane. Dialysis fluid circulates on the other side. The membrane allows wastes to move from the blood into the dialysis fluid. Patients typically must use the machine for four to six hours three times a week.

Workbook Activity 20

Lesson 4: Coordinating Bodily Activities

Objectives

After reading this lesson, you should be able to
- explain how the endocrine system coordinates activities.
- explain how messages travel through the nervous system.
- describe the features of some invertebrate nervous systems.
- describe the organization of the vertebrate nervous system.

Coordinate
Work together

Hormone
A chemical signal that glands produce

All of the parts of an animal have to work together. Animals must **coordinate** the activities of their cells, tissues, and organs. To coordinate means to work together. Most animals have two systems for coordinating the activities of their parts. Both systems use chemicals. The chemicals act as signals to tell tissues and organs what to do.

The Endocrine System

One system that coordinates some activities is the endocrine system. This system is made of glands that secrete chemicals called **hormones**. The circulatory system carries the hormones all through the body. However, the hormones affect only certain cells. The cells must have a particular type of protein. The hormone binds to the protein. This causes the cell to change its activity. There are many different hormones. Each hormone may work on different kinds of cells. Hormones are found in vertebrates and invertebrates.

Let's look at an example of how the endocrine system works. Mammals produce a hormone called vasopressin. This hormone is made when a mammal's body does not have enough water. A gland at the base of the brain secretes vasopressin. The blood transports this hormone all through the body. The hormone binds to proteins in kidney cells. This causes the kidney to excrete urine with a low water concentration. Less water leaves through the urine. More water stays in the body.

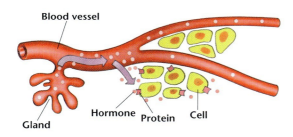

How Animals Stay Alive   Chapter 6   **133**

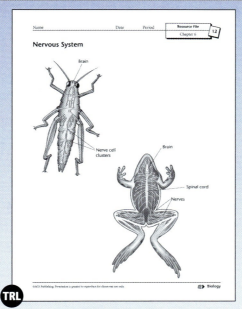

Resource File 12

Lesson at a Glance

Chapter 6 Lesson 4

Overview In this lesson, students explore the endocrine and nervous systems of animals and compare the brains of different types of vertebrates.

Objectives
- To explain how the endocrine system coordinates activities
- To explain how messages travel through the nervous system
- To describe the features of some invertebrate nervous systems
- To describe the organization of the vertebrate nervous system

Student Pages 133–138

Teacher's Resource Library
Workbook Activity 21
Alternative Activity 21
Resource File 12

Vocabulary

coordinate	central nervous
hormone	system
impulse	peripheral
neurotransmitter	nervous system
nerve net	cerebrum

Science Background
Endocrine System

All animals coordinate the functions of their cells by producing hormones. When a hormone binds to a specific protein on a cell, the binding triggers a cascade of reactions within the cell. These reactions may amplify the effect of each hormone molecule. This is why hormones can have major effects on the body though they are secreted in minute amounts.

Nervous System

Nerve cells release chemical signals close to the cells that are being controlled. The effect that a neurotransmitter produces on a cell depends partly on the type of cell. A nerve cell may produce an impulse and release its own neurotransmitter. A muscle cell may shorten, causing a limb to move. A glandular cell may release a hormone into the blood.

How Animals Stay Alive   Chapter 6   **133**

Warm-Up Activity

Surprise students with a sudden loud noise such as a whistle or air horn blast. Ask students to describe their reactions. Explain that the noise triggered their "fight-or-flight" response. It pumps adrenaline into the bloodstream, causes the heart to race, and gives a rush of energy. This response prepares an animal to react to a threatening situation by defending itself or running away. Inform students that this lesson will explain how two body systems coordinate this response and other bodily activities.

Teaching the Lesson

Show students how to set up a Venn diagram to compare and contrast two subjects. As they read, have them make notes on the diagram to show how the endocrine and nervous systems are alike and different.

Have students recall their response in the Warm-Up Activity. Ask them to explain what signals were sent by the nervous and endocrine systems. (*Neurotransmitters made the heart beat faster and stimulated glands. Hormones caused more blood to go to the muscles and brain.*)

TEACHER ALERT

Explain that in humans, endocrine glands control growth and sexual development, among other functions. The pituitary, for example, produces growth hormone, which regulates the growth of bones. If too much or too little is produced, the disorders gigantism and dwarfism result. The pancreas produces insulin, which stimulates the liver to store glucose. If too little is produced, cells are unable to use glucose properly, and diabetes results.

Impulse
A message that travels along nerve cells

Neurotransmitter
A chemical signal that a nerve cell releases

The movement of hormones through the body takes awhile. Hormones must get from glands to cells or organs. The endocrine system is suited to control activities that happen slowly. For example, hormones control the metamorphosis of a tadpole into a frog.

The Nervous System

The second system that coordinates activities in animals is the nervous system. The nervous system carries its messages directly to parts of the body. It does not need the circulatory system. Instead, nerve cells carry the messages. Nerve cells have long, thin branches at the ends. Some nerve cells are very long. For example, some nerve cells reach from your lower back to the tips of your toes.

Messages that travel along nerve cells are called **impulses**. One end of a nerve cell starts an impulse. The impulse travels across the cell to the other end. At this end, the impulse causes the cell to release a chemical signal. This signal is called a **neurotransmitter**. It binds to proteins on nearby nerve cells. The impulses move from cell to cell. This is the way messages travel from one nerve cell to another.

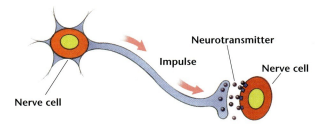

Impulses travel quickly along nerve cells. The fastest impulses can reach speeds of 120 meters per second. The nervous system is suited to control activities that happen quickly. For example, your nervous system directs the movements of your fingers when you type at a keyboard.

Nerve net
A bunch of nerve cells that are loosely connected

Did You Know?

Even simple invertebrates can be trained. For example, some flatworms can be trained to remain still when a light is shone on them.

Invertebrate Nervous Systems

Except for sponges, all animals have a nervous system. Cnidarians, such as hydras, have the simplest nervous system. They do not have brains. They have a bunch of nerve cells that are loosely connected. This is called a **nerve net**. This is all cnidarians need to control their simple activities. The nerve net causes a hydra to shrink if it is touched. It controls the movement of tentacles when a hydra feeds.

Flatworms, segmented worms, and arthropods are more highly developed than cnidarians. They have structures to sense their environment. The front end of the animal contains eyes and sometimes antennae. The front end also has clusters of nerve cells. These serve as a simple brain. The brain receives information from the sense structures. In segmented worms and arthropods, each body segment also has a cluster of nerve cells. The clusters are connected to each other and to the brain. The nervous system of these invertebrates coordinates movement, feeding, reproduction, and other activities.

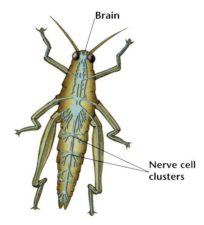

Of all the invertebrates, squids and octopuses have the most highly developed nervous systems. Their brains contain millions of nerve cells. Octopuses can be taught to solve simple problems. They can learn to recognize objects based on their shape or feel.

 Reinforce and Extend

LEARNING STYLES

Body/Kinesthetic
Have students form a line and hold hands. Each person represents a nerve cell. The left arm picks up a neurotransmitter and their right arm carries impulses and releases a neurotransmitter. Initiate an "impulse" in the first student by squeezing his or her left hand. Students should gently squeeze the hand of the person to their right as soon as they feel their left hand being squeezed. Have a separate student time the "impulse" by having the last person signal receiving the impulse by raising the right hand. Compare this travel speed with that of a neurotransmitter, which can travel up to 120 m/sec.

CAREER CONNECTION

Invite an animal trainer, or a person who has trained a dog effectively, to speak to the class. The expert can tell students how he or she teaches an animal to perform a desired behavior, and how capable different species are of learning. Encourage students to explore methods used by animal trainers.

SCIENCE INTEGRATION

 Physical Science
Display a battery and have students point out the positive and negative ends. Review how electricity flows when the battery is connected to a circuit. Ask students to consider how the nervous system is like the positive side and negative side of the battery. (*Nerve cells send signals through the body as a battery sends electricity through its circuit.*)

Did You Know?

Explain to students the difference between innate, or instinctive, behaviors and acquired behaviors. The flatworms that were trained to be still when a light shone had responded to a condition, much as a fish that is fed by people will swim to the surface for any people who come by. Discuss which animals are likely to exhibit the most acquired behaviors. (*those with long life spans, such as mammals, because learned behaviors take time to develop*)

LEARNING STYLES

Visual/Spatial
Have pairs of students make a crossword puzzle using the vocabulary words for the lesson and other important terms, such as *endocrine system, brain,* and *spinal cord.* Pairs can exchange and solve each other's crossword puzzles.

CROSS-CURRICULAR CONNECTION

Physical Education
Pair students and give each pair a cm ruler. One partner holds the ruler vertically, with the zero end down. The other places thumb and forefinger on either side of the bottom of the ruler, without touching it. When partner 1 drops the ruler, partner 2 catches it as soon as he or she sees it falling. Record how many cm past zero the ruler falls before it is caught. After repeating this five times, partners calculate average reaction time. Have students compare their reaction times and discuss the importance of reaction time.

ONLINE CONNECTION

Students will find links to articles on many subjects relating to the brain and spinal column at faculty.washington.edu/chudler/introb.html. Articles are grouped into common categories. Especially useful categories include Brain Basics and Sensory Systems.

Central nervous system
The brain and spinal cord

Peripheral nervous system
The nerves that send messages between the central nervous system and other body parts

The Vertebrate Nervous System

The nervous system is organized the same way in all vertebrates. The nervous system is divided into two parts. One part is the **central nervous system**. It includes the brain and spinal cord. The second part is the **peripheral nervous system**. *Peripheral* means "outer." This part is made up of nerves. The nerves connect the central nervous system with the rest of the body. The diagram below shows how these parts are arranged in a frog.

Bones protect the brain and the spinal cord. The skull protects the brain. The backbone protects the spinal cord. Nerves pass through holes in the skull and backbone.

The brain is the control center for the vertebrate nervous system. The brain interprets messages from the sense organs. It directs the movement of muscles. It controls how fast the heart beats. It controls how fast an animal breathes. It helps maintain balance when an animal walks. In some vertebrates, the brain is the center of emotions and reasoning.

The spinal cord links the brain and the body below the neck. It relays information from the body to the brain. It carries the brain's commands back to the body.

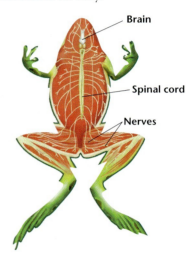

136 Chapter 6 How Animals Stay Alive

Cerebrum

The largest part of the brain that controls thought, memory, learning, feeling, and body movement

Comparing Vertebrate Brains

The brains of different types of vertebrates are different. One difference is the size of the brains. The brains of fish, amphibians, and reptiles are small compared to the rest of their body. Birds and mammals have much larger brains for their body size. The brain of a 100-gram mammal is about ten times bigger than the brain of a 100-gram fish.

Another difference is the size of the different parts of the brain. A part called the **cerebrum** is especially large in birds and mammals. This part controls thought, memory, and learning. These functions are more highly developed in birds and mammals than in other vertebrates.

In some mammals, the surface of the cerebrum is highly folded. The folds increase the surface area of the cerebrum. Since nerve cells are located near the surface, folding makes room for more nerve cells. The more nerve cells an animal has, the more it can learn and remember. Porpoises and humans have brains with the greatest number of folds.

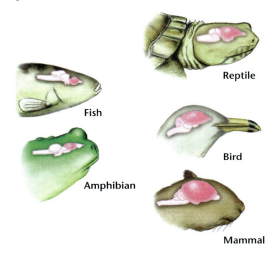

Brains of Different Vertebrates

SCIENCE INTEGRATION

Technology
Ask interested students to research computerized tomography technology, which allows CAT scans and PET scans to be made. Have them explain how the beams differ from X rays and how they create pictures of a "slice" of the brain.

SCIENCE JOURNAL

Have students write a paragraph telling how the nervous system permits them to carry out an action such as catching a ball or speaking.

LEARNING STYLES

LEP/ESL
Use pictures of animals labeled with their names to reinforce animal names. Students might label the pictures with the names in their first language as well as in English.

How Animals Stay Alive   Chapter 6   **137**

Lesson 4 Review Answers

1. They are secreted by glands, enter the bloodstream, and are pumped throughout the body by the circulatory system. **2.** These chemical signals are set in motion by a nerve impulse; they move to nearby cells and bind with proteins, causing the impulse to move on to the next cell. **3.** loosely connected nerve cells that coordinate responses and control activities in cnidarians **4.** The two-part system is comprised centrally of the brain and spinal cord, which are connected to the rest of the body by nerves. **5.** Brain folds give more surface area, allowing for more nerve cells; the more nerve cells the vertebrate brain has, the more the animal is able to learn and remember.

Portfolio Assessment

Sample items include:
- Venn diagram from Teaching the Lesson
- Lesson 4 Review answers

Science at Work

Invite volunteers to tell about the kinds of jobs they have observed at a veterinary clinic. Read the Science at Work feature on page 138 and ask students to predict the qualities a veterinarian assistant might need. If possible, have a veterinary assistant speak to the class about his or her job.

Lesson 4 REVIEW

Write your answers to these questions on a separate sheet of paper. Write complete sentences.

1. How do hormones move through the body?
2. What is the function of neurotransmitters?
3. What is a nerve net?
4. What are the parts of the vertebrate central nervous system?
5. What is the importance of having a highly folded cerebrum?

Science at Work

Veterinary Assistant

A veterinary assistant must be able to communicate effectively and use correct medical terms. Being able to observe, measure, and record is important. Working as part of a team and following procedures are part of this job. A veterinary assistant may receive on-the-job training or take classes at a trade school or community college.

Veterinary assistants help a veterinarian examine and care for animals. They also help keep the office running smoothly. They should like working with animals and be interested in medical science. Duties might include holding the animal while a veterinarian examines and treats it. A veterinary assistant may also prepare an animal for surgery and help during surgery. This person may also perform laboratory tests and provide for the needs of animals that stay during the day or overnight. Veterinary assistants may answer phones, and make appointments. They work not only with the veterinarian but also with the pet owners.

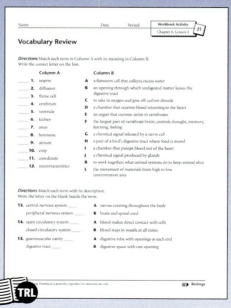

Workbook Activity 21

Chapter 6 SUMMARY

- Animals feed by filtering food, sucking fluids, or consuming large pieces of food.
- Animals digest food in their cells, in a gastrovascular cavity, or in a digestive tract.
- Most animals that live in water use gills to exchange oxygen and carbon dioxide. Land animals use lungs or tubes.
- In an open circulatory system, blood leaves the vessels and makes direct contact with cells. In a closed circulatory system, blood stays inside vessels.
- Bony fish in the sea drink seawater to replace the water they lose by osmosis. Freshwater bony fish use their kidneys to excrete water gained by osmosis.
- Land animals lose water by respiration and in their sweat, urine, and feces. They replace lost water by drinking or eating.
- Animals must excrete the ammonia they produce when they break down proteins.
- The endocrine system uses hormones, which are carried by the blood.
- The nervous system uses chemical signals called neurotransmitters, which are released by nerve cells.
- Cnidarians have a simple nervous system called a nerve net. Most other invertebrates have a brain that is connected to nerve cells.
- The vertebrate nervous system is made of the central nervous system and the peripheral nervous system.

Science Words

annelid, 125	closed circulatory system, 125	flame cell, 129	neurotransmitter, 134
anus, 119	coordinate, 133	gastrovascular cavity, 118	open circulatory system, 125
atrium, 126	crop, 119	gizzard, 119	peripheral nervous system, 136
carnivore, 117	diffusion, 123	herbivore, 117	respire, 123
central nervous system, 136	digestive tract, 119	hormone, 133	secrete, 118
cerebrum, 137	enzyme, 118	impulse, 134	ventricle, 126
circulatory, 125	excrete, 128	kidney, 130	
	filter feeding, 116	nerve net, 135	

Chapter 6 Review

Use the Chapter Review to prepare students for tests and to reteach content from the chapter.

Chapter 6 Mastery Test

The Teacher's Resource Library includes two parallel forms of the Chapter 6 Mastery Test. The difficulty level of the two forms is equivalent. You may wish to use one form as a pretest and the other form as a posttest.

Review Answers

Vocabulary Review

1. hormones 2. digestive tract
3. ventricle 4. flame cells 5. closed circulatory system 6. herbivore 7. nerve net 8. gastrovascular cavity 9. gizzard
10. filter feeding

Concept Review

11. B 12. A 13. C 14. D 15. B 16. D
17. A 18. B

TEACHER ALERT

In the Chapter Review, the Vocabulary Review activity includes a sample of the chapter's vocabulary terms. The activity will help determine students' understanding of key vocabulary terms and concepts presented in the chapter. Other vocabulary terms used in the chapter are listed below:

annelid	enzyme
anus	excrete
atrium	impulse
carnivore	kidney
central nervous system	neurotransmitter
cerebrum	open circulatory system
circulatory	peripheral nervous system
coordinate	respire
crop	secrete
diffusion	

Chapter 6 REVIEW

Vocabulary Review

Word Bank
- closed circulatory system
- digestive tract
- filter feeding
- flame cells
- gastrovascular cavity
- gizzard
- herbivore
- hormones
- nerve net
- ventricle

Choose the word or words from the Word Bank that best complete each sentence. Write the answer on a sheet of paper.

1. The circulatory system carries chemical signals known as _____.
2. A tubelike digestive space with an opening at each end is called a _____.
3. A chamber that pumps blood out of a vertebrate heart is called a _____.
4. Flatworms remove excess water through a system of tubes connected to _____.
5. Blood stays inside vessels at all times in a _____.
6. An animal that eats plants is called a _____.
7. Cnidarians have a simple nervous system known as a _____.
8. A _____ is a digestive space with a single opening.
9. In most animals, a _____ grinds food into a watery paste.
10. A sponge gets food by _____.

Concept Review

Choose the answer that best completes each sentence. Write the letter of the answer on your paper.

11. Leeches get food by _____.
    A filter feeding
    B sucking fluids
    C eating solid pieces of food
    D osmosis

12. Animals break down food by secreting _____.
    A digestive enzymes
    B hormones
    C neurotransmitters
    D impulses

140   Chapter 6 How Animals Stay Alive

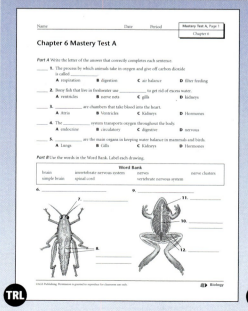
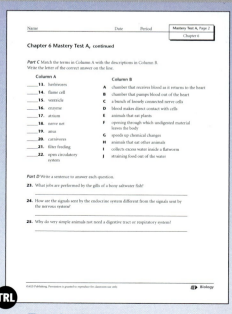

Chapter 6 Mastery Test A

13. Insects use a system of _____ to carry air into the body.
    A gills   B lungs   C tubes   D flame cells

14. In the heart of a bird or mammal, blood goes from the left atrium to the _____.
    A body   C lungs
    B right atrium   D left ventricle

15. Most sea invertebrates have an inside water concentration that is the same as the water concentration in _____.
    A freshwater   B seawater   C bony fish   D flatworms

16. When proteins are broken down in the body, _____ is formed as a waste product.
    A oxygen   B fat   C sugar   D ammonia

17. The endocrine system works best to control activities that happen _____.
    A slowly   B quickly   C only once   D often

18. The brain and spinal cord make up the _____ nervous system.
    A peripheral   C invertebrate
    B central   D impulse

Critical Thinking

Write the answer to each of the following questions.

19. What method of feeding is used by the animals shown in the photo? Explain your reasoning.

20. Some animals live in water that has a low oxygen concentration. They often have large gills. Why would large gills be helpful to these animals?

Test-Taking Tip Before you start a test, look it over so you can plan how to use your time.

Critical Thinking

19. The animals are attached and stationary and live in water. They must be filter feeders, for this way they can take in tiny bits of food and do not have to move to get it. 20. Large gills would have a large surface area. A greater volume of water could be drawn over the gills, so more oxygen could diffuse into the animal's body.

Alternative Assessment

Alternative Assessment items correlate to the student Goals for Learning at the beginning of this chapter.

- Have students refer to Resource File 11. Ask them to explain how each of these animals obtains and digests food.

- Provide students with red and blue pencils and a diagram of the circulatory system of a bird or mammal. Ask them to trace the path of blood through the heart, lungs, and rest of the body. They should use red for oxygen-rich blood and blue for oxygen-poor blood.

- Have students draw two simple diagrams of a bony fish and label the mouth, gills, and kidneys. They can label one diagram "Sea" and the other "Freshwater." Ask them to draw arrows that indicate where water and salt move into and out of each fish's body.

- Have students use the illustration on page 133 to explain how hormones are transported and work.

- Provide students with simple diagrams of a nerve cell, an invertebrate nervous system, and a vertebrate nervous system. Have them label the parts and explain how impulses move through each.

Chapter 7

Planning Guide
How Plants Live

	Student Pages	Vocabulary	Lesson Review
Lesson 1 The Vascular System in Plants	144–147	✔	✔
Lesson 2 How Plants Make Food	148–151	✔	✔
Lesson 3 How Plants Give Off Oxygen	152–155	✔	✔
Lesson 4 How Plants Reproduce	156–162	✔	✔

Chapter Activities

Student Text
Science Center

Teacher's Resource Library
Community Connection 7: Plants in Your Community

Assessment Options

Student Text
Chapter 7 Review

Teacher's Resource Library
Chapter 7 Mastery Tests A and B

Student Text Features								Teaching Strategies						Learning Styles						Teacher's Resource Library				
Achievements in Science	Science at Work	Science in Your Life	Investigation	Science Myth	Note	Technology Note	Did You Know?	Science Integration	Science Journal	Cross-Curricular Connection	Online Connection	Teacher Alert	Applications (Home, Career, Community, Global, Environment)	Auditory/Verbal	Body/Kinesthetic	Interpersonal/Group Learning	Logical/Mathematical	Visual/Spatial	LEP/ESL	Workbook Activities	Alternative Workbook Activities	Lab Manual	Resource File	Self-Study Guide
	147		145							146			146						145	22	22	19		✓
151							150			150	149	149			149		150			23	23			✓
		155					154	153	154	154			154	153						24	24		13	✓
			161	✓	✓			158	159	157	158	157, 158	159			158, 160		157		25	25	20, 21	14	✓

Pronunciation Key

a	hat	e	let	ī	ice	ô	order	ù	put	sh	she
ā	age	ē	equal	o	hot	oi	oil	ü	rule	th	thin
ä	far	ėr	term	ō	open	ou	out	ch	child	ṮH	then
â	care	i	it	ȯ	saw	u	cup	ng	long	zh	measure

ə { a in about / e in taken / i in pencil / o in lemon / u in circus }

Alternative Workbook Activities

The Teacher's Resource Library (TRL) contains a set of lower-level worksheets called Alternative Workbook Activities. These worksheets cover the same content as the regular Workbook Activities but are written at a second-grade reading level.

Skill Track Software

Use the Skill Track Software for Biology for additional reinforcement of this chapter. The software program allows students using AGS textbooks to be assessed for mastery of each chapter and lesson of the textbook. Students access the software on an individual basis and are assessed with multiple-choice items.

Chapter at a Glance

Chapter 7: How Plants Live
pages 142–165

Lessons

1. **The Vascular System in Plants**
   pages 144–147
2. **How Plants Make Food**
   pages 148–151
3. **How Plants Give Off Oxygen**
   pages 152–155
4. **How Plants Reproduce**
   pages 156–162

Investigation 7 pages 161–162

Chapter 7 Summary page 163

Chapter 7 Review pages 164–165

Skill Track Software for Biology

Teacher's Resource Library

Workbook Activities 22–25

Alternative Workbook Activities 22–25

Lab Manual 19–21

Community Connection 7

Resource File 13–14

Chapter 7 Self-Study Guide

Chapter 7 Mastery Tests A and B

Chapters 1–7 Midterm Mastery Test

(Answer Keys for the Teacher's Resource Library begin on page 420 of the Teacher's Edition. A list of supplies required for Lab Manual Activities in this chapter begins on page 449.)

Science Center

Provide (or ask students to bring) tissue paper, crepe paper, pipe cleaners, paper tubes, polystyrene forms, polystyrene peanuts, egg cartons, clay markers, paint, glue, tape, and wire. As students study the chapter, have them make models of different plant parts, such as roots, stems, leaves, and flowers.

Community Connection 7

Chapter 7

How Plants Live

A sunflower seed grew into the plant you see in the photo. Notice the roots, stem, and leaves of the plant. These plant parts all have jobs to do that help the plant survive. If the plant gets everything it needs, it will grow into an adult plant, have sunflowers, and produce seeds that can grow into new plants. In this chapter, you will learn about the parts of a plant. You will also learn how plants make and transport food and water, how plants produce oxygen, and how plants reproduce.

Organize Your Thoughts

Plants
- Plant structures
  - Roots
  - Stems
  - Leaves
  - Flowers
  - Vascular tissue
  - Seeds and spores
- Life processes
  - Photosynthesis
  - Oxygen production
  - Reproduction
  - Respiration

Goals for Learning

◆ To identify the main parts of a plant
◆ To explain how plants make food, transport food and water, and produce oxygen
◆ To describe how plants reproduce

Introducing the Chapter

Bring in a plant, such as an upright flowering plant, or different parts of different plants. If the plants do not have fruits forming, also bring in some fruits. Ask the class how many different parts of the plants they can identify. Write the names of the parts across the top of the board. Then ask how many different functions they can think of for each plant part. Have the students copy the information. When they have finished the chapter, have them revise and extend their lists as necessary.

Notes and Technology Notes

Ask volunteers to read the notes that appear in the margins throughout the chapter. Then discuss them with the class.

TEACHER'S RESOURCE

The AGS Teaching Strategies in Science Transparencies may be used with this chapter. The transparencies add an interactive dimension to expand and enhance the *Biology* program content.

CAREER INTEREST INVENTORY

The AGS Harrington-O'Shea Career Decision-Making System-Revised (CDM) may be used with this chapter. Students can use the CDM to explore their interests and identify careers. The CDM defines career areas that are indicated by students' responses on the inventory.

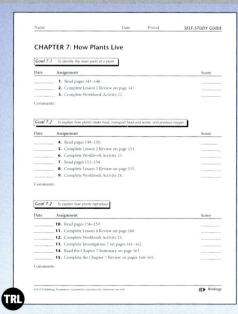

Chapter 7 Self-Study Guide

Lesson at a Glance

Chapter 7 Lesson 1

Overview In this lesson, students learn about the vascular system and its functions in the roots, stems, and leaves of plants.

Objectives

- To identify the main parts of a plant
- To describe the structure and functions of roots
- To describe the structure and functions of stems
- To describe the structure and functions of leaves

Student Pages 144–147

Teacher's Resource Library

Workbook Activity 22

Alternative Workbook Activity 22

Lab Manual 19

Vocabulary

xylem   petiole
phloem stoma
annual growth ring

Science Background
The Vascular System in Plants

The vascular system in vascular plants is made up of xylem and phloem. This system is composed of dead cells that are joined to form tubes. Water moves up to the leaves from the roots mainly through the xylem tissue. Phloem conducts food, in the form of sugars, from the leaves to all plant parts.

Cohesion, the attraction of water molecules for each other, and adhesion, the attraction of water molecules to the sides of the tubes, are important in the movement of water from the roots. Water exits the plant through the leaves in a process called transpiration. When this occurs, molecules of water escape from the plant. Cohesion and adhesion cause more water molecules to be pulled up to replace them. This helps move water through the plant.

144 Chapter 7 How Plants Live

Lesson 1: The Vascular System in Plants

Objectives

After reading this lesson, you should be able to

- identify the main parts of a plant.
- describe the structure and functions of roots.
- describe the structure and functions of stems.
- describe the structure and functions of leaves.

A vascular system of tiny tubes runs through the roots, leaves, and stems of most plants. It connects all parts of the plant. To make food and to survive, plant roots take water and minerals from the soil. Plant leaves collect light from the sun and carbon dioxide from the air. Without this system, the parts of the plant could not do their jobs.

What Roots Do

Have you ever tried to pull a weed out of the ground? You were probably surprised by how hard you had to pull. You discovered an important function of roots. They hold plants firmly in the ground. Roots also have three other functions. First, they absorb water and minerals from the soil. Roots push their way through the soil to reach the water and minerals they need. Second, roots store water and minerals. They can also store food that is made in leaves. Third, the root vascular system brings water and minerals to other parts of the plant.

Roots hold a plant in the ground, absorb water and minerals, and store food.

144 Chapter 7 How Plants Live

Lab Manual 19, pages 1–2

Xylem
The vascular tissue in plants that carries water and minerals from roots to stems and leaves

Phloem
The vascular tissue in plants that carries food from leaves to other parts of the plant

Annual growth ring
Ring in a tree trunk formed by the growth of wood in layers

The Parts of a Root

The tip of a root is always growing. As it grows, it pushes its way through the soil.

Millions of tiny root hairs cover the tip of each root. It is the root hairs that absorb water and minerals from the soil. Roots can store the water and minerals until needed. Water and minerals can also move to the stems and leaves through the root's vascular tissue. **Xylem** vascular tissue forms tubes that carry water and minerals from roots to stems and leaves. The leaves use the water and minerals to make food. **Phloem** vascular tissue forms tubes that carry food from leaves to stems and roots. The roots can also store food.

What Stems Do

Stems are the parts of plants that connect the leaves with the roots. Most stems are above the ground. Stems have three functions. First, stems support the leaves. They hold the leaves up so that they can receive sunlight. Second, stems transport food, water, and minerals through the plant. Third, stems can store food.

Xylem and phloem form the annual growth rings of trees.

The Parts of a Stem

Like roots, stems contain xylem and phloem. They also contain a special layer of growth tissue. It produces new layers of xylem and phloem cells. These layers build up in some plants, so stems become thicker as they get taller. In trees, these layers become wood. In a tree trunk, one layer forms a new ring each year. You can count these rings, called **annual growth rings**, to tell the tree's age.

Science Myth

The stems of plants are always above ground.
Fact: Some plants have underground stems. Tubers, rhizomes, and bulbs are some kinds of underground stems. A potato is a tuber that swells to store food. Many ferns have rhizomes beneath the soil that spread out and form new plants. An onion is a bulb. The layers of the onion are modified leaves attached to the stem part of the bulb.

How Plants Live   Chapter 7   **145**

 Warm-Up Activity

Make a cutting of a houseplant that will grow roots in water. Let the students examine the root closely so that they can see the root hairs.

 Teaching the Lesson

Tell students that when they finish reading the lesson, they should make a diagram of a plant that they could use to explain to another person part or all of the lesson's information. Suggest that they work on rough sketches as they read.

Bring in (or have students bring in) leaves. Have students trace one of the leaves, add details, and label the three main parts.

Ask students why carving their initials in a tree might damage the tree. Students should infer that the cuts might interfere with the tree's vascular tissue. Some students might mention that disease organisms can enter the vascular system through the cuts.

Science Myth

Have one student read aloud the myth portion and one the fact portion of the Science Myth on page 145. If possible, bring a potato, a fern, and an onion to class. Point out the stem portion of each plant. Point out that potatoes and onions are stems that we use for food. Explain that sugar cane is also a stem used for food.

 Reinforce and Extend

LEARNING STYLES

 LEP/ESL
Write the headings "Roots," "Stems," and "Leaves" on the board. Name several functions and characteristics and ask students to choose the correct heading for each. Write the information on the board. Then ask several students to state one fact about a root, stem, or leaf. Write the information under the appropriate heading.

CROSS-CURRICULAR CONNECTION

Social Studies

Obtain a cross section of a tree. Have students count the annual growth rings to determine the tree's age. Encourage students to make a timeline showing significant events that have occurred throughout the tree's lifetime.

AT HOME

Have students observe various plants outside their homes. Instruct them to find examples of at least two different stems and two different leaves. Have them make a sketch of each example and label each part of the leaves. Have them display their sketches as they discuss interesting differences in their specimens and reasons for these differences. For example, they might compare the stems of a dandelion and a tree, discussing why the tree trunk is so much larger than the dandelion stem. Have students share names of the plants they have sketched.

Petiole
The stalk that attaches a leaf to a stem

Stoma
A small opening in a leaf that allows gases to enter and leave (plural is stomata)

What Leaves Do

Leaves are the parts of the plant that trap sunlight. Leaves have four functions. First, they make food. Second, they store food. Third, they transport food to stems. Fourth, they allow gases to enter and leave the plant.

The Parts of a Leaf

Leaves have three main parts: the **petiole**, the blade, and the veins. The petiole, or stalk, attaches the leaf to a stem or a branch. The blade is the main part of the leaf. It collects light from the sun to make food. Many leaves are thin and have flat surfaces. A tree full of leaves can gather large amounts of energy from the sun.

The veins are part of the plant's vascular system. They are thin tubes that are arranged in a pattern. Veins run throughout the blade. They also run through the petiole to the stem. The veins of leaves transport food and water between the stem and the leaf.

The underside of each leaf has many small openings called **stomata**. Each opening is called a stoma. Stomata allow gases, such as carbon dioxide and oxygen, to enter and leave the leaf. Water vapor also leaves through stomata.

The parts of a leaf are the petiole, the blade, and the veins.

Lesson 1 REVIEW

Write your answers to these questions on a separate sheet of paper. Write complete sentences.

1. What are the functions of roots?
2. What is the difference between xylem and phloem tissue?
3. How are annual growth rings made?
4. What are the main parts of a leaf?
5. What do stomata do?

Science at Work

Florist

A florist must be creative and have a good sense of design. Good business management, marketing, and communication skills are also important. A florist can get on-the-job training or take classes at a floral design school, community college, or university.

Florists sell cut flowers and flower arrangements. Florists create beautiful arrangements for weddings, funerals, and other special occasions. In addition to being artistic, florists need to be able to communicate well with customers. Florists also need to know how to prepare and store flowers and plants so they stay fresh. Besides knowing about flowers, florists need to know how to run a business, market their business, and work with other people.

Lesson 1 Review Answers

1. Roots hold the plant in the ground, absorb water and minerals; store water, minerals, and food; and transport them to different parts of the plant. 2. Xylem carries water and minerals from roots to stems and leaves. Phloem carries food from leaves to stems and roots. 3. by the buildup of new layers of xylem and phloem cells each year 4. the petiole, the blade, and the veins 5. Stomata allow water vapor and gases, such as carbon dioxide and oxygen, to enter and leave the leaf.

Portfolio Assessment

Sample items include:
- Diagram of a plant from Teaching the Lesson
- Drawing from Teaching the Lesson
- Lesson 1 Review answers

Science at Work

Have volunteers read aloud Science at Work on page 147. Point out that florists must also be familiar with various kinds of flowers and plants and their traits. Suggest that interested students bring in pictures of attractive flower arrangements from magazines, identify the flowers and plants in them, and display them on a bulletin board.

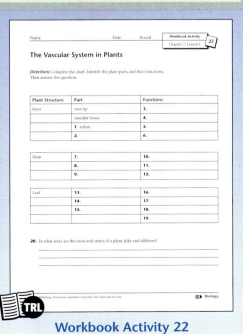

Workbook Activity 22

Pigment
A chemical that absorbs certain types of light

Chlorophyll
The green pigment in plants that absorbs light energy for photosynthesis

Plants get the energy they need when light shines on their chloroplasts. Chloroplasts are organelles in plant cells where photosynthesis takes place. Chloroplasts contain a green **pigment** called **chlorophyll**. A pigment is a chemical that absorbs certain types of light. The cells of the green parts of plants, such as leaves, contain many chloroplasts. When sunlight hits the chloroplasts in the leaves, the chlorophyll absorbs light. The sunlight then supplies the energy for photosynthesis.

Photosynthesis takes place in the chloroplasts of plant cells.

Plants use the energy to split water into hydrogen and oxygen. The oxygen leaves the plant through the stomata and goes into the air. The hydrogen combines with the carbon dioxide to make simple sugar. Plants store the energy of sunlight in the sugar as chemical energy.

 1 Warm-Up Activity

Remind students that, unlike animals, plants make their own food. Ask students what plants need to live. They will probably answer that they need sunlight and water. Have students use these answers and other prior knowledge to form a hypothesis about how plants make food. After the lesson, have them revise the hypotheses according to what they have learned.

 2 Teaching the Lesson

Before beginning this lesson, ask students to scan the lesson, looking for unfamiliar words. Have students work in pairs to define the words on paper.

In discussing photosynthesis, explain that it involves complex chemical reactions that they will not go into now. However, you can simplify the process by explaining that it essentially has two major steps. In the first step, light energy separates the hydrogen and oxygen in water. Oxygen is discarded. In the second step, the hydrogen combines with carbon dioxide to form glucose.

Teacher Alert

 Students may think that photosynthesis takes place only in the leaves of plants. Photosynthesis occurs in all of the green parts of a plant, including stems. In cacti, most of the photosynthesis takes place in the fleshy stems.

Learning Styles

Body/Kinesthetic
Have students act out photosynthesis. You may wish to write the chemical equation for photosynthesis on the board. ($6CO_2 + 6H_2O + $ light energy $\rightarrow C_6H_{12}O_6 + 6O_2$) Explain that the numbers in the demonstration will be different but that the molecules will recombine in a similar way. Give an index card with a chemical symbol for an element to each participant. Have each of two "carbons" join hands with two "oxygens" to form carbon dioxide molecules. Have each of two "oxygens" join hands with two "hydrogens" to form water molecules. Now have the "sun" rearrange the students as follows. Remove the "oxygens" from the water and have them join hands to form an oxygen molecule. Join the remaining "hydrogens" to the carbon dioxides to form sugar molecules.

 3 Reinforce and Extend.

Online Connection

 Students who want to study photosynthesis further can consult the Web site below. It offers a more detailed explanation of photosynthesis and photographs, drawings, and diagrams.
http://photoscience.la.asu.edu/photosyn/education/photointro.html .

LEARNING STYLES

Logical/Mathematical
Explain that the number of atoms on the left side of a chemical equation must equal the number of atoms on the right. Show students how to count the number of atoms in each molecule. Multiply the number in front of the chemical formula by the subscript or implied subscript (1). For example, in $6CO_2$, multiply 6 by the implied subscript to get 6 carbon atoms. Multiply 6 by the subscript for oxygen, 2, to get 12 oxygen atoms. Then have them add up the number of carbon, hydrogen, and oxygen atoms on both sides of the equation to see that they are equal.

CROSS-CURRICULAR CONNECTION

Language Arts
Discuss the origin of the word *photosynthesis*. Ask students to name other words that contain the parts *photo* and *synthesis* or sound similar to those parts. Examples include *photography*, *photograph*, *photoelectric cell*, *photocopy*, *synthesize*, and *synthetic*. Then ask the students what they think *photo* and *synthesis* mean. (*Photo* means light. *Synthesis* means *the combination of parts to make a whole.*) Point out that in plants, photosynthesis involves the combination of carbon dioxide and water, using light energy to form sugars.

 Did You Know?
During the summer, leaves contain a lot of chlorophyll. You can't see the other pigments in the leaves. In autumn, the pigments break down. Chlorophyll breaks down first. Then you can see the red, yellow, and orange colors of the other pigments.

Chemical Energy

Chemical energy is energy stored in the bonds that hold a chemical's molecules together. When the chemical breaks apart, the energy is released. Glucose is the simple sugar that plants make during photosynthesis. Glucose contains stored chemical energy. Plants and animals that eat plants use that stored energy. In Lesson 3, you will learn how plants and animals use this energy.

The Chemical Equation for Photosynthesis

You can write a chemical equation that shows how photosynthesis works. In an equation, the left side and the right side are equal. Each side of this equation has the same number of oxygen, hydrogen, and carbon atoms. The chemical equation for photosynthesis looks like this:

$$6CO_2 + 6H_2O + \text{light energy} \longrightarrow C_6H_{12}O_6 + 6O_2$$

$$\text{carbon dioxide} + \text{water} + \text{light energy} \longrightarrow \text{makes sugar} + \text{oxygen}$$

The substances to the left of the arrow are those needed for photosynthesis: carbon dioxide (CO_2), water (H_2O), and light from the sun. The substances to the right of the arrow are the products of photosynthesis: sugar ($C_6H_{12}O_6$) and oxygen (O_2). In photosynthesis, six molecules of carbon dioxide ($6CO_2$) join with six molecules of water ($6H_2O$). They form one molecule of sugar ($C_6H_{12}O_6$) and six molecules of oxygen ($6O_2$).

Did You Know?

Ask a volunteer to read aloud Did You Know? on page 150. Ask students why they think the leaves of evergreen trees do not turn colors in the autumn. Explain that although evergreen leaves also contain chlorophyll, the leaves' thin shape prevents them from being affected by temperature as much as broad leaves from deciduous trees are.

Lesson 2 REVIEW

Write your answers to these questions on a separate sheet of paper. Write complete sentences.

1. Why are leaves green?
2. What is photosynthesis and where does it occur?
3. What do plants need to make food?
4. What is the source of the chemical energy stored in plants?
5. What is the chemical equation for photosynthesis?

Achievements in Science

The Discovery of Photosynthesis

Today you can study about photosynthesis because of many scientific experiments performed over the years. You know that in photosynthesis, plants use carbon dioxide, water, and light to make sugars (food). Plants give off oxygen during photosynthesis. The discovery of the way photosynthesis works was a process. It did not happen all at once.

The process started in the 1770s, when a scientist discovered that plants give off oxygen. His experiments involved burning a candle in a closed container with a plant. When the burning used all the oxygen in the container, the flame went out. After a few days, the plant replaced the oxygen, and the candle could be burned again.

Several years later, another scientist heard about this discovery and decided to experiment further. His experiments showed that plants need light to produce oxygen. Other scientists went on to discover that plants need carbon dioxide and water for photosynthesis. They also discovered that plants convert solar energy to chemical energy. When scientists share information, other scientists can build on it to plan their own experiments. Even today, research on plants and photosynthesis continues.

How Plants Live   Chapter 7   **151**

Lesson 2 Review Answers

1. Leaves contain chloroplasts, which contain a green pigment called chlorophyll. 2. Photosynthesis is the process that uses the energy of sunlight to change carbon dioxide and water into simple sugars and oxygen. It occurs mainly in the leaves of plants. 3. To make food, plants need energy from the Sun, carbon dioxide, and water. 4. The Sun is the source of the chemical energy stored in plants. 5. $6CO_2 + 6H_2O + \text{light energy} \rightarrow C_6H_{12}O_6 + 6O_2$

Portfolio Assessment

Sample items include:
- Hypotheses from the Warm-Up Activity
- Definitions from Teaching the Lesson
- Lesson 2 Review answers

Achievements in Science

Read aloud Achievements in Science on page 151. Tell students that Joseph Priestley, a chemist, was the scientist who noticed that plants give off oxygen in 1771. In 1779, Jan Ingerhausz, a doctor, discovered that plants needed light to produce oxygen. Their work was the basis for later studies that resulted in the scientific understanding of photosynthesis.

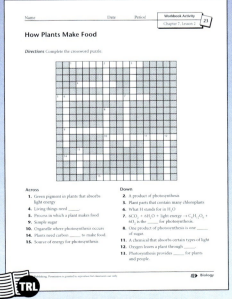

Workbook Activity 23

How Plants Live   Chapter 7   **151**

Lesson at a Glance

Chapter 7 Lesson 3

Overview In this lesson, students learn why oxygen is important to living things. They also learn how plants produce and release oxygen.

Objectives
- To identify the importance of oxygen for living things
- To describe how plants produce and release oxygen

Student Pages 152–155

Teacher's Resource Library TRL
- Workbook Activity 24
- Alternative Activity 24
- Resource File 13

Vocabulary
cellular respiration
guard cell

Science Background
Oxygen

Oxygen gas is made up of molecules that contain two oxygen atoms. About one-fifth of the volume of air is oxygen gas. Oxygen is used by the cells of most organisms to break down food and release its energy. This process is called cellular respiration. Plants produce oxygen when they split water molecules during photosynthesis. The plant uses some of the oxygen for cellular respiration. The rest is released into the atmosphere where it can be used by animals and other organisms.

1 Warm-Up Activity

Have students inhale deeply and then exhale. Ask them if they know the difference between the air they inhale and the air they exhale. Record their ideas on the board. Then explain that inhaled and exhaled air contain both oxygen and carbon dioxide. However, inhaled air contains more oxygen and less carbon dioxide than exhaled air. Ask how oxygen gets into the air. (*Plants release oxygen.*) Ask how plants use the carbon dioxide that we exhale (*to make food*).

Lesson 3: How Plants Give Off Oxygen

Objectives
After reading this lesson, you should be able to
- identify the importance of oxygen for living things.
- describe how plants produce and release oxygen.

Cellular respiration
The process in which cells break down food to release energy

Oxygen is a gas that living things need. You breathe in oxygen thousands of times each day. Most of that oxygen was released by plants during photosynthesis.

Properties of a Gas

Gases make up the air around you. Fan yourself with a sheet of paper. The breeze you feel on your face is moving air. Why can't you see or hold air? The tiny, invisible particles of the gases in air are far apart. There is a lot of space between them. The particles move around quickly. You cannot hold or touch a gas such as air. In a solid, such as your desk, the particles are packed tightly together. They hardly move. That is why you can touch your desk.

The Importance of Oxygen

Two of the most important gases in the air that you breathe are carbon dioxide and oxygen. Oxygen is important to most living things. They use oxygen to break down food to release the chemical energy stored in it. This process is called **cellular respiration**.

Photosynthesis happens only in plants. Respiration happens in both plants and animals. Cellular respiration is a special low-temperature kind of burning that breaks down glucose. Glucose is the simple sugar that plants make during photosynthesis. Glucose is also your body's main source of energy. You get that energy when your cells burn sugars and starches that come from the plants you eat. Your body cells use oxygen to break apart the sugar molecules. During cellular respiration, oxygen combines with hydrogen to make water. Carbon dioxide is released as a waste product. Does this sound familiar? It is the opposite of photosynthesis.

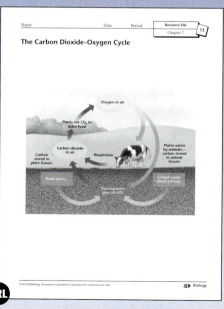

Resource File 13

Photosynthesis and respiration are part of the carbon dioxide–oxygen cycle. Plants take in carbon dioxide and water and give off oxygen during photosynthesis. Plants and animals take in oxygen and give off carbon dioxide and water during respiration. This cycle is necessary for life on Earth.

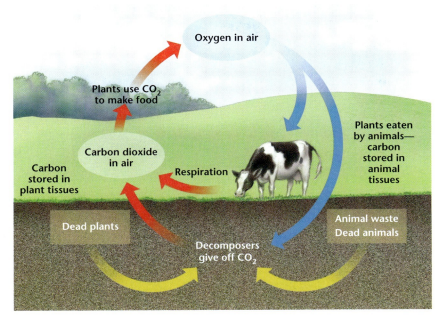

Producing Oxygen

The oxygen that plants produce comes from water. During photosynthesis, plants use water and carbon dioxide to make sugars. Photosynthesis splits water into hydrogen and oxygen. The hydrogen combines with the carbon dioxide to make sugar and more water. The oxygen forms into oxygen gas. The plant uses some of the oxygen for cellular respiration. A plant, however, makes more oxygen than it needs. The rest of the oxygen leaves the plant and goes into the air.

2 Teaching the Lesson

Have students write a question about what they think they will learn in the text under each subhead. For example, the subhead Producing Oxygen might suggest the question "What produces oxygen?" Have students write the answer to each question as they read.

Tell about aquariums developed by NASA that are completely sealed in glass. They contain tiny brine shrimp, brine algae, snails, and bacteria. Ask students how they think the organisms survive without air from the outside. (*The algae use light energy and carbon dioxide produced by the other organisms to carry out photosynthesis. The shrimp and snails breathe oxygen produced by the algae.*)

3 Reinforce and Extend

LEARNING STYLES

Auditory/Verbal
Have students work in pairs. Have each pair look at the illustration on page 153 and take turns explaining the carbon dioxide-oxygen cycle.

SCIENCE INTEGRATION

Physical Science
Chemical equations represent chemical reactions in processes such as photosynthesis and respiration. The substances that react, called reactants, are on the left side of the equation. The substances that are produced, called products, are on the right side of the equation. Write the chemical equation for the production of sugars in photosynthesis. Label the molecules and point out how photosynthesis is the opposite of respiration.

$6CO_2 + 6H_2O + \text{light energy} \rightarrow C_6H_{12}O_6 + 6O_2$

Did You Know?
Read aloud Did You Know? on page 154. Discuss how the release of oxygen by plants is beneficial to animals.

CROSS-CURRICULAR CONNECTION

Health
Explain that emphysema is a lung disease in which carbon dioxide does not flow out of the body as it should. Therefore, sufferers do not get all the oxygen they need in their cells. Have students find and report on more information on emphysema, such as how many Americans it affects and its treatments.

IN THE ENVIRONMENT

Carbon dioxide in the atmosphere traps some of the heat energy from the sun. Without this greenhouse effect, Earth would be too cold for life to exist. However, burning fossil fuels has released more carbon dioxide into the air than plants can use in photosynthesis. Have students find newspaper articles about how an increased greenhouse effect may affect Earth's climate.

SCIENCE JOURNAL

Encourage students to explain in their own words the importance of photosynthesis and the carbon dioxide-oxygen cycle to human life. Have them discuss why it is important to care for our forests as a result.

Guard cell
A cell that opens and closes stomata

Did You Know?
Most of the oxygen that we breathe comes from plants.

Releasing Oxygen

The oxygen that goes out of the plant into the air leaves through the stomata. Remember that stomata are small openings on a leaf. Each stoma has two special cells called **guard cells**. The size and shape of the guard cells change as they take up and release water. When the guard cells take up water and swell, the stomata open. Oxygen, carbon dioxide, and water vapor can move in and out of the leaf through the openings. When the guard cells lose water, the stomata close.

The amount of light affects the opening and closing of stomata. The stomata of most plants close at night. They open during the day when photosynthesis takes place. The amount of water also affects the opening and closing of stomata. When the soil and air are dry, stomata close, even during the day. This prevents the plant from losing water during short dry periods.

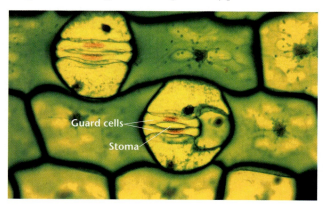

Guard cells open and close a stoma.

Lesson 3 REVIEW

Write your answers to these questions on a separate sheet of paper. Write complete sentences.

1. Why can't you see or hold a gas?
2. Why do living things need oxygen?
3. How is respiration the opposite of photosynthesis?
4. What happens to water during photosynthesis?
5. What do guard cells do?

Science in Your Life

What's that plant?

Do you eat genetically engineered foods? The answer is most likely yes. A plant that contains some DNA from another species is a genetically engineered (GE) plant or a genetically modified organism (GMO).

When a scientist puts genes from one living thing into another living thing, the result is a GMO. One example is taking a gene from a fish that lives in very cold water and putting it into a strawberry. The strawberry would then be able to stand colder temperatures without freezing. We may never be able to buy this kind of strawberry at the supermarket. However, about 60 percent of the foods at the supermarket contain at least some genetically engineered ingredients. Most processed foods, such as chips, cereals, and peanut butter, contain these ingredients. The ingredients they contain are most likely from GE soybeans, corn, or canola.

Scientists are researching many different kinds of GE plants. These plants may help make more food crops stay free of disease. However, many people object to genetically engineered foods. They are concerned about environmental and health effects that might not be known for years. As research continues, genetically engineered organisms will continue to be controversial.

Lesson 3 Review Answers

1. The molecules in a gas are far apart and move around quickly. **2.** to release food energy, which is stored in chemicals in cells **3.** Plants take in carbon dioxide and water and give off oxygen during photosynthesis. Animals breathe in oxygen and breathe out carbon dioxide and water during respiration. **4.** The hydrogen combines with the carbon dioxide to make sugar and more water. The oxygen becomes oxygen gas. **5.** control the opening and closing of stomata

Portfolio Assessment

Sample items include:
- Questions and answers from Teaching the Lesson
- Lesson 3 Review answers

Science in Your Life

Have volunteers read aloud Science in Your Life on page 155. Discuss the ways in which genetic engineering can improve food. Explain that it may make some foods, such as strawberries, hardier in the growing stage. It may make other foods more nourishing or less harmful, as when unhealthy fats are decreased in a food. It may also protect foods from spoilage. Ask how genetic engineering could help people all over the world. (*It could decrease hunger by providing a greater quality and quantity of foods.*)

Workbook Activity 24

Lesson at a Glance

Chapter 7 Lesson 4

Overview In this lesson, students learn how seed plants and seedless plants reproduce.

Objectives
- To identify the difference between sexual and asexual reproduction
- To describe how mosses and ferns reproduce
- To discuss sexual reproduction in angiosperms and gymnosperms

Student Pages 156–162

Teacher's Resource Library

Workbook Activity 25
Alternative Activity 25
Lab Manual 20, 21
Resource File 14

Vocabulary
zygote   ovary
stamen   nectar
pollen   pollination
pistil   germinate
stigma

 Warm-Up Activity

Bring in a fern frond (with sori containing spores on the underside of the leaves), a pinecone (with seeds still attached), and a fruit (with seeds). Have students describe how the plant parts are alike and different. Students should recognize that they all have seeds or seedlike structures (spores). Explain that these structures are all ways in which plants reproduce.

 Teaching the Lesson

Recall the discussion of asexual and sexual reproduction of protists on page 97. Have students make a chart that compares reproduction in seed plants and seedless plants. Label columns "Seed Plants" and "Seedless Plants." Label rows "Asexual" and "Sexual." As they read, have students write a brief description in the appropriate spaces.

156 Chapter 7 *How Plants Live*

Lesson 4 How Plants Reproduce

Objectives

After reading this lesson, you should be able to
- identify the difference between sexual and asexual reproduction.
- describe how mosses and ferns reproduce.
- discuss sexual reproduction in angiosperms and gymnosperms.

Zygote
A fertilized cell

Plants can reproduce by sexual reproduction or by asexual reproduction. Sexual reproduction involves two parents. The female parent provides the egg. The male parent provides the sperm. The sperm and egg cells join to form a new plant. Asexual reproduction involves only one parent and no egg or sperm. Many plants can reproduce both sexually and asexually.

Reproduction in Seedless Plants

Mosses and ferns are seedless plants. They reproduce asexually and sexually. Asexual reproduction happens in mosses when a small piece of the parent plant breaks off. That piece forms a new plant. Asexual reproduction happens in ferns when a new plant grows from an underground stem.

Seedless plants reproduce sexually from spores. A spore is a reproductive cell with a thick protective coating. Spores develop into tiny plants that are male, female, or both male and female. The plants produce sperm and eggs. A sperm cell swims to an egg cell through the moisture around the plants. The egg and sperm come together during fertilization. The fertilized cell is called a **zygote**. The zygote is the beginning of a new plant. When the plant matures, it produces spores that create a new generation of plants.

Seedless plants such as ferns reproduce from spores.

156 Chapter 7 *How Plants Live*

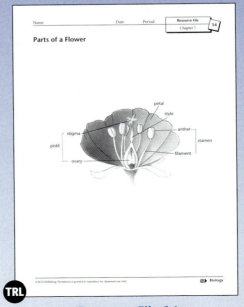

Resource File 14

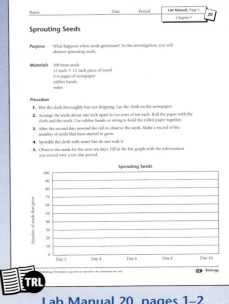

Lab Manual 20, pages 1–2

Stamen
The male organ of reproduction in a flower, which includes the anther and filament

Pollen
Tiny grains containing sperm

Pistil
The female organ of reproduction in a flower

Stigma
The upper part of the pistil, on the tip of the style

Ovary
The lower part of the pistil that contains eggs

Some foods people think of as vegetables are actually fruits. If a food contains seeds, it was made by the ovary of a flower and is a fruit. Tomatoes, cucumbers, and green beans are fruits.

Reproduction in Seed Plants

There are two types of seed plants: flowering plants and nonflowering plants. Seed plants can reproduce asexually. New plants can grow from a piece of a plant called a cutting. A single leaf or stem can grow roots and become a new plant. However, seed plants usually reproduce sexually.

Sexual Reproduction in Angiosperms

Flowering plants are angiosperms. The flower is the part of an angiosperm that contains eggs and sperm. In a flower, the **stamens** are the male organs of reproduction. The stamen includes the anther and filament. They produce **pollen**, which are tiny grains containing sperm. The **pistil** is the female organ of reproduction. The upper part of the pistil is the **stigma**, on the tip of the style. The lower part of the pistil is the **ovary**, which contains eggs.

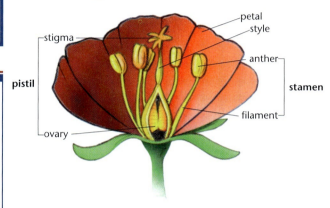

Have students work in pairs to take apart flowers and identify as many structures as they can. The parts of large flowers, such as tulips and lilies, will be easiest to see. You can refer students to page 158 for a brief discussion of fertilization.

Try to get an apple fresh from a tree. Point out the dried-up stamens on the "bottom" of the apple, opposite the stem. These stamens remain from the withered apple flower. Cut the apple in half. Show students the core, which is the ripened ovary of the flower, and the seeds. Ask what part of the apple plant protects the seeds. (*the fruit*)

TEACHER ALERT

Students may think that all flowers have petals. However, some flowers, such as those of some grasses and trees, have no petals. They depend on wind for pollination and do not need brightly colored petals to attract pollinators.

3 Reinforce and Extend

LEARNING STYLES

Visual/Spatial
Make transparencies showing the life cycles of a fern, an angiosperm, and a gymnosperm. Refer to the transparencies as you talk about reproduction.

CROSS-CURRICULAR CONNECTION

Art
Have students draw and color a cross-section picture of a flower like the one on page 157. Instruct them to draw each part shown in the diagram on page 157 and to draw and color it as realistically as possible.

Teacher Alert

Students may think that all brightly colored parts of a plant, such as the red part of a poinsettia, are flower petals. However, this is not always true. On poinsettias, the red parts are colored leaves, called bracts. The tiny flowers are clustered in the center of the bracts.

Science Integration

Physical Science

Flowers are usually brightly colored in the visible spectrum, which we can see. They reflect visible light. Some flowers also reflect ultraviolet light. We cannot see ultraviolet light, but some insects, such as bees, can. Have students research the differences between visible and ultraviolet light. Show photographs from a college text or ultraviolet markings on flowers.

Learning Styles

Interpersonal/Group Learning

Point out that plant pollen is a cause of allergic reactions in some people. Have groups of students research different kinds of information about pollen and allergies. For example, they could research the most common pollen allergies, the most common pollens in your area, causes and effects of pollen allergies, and treatment of pollen allergies.

Nectar
A sweet liquid that many kinds of flowers produce

Pollination
The process by which pollen is transferred from the stamen to the pistil

Germinate
Start to grow into a new plant

For reproduction to take place, the sperm in pollen must fertilize the egg. Flowers have many colors and shapes to attract insects and birds. They land on flowers to drink **nectar**, which is a sweet liquid that many kinds of flowers produce. While insects and birds drink, pollen sticks to their bodies. They carry the pollen to the pistil of other flowers or to the same flower. Wind also spreads pollen. The process by which pollen is transferred from the stamen to the pistil is called **pollination**.

Fertilization

After pollination, the pollen grain grows a tube. The tube reaches down through the pistil to the eggs in the ovary. When the pollen inside the tube meets an egg, fertilization takes place. The ovary grows and becomes a fruit with seeds inside. The fruit protects the seeds.

Seeds

Seeds contain the embryo, or beginning stages of a new plant. If the temperature and amount of water are just right, the seed **germinates**. That means it starts to grow into a new plant. Seeds also contain stored food. The young plant uses this food until it can make its own. If the new plant is fertilized, a new set of seeds develops in the ovary.

Online Connection

The California Native Plant Society has a colorful Web site that discusses a variety of plants topics including reproduction. Have students visit the site at www.cnps.org. To find out about pollination, they should click on Kid's Stuff.

Sexual Reproduction in Gymnosperms

Gymnosperms are nonflowering plants. The largest group of gymnosperms are conifers, or cone-bearing plants. Most evergreen trees are gymnosperms. The reproductive organs of gymnosperms are in cones, not flowers. Some cones are male. Some are female. Male cones are usually smaller than female cones.

During reproduction, male cones release millions of pollen grains into the air. Some of the pollen reaches female cones. As in flowering plants, the pollen grain grows a tube that reaches eggs in the ovary. When the pollen and egg meet, fertilization takes place. But unlike angiosperms, a fruit does not cover gymnosperm seeds. The uncovered seeds are under the scales of the cones.

How do male and female cones differ from each other?

IN THE COMMUNITY

Have students find out about garden clubs and other organizations in their community that plant and care for flowers, trees, and other plants in public areas. Have them find out one or two specific projects that the club has participated in to beautify the community. Interested students can research and participate in volunteer opportunities with these organizations.

CAREER CONNECTION

Forest rangers are employees of national forests. They have a wide range of jobs including protecting forests from fires, diseases, and pests; planting new trees; supervising tree cutting; and rescuing forest visitors in trouble. Some forestry positions require advanced degrees in biology or forestry. Others require college or high school diplomas. Some colleges offer programs in forestry. Students can find out more about working for the U.S. Forest Service on their Web site at www.fs.fed.us.

SCIENCE JOURNAL

Have students keep a record of all of the plant reproductive parts they eat in a day or a week. This can include fruits, certain vegetables, and seeds.

GLOBAL CONNECTION

When studying gymnosperms, students may think first of conifers, such as pines, firs, and cedars. Conifers are the largest group of gymnosperms, but another kind of gymnosperm is the ginkgo. The ginkgo is an ornamental tree that was first found in temple gardens in China. It is now frequently planted in many urban areas around the world because it thrives even in places plagued by air pollution. Ginkgoes have fan-shaped leaves and do not grow in the wild. Have students find an illustration of a ginkgo and make their own sketch of one.

LEARNING STYLES

Interpersonal/Group Learning
Have groups of students collect and bring to class examples of a moss, a fern, an angiosperm, and a gymnosperm. Have each group display their samples, identify each, and explain how it reproduces.

Lesson 4 Review Answers

1. Sexual reproduction involves two parents. One parent provides the egg cell, and the other parent provides the sperm cell. Asexual reproduction involves only one parent and no egg or sperm.
2. mosses and ferns **3.** After pollination, the pollen grain grows a tube that reaches down to the ovary. The ovary contains the eggs. When the pollen inside the pollen tube meets an egg, fertilization takes place. **4.** The seed begins to grow into a new plant. **5.** Male cones release pollen, some of which reach the female cones. The pollen grain grows a tube that reaches eggs in the ovary, where fertilization takes place.

Portfolio Assessment

Sample items include:
- Chart from Teaching the Lesson
- Lesson 4 Review answers

Lesson 4 REVIEW

Write your answers to these questions on a separate sheet of paper. Write complete sentences.

1. What is the difference between sexual reproduction and asexual reproduction?
2. Which type of plants uses spores to reproduce?
3. Describe the process of fertilization in angiosperms.
4. What happens when a seed germinates?
5. Describe the process of reproduction in a conifer.

Technology Note

People have been changing the traits of organisms for centuries. They do this by breeding organisms that have the traits they want. Today, scientists use genetic engineering to change the DNA of some plants used for food. Scientists have learned how to take pieces of DNA from the genes of one kind of organism, such as a bacteria or virus, and combine them with a plant's DNA. The result is a plant with a trait it doesn't normally have, such as resistance to disease, weed killers, or spoilage.

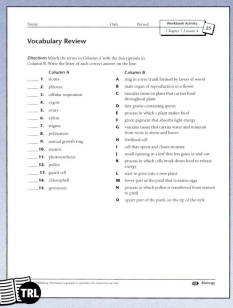

Workbook Activity 25

INVESTIGATION 7

Growing an African Violet from a Leaf

Purpose
Can one leaf grow into a whole new plant? In this investigation, you will grow a plant, using asexual reproduction.

Materials
- African violet plant
- water
- 2 paper cups
- aluminum foil
- potting soil

Procedure

1. Have your teacher cut a leaf with a long stem from the African violet plant.

2. Fill a cup with water. Then cover it with aluminum foil. With a pencil, poke a hole in the center of the foil.

3. Insert the leaf into the hole. The end of the stem should be in the water.

4. Place the leaf and cup in a window where the leaf will get sunlight.

5. Change the water in the cup every few days. As you do, observe the end of the stem. Observe and record any changes.

6. When roots appear and begin to grow, plant your leaf in a cup of potting soil. Bury the roots and part of the stem in the soil. Water the soil.

7. Place the potted leaf on a windowsill. Keep the soil moist. What eventually happens?

How Plants Live   Chapter 7   **161**

Investigation 7

This investigation will take approximately 20 minutes to set up and then 5 minutes every other day until the investigation is completed. Students will use these process and thinking skills: observing, describing, sequencing, making inferences.

Preparation
- You may wish to tear the aluminum foil into pieces ahead of time so students do not use too much.
- Students may use Lab Manual 21 to answer the questions.

Procedure
- Several plants can be easily grown using cuttings of stems and leaves. Examples are African violets, begonias, ivies, impatiens, and geraniums. Refer to a houseplant reference for other ideas.
- You may use a commercial root growth stimulator to speed up root growth. Follow the directions on the package. Or dip the cut stems in the root growth stimulator before placing them in the water.
- Allow several weeks for this investigation. Students can observe development of roots and root hairs over time. At the end, every student will have a new plant.
- If you use root growth stimulator, you might want to root some leaves with and without it. You can compare the rate at which the roots grow.
- Have students measure the growth in millimeters of the longest root or the first root that appears every few days or weekly. Have them graph the growth.
- Keep the original plant. Have the students compare it to the new plants. Because the new plants were produced by asexual reproduction, they will be exactly like the original plant.

Results
Students will observe the growth of roots and root hairs. After the stem is planted, the new plant should grow new leaves and perhaps flowers.

SAFETY ALERT
- Do not allow students to use the knife. Remove the leaves from the plant yourself.
- Warn students not to poke themselves with the pencil.
- Follow the safety directions on the package of root growth stimulator.

Lab Manual 21

Questions and Conclusions Answers

1. Students should observe that new roots grow from the stem. The new roots are white and covered with root hairs.
2. The leaf can make food because it has chloroplasts, but it needs roots to be able to absorb nutrients from the soil.
3. New leaves appear.
4. This is asexual reproduction because a new plant grew from a part of the original plant.

Explore Further Answers

Encourage students to share their results with the class.

Assessment

Check to be sure students are accurately measuring and recording root growth. You might include the following items from this investigation in student portfolios:

- Graph of root growth
- Answers to Questions and Conclusions and Explore Further

Questions and Conclusions

1. What was the first change that you observed in the leaf? Describe your observation.
2. Why do you think the plant produced this type of new growth?
3. How does the plant change after the leaf is planted in soil?
4. What type of reproduction occurred in this investigation? Explain your answer.

Explore Further

Many plants will grow from leaf or stem cuttings. Try to grow some other plants this way. You may want to use a book on houseplants as a reference.

Chapter 7 SUMMARY

- Roots hold plants in the ground and absorb water and minerals.
- Food is moved from the leaves and stems to the roots through phloem vascular tissue. Food can be stored in the roots.
- Stems support the leaves, store food, and transport food, water, and minerals through the plant.
- Most leaves are thin and flat with three main parts: the petiole, the blade, and the veins.
- Plants make food in the green parts of the plant. The cells in these parts contain chloroplasts, where photosynthesis takes place.
- Plants need carbon dioxide and water to make food. Food stores chemical energy. The process of releasing energy from food is cellular respiration.
- Plants give off oxygen gas. Most living things need oxygen for cellular respiration.
- The oxygen a plant produces leaves the plant through the stomata.
- Plants can reproduce by sexual reproduction, which involves two parents, or by asexual reproduction, which involves only one parent.
- Plants reproduce asexually by growing new plants from parts or cuttings. Mosses and ferns reproduce sexually by forming spores. Seed plants reproduce sexually by producing seeds.
- Angiosperms use flowers to reproduce sexually. Pollen from the stamen fertilizes an egg in the pistil. A fruit surrounds the seed.
- In gymnosperms, male and female organs are in different cones or in different trees. A fruit does not surround the seed.

Science Words

annual growth ring, 145	guard cell, 154	pigment, 149	stoma, 146
cellular respiration, 152	nectar, 158	pistil, 157	xylem, 145
chlorophyll, 149	ovary, 157	pollen, 157	zygote, 156
germinate, 158	petiole, 146	pollination, 158	
	phloem, 145	stamen, 157	
	photosynthesis, 148	stigma, 157	

How Plants Live Chapter 7

Chapter 7 Review

Use the Chapter Review to prepare students for tests and to reteach content from the chapter.

Chapter 7 Mastery Test

The Teacher's Resource Library includes two parallel forms of the Chapter 7 Mastery Test. The difficulty level of the two forms is equivalent. You may wish to use one form as a pretest and the other form as a posttest.

Chapters 1–7 Midterm Mastery Test

The Teacher's Resource Library includes the Midterm Mastery Test. This test is pictured on page 417 of this Teacher's Edition. The Midterm Mastery Test assesses the major learning objectives for chapters 1–7.

Review Answers

Vocabulary Review
1. nectar 2. guard cell 3. phloem
4. xylem 5. stomata 6. photosynthesis
7. pollination 8. stamen 9. pollen
10. petiole 11. cellular respiration
12. chlorophyll 13. pistil 14. germinate

Concept Review
15. A 16. C 17. D 18. B

TEACHER ALERT

In the Chapter Review, the Vocabulary Review activity includes a sample of the chapter's vocabulary terms. The activity will help determine students' understanding of key vocabulary terms and concepts presented in the chapter. Other vocabulary terms used in the chapter are listed below:

annual growth ring
ovary
pigment
stigma
zygote

Chapter 7 REVIEW

Vocabulary Review

Choose the word or words from the Word Bank that best complete each sentence. Write your answer on a sheet of paper.

Word Bank
cellular respiration
chlorophyll
germinate
guard cell
nectar
petiole
phloem
photosynthesis
pistil
pollen
pollination
stamen
stomata
xylem

1. Many kinds of flowers produce a sweet liquid called _____.
2. The type of cell that opens and closes a stoma is a(n) _____.
3. Food moves around in a plant through _____.
4. Water and minerals move in a plant through _____.
5. Gases move in and out of a leaf through _____.
6. Plants make simple sugars during _____.
7. Pollen is transferred from the stamen to the top of the pistil in the process of _____.
8. The male part of a flower is the _____.
9. The sperm of seed plants is contained in grains of _____.
10. The _____ attaches a leaf to a stem.
11. Oxygen is used and carbon dioxide is given off during _____.
12. The green pigment, _____, absorbs light energy for photosynthesis.
13. The female part of a flower is the _____.
14. When conditions are right, a seed will _____, or start growing into a new plant.

164   Chapter 7   How Plants Live

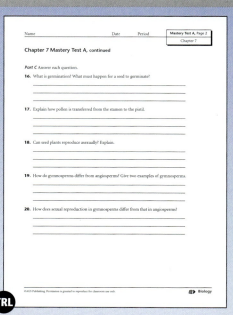

Chapter 7 Mastery Test A

Concept Review

Choose the answer that best completes each sentence. Write the letter of the answer on your paper.

15. The main parts of a plant are roots, stems, and _____.
- A leaves
- B stomata
- C pollen
- D chlorophyll

16. Water vapor, carbon dioxide, and oxygen enter and leave a plant through the _____.
- A roots
- B flowers
- C stomata
- D stems

17. The food that is made during photosynthesis contains _____ energy.
- A little
- B light
- C cellular
- D chemical

18. The process in which a sperm joins with an egg is _____.
- A pollination
- B fertilization
- C respiration
- D asexual reproduction

Critical Thinking

Write the answer to each of the following questions.

19. The chemical formula for glucose is $C_6H_{12}O_6$. How many carbon, hydrogen, and oxygen atoms does one molecule of glucose have?

20. Explain how energy for animals comes originally from the Sun.

> **Test-Taking Tip** Prepare for a test by making a set of flash cards. Write a word or phrase on the front of each card. Write the definition on the back. Use the flash cards in a game to test your knowledge.

Critical Thinking

19. 6 carbon, 12 hydrogen, and 6 oxygen

20. Plants use energy from the Sun to make simple sugars and other molecules. Animals eat plants or other animals that eat plants. When animals break down the food, they release the energy that was stored in the chemicals from the Sun.

ALTERNATIVE ASSESSMENT

Alternative Assessment items correlate to the student Goals for Learning at the beginning of this chapter.

- Give each student a plant, possibly a weed. Have them identify the roots, stems, and leaves and trace the paths of xylem and phloem.
- Have students make a graphic organizer for photosynthesis in plants. Their organizers should identify steps in the process as well as how plants transport food and water.
- Have students take apart fruits and compare where the seeds are found and how they are arranged. As they do so, they should describe the reproduction process and the role plant parts play in it. Show students a plant cutting and have them tell how the cutting can be used in reproduction.

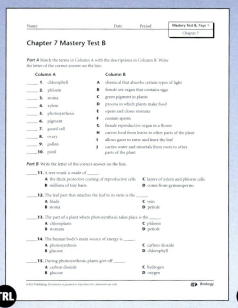

Chapter 7 Mastery Test B

Chapter 8

Planning Guide
Human Body Systems

	Student Pages	Vocabulary	Lesson Review
Lesson 1 How the Body Digests Food	168–172	✔	✔
Lesson 2 How Materials Move to and from Cells	173–180	✔	✔
Lesson 3 How We Breathe	181–183	✔	✔
Lesson 4 How the Body Gets Rid of Wastes	184–186	✔	✔
Lesson 5 How the Nervous System Controls the Body	187–191	✔	✔
Lesson 6 The Sense Organs	192–195	✔	✔
Lesson 7 How the Endocrine System Controls the Body	196–198	✔	✔
Lesson 8 How the Body Moves	199–204	✔	✔

Student Text Lesson

Chapter Activities

Student Text
Science Center

Teacher's Resource Library
Community Connection 8: The Dominant Eye

Assessment Options

Student Text
Chapter 8 Review

Teacher's Resource Library
Chapter 8 Mastery Tests A and B

	Student Text Features								Teaching Strategies						Learning Styles						Teacher's Resource Library				
	Achievements in Science	Science at Work	Science in Your Life	Investigation	Science Myth	Note	Technology Note	Did You Know?	Science Integration	Science Journal	Cross-Curricular Connection	Online Connection	Teacher Alert	Applications (Home, Career, Community, Global, Environment)	Auditory/Verbal	Body/Kinesthetic	Interpersonal/Group Learning	Logical/Mathematical	Visual/Spatial	LEP/ESL	Workbook Activities	Alternative Workbook Activities	Lab Manual	Resource File	Self-Study Guide
	172						✔	170	171			169	169, 170				171	171	170		26	26			✔
	178			179				175			175, 177		177	174	174, 176	175			176	27	27	22	15	✔
					183					182	182		182			182					28	28	23		✔
						✔			185, 186		185		185								29	29			✔
			190						188		189	190	188	189							30	30			✔
										194	193			194	193	194					31	31	24		✔
						✔			196				197		197						32	32			✔✔
		204							201	200	202	203	201	200, 203		201			202		33	33		16	✔

Pronunciation Key

a	hat	e	let	ī	ice	ô	order	ù	put	sh	she	ə { a in about
ā	age	ē	equal	o	hot	oi	oil	ü	rule	th	thin	e in taken
ä	far	ėr	term	ō	open	ou	out	ch	child	ᴛʜ	then	i in pencil
â	care	i	it	ò	saw	u	cup	ng	long	zh	measure	o in lemon
												u in circus }

Alternative Workbook Activities

The Teacher's Resource Library (TRL) contains a set of lower-level worksheets called Alternative Workbook Activities. These worksheets cover the same content as the regular Workbook Activities but are written at a second-grade reading level.

Skill Track Software

Use the Skill Track Software for Biology for additional reinforcement of this chapter. The software program allows students using AGS textbooks to be assessed for mastery of each chapter and lesson of the textbook. Students access the software on an individual basis and are assessed with multiple-choice items.

Chapter at a Glance

Chapter 8: Human Body Systems
pages 166–207

Lessons

1. How the Body Digests Food pages 168–172
2. How Materials Move to and from Cells pages 173–180

Investigation 8 pages 179–180

3. How We Breathe pages 181–183
4. How the Body Gets Rid of Wastes pages 184–186
5. How the Nervous System Controls the Body pages 187–191
6. The Sense Organs pages 192–195
7. How the Endocrine System Controls the Body pages 196–198
8. How the Body Moves pages 199–204

Chapter 8 Summary page 205

Chapter 8 Review pages 206–207

Skill Track Software for Biology

Teacher's Resource Library TRL

- Workbook Activities 26–33
- Alternative Workbook Activities 26–33
- Lab Manual 22–24
- Community Connection 8
- Resource File 15–16
- Chapter 8 Self-Study Guide
- Chapter 8 Mastery Tests A and B

(Answer Keys for the Teacher's Resource Library begin on page 420 of the Teacher's Edition. A list of supplies required for Lab Manual Activities in this chapter begins on page 449.)

Science Center

Write the name of an organ or other significant body part (such as artery) on separate pieces of paper. Randomly hand one to each student. Tell students they are going to make a "Body Gallery" (similar to an art gallery) for the classroom. Have them research the body part, make a drawing or model, and write three interesting facts to go with the drawing. As students study the chapters, have them periodically add to their fact list. Alternatively, display several models of body systems and organs. Students may enjoy putting together such models from kits. Encourage students to view the "Body Gallery" often.

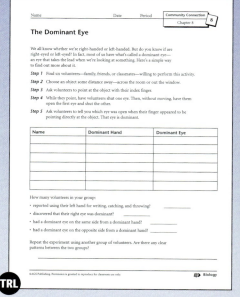

Community Connection 8

166 Chapter 8 Human Body Systems

Chapter

8 Human Body Systems

Imagine being the runner in the photo. The muscles in your legs spring into action when you hear the starting pistol. You feel your feet hit the ground as you move ahead of the other runners. You breathe deeply as air moves into and out of your lungs. You feel your heart pounding. Before long, you begin to sweat. Different systems in your body work together to make all this possible. In this chapter, you will learn about these body systems and what they do. In Chapter 9, you will learn about the reproductive system.

Organize Your Thoughts

- Skeletal — Protects and supports
- Endocrine — Controls activities
- Nervous — Controls activities
- Muscular — Allows movement
- **Body Systems**
- Excretory — Gets rid of waste
- Digestive — Breaks down food
- Circulatory — Carries materials
- Respiratory — Gets oxygen Releases CO_2

Goals for Learning

- ◆ To identify eight systems of the human body
- ◆ To describe the structure and function of each human body system
- ◆ To recognize that body systems work together to carry out basic life activities

167

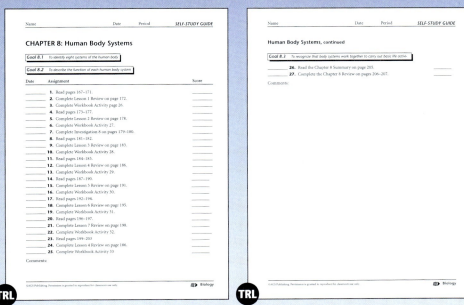

Chapter 8 Self-Study Guide

Introducing the Chapter

Have students look at the runner in the photograph. Ask: Why do you think this photo was chosen to illustrate a chapter entitled "Human Body Systems?" (*Sample answer: A person in action demands more of his or her body than a person at rest.*) Now direct students' attention to the diagram of the eight body systems below the opening paragraph. Invite volunteers to read the name and description of each system, helping them with the pronunciation as necessary. Challenge students to identify which systems are most important to the runner in the photograph as he runs. List their ideas on the board. Point out that this chapter will identify how each body system is essential to human life.

Notes and Technology Notes

Ask volunteers to read the notes that appear in the margins throughout the chapter. Then discuss them with the class.

TEACHER'S RESOURCE

The AGS Teaching Strategies in Science Transparencies may be used with this chapter. The transparencies add an interactive dimension to expand and enhance the *Biology* program content.

CAREER INTEREST INVENTORY

The AGS Harrington-O'Shea Career Decision-Making System-Revised (CDM) may be used with this chapter. Students can use the CDM to explore their interests and identify careers. The CDM defines career areas that are indicated by students' responses on the inventory.

Lesson at a Glance

Chapter 8 Lesson 1

Overview In this lesson, students learn how food moves through the digestive system and how digestive chemicals change food into usable molecules.

Objectives

- To trace the path of food through the digestive system
- To tell how and where digestive chemicals act on food
- To describe what happens to undigested food

Student Pages 168–172

Teacher's Resource Library

Workbook Activity 26

Alternative Workbook Activity 26

Vocabulary

peristalsis	villi
chyme	feces
gallbladder	rectum
bile	

Science Background
The Digestive System

The cells that line the stomach secrete mucus that protects the stomach lining from damage by acid and enzymes. Hydrochloric acid kills most bacteria, softens connective tissue in foods, and makes an acidic environment needed for pepsin (an enzyme) to function. Pepsin digests collagen, which is a protein in meat. When food enters the small intestine, enzymes from the pancreas secrete sodium bicarbonate, which neutralizes the pH in the small intestine. The large intestine is home to millions of bacteria. Some of them produce vitamin K, which is needed for blood to clot.

168 Chapter 8 Human Body Systems

Lesson 1: How the Body Digests Food

Objectives

After reading this lesson, you should be able to

- trace the path of food through the digestive system.
- tell how and where digestive chemicals act on food.
- describe what happens to undigested food.

Digestion

Your body is made up of systems that work together to maintain your good health. One of these systems is your digestive system, which breaks down food for your body to use. Food contains energy for your body's cells. However, this food is too large to enter cells. The food must be broken down into smaller pieces. This process is called digestion. Food contains carbohydrates, proteins, and fats. In digestion, these nutrients are broken down into a form that your cells can use for energy. As you read about digestion, refer to the diagram of the digestive system shown below.

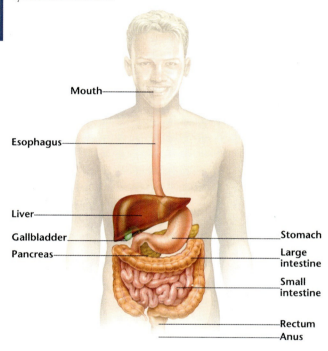

The Digestive System

168 Chapter 8 Human Body Systems

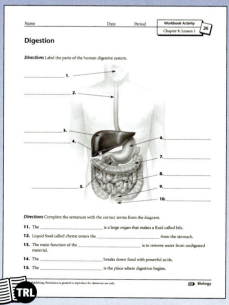

Workbook Activity 26

Peristalsis
The movement of digestive organs that pushes food through the digestive tract

Digestion Begins Inside Your Mouth

Your teeth and jaws chew and crush food while your tongue turns it over. This mechanical action makes pieces of food smaller. As you chew, salivary glands secrete saliva, a fluid that has a digestive enzyme. An enzyme is a protein that causes chemical changes. The enzyme in saliva changes carbohydrates into sugars as you chew. Digestive enzymes help to break down food. Different parts of the digestive system have their own special digestive enzymes.

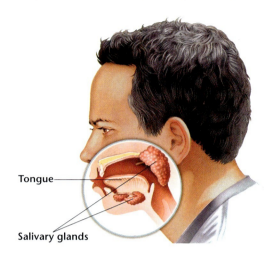

Digestive Organs in the Mouth

The Esophagus

As you chew, food moves around in your mouth. When you swallow, the food moves into your pharynx, or throat. From there, it moves into the esophagus. This long tube connects the mouth to the stomach. Smooth muscles in the esophagus contract, or squeeze together, to push food toward the stomach. This movement is called **peristalsis**.

---

1 Warm-Up Activity

Ask students to recall the definition of digestion from Chapter 1 (the process of breaking down food). Ask where digestion begins in the human body. Students probably will say the stomach. Have a volunteer hold a piece of cracker in his or her mouth for one minute before swallowing it. Have the volunteer describe what happened before he or she swallowed. (*The cracker began to dissolve.*) Explain that the cracker dissolved because digestion begins in the mouth.

2 Teaching the Lesson

Have students divide a sheet of paper in half vertically. They should write *esophagus, stomach, small intestine,* and *large intestine* in the left column. Tell them to leave several lines between each term. As students read, they should write in the right column what happens to food in each organ.

TEACHER ALERT

Students may think that food moves from the mouth to the stomach because of gravity. Explain that peristalsis causes food to move to the stomach. Astronauts who are weightless can eat and digest food.

3 Reinforce and Extend

ONLINE CONNECTION

Students will find useful supplementary information on material discussed in this lesson by going to www.stcms.si.edu, clicking on *Human Body Systems*, and then on *digestive system*. This site, developed specially for middle schools by the National Science Resources Center, refers visitors to lively, reliable Internet sources for all topics covered in this chapter.

LEARNING STYLES

Visual/Spatial
Pour one-half cup of water into each of two bowls. Add one spoonful of cooking oil to each bowl. Add several drops of dishwashing detergent to one of the bowls. Using separate spoons, stir the liquid in each bowl and ask students to describe what they see. (*The oil with the detergent breaks up into tiny globules. The oil without the detergent breaks up into large globules.*) Explain that the detergent acts on the oil as bile acts on fat. Add one drop of food coloring to each bowl to make the globules easier to see.

TEACHER ALERT

Point out that food does not pass through the liver, gallbladder, or pancreas.

Did You Know?

After reading Did You Know? on page 170, show students a sheet of aluminum foil. Fold parts of the foil back and forth in an accordion fashion. Point out how the space taken up by the original sheet has shrunk, but the sheet has the same surface area. Compare the folds to villi.

Chyme
Partly digested liquid food in the digestive tract

Gallbladder
The digestive organ that stores bile

Bile
A substance made in the liver that breaks down fats

Did You Know?
If the villi in the small intestine could be laid out flat, the area would be as big as a baseball diamond.

The Stomach

Digestion continues in the stomach. Strong muscles of the stomach walls contract. This action churns and mixes the food. The stomach walls secrete digestive juices. These juices are hydrochloric acid and digestive enzymes. A special moist lining protects the stomach from being eaten away by the acids. The acid and enzymes break down large molecules of food. Solid food becomes liquid. This liquid food is called **chyme.**

The Small Intestine

Peristalsis squirts chyme from the stomach into the small intestine. The small intestine is a coiled tube that is about 4 to 7 meters long. This is where most digestion takes place.

Two organs and a gland close to the small intestine aid digestion. These are the liver, the **gallbladder**, and the pancreas. The liver makes a fluid called **bile**. Bile breaks apart fat molecules. The gallbladder stores the bile. The bile enters the small intestine through a tube called a bile duct. Glands are organs that produce chemicals for the body to use. The pancreas is a gland that secretes enzymes that complete the digestion of carbohydrates, proteins, and fats.

Villi provide a large surface area through which food molecules can pass into the blood.

Villi
Tiny fingerlike structures in the small intestine through which food molecules enter the blood

Feces
Solid waste material remaining in the large intestine after digestion

Rectum
Lower part of the large intestine where feces are stored

By this time, the food molecules are ready to be absorbed, or taken in, by body cells. They are absorbed through tiny, fingerlike structures called **villi**. Thousands of villi line the small intestine. Villi make the surface area of the intestine larger. Many food molecules can be absorbed through the blood vessels of the villi. Blood carries the food molecules to cells all through the body.

The Large Intestine

Peristalsis moves material that cannot be digested to the large intestine. The main function of the large intestine is to remove water from undigested material. The water is returned to the body. The undigested material forms a solid mass called **feces**. Feces are stored in the **rectum** for a short time. The rectum is the last part of the large intestine. Smooth muscles line the large intestine. They contract and push the feces out of the body through an opening called the anus. The journey of food through your digestive system takes about 24 to 33 hours.

Medical technology is often used to study the digestive system. One test involves taking an X ray of the esophagus, stomach, and small intestine. The person being tested swallows a liquid that coats the lining of these organs. The barium in the liquid makes these organs show up on the X ray. This test, along with others, can help doctors determine the health of the digestive system.

LEARNING STYLES

Logical/Mathematical
In groups or as a class, have students measure out 9 meters of yarn. Have students arrange the yarn to resemble a digestive tract. Use these figures as a guide: esophagus, 23–25 cm; stomach, 20 cm; small intestine, 4–7 meters; large intestine, 1.5 meters.

LEARNING STYLES

Interpersonal/Group Learning
Give small teams of students index cards and ask them to label each card with one part of the digestive system. Have them put all their cards together, face downward, on a flat surface. Each team should designate a member to turn over a card and show it to his or her teammates. Allow members to confer and come up with one fact about that term. Have students continue turning over cards until they have identified all parts of the digestive system and recalled their major functions.

SCIENCE INTEGRATION

Physical Science
Tell students that X rays are a part of the electromagnetic spectrum. The ability of X rays to penetrate soft tissues helps make them useful in the medical field. Bones and dense tumors, which the rays cannot easily penetrate, appear white on an X-ray photograph. Barium sulfate, used to reveal the digestive system, is a metallic compound that will not allow X rays to pass through.

Lesson 1 Review Answers

1. The digestive system breaks down food for your body to use. **2.** Digestion begins inside the mouth. Teeth and jaws chew and crush food, while the tongue turns it over. Enzymes in saliva begin to break down carbohydrates into sugars. **3.** Food moves from the mouth, to the pharynx, to the esophagus, to the stomach, and to the small intestine. Undigested food moves to the large intestine. **4.** The shape of villi makes the surface area of the intestine larger. Many food molecules can be absorbed through the blood vessels of the villi. **5.** Smooth muscles in the large intestine contract and push feces out of the body through the anus.

Portfolio Assessment

Sample items include:
- Table from Teaching the Lesson
- Lesson 1 Review answers

Achievements in Science

Read Achievements in Science on page 172. Some students may be interested to hear more about the curious story of William Beaumont and Alexis St. Martin.

- St. Martin was a nineteen-year-old French-Canadian trapper. He was accidentally shot at close range in the American Fur Company Store.
- The healed wound on St. Martin's stomach was usually closed, but it opened up like a mouth when he ate too much.
- Beaumont paid St. Martin to observe his stomach in action.
- St. Martin outlived the doctor. He died at age 83.

Lesson 1 REVIEW

Write your answers to these questions on a separate sheet of paper. Write complete sentences.

1. What does the digestive system do?
2. Where and how does digestion begin?
3. Describe the path that food takes through the digestive system.
4. Why are villi shaped the way they are?
5. How do feces leave the body?

Achievements in Science

Human Digestion Observed

Today, doctors can view the human stomach with a tiny camera. They understand the digestive process and the way the stomach works. But many years ago, doctors didn't know what the stomach did. Some thought it cooked the food and others thought it ground up the food. The first opportunity for a doctor to actually observe digestion taking place in the stomach occurred by accident.

In 1822, Alexis St. Martin was badly injured by a shotgun. Dr. William Beaumont, an Army doctor assigned to Fort Mackinac, Michigan, took care of him. Even after St. Martin was treated, he had a permanent hole in his stomach. With St. Martin's permission, Dr. Beaumont began experiments.

Beaumont was the first person to observe human digestion as it was occurring. His experiments showed that digestion is a chemical process and that digestive juices need heat to work. Beaumont carefully recorded his observations. He published a book containing the results of his work.

Lesson 2: How Materials Move to and from Cells

Objectives

After reading this lesson, you should be able to
- identify the major parts of the circulatory system and their functions.
- tell how arteries and veins are alike and different.
- trace the flow of blood through the heart.
- describe the parts of the blood and explain their functions.

Cardiac
Relating to the heart

Body cells must have a way to get oxygen and nutrients. They must get rid of wastes. The circulatory system performs these functions. The circulatory system consists of the heart and blood vessels. The heart pumps blood throughout the body through blood vessels. Blood carries food and oxygen to all the body cells. Blood also carries away wastes from the body cells.

The Heart

The main organ of the circulatory system is the heart. The heart is about the size of a human fist. It is located between the lungs in the chest cavity. The heart is the most powerful organ in the body. It is made mostly of muscle tissue called **cardiac** muscle. *Cardiac* means "of the heart" or "relating to the heart." The heart contracts and relaxes in a regular rhythm known as the heartbeat. The heartbeat is automatic. A person does not have to think about it to make it happen.

The heart beats about 70 times a minute in adults who are sitting or standing quietly. That is over 100,000 beats per day pumping more than 7,000 liters per day. The heart beats faster in children and teenagers. The heart beats even faster when a person runs, swims, or does other physical activities. Each time the heart contracts, it squeezes blood out of itself and into blood vessels. Pressure increases inside the walls of certain blood vessels, and they bulge. This bulge can be felt at the wrist and on the side of the neck as the pulse.

Hold two or three fingers of one hand on the thumb side of your other wrist. You can count the number of times your heart beats each minute.

Human Body Systems Chapter 8 173

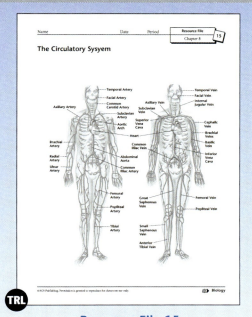

Resource File 15

Lesson at a Glance

Chapter 8 Lesson 2

Overview In this lesson, students learn the parts and function of the circulatory system. They learn how blood flows through the heart and body. They explore the functions and components of blood.

Objectives

- To identify the major parts of the circulatory system and their functions
- To tell how arteries and veins are alike and different
- To trace the flow of blood through the heart
- To describe the parts of the blood and explain their functions

Student Pages 173–180

Teacher's Resource Library TRL
- Workbook Activity 27
- Alternative Activity 27
- Lab Manual 22
- Resource File 15

Vocabulary

cardiac	blood pressure
aorta	plasma
artery	antibody
capillary	hemoglobin
vein	platelet

Science Background
The Circulatory System

Heart muscle does not get its blood supply from blood within the heart. Coronary arteries branch off from the aorta and supply the heart with blood. Diagrams of the circulatory system appear in Appendix C on page 391.

The proteins that determine blood type are antigens, which stimulate antibody production. If type B blood is given to a person with type A blood, the person's body will recognize the type B blood as "foreign" and attack it. This causes the red blood cells to clump. Laboratories mix samples of the donor's and recipient's blood before transfusions, to check for this reaction.

Human Body Systems Chapter 8 173

1 Warm-Up Activity

Tell students to locate a pulse by gently pressing the side of their neck just below the jaw. Ask what causes the bulging movement. (*blood surging through the vessel*)

2 Teaching the Lesson

Have students write a question about what they are to learn in each section following a subhead. As students read, they should write the answer to each question.

Explain that the inner walls of arteries can be narrowed by a buildup of fats. This can lead to an increase in blood pressure. The blood is forced through less space. Tell students to picture water coming from a garden hose. Ask them what happens when they partially cover the end of the hose with their thumb. (*The water sprays out more forcefully and goes farther.*) Demonstrate this if you have access to a faucet and garden hose.

3 Reinforce and Extend.

LEARNING STYLES

Auditory/Verbal
Explain that blood flows through the heart in certain directions because the heart contains one-way valves. The heartbeat can be heard as two separate sounds. The first sound, "lub," is low-pitched and is caused by the contraction of the ventricles and the closing of valves. The second sound, "dub," is shorter and higher pitched. This sound results from the closing of other valves as the ventricles relax. Borrow a stethoscope from the school nurse and have each student listen to his or her heartbeat.

Aorta
A large vessel through which the left ventricle sends blood to the body

How Blood Circulates

Look at the diagram of the heart below. Notice that it has two sides, left and right. Each side has an upper chamber called the atrium and a lower chamber called the ventricle. Use your finger to trace over the diagram as you read how blood moves through the heart.

The right atrium receives blood from the rest of the body. The blood is low in oxygen and high in carbon dioxide. The blood moves into the right ventricle. The right ventricle pumps the blood to the lungs. In the lungs, the blood is filled with oxygen. Carbon dioxide leaves the blood and is exhaled, or breathed out. From the lungs, blood that is high in oxygen goes back to the heart. It goes first to the left atrium and then to the left ventricle. The left ventricle is a thick, powerful muscle. It sends the blood surging out through a large vessel called the **aorta**. The blood then moves to the rest of the body.

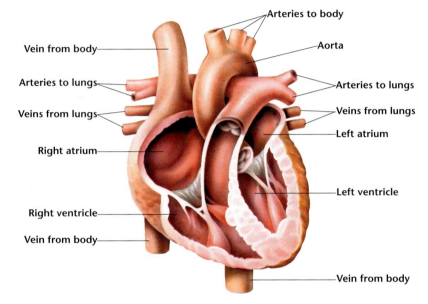

The Human Heart

LEARNING STYLES

Body/Kinesthetic
Provide pairs of students with modeling dough and red and blue pipe cleaners. Have the partners work to construct a heart and lungs from the dough. They should attach the pipe cleaners (red for arteries and blue for veins) to the organs and connect them to one another. Emphasize that the model should be accurate but does not have to be artistic.

Artery
A blood vessel that carries blood away from the heart

Capillary
A blood vessel with a wall one cell thick through which oxygen and food molecules pass to body cells

Vein
A blood vessel that carries blood to the heart

Did You Know?
If all of your blood vessels could be lined up end to end, they would measure about 96,000 kilometers. That is more than twice around the world!

Blood Vessels

Blood travels in only one direction to form a circle pattern. Blood vessels that carry blood away from the heart are called **arteries**. Arteries carry blood full of oxygen. The aorta is the largest artery. The aorta leads to the rest of the body. The arteries become smaller as they move away from the heart. Only one artery carries blood high in carbon dioxide. This artery carries blood from the right ventricle to the lungs.

Tiny arteries branch into blood vessels called **capillaries**. The walls of the capillaries are only one cell thick. Oxygen and food molecules pass easily through the capillary walls to cells. Wastes, such as carbon dioxide, move into the capillaries from cells. The body has so many capillaries that each of its millions of cells is next to a capillary wall.

Capillaries that branched out from arteries join up again to form **veins**. Veins are blood vessels that carry blood to the heart. They carry carbon dioxide from the cells. Veins become larger as they move toward the heart. Pressure from the heart is not as great in veins as in arteries. Veins have one-way valves that keep the blood from flowing backward. Some veins are squeezed by muscles. This helps blood flow back to the heart. For example, when you walk, the muscles in your legs help blood to flow from your legs back to your heart. Only one vein carries blood high in oxygen. This vein carries blood from the lungs into the heart.

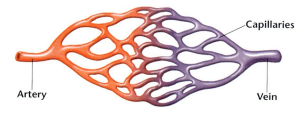

Blood Vessels

Human Body Systems   Chapter 8   175

CROSS-CURRICULAR CONNECTION

 Health
Discuss with students how the heart is vital to human health. Point out that it is also subject to many conditions, making heart disease the number one cause of death in the United States. Have students research the subject of the heart and health. Suggest that they conduct an Internet search using the keyword phrases *heart disease* or *heart health*. Have them print out information from a site and share their findings with classmates.

LEARNING STYLES

 Interpersonal/ Group Learning
Cover the labels on a copy of the figure on page 174 with sticky notes. Have students identify parts of the heart. Invite volunteers to draw a master diagram on the board and to draw arrows showing the flow of blood to and from the heart.

Did You Know?
Read Did You Know? on page 175. Explain to students that an English doctor named William Harvey discovered the principles of human circulation. In 1628, after years of experimenting on living animals and cadavers, Harvey published his findings. The heart was a pump, he claimed, sending the blood in a continuous loop through arteries and veins. One key aspect of the circulatory system Harvey did not detect: The microscopes he used were not powerful enough for him to detect the tiny capillaries.

LEARNING STYLES

Body/Kinesthetic
Invite the school nurse to explain the importance of healthy blood pressure levels and to demonstrate the use of the blood pressure monitor. Have student volunteers, under supervision, practice taking each other's blood pressure.

LEARNING STYLES

LEP/ESL
Write the word *antibodies* on the board. Circle the prefix *anti-* and ask students what they think it means. (*against; opposed to*) Explain that antibodies act *against* foreign or harmful bodies in the blood. Divide the class into small groups. Ask them to find three words that use the prefix *anti-* in this manner in a dictionary. Encourage students learning English to act as group scribes, writing down the words and their definitions. Invite groups to present their words to the class.

Blood pressure
The force of blood against the walls of blood vessels

Plasma
The liquid part of blood

Antibody
A protein in plasma that fights disease

Hemoglobin
A substance in red blood cells that carries oxygen

Blood Pressure

Blood pressure is the force of blood against the walls of blood vessels, usually arteries. When your heart beats, blood is pushed into the arteries. This forces the artery walls to bulge for a moment. The heart works hard to pump blood so that it reaches every part of the body. If blood vessels lose their ability to stretch, blood pressure rises. If blood pressure goes up too high, the heart can be damaged.

Blood and Its Parts

Blood does more than deliver oxygen and carry wastes to the lungs. As you recall from Lesson 1, it also delivers nutrients from the digestive system to cells. Food molecules pass through blood vessels in the villi. Blood carries waste products to the kidneys. You will learn more about that in Lesson 4. Blood also contains materials that fight infections and heal wounds.

Blood has a liquid part and a part that contains cells. The liquid part of blood is called **plasma**. Plasma is mostly water. It makes up half of your blood. Plasma contains dissolved substances including oxygen, food molecules, minerals, and vitamins.

Some proteins in plasma fight disease. These proteins are called **antibodies**. Antibodies help a person to be immune to, or resist, disease. Antibodies fight harmful microorganisms and cancer cells. Your body makes antibodies against these foreign substances. It makes one kind of antibody for each different foreign substance.

Red blood cells make up almost half of the blood. These cells are filled with **hemoglobin**, which is a protein that carries oxygen in the blood. Oxygen plus hemoglobin gives blood its bright red color. Hemoglobin easily gives up oxygen to the cells. It picks up carbon dioxide. In the lungs, hemoglobin exchanges carbon dioxide for oxygen.

White blood cells are larger than red blood cells. However, they are fewer. There is only about one white blood cell for every 700 red blood cells. Like antibodies, white blood cells protect

Platelet
A tiny piece of cell that helps form clots

the body against foreign substances. White blood cells move through the walls of capillaries to where they are needed. Some wrap themselves around invaders and trap them. Others make chemicals that kill harmful germs.

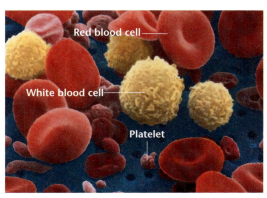

Find the red blood cells, white blood cells, and platelets.

Platelets are tiny cell pieces that help blood to clot. A clot is a thick mass of blood. Platelets do not have a regular shape like red blood cells and white blood cells. Platelets collect at the place where skin is cut. They stick to each other and to the broken blood vessel. Red blood cells stick to the platelets. This mass of platelets and red blood cells forms a clot.

Blood Types

People have one of four blood types. These are type A, type B, type AB, and type O. Different blood types are caused by different proteins in the red blood cells. Sometimes people who are injured or who have certain illnesses need blood. They may receive a blood transfusion to replace lost blood. Blood banks are set up to provide all types of blood. Before a transfusion, health care workers must identify the patient's blood type. If the wrong blood type is given, the person's blood clumps. This blocks the tiny capillaries. Oxygen does not get to cells. Without oxygen, cells die.

CROSS-CURRICULAR CONNECTION

Language Arts

Ask students to imagine that they have been shrunk to the size of blood cells and given a tiny submarine in which to ride through the body's veins and arteries. Have them use the information in this lesson to write an account of their adventures. Encourage them to write vividly, describing the dangers and challenges they encounter. Invite volunteers to share their fantastic voyages with the class.

IN THE ENVIRONMENT

Mosquito-borne diseases include encephalitis, malaria, West Nile Virus, and yellow fever. They enter the body through the bloodstream. Because mosquitoes can be carriers of diseases and spread them to animals they bite, it is important to control the mosquito population. Encourage students to find out ways mosquito populations can be controlled, such as using larvicides and pesticides, draining standing water, and cleaning areas to prevent standing water. They can use reference books or contact a mosquito abatement district in their area for information.

Lesson 2 Review Answers

1. The two main parts are the heart and the blood vessels. 2. Arteries carry blood away from the heart. Veins carry blood to the heart. 3. Blood delivers oxygen and carries wastes to the lungs; delivers nutrients from the digestive system to cells; carries waste products to the kidneys; and contains materials that fight infections and heal wounds. 4. Red blood cells carry oxygen. White blood cells fight disease. Platelets help blood to clot. 5. If you need blood and are given the wrong blood type, your blood may clump.

Portfolio Assessment

Sample items include:
- Answers to questions in Teaching the Lesson
- Stories from Cross-Curricular Connection, page 177
- Lesson 2 Review answers

Achievements in Science

Students may be interested to learn that the first blood transfusion known to be successful occurred 1665 in England. The doctor was Richard Lower; his "patient" was a dog. Lower first drew off so much blood that the dog nearly died. Then he transferred blood, using a silver tube, from another dog fastened beside the first one. The weak dog regained strength "with no sign of discomfort or of displeasure," according to Lower.

Lesson 2 REVIEW

Write your answers to these questions on a separate sheet of paper. Write complete sentences.

1. What are the two main parts of the circulatory system?
2. What is the difference between arteries and veins?
3. What does blood do?
4. What are the functions of red blood cells, white blood cells, and platelets?
5. Why is it important to know your blood type?

Achievements in Science

Human Blood Groups Discovered

When people were given blood transfusions before the early 1900s, they often died because they were not given the correct blood. Doctors didn't know about blood groups. In 1901, an important discovery was made that has helped to save lives.

Austrian doctor Karl Landsteiner discovered that people have different blood groups. He identified these groups as A, B, and O. Later the blood group AB was discovered. With this information, doctors could make sure that blood types matched during blood transfusions. In 1930, Landsteiner received a Nobel Prize for his discovery of blood groups.

Identifying the different blood groups has increased the safety of transfusions and has saved many lives. Today, there are more than four million people in the United States who get blood transfusions each year.

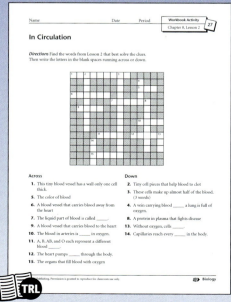

Workbook Activity 27

INVESTIGATION 8

How Does Exercise Change Heart Rate?

Purpose
Does your heart beat faster when you're very active? In this investigation, you will observe the changes in heart rate during different amounts of activity.

Materials
- watch or clock with a second hand
- graph paper

Procedure

1. Copy the table below on a sheet of paper.

2. Sit quietly for three minutes. Then find your pulse as shown on page 173.

3. Take your pulse. To do this, count the number of times you feel your pulse for 15 seconds. Multiply this number by 4. Your answer is your pulse for one minute. Record this number as your resting heart rate.

Activity	Heart Rate
Sitting (resting heart rate)	
Standing up	
After running in place	
Resting 30 seconds	
Resting 1 minute	
Resting $1\frac{1}{2}$ minutes	
Resting 2 minutes	
Resting $2\frac{1}{2}$ minutes	
Resting 3 minutes	

Investigation 8

This investigation will take approximately 30 minutes to complete. Students will use these process and thinking skills: measuring, observing, collecting and interpreting data, inferring, and recognizing time relationships

Preparation
- Make sure stopwatches or watches or clocks with second hands are available for students' use.
- Students may use Lab Manual 22 to record their data and answer the questions.

Procedure
- Students should individually conduct this investigation.
- Show students how to feel a pulse on the wrist.
- Make sure students do not use their thumbs to take their pulse. The thumb also has a pulse.
- If students have difficulty feeling a pulse on their wrist, have them gently press the carotid artery on the side of the neck.
- Tell students not to count aloud. This distracts others.

SAFETY ALERT
- Caution students not to press too hard on the carotid artery.
- Students who have a health condition should not perform step 5.

Lab Manual 22

Results
Heart rates will vary, but students should find that their heart rate is lower when at rest than immediately after exercise. After several minutes following the exercise, the heart rate should return to resting rate.

Questions and Conclusions Answers
1. Heart rate increases with amount of activity.
2. When the heart rate was lowest, the heart was not beating as fast. When the heart rate was highest, the heart was beating faster than at any other time. When the heart rate was lowest, the activity level was low.
3. The heart rate increases as activity increases because your body needs more oxygen delivered to cells. Cells need more oxygen to release energy from glucose. The energy is needed for movement.

Explore Further Answers
Investigations will vary, but make sure students have a clear purpose and can obtain measurable results.

Assessment
Check data tables to be sure the different pulse rates correspond to different amounts of activity. Graphs should show a direct relationship between amount of activity and pulse rate. You might include the following items from this investigation in student portfolios:
- Graph
- Answers to Questions and Conclusions and a written proposal for the Explore Further investigation

---

4. Stand up and immediately take your pulse again. Record this number.
5. Run in place for 200 steps. Then immediately take your pulse and record it.
6. Sit quietly for 30 seconds and take your pulse again. Record this number.
7. Repeat step 6 until you have taken your pulse every 30 seconds for three minutes. Remember to record your data.
8. Graph all of your data. Set up the graph so that time is on the *x*-axis (the line that runs the same direction as the bottom of the page). Put your heart rate on the *y*-axis (which runs the same direction as the side of the page).

Questions and Conclusions
1. How does the amount of activity affect heart rate?
2. Describe the demands on your heart when the heart rate was lowest and when it was highest.
3. Why does the heart rate change as the amount of activity changes?

Explore Further
Design an investigation about heart rate. You could investigate one of the following questions or one of your own: What activities increase heart rate the most? How much do people's resting heart rates differ? Once you choose a purpose for your investigation, write a procedure and do the investigation.

Lesson 3: How We Breathe

Objectives

After reading this lesson, you should be able to
- identify the function and parts of the respiratory system.
- describe the process of gas exchange in the lungs.
- explain how the diaphragm moves when a person breathes.

Pharynx
The passageway between the mouth and the esophagus for air and food

Larynx
The voice box

Trachea
The tube that carries air to the bronchi

Bronchus
A tube that connects the trachea to a lung (plural is bronchi)

Bronchiole
A tube that branches off the bronchus

The Respiratory System

The function of the respiratory system is to get oxygen into the body and to get rid of carbon dioxide. Lungs are the most important organs in the respiratory system. They connect your body to the outside air. The circulatory system then carries the oxygen from your lungs to the rest of your body. You have two lungs. One is in the right side of your chest, and one is in the left side. Your heart lies between them.

How Air Moves to the Lungs

Look at the diagram on page 182. Air comes into your body through your nose and mouth. It travels through the **pharynx**. Air and food share this passageway. From the pharynx, the air moves through the **larynx**, or voice box. A flap of tissue covers the larynx when you swallow. This tissue prevents food from going into your airways. From the larynx, air moves into a large tube called the **trachea**, or windpipe. The trachea branches into two smaller tubes called **bronchi**. One bronchus goes into each lung. In the lungs, each bronchus branches into smaller tubes called bronchial tubes. The bronchial tubes continue to branch and become **bronchioles**.

Respiration

At the end of each bronchiole are tiny sacs that hold air. They are so small that you need a microscope to see them. These microscopic air sacs are called alveoli. The walls of alveoli are only one cell thick. They are always moist. Many tiny capillaries act like nets and wrap around the alveoli.

Blood returning to the heart from the rest of the body is full of carbon dioxide. The right ventricle pumps blood through an artery to the lungs. Recall that this is the only artery in the body that is high in carbon dioxide instead of oxygen. It is called the pulmonary artery. *Pulmonary* means "lung." The carbon dioxide passes out of the capillaries around the alveoli. Carbon dioxide leaves the body when you exhale, or breathe out.

Human Body Systems Chapter 8 181

Lesson at a Glance

Chapter 8 Lesson 3

Overview In this lesson, students learn the parts and function of the respiratory system. They explore how gas exchange occurs and what causes a person to breathe.

Objectives
- To identify the function and parts of the respiratory system
- To describe the process of gas exchange in the lungs
- To explain how the diaphragm moves when a person breathes.

Student Pages 181–183

Teacher's Resource Library TRL
- Workbook Activity 28
- Alternative Activity 28
- Lab Manual 23

Vocabulary

pharynx	bronchus
larynx	bronchiole
trachea	alveolus

Science Background
The Respiratory System

The cells of alveoli are only one cell layer thick. Oxygen and carbon dioxide diffuse in and out, respectively, through passive diffusion. If the level of carbon dioxide in the blood becomes too high, the brain stem automatically increases the rate of breathing. The diaphragm is not the only muscle that helps in breathing. Intercostal muscles extend from the lower border of each rib to the upper border of the rib below. They assist the diaphragm.

1 Warm-Up Activity

Ask students to count the number of times they breathe per minute. Ask why they would breathe faster if they went up a mountain about 2,400 meters. (*The amount of oxygen decreases with increasing altitude. A breath at high altitude does not deliver the same amount of oxygen to the lungs.*)

Lab Manual 23, pages 1–2

2 Teaching the Lesson

Explain to students that when they finish reading the lesson, they should make a drawing they could use to explain part or all of the lesson's information to another person.

TEACHER ALERT

 Students may confuse cellular respiration, respiration, and breathing. In cellular respiration, oxygen joins food molecules to release energy and carbon dioxide. This process occurs in cells. Respiration is the exchange of gases between the organism and its environment. Breathing is drawing air into the airways and expelling air.

 ## 3 Reinforce and Extend.

LEARNING STYLES

 Body/Kinesthetic
Write the following on separate index cards: *nose, pharynx, larynx, trachea, bronchus, bronchiole, alveolus, capillary, pulmonary vein, left atrium, left ventricle, body, right atrium, right ventricle, pulmonary artery, lungs.* In random order, give one card to each student. Beginning with the nose, have students line up in the order that gases travel through the body.

SCIENCE JOURNAL

 Have students discuss how oxygen travels through the body. Ask them to describe in their Science Journal how a molecule of oxygen travels from the air through the body to cells.

Alveolus
A tiny sac at the end of each bronchiole that holds air (plural is alveoli)

When you inhale, or breathe in, oxygen comes into the lungs. The oxygen moves through the walls of the **alveoli**. Oxygen then moves through the walls of the tiny capillaries and into the blood. The exchange of carbon dioxide and oxygen is called respiration.

A vein carries the blood that is high in oxygen to the left atrium. This vein is called the pulmonary vein. Recall that it is the only vein that carries blood high in oxygen.

Breathing

At rest, you usually breathe about 12 times a minute. With each breath, the lungs stretch and you take in about half a liter of air. A strong muscle below your lungs helps you breathe. This muscle is called the diaphragm. It separates the lung cavity from the abdominal cavity.

Breathing happens partly because the pressure inside the chest cavity changes. When the diaphragm contracts, or tightens, it moves down. The ribs move upward. This movement increases the volume of the chest cavity. Air is inhaled, or pulled in, to fill this larger volume. When the diaphragm relaxes, it moves up. The ribs move downward. This movement reduces the volume of the chest cavity. Air is exhaled, or forced out of the lungs. Just like the heartbeat, breathing is automatic.

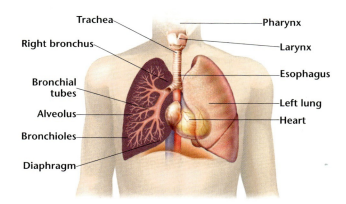

The Respiratory System

CROSS-CURRICULAR CONNECTION

 Health
Have students research information about asthma. Suggest that students find out what can trigger asthma attacks, how asthma is treated, and its effects on breathing.

PRONUNCIATION GUIDE

Use this list to help students pronounce difficult words in this lesson. Refer to the pronunciation key on the Chapter Planning Guide for the sounds of these symbols.

diaphragm (dī ə fram´)
abdominal (ab dä´ mə nəl)

Lesson 3 REVIEW

Write your answers to these questions on a separate sheet of paper. Write complete sentences.

1. What is the function of the respiratory system?
2. Describe the path of air from the nose to the alveoli.
3. What is respiration?
4. Where and how does respiration take place?
5. How does the diaphragm move when you inhale and exhale?

Science Myth

Air exhaled from the lungs contains carbon dioxide but no oxygen.

Fact: The air a person breathes in usually contains about 20 percent oxygen. Only about 5 percent of the oxygen in the air moves through the walls of alveoli in the lungs. The air that is exhaled is about 15 percent oxygen.

Human Body Systems Chapter 8 183

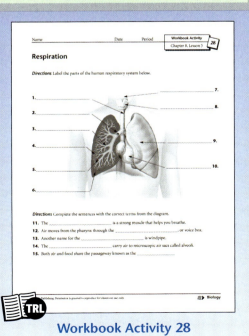

Workbook Activity 28

Lesson 3 Review Answers

1. to get oxygen into the body and get rid of carbon dioxide 2. Air moves from the nose to the pharynx, to the larynx, to the trachea, to the bronchi, to the bronchial tubes, to the bronchioles, and to the alveoli. 3. the exchange of carbon dioxide and oxygen 4. Respiration takes place in the capillaries that surround the alveoli. Oxygen that is breathed in moves through the walls of the alveoli and into the blood. Carbon dioxide from the blood passes out of the capillaries into the alveoli and is exhaled. 5. When you inhale, the diaphragm contracts and moves down. When you exhale, the diaphragm relaxes and moves up.

Portfolio Assessment

Sample items include:
- Drawing from Teaching the Lesson
- Science Journal description
- Lesson 3 Review answers

Science Myth

Ask students if they know which gas comprises the greatest percentage of the air we breathe. (*nitrogen, about 78 percent*) Explain that the percentage of nitrogen in the air is unchanged by being taken into our lungs. Plants and animals are unable to use nitrogen directly from air. Bacteria in soil convert nitrogen gas to compounds that plants can use.

Lesson at a Glance

Chapter 8 Lesson 4

Overview In this lesson, students explore the ways the body gets rid of wastes. They learn the functions of perspiration. They also learn how the excretory system filters and removes wastes.

Objectives

- To explain how perspiration gets rid of wastes
- To describe the function of kidneys
- To explain how urine leaves the body

Student Pages 184–186

Teacher's Resource Library

Workbook Activity 29
Alternative Activity 29

Vocabulary

epidermis	excretory system
dermis	urine
fatty layer	ureter
perspiration	urethra

Science Background
The Excretory System

Kidneys filter blood to eliminate nitrogen wastes from the breakdown of proteins and nucleic acids. Most of these nitrogen wastes are excreted as urea. Kidneys also help regulate blood pressure by balancing the amounts of water and salt in the body. Each kidney has about one million filtering units called nephrons.

Besides producing perspiration that evaporates and cools the body, the skin contains blood vessels that also help regulate body temperature. These blood vessels constrict to prevent heat loss when the temperature is cold. They dilate (enlarge) to release heat when a person becomes hot.

184 Chapter 8 Human Body Systems

Lesson 4 How the Body Gets Rid of Wastes

Objectives

After reading this lesson, you should be able to
- explain how perspiration gets rid of wastes.
- describe the function of kidneys.
- explain how urine leaves the body.

Epidermis
The thin outer layer of skin

Dermis
The thick layer of cells below the epidermis

Fatty layer
Protects organs, keeps in heat

Perspiration
Liquid waste made of heat, water, and salt released through the skin

You know that when you exhale, you get rid of carbon dioxide. Carbon dioxide is one of the wastes that cells make when they use oxygen and food to release energy. Other wastes that your cells make include water, heat, salt, and nitrogen. Exhaling releases some of the extra water and heat.

Perspiration

Many wastes leave your body through its largest organ, the skin. There are three layers of skin. The outer layer, the **epidermis**, is the thinnest. It protects the deeper layers of the skin. The **dermis** is a thicker layer under the epidermis. It contains blood vessels, nerves, and glands. The **fatty layer** protects the body's organs and keeps in heat. Your blood carries heat, water, and salt to sweat glands in your skin. These wastes form a salty liquid called **perspiration**. Thousands of sweat glands in the skin release perspiration through pores onto the skin's surface. Perspiration cools your body as the water evaporates from the skin.

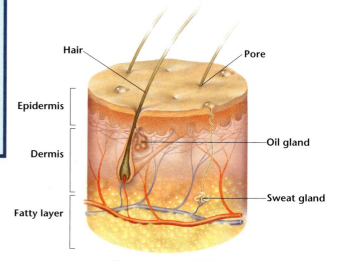

Three Layers of Skin

184 Chapter 8 Human Body Systems

Excretory system
A series of organs that get rid of cell wastes in the form of urine

Urine
Liquid waste formed in the kidneys

Ureter
A tube that carries urine from the kidney to the urinary bladder

Urethra
The tube that carries urine out of the body

The Excretory System

Your cells make nitrogen wastes, which are poisonous. The **excretory system** gets rid of these wastes. The kidneys are the main organs of the excretory system. The body has two kidneys, located in the lower back. Kidneys filter nitrogen wastes out of the blood. The kidneys also remove some extra water and salt from the blood.

The filtered wastes form a liquid called **urine**. Tubes called **ureters** carry urine from each kidney. The urine collects in the urinary bladder. This muscular bag stretches as it fills. When the urinary bladder is almost full, you feel the need to urinate. When you do, the urinary bladder squeezes urine out of your body through a tube called the **urethra**. Follow the path of urine through the excretory system in the diagram.

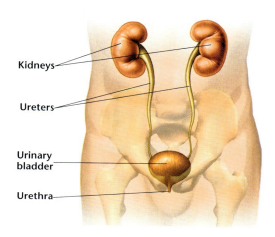

The Excretory System

Science Integration

Earth Science

The heat index describes how hot the air feels. It takes into account the temperature and the relative humidity. Relative humidity is the ratio of the water vapor in the air compared to what it can hold. For example, when the relative humidity of air is 100%, air cannot hold any more moisture. When the air temperature is higher than the body temperature but the air is dry, heat can leave the body through evaporation. However, at a high temperature and relative humidity, a person can become overheated. Perspiration will not evaporate, and body heat will not move into the surrounding hotter air. Have students find a table of the heat index, calculate the heat index for today, and list precautions when the heat index is hazardous.

1 Warm-Up Activity

Dampen a student's forearm. Ask the student to blow on the arm and describe the sensation. (*feels cool*) Ask students what happens when they perspire and a breeze blows on them. (*They feel cooler.*) Explain that perspiration is one way the body cools itself.

2 Teaching the Lesson

Ask students to write this lesson's title and objectives on paper. Tell students that as they read the lesson, they should write sentences that meet the objectives.

Have students compare the removal of nitrogen wastes by the kidneys with the removal of carbon dioxide wastes by the lungs.

Explain that urine moves from the kidney to the urinary bladder by means of peristalsis. Ask them to recall the definition of peristalsis from their study of the digestive system.

TEACHER ALERT

Students may think that they can get water only from beverages. Tell them that many fruits and vegetables contain 80 to 90 percent water.

3 Reinforce and Extend

CROSS-CURRICULAR CONNECTION

Math

Write the following information on the board: Exhaled air—350 mL; Feces—150 mL; Perspiration—500 mL; Urine—1,800 mL. These numbers represent the amount of water lost from the body each day. Have students use the data to construct a circle graph. Have them calculate the number of 8-ounce glasses of water this water loss represents. (*Total of 2,800 mL or 2.8 liters × 1.06 = 3 qt. 3 qt × 32 oz = 96 oz. 96 oz ÷ 8 oz = 12 glasses.*)

Lesson 4 REVIEW

Write your answers to these questions on a separate sheet of paper. Write complete sentences.

1. List four wastes that your cells produce.
2. What is perspiration?
3. What is the function of kidneys?
4. What is urine made of?
5. Describe how urine travels through the excretory system.

Technology Note

Biometrics is a new type of technology that provides unique ways to identify human characteristics. There are a number of biometric technologies, including fingerprint identification, voice identification, and even body odor identification. One company is working on a product that can record your body odor. A sensor is used to "capture" your body odor from your hand.

SCIENCE INTEGRATION

Technology

Write the following equations on the board: bios = life; metrikos = measurement. Point out that these two Greek words combine to make the word *biometrics*, the technology that identifies individuals by measuring and recognizing certain features. Encourage students to discuss the uses (personal and national security) and misuses (invasion of privacy; abuse of power) that are potential in this technology.

Lesson 4 Review Answers

1. carbon dioxide, water, heat, salt, and nitrogen **2.** liquid waste made of heat, water, and salt released through the skin **3.** Kidneys filter nitrogen wastes out of the blood and remove some extra water and salt from the blood. **4.** nitrogen wastes, water, and salt **5.** Urine moves from the kidneys to the ureters, to the urinary bladder, to the urethra, and to the outside.

Portfolio Assessment

Sample items include:
- Sentences from Teaching the Lesson
- Lesson 4 Review answers

Workbook Activity 29

Lesson 5 How the Nervous System Controls the Body

Objectives

After reading this lesson, you should be able to
- identify the structures and functions of the nervous system.
- identify the function of the spinal cord.
- describe how impulses travel.
- explain the purpose of reflex actions.

Your body systems constantly work together to keep you healthy and functioning. For your systems to work, however, they have to be coordinated. All the different parts have to know what to do and when to do it. Your body has to respond to changes in the environment. For example, if you run, the heart has to know to pump faster. Your nervous system coordinates all of your body parts. It is your body's communication network.

The Nervous System

The nervous system is divided into two main parts. The central nervous system is made of the brain and the spinal cord. This system controls the activities of the body. The peripheral nervous system is made of nerves outside the central nervous system. This system carries messages between the central nervous system and other parts of the body.

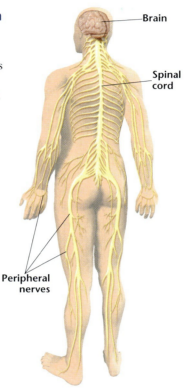

The Central and Peripheral Nervous Systems

Lesson at a Glance

Chapter 8 Lesson 5

Overview In this lesson, students learn how the brain and spinal cord control the activities of the body. They also learn how neurons send messages and control reflex actions.

Objectives
- To identify the structures and function of the nervous system
- To identify the function of the spinal cord
- To describe how impulses travel
- To explain the purpose of reflex actions

Student Pages 187–191

Teacher's Resource Library
  Workbook Activity 30
  Alternative Activity 30

Vocabulary
cerebellum   neuron
brain stem   synapse

Science Background
The Nervous System

Gray and white matter make up the brain. Gray matter is composed of nerve cell bodies. White matter is composed of nerve tracts that enable parts of the brain to communicate with one another. The cerebrum is divided into two hemispheres. Each can function separately. The right hemisphere controls primarily the left side of the body and vice versa. A nerve tract called the corpus callosum connects the two hemispheres.

1 Warm-Up Activity

Ask students if they play team sports or are in a band. Ask them who helps coordinate the players' movements and music. (*coach or conductor*) Ask what would happen if they did not have someone to lead them. Point out that the systems of the body must also be coordinated. Ask if they know how this is done.

Teaching the Lesson

Have students work in groups to draw a human body. Include the brain and other organs. Have them use reference books to map parts of the brain according to the parts of the body they control. They should color the areas of the brain and the corresponding parts of the body the same.

Suggest that students write the labels for each illustration on paper. Encourage them to write the description of each term as they read.

TEACHER ALERT

Make sure students understand that the spinal cord is a soft bunch of nerves protected by the backbone. It is not the backbone itself.

 Reinforce and Extend

SCIENCE INTEGRATION

 Technology

Encourage students to compare the human brain to the computer. Challenge them to find information online or at the library about our natural intelligence and the machines we have created to "think" for us. Ask students to bring to class a list of things that the brain can do better than the computer and vice versa. Discuss their findings and list them on the board. Then have students write a few sentences agreeing or disagreeing with the following statement: *It is possible that computers will control the world in the near future.* Invite students to share their ideas with the class.

Cerebellum
The part of the brain that controls balance

Brain stem
The part of the brain that controls automatic activities and connects the brain and the spinal cord

The Brain

The three parts of the brain are the cerebrum, the **cerebellum**, and the **brain stem**. The largest part is the cerebrum, as the diagram below shows. The cerebrum controls the way you think, learn, remember, and feel. It controls muscles that let you move body parts, such as your arms and legs. It interprets messages from the sense organs, such as the eyes and ears. The cerebrum is divided into two halves. The left half controls activities on the right side of the body. The right half controls activities on the left side of the body.

The cerebellum lies beneath the cerebrum. The cerebellum controls balance. It helps muscles work together so that you walk and write smoothly.

Under the cerebellum is the brain stem. It connects the brain and the spinal cord. The brain stem controls the automatic activities of your body. This includes heart rate, gland secretions, digestion, respiration, and circulation. The brain stem coordinates movements of muscles that move without your thinking about them, such as your stomach muscles.

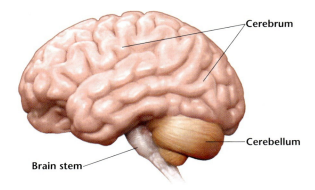

The Brain

Neuron
A nerve cell

Synapse
A tiny gap between neurons

The Spinal Cord

The spinal cord is a thick bunch of nerves that starts at the brain stem and goes down the back. The spinal cord is protected inside a backbone. The brain sends and receives information through the spinal cord. Thirty-one pairs of spinal nerves branch off from this cord. The spinal nerves send nerve messages all over the body. The spinal cord and brain are the central controls of the sense organs and body systems.

Neurons

Nerve cells are called **neurons**. They send messages in the form of electrical signals all through the body. These messages are called impulses. An impulse rapidly carries information from one nerve cell to the next. Neurons do not touch each other. Impulses must cross a small gap, or **synapse**, between neurons. This happens when an impulse travels from one end of a neuron to the other. When the impulse reaches the end of the cell, a chemical is released. The chemical moves out into the synapse and touches the next neuron. This starts another impulse. Information moves through your body by traveling along many neurons.

There are three kinds of neurons in your nervous system. Sensory neurons carry impulses from sense organs to the spinal cord or the brain. Motor neurons carry impulses from the brain and spinal cord to muscles and glands. Association neurons carry impulses from sensory neurons to motor neurons. Dendrites carry messages to the cell body. The axon carries messages away from the cell body.

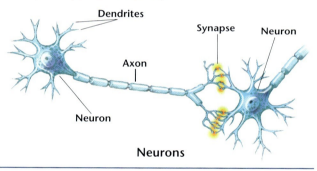

Neurons

GLOBAL CONNECTION

Acupuncture is a form of Chinese traditional medicine. Needles stuck into the skin relieve pain. Acupuncture is increasingly used by Western physicians. Have students research how acupuncture works.

CROSS-CURRICULAR CONNECTION

Art
Have students make models of the brain or neurons. Encourage them to be creative in their use of materials. Apart from clay, they might use foam, beads, pipe cleaners, string, or paper. Point out that many pictures are available in print or on the Internet to supplement the illustrations in the text. Invite students to present their models to the class.

ONLINE CONNECTION

"Created for all students and teachers who would like to learn about the nervous system," *Neuroscience for Kids* is maintained by Professor Eric H. Chudler of the University of Washington in Seattle. Students can find it at http://faculty.washington.edu/chudler/neurok.html. Have students click on *Explore the Nervous System* for information on topics covered in this lesson.

Science in Your Life

Read Science in Your Life on page 190. Ask students to consider why there should be such a number of "smart drugs" on the market. (*People are attracted to the idea of greater brainpower or are afraid of losing their memory as they grow older.*) Extend the discussion by inviting students to describe print or television ads for drugs that they have seen. Lead students to understand that manufacturers of pharmaceuticals often sell their products by showing images of people leading happy, successful lives. Then have students form small groups to create an advertisement that sells a product promising more brains, better looks, or greater strength.

Reflex Actions

Sneezing, coughing, and blinking are reflex actions. They happen automatically. What happens if you touch a hot frying pan? Sensory neurons send the "It is hot!" message to the spinal cord. Inside the spinal cord, association neurons receive the impulses and send them to the motor neurons. All of this happens in an instant, as you feel the heat. You pull your hand away quickly. You have been saved from a serious burn. Many other reflex actions protect the body from injury. For example, if an object comes flying toward your eyes, you blink without thinking.

Science in Your Life

What can smart drugs do?

You may have seen advertisements for "smart drugs" in stores, in magazines, or on the Internet. The drugs may be in the form of pills, drinks, or powders. They may be advertised as herbs, natural ingredients, nutritional supplements, or food additives. The ads might promise improved memory, more alertness, and better performance in school or on the job. But do smart drugs work?

Smart drugs are supposed to work by increasing the amount of blood that flows to the brain. They may also increase the level of neurotransmitters involved in learning and memory. These ideas originated with research on people who had strokes and people who have Alzheimer's disease. People who had a stroke often suffer from memory loss. This is because the blood supply to some parts of the brain is reduced. In Alzheimer's disease, memory loss is due to destruction of neurons in the brain. Research on Alzheimer's continues. So far, a few drugs have been developed that may slow symptoms but can't cure the disease.

Some drugs may help people with diseases affecting the brain. There are few scientific studies that have been done on healthy people. The studies that have been done provide conflicting information. Most scientists agree that, in healthy people, the brain receives enough blood.

Many smart drugs have not been part of reliable scientific experiments. There is no evidence that these drugs will improve memory. Claims are exaggerated or false and do not mention possible side effects of using the drugs. Sorry, but taking a smart drug will not improve your memory or help you pass your next test.

Retina
The back part of the eye where light rays are focused

Optic nerve
A bundle of nerves that carry impulses from the eye to the brain

Eardrum
A thin tissue in the middle ear that vibrates when sound waves strike it

Cochlea
The organ in the ear that sends impulses to the auditory nerve

Auditory nerve
A bundle of nerves that carry impulses from the ear to the brain

Behind the pupil is a lens that focuses light. The lens focuses light rays onto the **retina** at the back of the eye. Receptor cells on the retina send impulses to a nerve bundle called the **optic nerve**. Nerve impulses travel along the optic nerve to the brain. The brain translates the impulses into images you can see. All of this happens faster than you can blink.

The Sense of Hearing

Just as your eyes collect light, your ears collect sound. Review the diagram below as you read about how the ears work. The outer ear acts like a funnel to collect sound waves. The waves travel through the ear canal to the middle ear. The middle ear, just behind the **eardrum,** contains three small bones. The eardrum is a thin tissue that vibrates, or shakes, when sound waves strike it. The sound waves then travel through each of the three bones. The sound waves enter the inner ear. They cause fluid in the **cochlea** to vibrate. The cochlea is a hollow coiled tube that contains fluid and thousands of receptor cells. These cells vibrate when sound waves strike them. The cells send impulses to the **auditory nerve**, which goes to the brain. The brain translates the impulses into sounds you can hear.

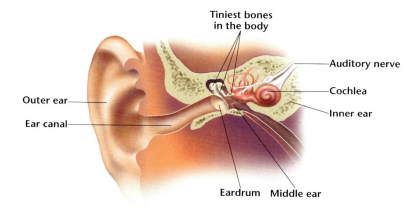

Help students understand that the camera is a form of mechanical eye. A lens lets light through a circular aperture (the pupil), the size of which is controlled by a mechanism (the iris). The image appears on a film (the retina).

Demonstrate how the sense of taste depends on the sense of smell. Provide small pieces of apple and onion for the class or a few volunteers. Do not let students know the identity of the foods. Have them close their eyes, hold their nose, and eat the foods. They should not be able to identify them.

3 Reinforce and Extend

CROSS-CURRICULAR CONNECTION

Language Arts
Have students make up a mnemonic to help them remember how energy travels through the eye and ear. For example, Cows in Portland leap right over bridges (cornea, iris, pupil, lens, retina, optic nerve, brain).

LEARNING STYLES

Auditory/Verbal
Have students form groups of six or seven and ask them to experiment with how we determine the direction of a sound. One student should wear a blindfold; the others should form a circle around him or her. The students in the circle should individually—in random order—make similar sounds (hand clapping or tapping pencils together). Without moving his or her head, the blindfolded student should point in the direction of the sound. Ask students to observe which directions the central student detected most accurately. Have groups repeat this experiment with different students in the center. Challenge groups to come to a conclusion about how we judge the direction of sound. (*It is easier to pinpoint sounds from the sides because our ears pick up the vibrations at slightly different times. Sounds from directly ahead or behind can confuse the listener because they reach the ears at the same time.*)

Learning Styles

Body/Kinesthetic Explain to students the principle of the Braille alphabet. (Patterns of up to six raised dots—arranged in rows like the dots on dominoes—represent letters, groups of letters, or numbers.) Then give groups of students a passage written in Braille and challenge them to take turns "reading" a word with their eyes closed. Ask them to describe the arrangement of dots in the letters they encounter. (If Braille text is unavailable, encourage students to test each other with their own Braille letters, using tiny lumps of modeling clay.)

In the Community

Have students find examples of Braille in public places in their community. Ask them why they think it was used in those locations. Invite students to think of other places where Braille information might be particularly useful to the visually impaired.

Science Journal

Ask students to keep a one-day "senses journal," describing the most memorable sights, sounds, touches, tastes, and smells that they experience during a twenty-four hour period. Have them share their notes in small groups. Encourage groups to comment on any similarities or differences they noticed in the observations students recorded.

The Sense of Touch

The skin receives messages about heat, cold, pressure, and pain. Receptor cells in the skin send nerve impulses to the brain. Then you can tell if something is cold, hot, smooth, or rough. Your fingertips and lips are most sensitive to touch because they have the most receptor cells.

The Senses of Taste and Smell

Taste buds are tiny receptor cells on the tongue that distinguish four basic kinds of tastes. The four tastes are sweet, sour, bitter, and salty. Notice in the diagram that certain parts of the tongue are sensitive to each taste. The taste buds send impulses to the brain. The brain uses these impulses along with impulses from the nose and interprets them as tastes.

Much of the sense of taste depends on the sense of smell. Receptor cells in the nose sense smells. If you hold your nose while you chew, much of your sense of taste goes away. Why does this happen? As you chew and swallow, air carrying the smell of the food reaches the nose. When you hold your nose, the air cannot flow freely. The smells never reach the receptor cells in your nose. The brain doesn't have impulses from the nose to use with impulses from the tongue to interpret taste.

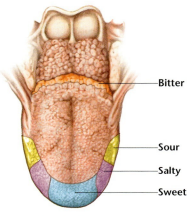

The Tongue

Lesson 6 REVIEW

Write your answers to these questions on a separate sheet of paper. Write complete sentences.

1. Name the three parts of the eye that light passes through in the order it travels from outside the eye to the retina.
2. What happens to light after it reaches the retina?
3. Describe the path of sound waves moving through the ear.
4. Name the four kinds of messages receptor cells in the skin receive.
5. Explain how holding your nose would affect how you taste food.

Lesson 6 Review Answers

1. cornea, pupil, lens 2. Special cells on the retina send information to the optic nerve. Impulses from the optic nerve go to the brain. 3. Sound waves move from the outer ear, through the ear canal, to the eardrum, to each of three bones, and to the cochlea. They are changed to impulses in the cochlea. 4. Receptor cells in the skin receive messages about heat, cold, pressure, and pain. 5. Holding your nose stops air from reaching the receptor cells in your nose. As a result the brain doesn't receive impulses from the nose, and you lose much of your sense of taste.

Portfolio Assessment

Sample items include:
- Mnemonics from Cross-Curricular Connection
- Entry from Science Journal
- Lesson 6 Review answers

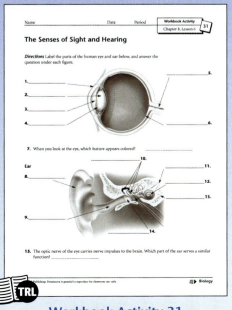

Workbook Activity 31

Lesson at a Glance

Chapter 8 Lesson 7

Overview In this lesson, students learn how hormones work. They also learn about the stress response.

Objectives
- To explain what hormones do
- To explain how a feedback loop works
- To describe the stress response

Student Pages 196–198

Teacher's Resource Library

Workbook Activity 32

Alternative Activity 32

Vocabulary
testis

Science Background
The Endocrine System

Not all glands are part of the endocrine system. Endocrine glands secrete hormones into the bloodstream. The secretions of exocrine glands move through ducts (e.g., sweat glands), or reach cells directly (e.g., pancreas). The pancreas also is considered an endocrine gland because it secretes the hormones insulin and glucagon. Insulin lowers blood glucose levels, while glucagon raises blood glucose levels.

Warm-Up Activity

Ask what comes to mind when you say *hormone*. Some students may answer "sex," "estrogen," or "testosterone." Tell students that some hormones do regulate reproductive activity. However, many other hormones control other body activities.

Teaching the Lesson

While reading this lesson, students should write one topical sentence for each paragraph.

Explain that hormones attach to cells like a key in a lock. Certain cells have molecules to which hormones bind. The molecules must be the right shape for the hormones to "fit."

196 Chapter 8 Human Body Systems

Lesson 7: How the Endocrine System Controls the Body

Objectives

After reading this lesson, you should be able to
- explain what hormones do.
- explain how a feedback loop works.
- describe the stress response.

Testis
The male sex organ that produces sperm cells (plural is testes)

Did You Know?

Scientists have changed some bacteria so that they produce growth hormone. This hormone is used to treat children who do not produce enough of it themselves.

The Endocrine Glands

The endocrine system works closely with the nervous system to control certain body activities. The endocrine system is made of glands. Some of them are circled in the illustration on the next page. These glands secrete substances called hormones. Hormones are chemical messengers. Glands release hormones into the bloodstream. The hormones then travel all through the body.

What Hormones Do

There are more than 30 different hormones. They affect everything from kidney function to growth and development. Hormones work by attaching to certain cells. They change the function of the cells. Some examples of hormones are aldosterone, insulin, and growth hormone. The adrenal glands secrete aldosterone. This hormone helps direct the kidneys to put more sodium and water back into the bloodstream. This may happen when a person has lost fluids. The pancreas secretes insulin. This hormone changes cells so that glucose can enter them. The pituitary gland secretes growth hormone. This hormone causes bones and muscles to grow. In females, the ovaries produce sex hormones. In males, the **testes** produce sex hormones.

The Feedback Loop

Glands must secrete the correct amounts of hormones for the body to work properly. After hormones reach the cells, the cells send a chemical signal back to the gland. That signal tells the gland to continue or to stop secreting the hormone. This process is called a feedback loop.

196 Chapter 8 Human Body Systems

Did You Know?

After reading Did You Know? on page 196, ask students how they think scientists might have "changed some bacteria." Lead them to understand that modifying the genetic makeup of plants, animals, and microorganisms is now an established, and often controversial, field known as biotechnology. Have students bring to class information about the promise—and the dangers—of biotechnology. Encourage them to share their findings with classmates and to discuss the future of this vast scientific frontier.

The pancreas is a part of both the endocrine system and the digestive system. As an endocrine gland, it releases the hormone insulin into the blood. As a gland of the digestive system, it secretes digestive enzymes into the small intestine.

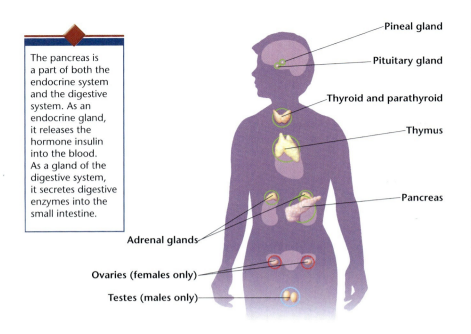

Endocrine Glands

Hormones and Stress

Some people get upset when they have to take a test. Their palms sweat, their heart rate goes up, and they breathe faster. These are signs of the stress response. When a person feels scared or excited, the adrenal glands secrete a hormone called adrenaline. Adrenaline causes these changes in the body. The stress response can be negative if it continues for a long time. A person can become depressed or ill.

However, the stress response can be positive. Suppose you are running a race. The increase in your heart rate causes more oxygen to be delivered to your muscles. Adrenaline also increases the amount of glucose to your muscles. After the race, your body returns to normal.

Students may think that stress is something external that causes them to feel "stressed out." Explain that stress is the response of the body. The source or cause of stress is a stressor. Emphasize that stressors cannot always be controlled. However, students can learn to control their response to stressors. Some ways students can deal with stressors include eating a healthful diet, exercising regularly, spending time with friends, and learning how to handle anger constructively. Ask students to share other ways they deal with stressors. (Coping with stress is discussed further in Chapter 10, page 258.)

3 Reinforce and Extend

CROSS-CURRICULAR CONNECTION

Health

Anabolic steroids are synthetic hormones that are illegal unless obtained by prescription. They are similar to the male hormone testosterone but have many negative effects on the body. Have students research the damage these drugs can do to the body.

AT HOME

The body uses iodine to make thyroid hormones. When the diet lacks adequate iodine, a person's thyroid gland becomes enlarged. The person may experience muscle tremors, increased heartbeat, and skin eruptions. Have students check the labels of their salt containers at home to see if "iodized salt" is listed. Students may wish to research how thyroid hormones affect the body.

Lesson 7 Review Answers

1. It helps to control certain body activities. **2.** chemical messengers secreted by glands **3.** Aldosterone helps direct the kidney to put more sodium and water into the bloodstream. Insulin changes cells so that glucose can enter them. Growth hormone causes bones and muscles to grow. **4.** After hormones reach the target cells, the cells send a chemical signal back to the gland. That signal tells the gland to continue or to stop secreting the hormone. **5.** Answers will vary. A positive example is that the stress response causes more blood and glucose to go to muscles when running a race. A negative example is that a person can become ill if the stress continues for a long time.

Portfolio Assessment

Sample items include:
- Sentences from Teaching the Lesson
- Lesson 7 Review answers

Lesson 7   REVIEW

Write your answers to these questions on a separate sheet of paper. Write complete sentences.

1. What is the function of the endocrine system?
2. What are hormones?
3. Name three hormones and how they affect the body.
4. How does the feedback loop work?
5. Give examples that show how the stress response can be positive and how it can be negative.

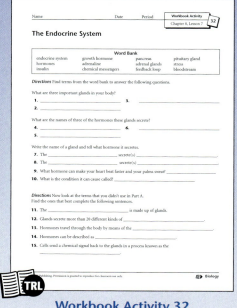

Workbook Activity 32

Lesson 8: How the Body Moves

Objectives

After reading this lesson, you should be able to
- identify five functions of bone.
- explain how bones and muscles work together to produce movement.
- describe the different kinds of muscles.

Skeletal system
The network of bones in the body

The Skeletal System

The 206 bones of the human body make up the **skeletal system**. Bones have several functions. First, bones support the body. They give the body a shape. Bones form a framework that supports the softer tissues of the body. Second, bones protect organs. For example, a rib cage protects the heart and lungs. Vertebrae protect the spinal cord. Vertebrae are the 33 bones that make up the backbone. The pelvis protects reproductive organs. The skull protects the brain. Find these bones in the diagram below.

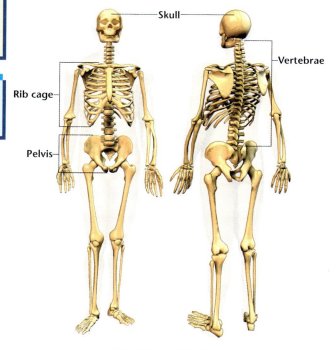

The Skeletal System

Human Body Systems Chapter 8 **199**

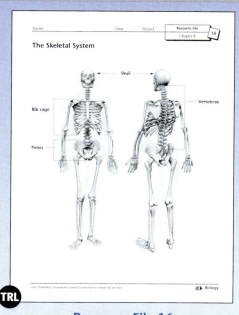

Resource File 16

Lesson at a Glance

Chapter 8 Lesson 8

Overview In this lesson, students explore how bones and muscles produce movement.

Objectives

- To identify five functions of bone
- To explain how bones and muscles work together to produce movement
- To describe the different kinds of muscles

Student Pages 199–204

Teacher's Resource Library

Workbook Activity 33
Alternative Activity 33
Resource File 16

Vocabulary

skeletal system	ligament
red marrow	voluntary muscle
osteoporosis	involuntary muscle

Science Background
The Skeletal System

Collagen fibers make bones flexible, and mineral salts make them hard. A long bone is made of compact bone and spongy bone. Compact bone contains canals through which blood vessels run. Spongy bone is lighter and contains red marrow. It is found at the ends of long bones.

The Muscular System

Nerve impulses initiate contraction of skeletal muscles by releasing a neurotransmitter called acetylcholine. Acetylcholinesterase is an enzyme that deactivates acetylcholine after the muscle contracts. Otherwise, the muscle would remain contracted. Muscles produce heat when they contract. This helps keep body temperature constant.

Human Body Systems Chapter 8 **199**

1 Warm-Up Activity

Bring in a newspaper article or Internet printout about a professional athlete's injury. Discuss how the injury affects the player's movement. Guide students to recall what they know about the bones or muscles involved in the injury.

2 Teaching the Lesson

Ask students to scan the lesson and look for unfamiliar words. Have students work in pairs to write definitions of the words.

Pass around a flashlight and have students briefly flash it (not too closely) into another student's eyes. Tell them to observe how the pupil becomes smaller and then larger. Explain that this is the result of the movement of muscles in the iris. These muscles involuntarily control the amount of light that enters the eye. Have students think of other muscles that are involuntary.

3 Reinforce and Extend

SCIENCE INTEGRATION

Physical Science
Draw students' attention to the fact that bones store calcium, phosphorus, and magnesium. Ask students what these minerals have in common. (If they need a hint, suggest that these substances often appear in a table including oxygen, gold, and uranium.) Lead students to understand that calcium, phosphorus, and magnesium are elements, the building blocks of all matter. Have students bring to class information on the elements that make up the human body. Help students create a pie graph on the board, showing the percentages of elements that combine to create the body.

Red marrow
The spongy material in bones that makes blood cells

Osteoporosis
A disease in which bones become lighter and break easily

Ligament
A tissue that connects bone to bone

A third function of bones is to allow movement. Muscles attach to bones and move them. The body has big bones, small bones, flat bones, wide bones, and bones that have unusual shapes. The variety of bones helps a person to move in different ways. Fourth, bones are the place where blood cells are formed. Bones contain spongy material called **red marrow**. Red marrow has special cells that make blood cells. Finally, bones store minerals, such as calcium, phosphorus, and magnesium.

Cartilage to Bone

Most bones start out as cartilage. Cartilage is a thick, smooth tissue. It is not as hard as bone. Before birth, the entire skeleton is cartilage. It is gradually replaced by bone. A baby is born with more than 300 bones. Over time, some of the bones join so that a person ends up with 206 bones. However, some parts of the body continue to have cartilage. Feel the end of your nose. It is cartilage and will never become bone. Your outer ear contains cartilage. Inside your body, cartilage surrounds your trachea.

How Bones Change

Bones are organs that are made of tissue. They are always changing. Bones are built up and broken down throughout life. This is a normal process. For example, enzymes break down bone tissue when the body needs calcium. Calcium is released into the bloodstream. However, if calcium is not replaced properly, a person can develop **osteoporosis**. With this disease, bones become lighter and break easily. It most often affects older people. Regular exercise and a diet higher in calcium can help prevent osteoporosis.

Joints

Bones come together at joints. Cartilage covers bones at the joints. This cartilage acts like a cushion. It protects bones from rubbing against one another. At the movable joints, strips of strong tissue called **ligaments** connect bones to each other. Ligaments stretch to allow the bones to move.

200 Chapter 8 Human Body Systems

TEACHER ALERT

Students may think that bones are hard and solid all the way through. Bring in a large beef bone that has been cut in half along the length and width by a butcher and boiled. Point out that bones contain soft marrow, blood vessels, and nerves.

There are several kinds of joints. The ball-and-socket joint allows the greatest range of motion. This type of joint is located at the hips and shoulders. It allows you to move your arms and legs forward, backward, side to side, and in a circular motion. The knee joint is the largest and most complex joint. The knee joint is a hinge joint. Some joints, such as your rib and spine joints, can move only a little. A few joints, such as those in your skull, do not move at all.

The Knee Joint

Did You Know?
Some leg muscles are half the length of your leg. Muscles that move tiny bones in your ear are the size of a pin.

The Muscular System

The muscular system consists of the more than 600 muscles in your body. The skeletal and muscular systems work together to produce movement. Tough strips of tissue called tendons attach muscles to bones.

Most muscles work in pairs. When a muscle contracts, or shortens, it pulls on the tendon. The tendon pulls on the attached bone, and the bone moves. A muscle cannot push. Therefore, a different muscle on the opposite side of the bone contracts to return the bone to its starting position.

LEARNING STYLES

Visual/Spatial
Explain that people have long been fascinated with the exterior muscles of the human body. Show students images of powerful men and women in ancient sculpture or fine art. Have students bring to class photographs of athletes, movie stars, or body builders. Discuss with students whether they think there is—or ought to be—an ideal body type for all people.

SCIENCE JOURNAL

Point out that we take for granted most of our body's routine muscular activity. Have students consider what their right (or left) arm would enter in a journal for one typical day in their lives. Remind them of the many activities their arms engage in, from brushing teeth and eating cereal to carrying books and throwing basketballs. Have students read their arm's journal to each other in small groups.

The diagram below shows an example of muscles working in pairs. When you bend your arm, the biceps muscle contracts. You can feel how the muscle shortens and hardens as it contracts. The biceps pulls on the tendon, which pulls your lower arm toward you. The triceps muscle on the underside of your arm is relaxed. It is long and thin. When you straighten your arm, the triceps contracts. It pulls the lower arm back to its starting position. Now the biceps muscle is relaxed.

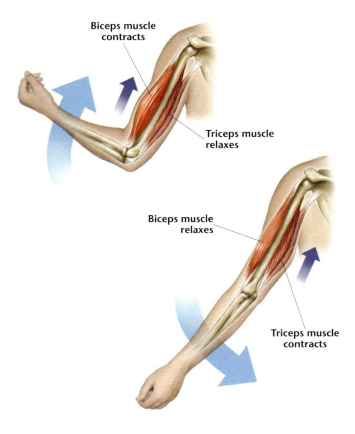

202   Chapter 8   Human Body Systems

Voluntary muscle
A muscle that a person can control

Involuntary muscle
A muscle that a person cannot control

Kinds of Muscle Tissue

The body has three kinds of muscle tissue—skeletal, smooth, and cardiac. Most muscle tissue is skeletal muscle. Skeletal muscles are attached to bones. Skeletal muscles are **voluntary muscles**. That is, you can choose when to use them. The muscles in your arms, legs, and face are voluntary.

The second kind of muscle tissue is smooth muscle. These muscles form layers lining the walls of organs. Smooth muscles are found in the esophagus, stomach, and intestines. These muscles move in wavelike actions to move food through the digestive system. The walls of the blood vessels also are lined with smooth muscles. These muscles contract and relax to maintain blood pressure. Smooth muscles are **involuntary muscles**. You cannot choose when to use them. They react to changes in the body.

The third kind of muscle tissue is cardiac muscle. These muscles make up the heart. They contract regularly to pump blood throughout your body. Cardiac muscles are involuntary.

CROSS-CURRICULAR CONNECTION

Physical Education

Have students go online or to the library to research the major skeletal muscles of the human body. Ask them to find out what exercises trainers recommend to develop these muscles. Invite volunteers to demonstrate some of these exercises for the class.

CAREER CONNECTION

Have a physical therapist visit the class and explain ways to take care of the skeletal and muscular systems. Ask this person to describe what he or she does and how to prepare for a career in the field.

Lesson 8 Review Answers

1. Bones support the body; protect organs; allow movement; form red blood cells; and store minerals. **2.** Before birth, the skeleton is cartilage. It changes to bone by the time the baby is born. Bones are built up and broken down all through life. Bones may get lighter in older people and break easily. **3.** Ligaments connect bones to one another. Tendons connect skeletal muscles to bones. **4.** As muscles on one side of the bone contract, they pull on the attached bone. **5.** skeletal, smooth, and cardiac

Portfolio Assessment

Sample items include:
- Definitions from Teaching the Lesson
- Entry from Science Journal
- Lesson 8 Review answers

Science at Work

Have a volunteer read aloud Science at Work on page 204. Invite the school nurse to talk to the class about the training, responsibilities, and rewards of a nursing career. Help students prepare questions before the visit. Encourage students to discuss their own experiences with nurses at clinics or hospitals.

Lesson 8 REVIEW

Write your answers to these questions on a separate sheet of paper. Write complete sentences.

1. What are the five functions of bone?

2. How does bone change during a person's lifetime?

3. What is the difference between ligaments and tendons?

4. How do muscles make bones move?

5. What are the three kinds of muscle tissue?

Science at Work

Nurse

A nurse needs good observation, decision-making, and communication skills. Nurses also need to keep accurate records, pay attention to detail, and perform medical tests and procedures carefully. The training for a nurse varies from about one year at a vocational school or a community college to four to five years at a university. All nurses must take and pass a state licensing exam.

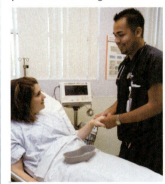

Nurses understand how all the human body systems work. They help people prevent disease and take care of people when they are sick, injured, or disabled. Nurses work in hospitals, doctors' offices, clinics, nursing homes, or other offices. The duties of a nurse include taking measurements such as blood pressure, giving injections, and monitoring.

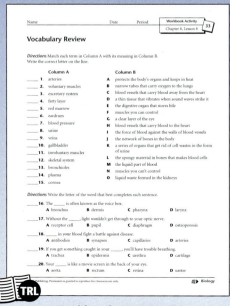

Workbook Activity 33

Chapter 8 SUMMARY

- During digestion, food changes into a form that can enter cells. The large intestine eliminates undigested food.
- The circulatory system moves materials to and from cells.
- Red blood cells carry oxygen in hemoglobin. White blood cells protect the body from disease. Platelets help blood to clot.
- The respiratory system brings oxygen into your lungs and releases carbon dioxide from your lungs.
- The kidneys filter the blood to get rid of toxic wastes. The filtered wastes form urine.
- The nervous system controls and coordinates body activities. Impulses carry information from one nerve cell to the next.
- The five main sense organs are the eyes, ears, skin, nose, and tongue. Special cells in each of the sense organs gather information for the brain.
- The glands of the endocrine system release hormones into the blood. Hormones help control body activities.
- Bones support and protect the body's soft tissues. Blood cells are made inside some bones. Bones also store minerals.
- Skeletal muscles work in pairs to pull on bones.

Science Words

alveolus, 182
antibody, 176
aorta, 174
artery, 175
auditory nerve, 193
bile, 170
blood pressure, 176
brain stem, 188
bronchiole, 181
bronchus, 181
capillary, 175
cardiac, 173
cerebellum, 188
chyme, 170
cochlea, 193
cornea, 192
dermis, 184
eardrum, 193
epidermis, 184
excretory system, 185
fatty layer, 184
feces, 171
gallbladder, 170
hemoglobin, 176
involuntary muscle, 203
iris, 192
larynx, 181
ligament, 200
neuron, 189
optic nerve, 193
osteoporosis, 200
peristalsis, 169
perspiration, 184
pharynx, 181
plasma, 176
platelet, 177
pupil, 192
receptor cell, 192
rectum, 171
red marrow, 200
retina, 193
skeletal system, 199
synapse, 189
testis, 196
trachea, 181
ureter, 185
urethra, 185
urine, 185
vein, 175
villi, 171
voluntary muscle, 203

Chapter 8 Review

Use the Chapter Review to prepare students for tests and to reteach content from the chapter.

Chapter 8 Mastery Test

The Teacher's Resource Library includes two parallel forms of the Chapter 8 Mastery Test. The difficulty level of the two forms is equivalent. You may wish to use one form as a pretest and the other form as a posttest.

Review Answers

Vocabulary Review

1. bile 2. Hemoglobin 3. alveoli 4. iris
5. villi 6. cochlea 7. chyme 8. platelets
9. ligament 10. neurons

TEACHER ALERT

In the Chapter Review, the Vocabulary Review activity includes a sample of the chapter's vocabulary terms. The activity will help determine students' understanding of key vocabulary terms and concepts presented in the chapter. Other vocabulary terms used in the chapter are listed below:

antibody	optic nerve
aorta	osteoporosis
artery	peristalsis
auditory nerve	perspiration
blood pressure	pharynx
brain stem	plasma
bronchiole	pupil
bronchus	receptor cell
capillary	rectum
cardiac	red marrow
cerebellum	retina
cornea	skeletal system
dermis	synapse
eardrum	testis
epidermis	trachea
excretory system	ureter
fatty layer	urethra
feces	urine
gallbladder	vein
involuntary muscle	voluntary muscle
larynx	

Chapter 8 REVIEW

Vocabulary Review

Word Bank
alveoli
bile
chyme
cochlea
hemoglobin
iris
ligament
neurons
platelets
villi

Choose the word or words from the Word Bank that best complete each sentence. Write the answer on a sheet of paper.

1. The liver produces _____, which breaks down fats.
2. _____ carries oxygen in the blood.
3. Oxygen from outside the body enters tiny air sacs in the lungs called _____.
4. The colored part of the eye is the _____.
5. Food molecules enter the blood through _____ that line the small intestine.
6. The _____ sends impulses to the auditory nerve so that you can hear sounds.
7. Partly digested liquid food called _____ passes from the stomach to the small intestine.
8. Tiny pieces of cells called _____ help blood clot.
9. A(n) _____ connects a bone to another bone.
10. Electrical signals travel along cells called _____.

Concept Review

Choose the answer that best completes each sentence. Write the letter of the answer on your paper.

11. The retina sends impulses to the _____ nerve.
    A optic   C motor
    B auditory   D receptor

12. The heart is made of _____ muscle.
    A smooth   C skeletal
    B cardiac   D voluntary

13. You think with your _____.
    A brain stem   C cerebellum
    B spinal cord D cerebrum

206 Chapter 8 Human Body Systems

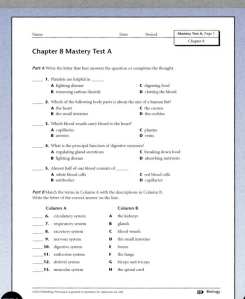
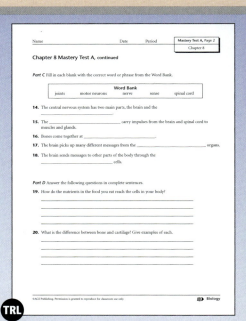

Chapter 8 Mastery Test A

14. Skeletal muscles work by _____.
    A pulling   C peristalsis
    B pushing   D relaxing

15. Perspiration leaves the body through the _____.
    A kidneys   C skin
    B lungs   D feces

16. You get oxygen into your lungs when you _____.
    A swallow   C exhale
    B relax   D inhale

17. Most digestion takes place in the _____.
    A esophagus   C small intestine
    B stomach   D large intestine

18. The _____ of the heart forces blood to the rest of the body.
    A right ventricle   C right atrium
    B left ventricle   D left atrium

Critical Thinking

Write the answer to each of the following questions.

19. The statement "All arteries carry blood that is high in oxygen" is incorrect. Why? Write a statement that describes all arteries.

20. Why do hormones travel in the blood?

Test-Taking Tip Make sure you have the same number of answers on your paper as the number of items on the test.

Concept Review

11. A **12.** B **13.** D **14.** A **15.** C **16.** D **17.** C **18.** B

Critical Thinking

19. The artery that carries blood from the heart to the lungs is low in oxygen. Correct statement: "All arteries carry blood away from the heart." **20.** to reach cells throughout the body

ALTERNATIVE ASSESSMENT

Alternative Assessment items correlate to the student Goals for Learning at the beginning of this chapter

- Encourage students to think of the human body and its eight systems as if it were a consumer product such as a new car or computer. Have students work with small groups of classmates to devise a television infomercial that would "sell" the body to someone unfamiliar with its systems.

- Write the names of the human body systems on scraps of paper and distribute these randomly to students. Students should form groups with classmates having the same topic and work together to come up with three or four reasons why their system is essential to human life.

- Write the heading "Body Systems" in the center of the board and arrange the names of the eight systems in a circle around it. Draw a line between two of the headings and ask, "How are these two systems related?" Help students make connections between the systems indicated. Encourage students to make their own connections and to create a web with more connecting lines.

- Have pair of students compose riddles consisting of five or six short statements that might be spoken by one of the body systems. (For example, the circulatory system might say, "I carry carbon dioxide from the cells.") Each riddle should end with the question, "Who am I?" Invite partners to read their riddles to the class.

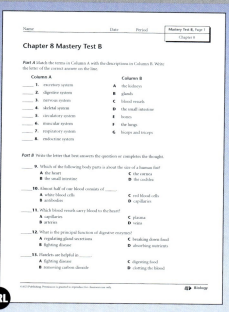

Chapter 8 Mastery Test B

Chapter 9

Planning Guide
Reproduction, Growth, and Development

	Student Text Lesson		
	Student Pages	Vocabulary	Lesson Review
Lesson 1 Where Life Comes From	210–213	✔	✔
Lesson 2 How Organisms Reproduce	214–219	✔	✔
Lesson 3 How Animals Grow and Develop	220–227	✔	✔
Lesson 4 How Humans Grow and Develop	228–236	✔	✔

Chapter Activities

Student Text
Science Center

Teacher's Resource Library
Community Connection 9:
 Observing Adaptations of Animals

Assessment Options

Student Text
Chapter 9 Review

Teacher's Resource Library
Chapter 9 Mastery Tests A and B

	Student Text Features							Teaching Strategies						Learning Styles						Teacher's Resource Library				
Achievements in Science	Science at Work	Science in Your Life	Investigation	Science Myth	Note	Technology Note	Did You Know?	Science Integration	Science Journal	Cross-Curricular Connection	Online Connection	Teacher Alert	Applications (Home, Career, Community, Global, Environment)	Auditory/Verbal	Body/Kinesthetic	Interpersonal/Group Learning	Logical/Mathematical	Visual/Spatial	LEP/ESL	Workbook Activities	Alternative Workbook Activities	Lab Manual	Resource File	Self-Study Guide
				213					211	211, 212			212	212				212		34	34			✔
219							218		217		215		217		216		216, 218	216	35	35	25	17	✔
			226			✔	223	222, 223	222	221, 224	224	223	224			221		222	222	36	36	26, 27	18	✔
	235	236				✔	232	231	234	233, 234			231, 232, 234		230, 234		232	230, 233		37	37			✔

Pronunciation Key

a	hat	e	let	ī	ice	ô	order	ù	put	sh	she		a	in about
ā	age	ē	equal	o	hot	oi	oil	ü	rule	th	thin		e	in taken
ä	far	ėr	term	ō	open	ou	out	ch	child	ŦH	then	ə	i	in pencil
â	care	i	it	ȯ	saw	u	cup	ng	long	zh	measure		o	in lemon
													u	in circus

Alternative Workbook Activities

The Teacher's Resource Library (TRL) contains a set of lower-level worksheets called Alternative Workbook Activities. These worksheets cover the same content as the regular Workbook Activities but are written at a second-grade reading level.

Skill Track Software

Use the Skill Track Software for Biology for additional reinforcement of this chapter. The software program allows students using AGS textbooks to be assessed for mastery of each chapter and lesson of the textbook. Students access the software on an individual basis and are assessed with multiple-choice items.

Chapter at a Glance

Chapter 9: Reproduction, Growth, and Development
pages 208–239

Lessons

1. **Where Life Comes From** pages 210–213
2. **How Organisms Reproduce** pages 214–219
3. **How Animals Grow and Develop** pages 220–227

   Investigation 9 pages 226–227
4. **How Humans Grow and Develop** pages 228–236

Chapter 9 Summary page 237

Chapter 9 Review pages 238–239

Skill Track Software for Biology

Teacher's Resource Library

   Workbook Activities 34–37

   Alternative Workbook Activities 34–37

   Lab Manual 25–27

   Community Connection 9

   Resource File 17–18

   Chapter 9 Self-Study Guide

   Chapter 9 Mastery Tests A and B

(Answer Keys for the Teacher's Resource Library begin on page 420 of the Teacher's Edition. A list of supplies required for Lab Manual Activities in this chapter begins on page 449.)

Science Center

Have students cut out magazine and newspaper pictures that show kinds of organisms and their offspring and place the pictures on a bulletin board. After students have studied the chapter, have them label each image with a term or phrase from the chapter. For example, if a dog is shown with puppies, students may write *sexual reproduction* or *internal fertilization* or *mammal with a placenta* beside the image. Have students try to relate each image to the chapter content.

208 Chapter 9 Reproduction/Growth/Development

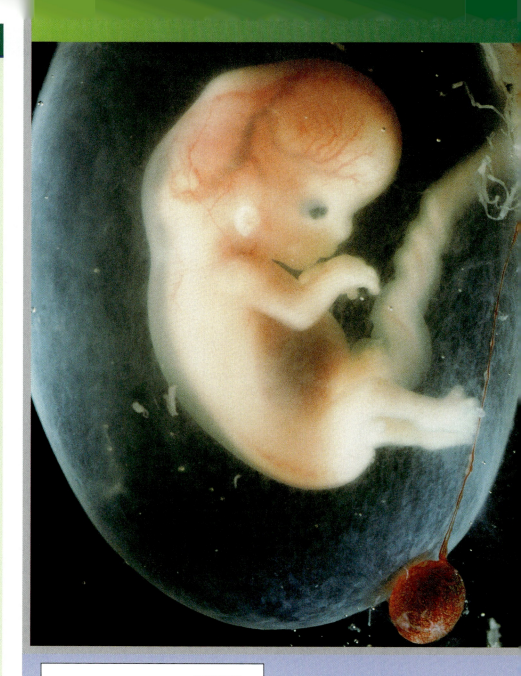

Community Connection 9

Chapter 9: Reproduction, Growth, and Development

You can tell by looking at the photo that the embryo is human. Even at about eight weeks old, you can see many of the features. This embryo and the embryos of other animals are a result of an egg and a sperm combining. Some organisms reproduce without eggs and sperm. In this chapter, you will learn about the advantages and disadvantages of different kinds of reproduction. You also will learn about reproduction and development in different groups of animals.

Goals for Learning

- ◆ To recognize that new life comes from existing life
- ◆ To compare and contrast asexual reproduction and sexual reproduction
- ◆ To describe the processes of reproduction and development in different groups of animals
- ◆ To trace the steps in human reproduction and development

Introducing the Chapter

Direct students' attention to the photograph on page 208. Ask them to identify what they see. When they identify the picture as a human baby, ask how they knew the photograph showed a human baby. Explain that the picture shows a fetus in the womb.

Draw a tiny dot on the chalkboard and tell students that this is about the size they were several days after conception. Tell students that at conception, a fertilized egg is about 0.14 mm in diameter and an average baby is about 30 cm (20 in.) long. Have students average the change in length over the 40 weeks of a typical pregnancy. (*about 7.5 mm per week*)

Have students read the Goals for Learning on page 209 and ask them to rewrite each goal as a question that they should answer as they read the chapter.

Notes and Technology Notes

Ask volunteers to read the notes that appear in the margins throughout the chapter. Then discuss them with the class.

TEACHER'S RESOURCE

The AGS Teaching Strategies in Science Transparencies may be used with this chapter. The transparencies add an interactive dimension to expand and enhance the *Biology* program content.

CAREER INTEREST INVENTORY

The AGS Harrington-O'Shea Career Decision-Making System-Revised (CDM) may be used with this chapter. Students can use the CDM to explore their interests and identify careers. The CDM defines career areas that are indicated by students' responses on the inventory.

Lesson at a Glance

Chapter 9 Lesson 1

Overview In this lesson, students learn that new life arises from existing life. They are also introduced to the broad concepts of genetic materials, reproduction, diversity, and adaptation. This information provides background for the lessons and chapters that follow.

Objectives

- To define spontaneous generation
- To describe how spontaneous generation was proved to be untrue
- To recognize the importance of DNA in reproduction
- To explain the importance of diversity in a population

Student Pages 210–213

Teacher's Resource Library
Workbook Activity 34
Alternative Workbook Activity 34

Vocabulary
spontaneous generation
trait
diversity

Science Background
DNA

Many people may think DNA (deoxyribonucleic acid) was a recent discovery. In fact, DNA was discovered by the Swiss biochemist Friederich Miescher in 1868. Yet, it wasn't until 1944 when a team headed by an American geneticist, Oswald T. Avery, discovered that DNA alone determined heredity.

Although scientists knew that a DNA molecule was made up of phosphate, deoxyribose, and four bases—adenine, cystosine, guanine, and thymine, they did not know how these units fit together. Then in 1953, James Watson, an American, and Francis H. C. Crick of Britain came up with a model of DNA that resembled a twisted ladder. Later experiments and research proved their twisted ladder model to be correct.

210 Chapter 9 Reproduction/Growth/Development

Lesson 1 Where Life Comes From

Objectives

After reading this lesson, you should be able to

- define *spontaneous generation*.
- describe how spontaneous generation was proved to be untrue.
- recognize the importance of DNA in reproduction.
- explain the importance of diversity in a population.

Spontaneous generation
The idea that living things can come from nonliving things

It may seem obvious that flies come from other flies and frogs come from other frogs. At one time, however, people were not sure where some living things came from.

Spontaneous Generation

Many years ago, some people thought that living things could come from nonliving things. This idea is called **spontaneous generation**. When people saw flies come out of rotten meat, they reasoned that rotten meat produces flies. When they saw frogs hop out of muddy ponds, they thought mud makes frogs.

The diagram shows a famous experiment that helped prove that spontaneous generation does not occur. In the 1600s, an Italian scientist named Francesco Redi put pieces of meat in jars. He covered some of the jars with netting. The netting allowed air, but no adult flies, to get in. Redi left the other jars open. After some time, flies appeared only in the open jars.

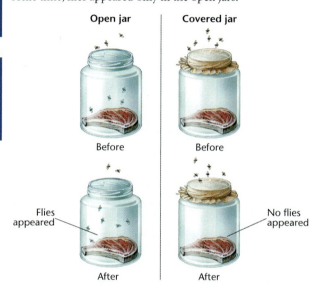

210 Chapter 9 Reproduction, Growth, and Development

Trait
A characteristic of an organism

Later, scientists discovered that fly eggs laid in meat had produced the flies. Scientists also found frog eggs on the edges of muddy ponds, which explained where the frogs came from. Through many experiments over more than 100 years, spontaneous generation was proved not to be true. Living things come only from other living things.

DNA

As you learned in Chapter 1, all living things are made of cells. Each organism's cells contain the information needed to make another organism just like itself. Cells store this information in the form of a chemical called DNA.

DNA stores information in a pattern of chemicals, much as a book stores information in a pattern of letters. The pattern in DNA provides information for making a new cell or a whole new organism. Your DNA contains all the information needed to make a human with your characteristics, or **traits**.

Each type of organism has its own unique, or special, DNA. All humans have information in their DNA about how to make a human, rather than a rabbit or a frog. Rabbit DNA has information about making a rabbit. A frog has DNA for making a frog.

DNA and Reproduction

When living things reproduce, they pass exact copies of their DNA to their offspring. In some organisms, such as bacteria, reproduction usually involves just one parent. The parent passes copies of all of its DNA to its offspring. The offspring is identical, or exactly the same as, the parent because its DNA is identical to the parent's DNA. Most one-celled organisms reproduce this way.

Reproduction in animals and plants usually involves two parents. You may resemble one of your parents more than the other. However, you have traits from both of your parents. That is because your DNA is not an exact copy of DNA from just one of your parents. Your DNA is a combination of both parents' DNA. For that reason, humans, like many other kinds of organisms, produce offspring that are unique.

Reproduction, Growth, and Development Chapter 9   211

CROSS-CURRICULAR CONNECTION

Language Arts

Have students locate and read examples of Greek or Native American myths that explain natural phenomena, such as day and night or stories about the constellations. Point out that myths provided explanations for observations people made. In the absence of scientific theories, the myths made sense. Let students take turns telling the myths they read. Have volunteers describe what observations were being described and what scientific theories might explain it.

 Warm-Up Activity

On the chalkboard write this question: *Where does life come from?* Ask students to write a brief explanation. Let volunteers read their explanation. Summarize each explanation in a phrase and write the phrase on the chalkboard under the question. Leave the question and the summaries on the chalkboard. Return to them at the end of the lesson and have students discuss which ideas were correct.

Teaching the Lesson

Have students scan the lesson and write down the headings. Ask students to write a question that they expect to have answered under each heading. As students read the text, have them look for answers to their questions.

Discuss with students what spontaneous generation is and why it might have made sense to people before experiments disproved it.

Review with students the importance of DNA and its function in the reproduction process. Let students describe instances when they have read or heard about DNA, such as in criminal investigations and in cloning.

Finally let students tell about times they have heard the topic of diversity discussed and in what context. Let students tell what diversity means to them. Then ask why diversity is important in a social context and why diversity is important in a biological context. Be sure students understand that diversity promotes survival of a *population*—not necessarily of an individual.

 Reinforce and Extend

SCIENCE JOURNAL

Ask students to write Redi's experiment as a scientific investigation. Tell them to state the problem, an hypothesis, and a conclusion. Also have students include what the control and the dependent and independent variables were.

Reproduction/Growth/Development Chapter 9   211

LEARNING STYLES

Auditory/Verbal
To reinforce the use of correct reproductive and anatomical terminology, prepare an audiotape to accompany the lessons in the chapter. Pronounce each vocabulary term carefully and give its definition. Place the tape in the science center and encourage students to listen to it as they read and discuss the lessons.

LEARNING STYLES

Visual/Spatial
Emphasize that people are similar and different in many ways. Discuss similarities in needs, feelings, and growth and development. Emphasize differences in interests, dispositions, and talents. Encourage students to make a two-column chart in which they identify how they are alike and different from their classmates.

GLOBAL CONNECTION

Let small groups of students use an almanac or other reference source to find out the current size of the world's population and how fast it is growing. Groups might also want to find out which countries are growing fastest.

The Advantage of Diversity

Diversity
The range of differences among the individuals in a population

Humans and other species that produce unique offspring are said to have **diversity**. Diversity is the range of differences found among the members of a population. You can see diversity in the group of children in the photo.

If a population's surroundings, or environment, change suddenly, the population's diversity can help it continue to survive. Suppose a disease sweeps through a population of deer, killing many of them. If none of the deer are able to fight, or resist, the disease, then the population will die off. However, in a diverse population, chances are that a few deer will be able to resist the disease. As a result, the population will survive. Resistance to the disease is a kind of adaptation. Adaptations are traits that allow organisms to survive in certain environments. Adaptations result from the information in an organism's DNA.

Humans show a great deal of diversity.

Chapter 9 Reproduction, Growth, and Development

CROSS-CURRICULAR CONNECTION

Logical/Mathematical
Have students use Punnett Squares to calculate the likelihood of their inheriting certain traits, such as blue eyes and dark hair. Students can enter a search for Punnett Squares on the Internet or go to the Web site Introducing Punnett Squares at this Web address: www.usoe.k12.ut.us/curr/science/sciber00/7th/genetics/sciber/punnett.htm.

Lesson 1 REVIEW

Write your answers to these questions on a separate sheet of paper. Write complete sentences.

1. What is spontaneous generation?
2. Explain how Redi's experiment helped prove that spontaneous generation does not occur.
3. What is DNA, and what is its function?
4. Why are human offspring unique?
5. What is the advantage of having diversity in a population?

Science Myth

If you look more like one of your parents than the other, you received your DNA for facial features from that parent.

Fact: Your body cells each contain the same 23 pairs of chromosomes. Each chromosome in one pair contains DNA for the same traits. One chromosome in each pair is from your mother, and the other is from your father. You received the same amount of DNA for facial features from each parent.

Lesson 1 Review Answers

1. Spontaneous generation is the idea that living things can arise from nonliving things. **2.** Redi showed that maggots came from eggs laid by adult flies, not from nonliving meat. **3.** DNA is a chemical found in an organism's cells and stores information about the organism. **4.** Human offspring are unique because they have a combination of DNA from each parent. **5.** Diversity increases the likelihood that the population can survive sudden changes in the environment.

Portfolio Assessment

Sample items include:
- Questions and answers from Teaching the Lesson
- Lesson 1 Review answers

Science Myth

Ask volunteers to read the Science Myth on page 213. You might wish to further explain that Gregor Mendel (1822–1884) experimented with garden peas and came up with the idea of dominant and recessive genes. So although humans are a combination of the genes of both parents, it is possible for some traits to be more obvious, or dominant.

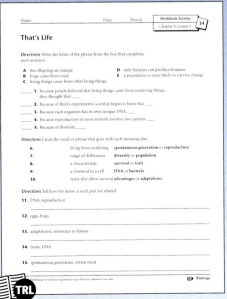

Workbook Activity 34

Lesson at a Glance

Chapter 9 Lesson 2

Overview In this lesson, students learn the advantages and disadvantages of asexual and sexual reproduction. They also learn the differences between mitosis and meiosis and how fertilization takes place.

Objectives

- To recognize the advantages and disadvantages of asexual reproduction
- To recognize the advantages and disadvantages of sexual reproduction
- To compare mitosis with meiosis
- To trace the steps in the fertilization of an egg

Student Pages 214–219

Teacher's Resource Library TRL
- Workbook Activity 35
- Alternative Activity 35
- Resource File 17
- Lab Manual 25

Vocabulary

mitosis	external fertilization
chromosome	
gamete	internal fertilization
meiosis	

Science Background
Mitosis and Meiosis

Cells divide forming new cells. In nucleated cells (eukaryotes), the nucleus divides first, followed by division of the entire cell into two new cells, each with its own nucleus. Mitosis and meiosis refer to the division of a cell's nucleus during cell division.

Mitosis occurs in cells that are multiplying as part of an organism's growth and repair process. Meiosis occurs during the formation of sex cells, or gametes. Mitosis includes the copying of a cell's DNA followed by cell division. Meiosis involves the copying of a cell's DNA followed by two cell divisions.

Lesson 2 How Organisms Reproduce

Objectives

After reading this lesson, you should be able to

- recognize the advantages and disadvantages of asexual reproduction.
- recognize the advantages and disadvantages of sexual reproduction.
- compare mitosis with meiosis.
- trace the steps in the fertilization of an egg.

In Lesson 1, you learned that some organisms need only one parent to reproduce. Other organisms reproduce with two parents. In this lesson, you will learn more about these two types of reproduction.

Asexual Reproduction

Some organisms pass an exact copy of all of their DNA to their offspring. As a result, the offspring are identical to the parent. This form of reproduction is called asexual reproduction. One-celled organisms, such as bacteria and yeasts, usually reproduce asexually. Their cell divides to form two identical cells. Protists, fungi, and some plants and animals can reproduce by means of asexual reproduction, too.

Many of the cells that make up your body also undergo asexual reproduction. For example, skin cells, bone cells, and muscle cells reproduce asexually to form new cells. The new cells allow your body to grow, heal, and replace dead cells.

Advantages and Disadvantages of Asexual Reproduction

One advantage of asexual reproduction is that an organism can reproduce alone. It does not have to find a mate. Another advantage is time. One-celled organisms can reproduce quickly. For example, some bacteria divide every 20 minutes under certain conditions.

A disadvantage of asexual reproduction is that the offspring are exact copies of the parent. The offspring lack diversity. Thus, they are likely to respond to changes in the environment in the same way as the parent. If a change kills one of the offspring, it will probably kill them all. Thus, asexual reproduction is favorable in environments that do not change much.

214 Chapter 9 Reproduction, Growth, and Development

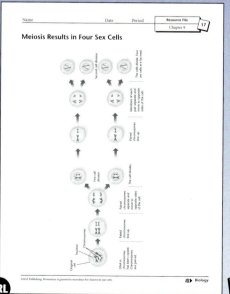

Resource File 17

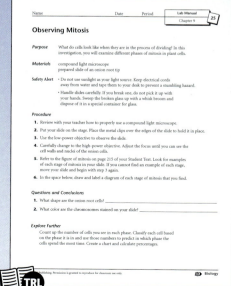

Lab Manual 25

214 Chapter 9 Reproduction/Growth/Development

Mitosis
The process that results in two cells identical to the parent cell

Chromosome
A rod-shaped structure that contains DNA and is found in the nucleus of a cell

Mitosis

In cells that have a nucleus, asexual reproduction occurs in the form of **mitosis**. Mitosis is the dividing of a cell's nucleus. Before a cell undergoes mitosis, it makes a copy of its DNA. The DNA is found in the nucleus in rod-shaped structures called **chromosomes**. When the DNA is copied, the chromosomes form pairs.

The diagram below shows the main steps of mitosis and cell division. During mitosis, the nucleus disappears. The pairs of chromosomes line up in the center of the cell. Then, the members of each pair separate and move to opposite ends of the cell. Next, the cell membrane pinches in between the two sets of chromosomes. A nucleus forms around each set. Two identical cells are formed. Each new cell has an exact copy of the chromosomes that were in the parent cell.

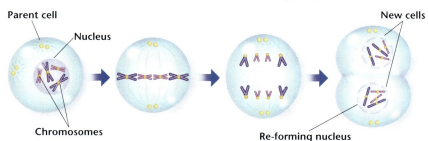

| DNA in chromosomes has been copied. Chromosomes are paired. | Nucleus disappears. Paired chromosomes line up in center of cell. | Members of paired chromosomes separate and move to opposite sides of the cell. | Nuclei re-form. Cell divides into two new cells. |

Mitosis

Sexual Reproduction

Humans and many other kinds of organisms have two parents. They use a form of reproduction called sexual reproduction. During sexual reproduction, a cell from one parent joins with a cell from the other parent. Most plants and animals reproduce by sexual reproduction.

Cell division that involves mitosis results in two new cells that are identical to the original cell. Both new cells contain a complete copy of the original cell's chromosomes; they are called diploid. Cell division that involves meiosis results in four new cells that have half the chromosome number of the original cell. The are said to be haploid.

 Warm-Up Activity

Have students turn back to the photograph on page 208. Ask them how the fetus is able to grow and develop. After a brief discussion, explain that cells must divide for growth to take place and that students will learn about the division of cells in this lesson.

 Teaching the Lesson

After students have read the sections on asexual and sexual reproduction, ask them to contrast the two by writing two statements that explain how the two processes are different.

Use Resource File 17 and page 215 to create an overhead transparency of mitosis and meiosis. Have students compare the two processes on a step-by-step basis, noting where the two differ.

Finally, discuss the two methods of fertilization. Challenge students to come up with a reason why external fertilization may be a good method of fertilization for some animals.

TEACHER ALERT

When analyzing diagrams of cell division, students may confuse chromosome mass with chromosome number. Point out, for example, that the cells resulting from the first cell division in meiosis have half the chromosome number, even though the chromosomes have twice the mass of the chromosomes in the original cell. The second cell division in meiosis restores the chromosome mass to that of the original cell.

Meiosis

| DNA in chromosomes has been copied. Chromosomes are paired. | Paired chromosomes line up. | Paired chromosomes separate and move to opposite sides of the cell. | The cell divides. |

Gamete
A sex cell, such as sperm or egg

Disadvantages and Advantages of Sexual Reproduction

A disadvantage of sexual reproduction is that an organism must find a mate to be able to reproduce. Also, sexual reproduction usually takes longer to produce offspring than does asexual reproduction. However, a big advantage of sexual reproduction is that it leads to greater diversity among offspring. That is because the DNA from two different parents is mixed. Each offspring is unique. Its combination of traits is different from the combination of traits of either parent.

Gametes

Sexual reproduction involves both a female parent and a male parent. The female produces egg cells. The male produces sperm cells. Sperm cells and egg cells are called sex cells, or **gametes**. In many animals, testes are the male sex organs that produce sperm cells. Ovaries are the female sex organs that produce egg cells.

216   Chapter 9   Reproduction, Growth, and Development

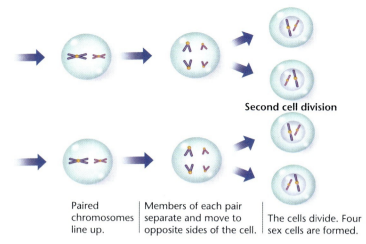

| Paired chromosomes line up. | Members of each pair separate and move to opposite sides of the cell. | The cells divide. Four sex cells are formed. |

Meiosis
The process that results in sex cells

The gametes of certain species contain one-half the number of chromosomes found in the species' nonsex cells. For example, human sex cells have 23 chromosomes in their nucleus. Human body cells have 46 chromosomes. The sex cells of a dog have 39 chromosomes. A dog's body cells have 78 chromosomes.

The testes produce millions of sperm cells. A sperm cell is about one-thousandth the size of an egg cell. A sperm cell usually has a tail that allows it to move toward an egg.

Meiosis

Gametes form by a division of the nucleus called **meiosis**. You can follow the steps of meiosis in the diagram above. As in mitosis, meiosis begins after the cell's chromosomes have been copied. During meiosis, the nucleus disappears. The pairs of chromosomes line up in the center of the cell and then separate. A nucleus forms around each set of chromosomes.

LEARNING STYLES

Auditory/Verbal
Play a game similar to *Jeopardy* with students. Prepare slips of paper with the steps in the process of mitosis and meiosis on them. Have students draw one slip at a time, read it aloud to the group, and respond with "What is mitosis?" or "What is meiosis?" or "What is mitosis and meiosis?"

CROSS-CURRICULAR CONNECTION

Health
Student may believe that human cells only reproduce sexually. Point out that during mitosis, human cells reproduce asexually to help the organism grow and repair itself. Have students work in pairs to list observable situations where asexual reproduction of cells is occurring. (*Possible answers: hair growth after a haircut, skin healing after a cut or scrape, formation of a scar after surgery, growth of fingernails, broken bones healing*)

SCIENCE INTEGRATION

Physical Science

During mitosis and meiosis, cells go through not only physical changes but also chemical changes. Let pairs of students research some of the substances found in DNA that undergo chemical changes. Have pairs share their findings with another pair of students.

LEARNING STYLES

Visual/Spatial

Ask students to read the description of external fertilization on page 218. Have them work in small groups to illustrate the process. Encourage group members to use their pictures as visual reinforcement of an oral presentation describing the process.

External fertilization
The type of fertilization that occurs outside the female's body

Internal fertilization
The type of fertilization that occurs inside the female's body

Next, the cell divides into two new cells. However, in meiosis, unlike mitosis, each new cell divides once again. In this way, one original cell produces four sex cells. Since cell division occurs twice, each sex cell contains one-half the number of chromosomes of the original cell.

Fertilization

Imagine that a female fish has laid her eggs under the water. A male fish swims above the eggs. He releases billions of sperm into the water. The sperm swim through the water toward the eggs. One sperm reaches an egg just ahead of dozens of other sperm. That sperm attaches itself to the outer membrane of the egg. Almost immediately, the egg's membrane changes so that no other sperm can attach.

Once a sperm is attached, its nucleus enters the egg. Then the nucleus of the sperm cell and the nucleus of the egg cell join. Fertilization is the process by which a sperm cell and an egg cell join to form one cell. The cell, called a zygote, begins to develop into a new organism. The zygote has a complete set of chromosomes. One-half of the chromosomes came from the sperm cell. The other half came from the egg cell.

The process described above is called **external fertilization** because the egg is fertilized outside the body of the female. Most fish and amphibians use external fertilization. All reptiles, birds, and mammals use **internal fertilization**. In those animals, the male places sperm inside the female's body, where fertilization occurs.

Lesson 2 REVIEW

Write your answers to these questions on a separate sheet of paper. Write complete sentences.

1. What are two advantages of asexual reproduction?
2. What is the result of the process of mitosis?
3. What is the main advantage of sexual reproduction?
4. What is the result of the process of meiosis?
5. Describe the process of fertilization.

Achievements in Science

Mammal Egg Discovered

Before the early 1800s, no one had observed a mammal egg. Scientists had many different beliefs about how mammals, including humans, reproduced. Some scientists believed that a woman's body contained tiny humans that were already formed. Others scientists thought that each sperm contained a tiny human.

In 1827, Karl Ernst von Baer discovered that mammals have eggs that develop into new organisms of the same kind. He made other discoveries, including identifying parts of an embryo that develop into the organs and systems of adult animals. He made many discoveries by comparing the embryos of different animals. His work established the new science of embryology.

Reproduction, Growth, and Development Chapter 9

Lesson 2 Review Answers

1. Asexual reproduction does not require a mate and produces offspring more quickly than sexual reproduction. **2.** The process of mitosis results in two cells that are identical to the parent cell. **3.** Sexual reproduction results in unique offspring. **4.** The process of meiosis results in four sex cells. **5.** A sperm cell inserts its nucleus into an egg cell. The nuclei of the sperm cell and the egg cell join, forming a zygote with a full set of chromosomes.

Portfolio Assessment

Sample items include:

- Contrast statements from Teaching the Lesson
- Graphic organizer from Learning Styles: Visual/Spatial
- Lesson 2 Review answers

Achievements in Science

Read the Achievements in Science feature on page 219 together. Point out that just as there were misconceptions about where life came from, resulting in the idea of spontaneous generation, there were also misconceptions about how humans reproduced. From as early as the 1600s, scientists searched for mammal eggs by looking in the uterus. Others mistook ovarian follicles for eggs. Baer found the first egg in a dog.

Workbook Activity 35

Lesson at a Glance

Chapter 9 Lesson 3

Overview This lesson provides an overview of animal growth and development. Students learn about cell differentiation and life cycles. This lesson also includes information about mammal development and gestation.

Objectives

- To describe how a zygote becomes an embryo
- To compare the development of different kinds of animals
- To compare the gestation times of different animals

Student Pages 220–227

Teacher's Resource Library TRL

　Workbook Activity 36
　Alternative Activity 36
　Resource File 18
　Lab Manual 26, 27

Vocabulary

embryo
cell differentiation
nymph
marsupial
uterus
placenta
gestation time

Science Background
Marsupials

Marsupials, such as kangaroos, are unique among mammals because their young are born extremely underdeveloped. After birth a newborn kangaroo wiggles along its mother's fur up to the pouch where it attaches itself to a nipple and stays attached until it no longer needs the mother's milk.

Marsupials are found only in Australia and the Americas. Marsupials in Australia include the kangaroo, koala, possums, wombats, bandicoots, and Tasmanian devils. Marsupials in the Americas include opossums and shrew opossums.

Lesson 3 How Animals Grow and Develop

Objectives

After reading this lesson, you should be able to

- describe how a zygote becomes an embryo.
- compare the development of different kinds of animals.
- compare the gestation times of different animals.

A new animal begins as a zygote. Recall that a zygote is a single cell that contains a complete set of chromosomes. Usually, the zygote divides to form two identical cells attached to each other. Then those two cells divide. This process is repeated many times, as shown in the diagram below. Eventually, the zygote divides into millions of cells that make up an **embryo**. An embryo is an early stage in the development of an organism. Remember that as the cells multiply to form a new organism, the same DNA is copied in each cell.

Differentiation

An embryo's cells gradually take on different shapes and functions. This process is called **cell differentiation**. In vertebrates, cells at one end of the developing embryo begin to form parts of the head. The cells form the eyes, mouth, gills, and other organs. Inside, the heart, stomach, and other organs develop. Eventually a complete organism forms.

Embryo
An early stage in the development of an organism

Cell differentiation
The process of cells taking on different jobs in the body

Development

The embryos of many animals, such as fish, reptiles, and birds, develop inside eggs. A fish may lay thousands of eggs. Turtles, birds, and other fish feed on those eggs. Laying large numbers of fish eggs increases the chances that a few eggs will live. This, in turn, increases the chances that the species of fish will survive.

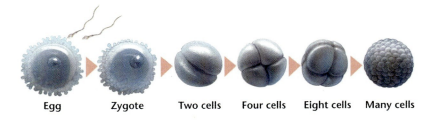

Egg → Zygote → Two cells → Four cells → Eight cells → Many cells

From Zygote to Embryo

220 Chapter 9 Reproduction, Growth, and Development

Nymph
A young insect that looks like the adult

In most species of fish, when the young hatch from the eggs, they look like the adult fish. Certain insects, such as grasshoppers and praying mantises, also produce young that look like the adults. These young insects are called **nymphs**. However, recall from Chapter 3 that butterflies produce young that are completely different from the adults. The young develop into adults in stages that vary greatly in form. In the diagrams, you can see the differences in the developments of a grasshopper and a butterfly. Grasshopper nymphs look like adult grasshoppers but cannot reproduce. Caterpillars look nothing like the butterflies into which they will develop.

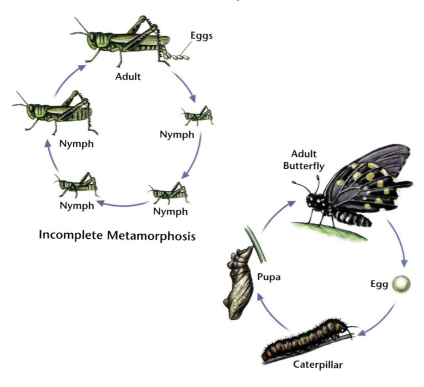

Incomplete Metamorphosis

Complete Metamorphosis

Reproduction, Growth, and Development Chapter 9 **221**

1 Warm-Up Activity

Ask students to make two lists. In the first list, have them identify at least five ways they are different now than they were at birth. In the second list, have them identify how they will be different fifty years in the future. When lists have been compiled, let volunteers share an item from each list.

2 Teaching the Lesson

Living things go through a cyclic development. Let students name some animals, including insects, and describe their stages of growth and development.

Write the question *Why do some animals produce many offspring and others produce few?* Have students write a hypothesis. Let volunteers read their ideas.

Ask students to speculate as to why the gestation period for some animals is short and for some animals long. Have students give examples of animals that that fit in each category. Be sure students tell why they think an animal belongs in a particular group.

3 Reinforce and Extend

CROSS-CURRICULAR CONNECTION

Language Arts
Have students work with a partner to write sentences that link the following pairs of terms from the lesson: *sperm/egg, fertilization/zygote, uterus/placenta, marsupial/mammal, gestation/development*. Have students share their sentences with the class. Let students help each other correct any misconceptions.

LEARNING STYLES

Interpersonal/Group Learning
Differentiation results in the formation of the various systems of the body. Divide the class into groups and assign each group one or more of these systems of the body: skeletal, nervous, circulatory, digestive, endocrine, lymphatic, respiratory, excretory, reproductive. Ask each group to list the major organs of each system. Let groups read their lists to the entire class.

Reproduction/Growth/Development Chapter 9 **221**

LEARNING STYLES

Visual/Spatial
Have partners choose two insects and research their life cycles. Then tell partners to draw or model the two life cycles. Partners should state whether their insects undergo complete or incomplete metamorphosis.

LEARNING STYLES

LEP/ESL
Tell students that English has several idioms are phrases that cannot be understood by knowing the meaning of the words used in them. Several idioms mention eggs. *Egg on one's face, all your eggs in one basket,* and *walking on eggs* are three examples. Encourage students to find out the meaning of these idioms and to use them in sentences.

SCIENCE JOURNAL

Ask students to compare a marsupial with another mammal. Tell students to describe how the two animals are alike and different in their development, focusing on the stage of development that is most different.

SCIENCE INTEGRATION

Earth Science
Australia has many unique animals. Let small groups of students research why. Suggest that they begin their research by reviewing the concept of plate tectonics and then relate that to the position of Australia. Ask each group to write and present a brief report on the topic.

Marsupial
A mammal that gives birth to young that are very undeveloped

Mammal Offspring

Unlike fish and many other kinds of animals, mammals produce few offspring. Mammal parents often take care of their young for long periods of time. This care protects the young from danger and increases their chances of survival. Thus, fewer offspring are needed to ensure the survival of a mammal species.

Most mammals do not lay eggs. Instead, the eggs develop and are fertilized inside the female's body. Only two kinds of mammals lay eggs: the spiny anteater and the duck-billed platypus. These two mammals live in Australia.

All other mammals carry their young inside their body, at least for some period of time. The female mammal gives birth to the young. **Marsupials** are mammals that give birth to young that are undeveloped. Kangaroos, koalas, and opossums are examples of marsupials. After being born, the tiny marsupial crawls into its mother's external pouch. There, the young continues to develop. All other mammals that do not lay eggs, including humans, cats, and whales, give birth to young ones that are more fully developed.

Marsupials complete their development in their mother's pouch.

Uterus
An organ in most female mammals that holds and protects an embryo

Placenta
A tissue that provides the embryo with food and oxygen from its mother's body

Did You Know?
There are about 4,500 species of mammals that exist today. Mammals live on every continent and in every ocean in the world.

Food for the Young Mammal

Young mammals may seem small and cuddly to us, but they are giants compared to baby fish. A fish embryo has little food in its egg, so it cannot grow much inside the egg. On the other hand, most mammal embryos, such as human embryos, get food from inside their mother's body. These mammals grow and develop a great deal before they are born.

Most female mammals have a **uterus**. The uterus is an organ that holds and protects a developing embryo. Inside the uterus, the embryo forms protective tissues around itself. Parts of these tissues form a **placenta**. The placenta provides food and oxygen from the mother's body to the developing embryo. When the young mammal is born, it feeds on milk produced by its mother's mammary glands.

The embryos of most marsupials do not obtain food through a placenta. When inside the mother, the embryo obtains food from the fertilized egg for a short time. After the young is born, it feeds on milk from inside the mother's pouch.

Recall that the spiny anteater and the duck-billed platypus are the only mammals that lay eggs. Mammal eggs are much smaller than a hen's egg. That is because the mammal eggs do not contain a large food supply. When the young are born, they feed on milk from their mother's mammary glands.

Mammals produce milk for their young.

TEACHER ALERT
Help students relate their navel to the placenta. Point out that a person's navel is the point of attachment of the umbilical cord, which in turn attached to the placenta, when they were in their mother's uterus.

SCIENCE INTEGRATION

Environment
The pesticide DDT had a serious effect on animal populations, causing some animals to become nearly extinct. Have students research DDT and the way it affected the growth and development of some animals. Be sure students find out what was done to halt the problem and to help animal populations that were affected. Have students write a short report on their research.

Did You Know?

Read aloud the Did You Know? feature on page 223. Point out that finding mammals in such a cold climate as Antarctica and its surrounding waters may seem surprising, but mammals, such as the fur seal, do exist there, spending time both on land and in the water. Whales, such as the blue whale, humpback whale, and right whale, migrate to Antarctica in the summer. Other animals that live on the continent include penguins, gulls, terns, mites, lice, and ticks. Krill, squid, and about 100 kinds of fish live in the waters surrounding Antarctica.

CROSS-CURRICULAR CONNECTION

Social Studies
Ask students to use a world map and to list the continents and oceans of the world. Beside each continent have them list five animals found on that continent.

ONLINE CONNECTION

Introduce students to the diversity of mammals past and present by having them investigate The Hall of Mammals Web site provided by the University of California Museum of Paleontology: www.ucmp.berkeley.edu/mammal/mammal.html. In addition to information about modern-day mammals, the site offers information about extinct mammals and also includes a helpful glossary. This site offers an audio section and links to recent fossil finds and classification diagrams.

AT HOME

Ask students to collect pictures of themselves at various ages—from birth (or before birth) until the present and to make a timeline showing their growth and development. Discuss the most striking changes that students have undergone.

Gestation time
The period of development of a mammal, from fertilization until birth

Gestation

Different mammals have different **gestation times**, depending on the size of the animal. Gestation time is the period of time from the fertilization of an egg until birth occurs. Gestation times for a variety of mammals are shown in the chart. As you can see, an elephant develops inside its mother for almost two years. But the gestation time of a mouse is only 20 days. In general, the larger the mammal, the longer is its gestation time.

Mammal	Approximate Number of Days
Mouse	20
Rabbit	31
Cat, dog	63
Monkey	210
Human	275
Cattle	281
Horse	336
Whale	360
Elephant	624

Lesson 3 REVIEW

Write your answers to these questions on a separate sheet of paper. Write complete sentences.

1. How does a zygote become an embryo?
2. Do young butterflies resemble their parents? Explain your answer.
3. What is the advantage of fish laying thousands of eggs at one time?
4. How is the development of a marsupial different from the development of other mammals that give birth to their young?
5. What is gestation time?

Technology Note

With new technology, scientists are developing genetically modified animals to help us in many ways. To increase food supplies, they are creating fish that grow faster. Some companies are bioengineering cows to produce medicines in their milk. Other scientists are modifying pigs to, one day, transplant their hearts into human patients. These modified animals could pass these features along to their offspring. There is still more research to be done. Scientists are studying the risks and the benefits of making these changes.

Reproduction, Growth, and Development Chapter 9 225

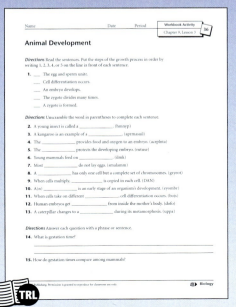

Workbook Activity 36

Lesson 3 Review Answers

1. The zygote undergoes many cell divisions and differentiation of cells, resulting in an embryo. **2.** No. The young go through different stages of development (egg, caterpillar, pupa) that look very different from the adult. **3.** Laying many eggs increases the chances of the species' survival. **4.** Marsupials give birth to young that are very underdeveloped and that complete their development outside the mother's body, in an external pouch. **5.** Gestation time is the period from fertilization of a mammal egg until birth occurs.

Portfolio Assessment

Sample items include:
- Lists from Warm-Up Activity
- Sentences from Cross-Curricular Connection: Language Arts
- Lesson 3 Review answers

Investigation 9

This investigation will take approximately 20 minutes to complete. Students will use these process and thinking skills: observing, comparing and contrasting, graphing, and drawing conclusions.

Preparation

- Find examples of bar graphs in science or math textbooks for students to use as models, if needed.
- If the Lab Manual is not being used, have graph paper available.
- Have available an assortment of colored pencils.
- Students may use Lab Manual 27 to record their data and answer the questions.

Procedure

- This investigation can easily be done independently.
- Remind students that the goal of the investigation is to determine how gestation times compare.
- Suggest that students begin by listing the animals from shortest gestation time to longest.
- Also suggest that students use a different colored pencil for each different gestation time.
- Challenge students to state a relationship between gestation time and animal size.

INVESTIGATION 9

Materials
- colored pencils
- graph paper

Graphing Gestation Times

Purpose
How do gestation times of different animals compare with each other? In this investigation, you will use a bar graph to compare different gestation times.

Procedure

1. Gestation time is the time from the fertilization of the egg until birth. Gestation times for a range of animals are shown below.

2. Copy the bar graph below on a sheet of graph paper. Use the information about gestation times to complete your bar graph. Arrange the information on the graph so that the shortest gestation time is shown on the first bar. Continue in this way until the longest gestation time is shown on the last bar. The first gestation time is done for you.

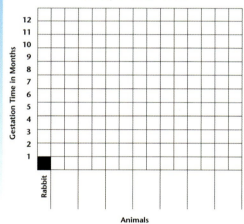

Animal	Months
Cat	2
Cow	10
Dog	2
Ewe	5
Goat	5
Mare	11
Rabbit	1
Sow	4

226 Chapter 9 Reproduction, Growth, and Development

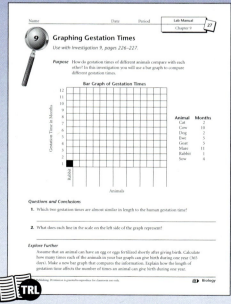

Questions and Conclusions

1. Which two gestation times are most similar in length to the human gestation time?
2. What does each line in the scale on the left side of the graph represent?

Explore Further

Assume that an animal can have an egg or eggs fertilized shortly after giving birth. Calculate how many times each of the animals in your bar graph can give birth during one year (365 days). Make a new bar graph that compares this information. Explain how the length of gestation time affects the number of times an animal can give birth during one year.

Reproduction, Growth, and Development Chapter 9 **227**

Results

The bars should represent the length of the gestation time. Bar lengths for the cat and dog should be equal, as should the lengths of the bars for the ewe and goat.

Questions and Conclusions Answers

1. the cow and the mare
2. one month

Explore Further Answers

The new bar graph should have a scale on the left side that represents the number of times an animal can reproduce in a year and should be numbered from 1 to 12. This is how many times each animal can reproduce in a year: rabbit, 12; cat and dog, 6; sow, 3; ewe and goat, $2\frac{2}{5}$; cow and mare, 1. The shorter the gestation time the more often an animal can reproduce.

Assessment

Check to be sure that students have made a bar for each animal and that the bar is the correct length. Also make sure the animals have been listed from shortest to longest gestation time. You might include the following items from this investigation in student portfolios:

- Investigation 9 bar graph
- Questions and Conclusions and Explore Further answers

Lesson at a Glance

Chapter 9 Lesson 4

Overview This lesson explains female and male reproductive systems, menstruation, pregnancy, birth, and puberty.

Objectives

- To identify the parts of the male and female reproductive systems in humans
- To define *ovulation* and *menstruation*
- To describe the main events of pregnancy
- To describe changes that occur in males and females during puberty

Student Pages 228–236

Teacher's Resource Library TRL

Workbook Activity 37

Alternative Activity 37

Vocabulary

scrotum	ovulation
testosterone	fallopian tube
penis	menstruation
vagina	pregnancy
prostate gland	umbilical cord
semen	fetus
estrogen	adolescence
progesterone	puberty

Science Background
Will It Be a Boy or a Girl?

What determines if a baby will be a boy or a girl? Special sex chromosomes do. Each female body cell contains two X chromosomes and each male body cell contains an X chromosome and a Y chromosome.

After meiosis, each sperm or egg cell has only one chromosome. All egg cells have an X chromosome. Half the sperm cells have an X chromosome and half have a Y chromosome. If a sperm with an X chromosome unites with an egg, the zygote has two X chromosomes and becomes a girl. If a sperm with a Y chromosome unites with an egg, the zygote will have an X and a Y chromosome and will become a boy.

Lesson 4 How Humans Grow and Develop

Objectives

After reading this lesson, you should be able to

- identify the parts of the male and female reproductive systems in humans.
- define *ovulation* and *menstruation*.
- describe the main events of pregnancy.
- describe changes that occur in males and females during puberty.

Scrotum
The sac that holds the testes

Like other mammals, humans reproduce sexually. A male parent and a female parent together produce a fertilized egg that develops inside the female's body.

The Male Reproductive System

The diagram shows the main reproductive organs of a human male. Recall that the testes produce sperm cells. The testes lie outside the body in a sac called a **scrotum**. Because the scrotum is outside the body, it is about 2°C cooler than the rest of the body. Sperm cells are sensitive to heat. The lower temperature of the scrotum helps the sperm to live.

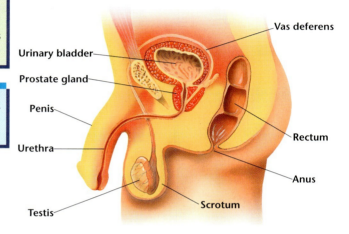

The Organs of the Male Reproductive System

228 Chapter 9 Reproduction, Growth, and Development

Testosterone
Male sex hormone

Penis
The male organ that delivers sperm to the female body

Vagina
The tube-like canal in the female body through which sperm enter the body

Prostate gland
The gland that produces the fluid found in semen

Semen
A mixture of fluid and sperm cells

The testes produce **testosterone**. This male sex hormone is important in the development of male sexual traits and in production of sperm. Beginning at puberty, the testes make more that 200 million sperm cells every day. Males are usually able to produce sperm from puberty through the rest of their life.

The external male organ, called the **penis**, delivers sperm to the female body. Before this happens, blood flows into the tissues of the penis. The blood causes the penis to lengthen and become rigid, or erect. The erect penis is inserted into a tube-like canal, called a **vagina**, in the female's body.

Sperm cells leave the male body through a tube in the penis called the urethra. The **prostate gland** connects to the urethra. The prostate gland produces fluid that mixes with the sperm cells and carries them through the urethra. The mixture of sperm and fluid is called **semen**. The semen flows through the urethra to the outside of the body. Urine also leaves the body through the urethra. However, urine and semen do not flow through the urethra at the same time.

Because half the sperm cells have X chromosomes and half have Y chromosomes, the chances of having a boy or a girl are approximately 50/50.

Warm-Up Activity

Draw the following chart on the chalkboard:

Physical	Mental	Emotional

Ask students to draw the chart on paper and to list at least two ways a person changes physically, mentally, and emotionally from birth to adolescence.

2 Teaching the Lesson

Have students scan the lesson and list unfamiliar terms. Let students work in pairs to define each word.

Ask students to compare the drawings on pages 230 and 231. Have them locate the ovary, fallopian tube, and uterus in each diagram.

Refer students to the diagram on page 233. Explain that when a woman is in labor, the process of giving birth is usually many hours long and painful. Ask students why, based on the diagram, this would be true.

Discuss with students how the care of a human infant compares with other mammals. Ask them to identify ways the care is similar and ways the care is different. Then ask students to make comparisons between humans and other types of animals, such as birds, reptiles, and insects.

3 Reinforce and Extend

LEARNING STYLES

Interpersonal/Group Learning
Some students may be uncomfortable during a discussion about the reproductive system and its organs. Emphasize that the reproductive system, like all others in the human body, is an important part of the human anatomy that makes human life possible.

LEARNING STYLES

Visual/Spatial
Some students may be confused about the location of internal reproductive organs in the body and in relationship to other organs. Use a large illustration of internal organs or a plastic model of the human body to help students identify the locations of the various organs.

Estrogen
Female sex hormone

Progesterone
Female sex hormone

Ovulation
The process of releasing an egg from an ovary

Fallopian tube
A tube through which eggs pass from an ovary to the uterus

Menstruation
The process during which an unfertilized egg, blood, and pieces of the lining of the uterus exit the female body

The Female Reproductive System

You can see the main female reproductive organs in the diagram below. Females are born with about 400,000 egg cells that are produced and stored in the ovaries. The ovaries also produce **estrogen** and **progesterone**. These female sex hormones regulate the reproductive development in females. One egg is usually released from one of the ovaries about every 28 days. The release of an egg is called **ovulation**.

After its release, an egg travels through one of the **fallopian tubes**. If sperm are present, a sperm cell may fertilize the egg cell in the fallopian tube. There, the fertilized egg will develop into an embryo, which travels to the uterus. If the egg is not fertilized, it eventually passes out of the female's body.

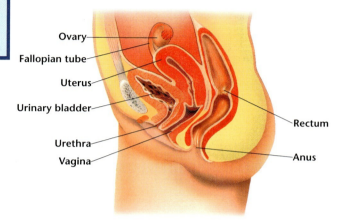

The Organs of the Female Reproductive System

Menstruation

Each month, the lining of the uterus thickens to form a blood-rich cushion. The lining can hold and nourish a developing embryo. If the egg is not fertilized and no embryo forms, the lining of the uterus breaks down. The unfertilized egg, blood, and pieces of the lining pass out of the female's body through the vagina. This process is called **menstruation**.

Pregnancy
The development of a fertilized egg into a baby inside a female's body

Umbilical cord
The cord that connects an embryo to the placenta

Pregnancy

When the male's penis releases semen into the female's vagina, the sperm cells swim to the uterus and fallopian tubes. If a sperm cell fertilizes an egg cell in a fallopian tube, the female begins a period of **pregnancy**. During pregnancy, the fertilized egg develops into a baby.

The diagram below shows the fertilized egg, or zygote, dividing in the fallopian tube. The zygote becomes an embryo. When the embryo reaches the uterus, it soon attaches to the blood-rich lining of the uterus.

Recall that the embryo forms a placenta. The **umbilical cord**, which contains blood vessels, connects the placenta to the embryo. The embryo's blood flows through blood vessels in the placenta. The mother's blood flows through blood vessels in the lining of the uterus. The two blood supplies usually do not mix. However, they come so close together that food and oxygen pass from the mother's blood to the embryo's blood. The embryo's waste products pass from the embryo's blood to the mother's blood. The embryo's wastes pass out of the mother's body along with her own wastes.

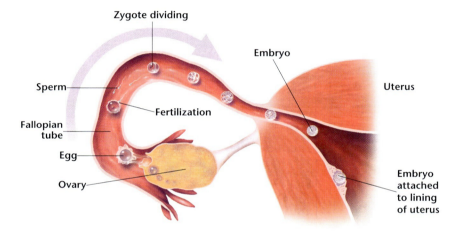

IN THE ENVIRONMENT

The growth and development of a baby within the uterus is very dependent on the baby's environment. A mother's habits, health, and exposure to chemicals and heat can affect the developing baby. Have students find out what environmental factors can affect a developing baby either positively or negatively. Ask them to identify what the mother can do to protect her child and to reduce the negative effects that factors, such as air pollution, smoke, chemicals, excessive heat, and exposure to viruses and bacteria, can have on the baby.

SCIENCE INTEGRATION

Technology

During pregnancy, the ob/gyn may give a woman an ultrasound, or sonogram. Let pairs of students find out what an ultrasound is, why it is administered, and what information it can provide for the doctor. Let partners write a short paragraph about what they found out and compare their information with that of another pair of students.

Did You Know?

Read aloud the feature on page 232. The structures referred to include the beginnings of the systems of the human body—skeletal, nervous, circulatory, digestive, endocrine, respiratory, excretory, muscular, and reproductive.

LEARNING STYLES

Logical/Mathematical
Provide students with a metric ruler and the following lengths: 0.5 mm, 1 mm, 4 mm, 8 mm, 19 mm, 38 mm, and 63 mm. Have students measure and draw the lengths on a sheet of paper. Then give students the following labels for each line they have drawn: 14 days, 18 days, 4 weeks, 6.5 weeks, 9 weeks, 11 weeks, and 15 weeks. Invite students to infer what the lines and labels refer to.
(*the average length of a developing human fetus*)

CAREER CONNECTION

Many health-care professionals are involved in the prenatal care of mother and child. These may include obstetricians, nutritionists, lab technicians, nurse-midwives, and others. Have students investigate the role of a professional who is involved in prenatal care. Encourage the students to find out what role the professional plays and how that role contributes to the health of the mother and child.

Fetus
An embryo after eight weeks of development in the uterus

Did You Know?

After eight weeks of development, a human fetus has all the major structures found in an adult but is only about 2.5 centimeters long.

Becoming a Baby

Inside the uterus, the embryo develops rapidly, using the food and oxygen provided by the mother. The embryo takes about nine months to become a fully developed baby.

The embryo first forms a hollow ball of cells. Soon, the cells differentiate. The body takes shape, and organs develop. After about three weeks, the embryo's heart begins to beat. Blood vessels form rapidly. The body has a head and has buds for arms and legs. At this point, the embryo is still smaller than a fingernail.

The human embryo at three weeks of development looks similar to the embryo of a fish, chick, or horse. All vertebrates look similar in their early stages of development. A human embryo soon begins to look more like a person.

At about four weeks, tiny hands begin to show fingers. Eyes appear as dark spots. After eight weeks, the embryo is called a **fetus**. It has all the major structures found in an adult. By the time pregnancy is half over, at $4\frac{1}{2}$ months, the fetus may suck its thumb. Sucking is an instinctive behavior that a baby uses to suck milk from its mother's mammary glands.

Technology Note

Some babies are born before they are fully developed. These premature babies rely on advances in technology to keep them alive. In a hospital's newborn intensive care unit, a baby might be kept in a special heated bed. This keeps the baby warm. At first, the baby might have a tube that puts liquids and nutrients directly into the blood. Later, the baby gets nutrients from a tube placed through the nose into the stomach. Wires attach the baby to devices that monitor blood pressure, heart rate, breathing, oxygen level, and temperature. A ventilator might be needed to help the baby breathe. A tube from the ventilator goes through the baby's mouth and into the trachea. This technology helps keep the baby alive and gives the baby time to develop.

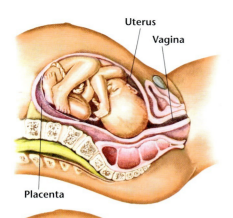

Uterus
Vagina
Placenta

Birth of a Baby

Usually, when the fetus reaches full size, the uterus begins to squeeze together, or contract. At first, the uterus contracts about every half hour. Then gradually, it contracts more often. As the uterus contracts, it pushes the baby out of the uterus and through the vagina.

After much work, the mother gives birth to the baby. The mother's body pushes out the placenta soon after the baby is born. The doctor clamps and cuts the umbilical cord. Now the baby can survive outside of the mother's body. The part of the umbilical cord that remains attached to the baby eventually falls off. A person's "belly button," or navel, is where the umbilical cord was once attached.

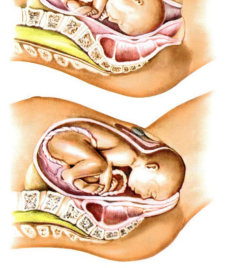

Birth

LEARNING STYLES

Visual/Spatial
Let pairs of students use books on child care to find out when babies start eating solid food. Encourage them to make a timeline showing what foods are eaten and when. Interested students might also find out how baby food can be made rather than purchased.

CROSS-CURRICULAR CONNECTION

Music
Challenge students to list tunes they might use to put a baby to sleep or to soothe a fussy baby. Let students share their list with the class, and if possible, play part of a song. Let the class express their opinions as to how effective the song would be.

Reproduction, Growth, and Development Chapter 9 **233**

Science Journal

After working through this lesson, students may have questions about topics that were discussed and not answered or about other related topics. Suggest that students write their questions in their Science Journal. Also suggest that they include a list of places they might look to find answers and people they might ask.

Cross-Curricular Connection

Art

Artists such as Mary Cassatt, Norman Rockwell, and Winslow Homer painted infants, children, and young adults. Ask students to locate paintings of children and adolescents by these artists or artists of their choice. Depending on the artwork selected, have them describe how the art shows diversity among humans, how infants or children are cared for, how children are the same and different today and in the past, or other topics related to the lessons in this chapter.

In the Community

In many communities, organizations offer social, recreational, and other services that help parents care for their children. Parenting classes, parent-child recreation activities, and day-care services help families provide for the needs of their children. Ask students to find out what services are available in your community. Some organizations may welcome student volunteers. Encourage interested students to volunteer their help. Ask them to share their experiences with the class.

Adolescence
The teenage years of a human

Puberty
The period of rapid growth and physical changes that occurs in males and females during early adolescence

Parental Care

Compared to a baby horse, which can stand just a few hours after birth, a human baby is born helpless. Human babies are not able to take care of themselves for many years.

Parents spend much time and effort caring for their babies. Some other animals have thousands of offspring at one time and then leave them without care. Humans usually have only one offspring at a time. The care that they give their offspring is one reason for the high survival rate of humans.

Adolescence

Humans care for their children through the teen years, called **adolescence**, and often into adulthood. Rapid growth and physical changes take place during **puberty**, which occurs at the beginning of adolescence.

During puberty in males, a boy's voice changes to a low pitch. Hair begins to grow on the face, under the arms, and in the area around the sex organs. The sex organs become more fully developed. During puberty in females, hair also begins to grow under the arms and around the sex organs. The breasts enlarge, and menstruation begins.

Adolescence is also a time when a teenager learns about becoming an adult. During this time, teenagers take on more responsibilities and learn more about themselves.

---

234   Chapter 9   Reproduction, Growth, and Development

Learning Styles

Interpersonal/Group Learning

Puberty can be an exciting, yet confusing, time for young people. Invite a pediatrician, nurse, or another specialist to talk to the class about the physical and psychological changes that puberty involves. Request that your guest include adequate time for a questioning period after his or her presentation.

Lesson 4 REVIEW

Write your answers to these questions on a separate sheet of paper. Write complete sentences.

1. Describe the path followed by sperm from the testes to the site of fertilization.
2. What happens during ovulation?
3. What is menstruation?
4. How does an embryo get nutrients during pregnancy?
5. What are some changes that occur in males and females during puberty?

Science at Work

Obstetrician/gynecologist

An obstetrician/gynecologist (ob/gyn) needs to have good observation, communication, problem-solving, and decision-making skills. An ob/gyn needs to be able to deal with emergencies and work with other medical professionals. A college degree followed by four years of medical school is required for ob/gyns. They are also required to do up to seven years of residency (caring for patients in a hospital under supervision). They must be licensed and pass an oral and written exam for certification.

An ob/gyn is a doctor who specializes in the health of women. The doctor focuses on the reproductive system, pregnancy, and birth. These doctors may see healthy patients for a checkup or women experiencing health problems. Listening to the patients, answering questions, and providing health information are important parts of every exam. Women who are pregnant visit an ob/gyn regularly to check on their health. The ob/gyn also checks the growth and development of the fetus. Ob/gyns help women deliver their babies. These doctors never stop learning because they need to keep up with new discoveries and new technology.

Reproduction, Growth, and Development Chapter 9   235

Lesson 4 Review Answers

1. Sperm leave the testes and move out from the penis through the urethra and into the female's body through the vagina. From the vagina, the sperm travel through the uterus and into the fallopian tubes where one sperm unites with an egg. **2.** An egg is released from an ovary into a fallopian tube. **3.** During menstruation, an unfertilized egg, blood, and pieces of the lining of the uterus pass out of the female body through the vagina. **4.** Food and oxygen pass from the mother's blood to the embryo's blood through the placenta. **5.** Answers will vary. Males and females begin to grow hair under the arms and in the area of the sex organs. Females' breasts enlarge and menstruation begins.

Portfolio Assessment

Sample items include:
- Chart from the Warm-Up Activity
- Words lists and definitions from Teaching the Lesson
- Lesson 4 Review answers

Science at Work

Read the Science at Work feature on page 235 together. Pronounce the words *obstetrician* and *gynecologist* for students. Then explain that the abbreviation *ob/gyn* is read and pronounced by saying the names of the letters: *o-b-g-y-n*. Further explain that an ob/gyn takes care of a woman during all the stages of her adult life by providing regular examinations when she begins menstruating, special care during pregnancy and delivery, and exams and help during and after menopause.

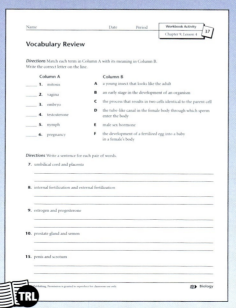

Workbook Activity 37   Reproduction/Growth/Development Chapter 9   235

Science in Your Life

Let volunteers read each section of the Science in Your Life feature on page 236. After each section is read, have another volunteer summarize the information found in the section.

Then lead a discussion about the effects of nutrition and drugs on a developing fetus. Ask: What effects are poor nutrition and drug use likely to have on a pregnant woman's body? What effect would there be on a developing baby?

Ask small groups to investigate the topic of birth defects and some things that can be done to try to prevent them. Suggest that students find out what some types of birth defects are and what advice is given to women before they get pregnant and after they become pregnant. Students might use encyclopedias or look up birth defects on the Web.

Science in Your Life

What substances are harmful during pregnancies?

An embryo and a fetus undergo many changes during development. During this time, the unborn baby is sensitive to substances in its environment. The baby's environment is its mother's body.

Any substance that a pregnant woman takes into her body can reach her unborn baby. A woman may take substances into her body by eating, drinking, breathing, and absorbing them through her skin. Just as food and oxygen can cross the placenta to the baby, so can drugs and other harmful substances.

What is wrong with drinking alcohol or smoking during pregnancy?

Alcohol has a more serious effect on a developing baby than it has on an adult. Even small amounts of alcohol drunk during pregnancy can cause fetal alcohol syndrome (FAS). Babies with FAS have various birth defects, including mental retardation. There is no safe amount of alcohol that a woman can drink during pregnancy.

The chemicals in tobacco also may harm an unborn baby. Pregnant women who smoke or are exposed to secondhand smoke are more likely to have babies with asthma, allergies, and other breathing problems.

How do other drugs affect unborn babies?

Pregnant women who take illegal drugs, such as cocaine, are likely to have babies who are addicted to the drugs. The babies also may have birth defects caused by the drugs. Even medicines, such as aspirin and cold medications, can harm a developing baby. It is important to read the information on the package of a medicine before taking it. Also, a woman should always tell her doctor if she is pregnant before taking a prescription drug.

Why is eating a balanced diet during pregnancy important?

To develop properly, an unborn baby needs certain nutrients. Without those nutrients, the baby may be born with physical or mental defects. That is why it is important for a pregnant woman to eat a healthy, balanced diet. Many doctors also recommend that pregnant women take vitamins to ensure that their developing baby gets all the nutrients needed.

Chapter 9 SUMMARY

- When organisms reproduce, they pass copies of their DNA to their offspring. The DNA determines the traits of the offspring.
- Asexual reproduction involves one parent and results in offspring identical to the parent.
- In cells that have a nucleus, asexual reproduction occurs by mitosis and cell division. Each new cell receives a copy of the DNA found in the parent cell.
- Sexual reproduction involves two parents and results in offspring that are unique. The offspring have a combination of the DNA found in the parents.
- Sexual reproduction allows for diversity among offspring.
- Sexual reproduction involves meiosis and cell division.
- When a sperm cell and an egg cell unite, a zygote is formed. The zygote develops into a new organism.
- In some types of animals, the female lays eggs that are fertilized outside her body. In mammals, the egg is fertilized inside the female's body.
- In most mammals, offspring develop inside the female's body. The uterus holds and protects the developing fetus.
- The fetus gets food and oxygen from the mother's body.
- After birth, mammal offspring are cared for by their parents.

Science Words

adolescence, 234	fetus, 232	ovulation, 230	spontaneous generation, 210
cell differentiation, 220	gamete, 216	penis, 229	testosterone, 229
chromosome, 215	gestation time, 224	placenta, 223	trait, 211
diversity, 212	internal fertilization, 218	pregnancy, 231	umbilical cord, 231
embryo, 220	marsupial, 222	progesterone, 230	uterus, 223
estrogen, 230	meiosis, 217	prostate gland, 229	vagina, 229
external fertilization, 218	menstruation, 230	puberty, 234	
fallopian tube, 230	mitosis, 215	scrotum, 228	
	nymph, 221	semen, 229	

Reproduction, Growth, and Development Chapter 9

Chapter 9 Review

Use the Chapter Review to prepare students for tests and to reteach content from the chapter.

Chapter 9 Mastery Test

The Teacher's Resource Library includes two parallel forms of the Chapter 9 Mastery Test. The difficulty level of the two forms is equivalent. You may wish to use one form as a pretest and the other form as a posttest.

Review Answers

Vocabulary Review

1. spontaneous generation 2. cell differentiation 3. gametes 4. uterus
5. nymph 6. fertilization 7. meiosis
8. chromosomes 9. diversity 10. placenta
11. marsupials 12. ovulation 13. fetus
14. puberty

TEACHER ALERT

In the Chapter Review, the Vocabulary Review activity includes a sample of the chapter's vocabulary terms. The activity will help determine students' understanding of key vocabulary terms and concepts presented in the chapter. Other vocabulary terms used in the chapter are listed below:

adolescence	penis
embryo	placenta
estrogen	pregnancy
external fertilization	progesterone
	prostate gland
gestation time	scrotum
internal fertilization	semen
	testosterone
menstruation	trait
mitosis	umbilical cord
nymph	vagina

Chapter 9 REVIEW

Vocabulary Review

Word Bank
cell differentiation
chromosomes
diversity
fertilization
fetus
gametes
marsupials
meiosis
nymph
ovulation
placenta
puberty
spontaneous generation
uterus

Choose the word or words from the Word Bank that best complete each sentence. Write the answer on a sheet of paper.

1. Redi's experiment helped disprove the idea of _____.
2. An embryo's cells take on different jobs in a process called _____.
3. Sperm cells and egg cells are _____.
4. The _____ holds and protects a developing embryo.
5. A(n) _____ is a young insect that looks like an adult.
6. The joining of an egg cell and a sperm cell is _____.
7. The process of _____ results in four sex cells.
8. DNA is found in structures called _____.
9. Sexual reproduction allows for _____ among offspring.
10. The _____ is a tissue that provides food and oxygen for an embryo.
11. Unlike most mammals, _____ are born undeveloped.
12. When an ovary releases an egg cell, _____ occurs.
13. After eight weeks of development, a human embryo is called a(n) _____.
14. The period of rapid growth at the beginning of adolescence is _____.

238 Chapter 9 Reproduction, Growth, and Development

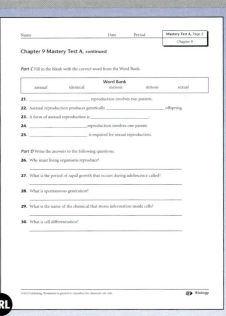

Chapter 9 Mastery Test A

Concept Review

Choose the answer that best completes each sentence. Write the letter of the answer on your paper.

15. Offspring resemble their parents because they have copies of their parents' _____.
    A DNA **B** gametes **C** cells **D** blood

16. The joining of a sperm cell and an egg cell results in a zygote that has _____ the number of chromosomes found in each sex cell.
    A one-half **B** four times **C** six times **D** twice

17. Mammals reproduce by _____.
    A sexual reproduction and external fertilization
    B asexual reproduction and internal fertilization
    C sexual reproduction and internal fertilization
    D sexual reproduction and external fertilization

18. The process of _____ occurs when a released egg cell has not been fertilized.
    A menstruation **C** pregnancy
    B ovulation **D** mitosis

Critical Thinking

Write the answer to each of the following questions.

19. How do you think the size of a *newborn* mammal might compare with the size of a *newly hatched* mammal? Explain your answer.

20. Women who use drugs during pregnancy are more likely to give birth to babies with birth defects. Explain why.

Test-Taking Tip Take time to organize your thoughts before answering a question that requires a written answer.

Reproduction, Growth, and Development Chapter 9 **239**

Concept Review
15. A 16. D 17. C 18. A

Critical Thinking
19. A newly hatched mammal is smaller than a newborn mammal because a mammal egg has a limited food supply and is too small to allow the young to grow much. 20. The drugs negatively affect the processes that occur during the development of an embryo and fetus.

Alternative Assessment

Alternative Assessment items correlate to the student Goals for Learning at the beginning of this chapter.

- Ask small groups to write an explanation of this statement: New life comes from existing life.

- Have students create a Venn diagram to compare and contrast asexual and sexual reproduction.

- Tell students to choose an animal and to answer each of these questions about the animal:
  - Where does fertilization occur?
  - Where does the offspring develop?
  - Where does the embryo and fetus get food before it is born?
  - Where does the offspring get food after it is born?

- Ask students to make a diagram that shows the path of a fertilized egg. Students should title their drawing, label its main parts, and write questions that can be answered by using the diagram.

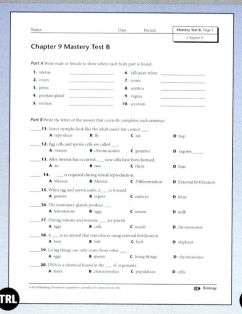

Chapter 9 Mastery Test B

Chapter 10

Planning Guide
Staying Healthy

	Student Pages	Vocabulary	Lesson Review
Lesson 1 How the Body Fights Disease	242–247	✔	✔
Lesson 2 Good Nutrition	248–255	✔	✔
Lesson 3 Healthy Habits	256–260	✔	✔

Student Text Lesson

Chapter Activities

Student Text
Science Center

Teacher's Resource Library
Community Connection 10:
   Opportunities to Build Healthy Habits

Assessment Options

Student Text
Chapter 10 Review

Teacher's Resource Library
Chapter 10 Mastery Tests A and B

	Student Text Features								Teaching Strategies						Learning Styles						Teacher's Resource Library				
	Achievements in Science	Science at Work	Science in Your Life	Investigation	Science Myth	Note	Technology Note	Did You Know?	Science Integration	Science Journal	Cross-Curricular Connection	Online Connection	Teacher Alert	Applications (Home, Career, Community, Global, Environment)	Auditory/Verbal	Body/Kinesthetic	Interpersonal/Group Learning	Logical/Mathematical	Visual/Spatial	LEP/ESL	Workbook Activities	Alternative Workbook Activities	Lab Manual	Resource File	Self-Study Guide
	247						✓	244	245		244, 246	246	243	244, 245	246			245	244		38	38	28	19	✓
		253		254		✓		249	250, 252		250		249, 250	251, 252	251	252				251	39	39	29	20	✓
			260		259					259	258	258		257	258	257	258				40	40	30		✓

Pronunciation Key

a	hat	e	let	ī	ice	ô	order	ù	put	sh	she
ā	age	ē	equal	o	hot	oi	oil	ü	rule	th	thin
ä	far	ėr	term	ō	open	ou	out	ch	child	ᴛʜ	then
â	care	i	it	ȯ	saw	u	cup	ng	long	zh	measure

ə { a in about, e in taken, i in pencil, o in lemon, u in circus }

Alternative Workbook Activities

The Teacher's Resource Library (TRL) contains a set of lower-level worksheets called Alternative Workbook Activities. These worksheets cover the same content as the regular Workbook Activities but are written at a second-grade reading level.

Skill Track Software

Use the Skill Track Software for Biology for additional reinforcement of this chapter. The software program allows students using AGS textbooks to be assessed for mastery of each chapter and lesson of the textbook. Students access the software on an individual basis and are assessed with multiple-choice items.

Chapter at a Glance

Chapter 10: Staying Healthy
pages 240–263

Lessons
1. How the Body Fights Disease pages 242–247
2. Good Nutrition pages 248–255

Investigation 10 pages 254–255

3. Healthy Habits pages 256–260

Chapter 10 Summary page 261

Chapter 10 Review
pages 262–263

Skill Track Software for Biology

Teacher's Resource Library

Workbook Activities 38–40

Alternative Workbook Activities 38–40

Lab Manual 28–30

Community Connection 10

Resource File 19–20

Chapter 10 Self-Study Guide

Chapter 10 Mastery Tests A and B

(Answer Keys for the Teacher's Resource Library begin on page 420 of the Teacher's Edition. A list of supplies required for Lab Manual Activities in this chapter begins on page 449.)

Science Center

Provide students with craft materials and discarded magazines. As students study the chapter, have them cut out pictures that illustrate good health and healthy behavior. Suggest that students focus on pictures that relate to the body's defenses and nutrition. Have students write information on an index card that relates the picture to the chapter content. For example, students might write *Good nutrition* and the names of nutrients in a pictured food. When they have finished reading the chapter, students can sort the images and their cards into a concept map. They could display the maps on a bulletin board.

Community Connection 10

Chapter 10: Staying Healthy

Imagine biting into one of the juicy, red apples in the photo. It not only tastes good, but it also contains some of the nutrients you need to stay healthy. The food choices you make every day provide you with energy and materials your body needs to grow and develop. In this chapter, you will learn what foods you need. You will learn how your body helps protect you from diseases. You also will learn about some habits that will help keep you healthy.

Organize Your Thoughts

- Staying Healthy
  - Preventing disease
    - Body's defenses
    - Sanitation
    - Vaccines
  - Good nutrition
    - Nutrients
    - Food Guide Pyramid
  - Healthy habits
    - Exercise
    - Avoid drug abuse
    - Cope with stress

Goals for Learning

- To define the term *infectious disease*
- To outline the body's defenses against germs
- To describe ways that infectious diseases can be prevented
- To identify nutrients the body needs to stay healthy
- To identify some good health habits

Introducing the Chapter

Direct students' attention to the photograph on page 240. Ask them what they think the apples have to do with staying healthy. Write *Health* in the center of the chalkboard or on a blank transparency and circle it. Draw lines from the circle as spokes. Invite students to list ways they can achieve good health. Write their suggestions on the spokes. (*Students may suggest eating good food, exercising, and getting enough rest.*) Encourage discussion of items related to nutrition and the body's defenses against disease.

Read the introductory paragraph together and have students study the Organize Your Thoughts chart. Explain any terms with which students are unfamiliar and ask them to predict what the terms have to do with health. As they read, students can check their predictions.

Notes and Technology Notes

Ask volunteers to read the notes that appear in the margins throughout the chapter. Then discuss them with the class.

TEACHER'S RESOURCE

The AGS Teaching Strategies in Science Transparencies may be used with this chapter. The transparencies add an interactive dimension to expand and enhance the *Biology* program content.

CAREER INTEREST INVENTORY

The AGS Harrington-O'Shea Career Decision-Making System-Revised (CDM) may be used with this chapter. Students can use the CDM to explore their interests and identify careers. The CDM defines career areas that are indicated by students' responses on the inventory.

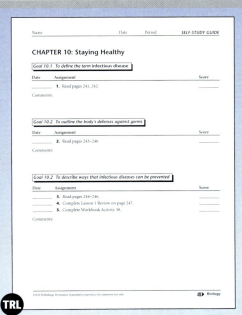

Chapter 10 Self-Study Guide

Lesson at a Glance

Chapter 10 Lesson 1

Overview In this lesson students learn how the body defends itself against pathogens and ways to prevent the spread of infectious disease.

Objectives

- To explain what an infectious disease is and how it is caused
- To list the body's main defenses against pathogens
- To describe how infectious diseases can be prevented

Student Pages 242–247

Teacher's Resource Library

Workbook Activity 38
Alternative Workbook Activity 38
Resource File 19
Lab Manual 28

Vocabulary

infectious disease	phagocyte
plague	lymphocyte
pathogen	immunity
virus	vaccine
immune system	sanitation

Science Background
Spread of Infectious Disease

Food, water, wind, and insects can spread infectious diseases. They also spread from person to person in several ways:

1. Touching, biting, kissing, or sexual intercourse are means of directly spreading disease. Sneezing and coughing within one meter of a person also are considered direct transmission.

2. People spread infectious diseases indirectly by contaminating objects and surfaces. When others touch the objects they may infect themselves by putting their fingers in their eyes, nose, or mouth. The Centers for Disease Control (CDC) notes that handwashing is the best way to prevent the spread of infectious diseases.

242 Chapter 10 Staying Healthy

Lesson 1: How the Body Fights Disease

Objectives

After reading this lesson, you should be able to
- explain what an infectious disease is and how it is caused
- list the body's main defenses against pathogens.
- describe how infectious diseases can be prevented.

Infectious disease
An illness that can pass from person to person

Plague
An infectious disease that spreads quickly and kills many people

Pathogen
A germ

Virus
A type of germ that is not living

If you have ever had a cold or the flu, you've had an **infectious disease**. An infectious disease is an illness that can pass from one person to another. In the past, many kinds of infectious diseases spread through populations. When an infectious disease spreads through a population quickly and kills many people, it is called a **plague**.

In the 1300s, two-thirds of the people in Europe caught the bubonic plague, or Black Death. In the 1700s, one out of every three English children came down with smallpox. In the 1800s, a cholera plague wiped out thousands of people in the United States. Even in the 1900s, an infectious disease called polio disabled and killed many children.

Today in the United States and many other parts of the world, these infectious diseases are rarely found. That is because steps have been taken to prevent them. However, one infectious disease that is of great concern today is AIDS. This disease destroys cells in the body that help fight other diseases. The body becomes unable to fight infection and cancer cells. Although there are drugs to help people with AIDS, research has yet to find a cure.

Pathogens

Germs, or **pathogens**, cause infectious diseases. Most pathogens are so small that they can be seen only with a microscope. Pathogens include protists, fungi, bacteria, and **viruses** that cause diseases.

A virus is a type of pathogen that is not living. However, living things are greatly affected by their presence. Viruses are smaller than bacteria. They are somewhere between living and nonliving things. Viruses are not cells, do not eat, and do not carry on respiration. Viruses do have DNA or RNA. When they are outside a cell, viruses are inactive. When they infect, or cause disease in, a cell, they insert their DNA or RNA into a cell. Viruses cannot reproduce on their own. They take over the cell's functions and cause the cell to make new viruses.

242 Chapter 10 Staying Healthy

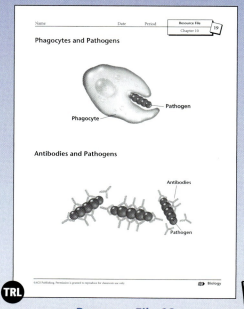

Resource File 19

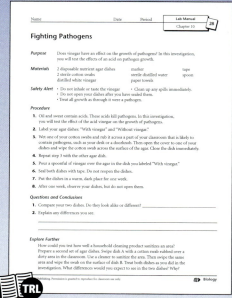

Lab Manual 28

Immune system
The body's most important defense against infectious diseases

Phagocyte
A white blood cell that surrounds and destroys pathogens

Pathogens are found in garbage, on unwashed dishes, and on the pages of this book. They are everywhere. Animals, wind, water, and other things carry pathogens from place to place.

Sometimes a virus can remain in a cell for long periods of time and not cause any problems. However, when the virus becomes active, the cell produces new viruses. They burst out of the cell, killing the cell and infecting other cells. Smallpox, the common cold, chickenpox, and AIDS are examples of diseases caused by viruses.

The Body's Defenses

If pathogens are everywhere, why aren't you sick all the time? You stay healthy because your body has ways to protect itself, called defenses. The body's first line of defense prevents pathogens from entering it. Your skin is a part of this defense. Skin forms a protective covering over your body and keeps pathogens out. The oil and sweat on your skin help kill pathogens.

When you breathe in pathogens, your nose has a sticky lining and tiny hairs that trap pathogens. When you sneeze or blow your nose, you get rid of many pathogens. If you swallow pathogens, acid in your stomach kills most of them.

If pathogens survive your body's first line of defense, your **immune system** fights them inside your body. The immune system is your body's most important defense against infectious diseases. White blood cells are specialized cells of your immune system. One kind of white blood cell is a **phagocyte**. Phagocytes surround pathogens and destroy them, as shown in the diagram.

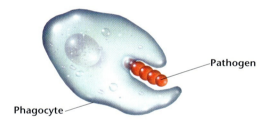

Staying Healthy Chapter 10 **243**

3. Tuberculosis, measles, influenza, and chicken pox are spread through the air over distances of more than one meter.

 Warm-Up Activity

Bring to class brochures and posters on infectious diseases from the American Red Cross, the CDC, your local health department, or other organizations. Use these materials to introduce and reinforce the concepts in Lesson 1.

 Teaching the Lesson

Review the terms *antibody, bacteria, protist,* and *fungi* with students. Ask students to list at least one example of each.

Have pairs of students write sentences linking the following pairs of terms: *infectious disease/plague, white blood cell/phagocyte, lymphocyte/pathogen, immunity/vaccine,* and *infectious disease/sanitation*. Allow students time to share sentences with another pair and have them correct each other's misconceptions.

Provide students with craft materials to make a model of a phagocyte engulfing a pathogen or an antibody binding to a pathogen. Students can refer to the drawings on Resource File 19. Ask students to use their models to explain the processes. Display the models in the classroom.

Ask students to describe ways they practice sanitation. (*washing hands, washing dishes, showering, bathing, avoiding rubbing eyes*) Discuss how each of these practices helps prevent the spread of infectious diseases.

PRONUNCIATION GUIDE

Use this list to help students pronounce difficult words in this lesson. Refer to the pronunciation key on the Chapter Planning Guide for the sounds of these symbols.

bubonic   [byü bä´ nik]

TEACHER ALERT

 Students may think that all bacteria, protists, and fungi are pathogenic. Explain that only a small percentage of these organisms cause infections. However, all viruses are pathogenic.

Staying Healthy Chapter 10 **243**

3 Reinforce and Extend

CROSS-CURRICULAR CONNECTION

Literature
Invite several students to read *Small Steps: The Year I Got Polio*, by Peg Kehret. Ask readers to relate the author's experience with polio to the content of the lesson.

IN THE ENVIRONMENT

An allergy is the result of the immune system reacting to foreign substances as if they were pathogens. For example, dust, pollen, and other nonpathogenic particles can cause a person's immune system to overreact. Students can find out what kinds of allergies are common in their community and how they are combated.

LEARNING STYLES

Visual/Spatial
Have students look up electron microscope photos or illustrations of viruses or bacteria in a reference book or on the Internet. Ask them to make a table for several of them, including diagrams of their shape, what disease each causes, and how each is spread. Allow students to present their information in small groups.

Lymphocyte
A white blood cell that produces antibodies

Immunity
The ability of the body to fight off a specific pathogen

Vaccine
A material that causes the body to make antibodies against a specific pathogen before that pathogen enters the body

Immunity

Your immune system includes white blood cells called **lymphocytes**. When a pathogen enters your body, the lymphocytes make antibodies to fight that pathogen. The diagram shows that antibodies fight specific pathogens by binding with them.

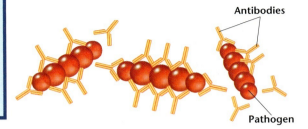

Antibodies help you get over a disease caused by a specific pathogen. In addition, the antibodies help prevent future attacks by that pathogen. That is because they remain in your body after the disease is gone. When this happens, your body is said to have **immunity** against the disease. If the pathogen that causes the disease enters your body again, it probably won't make you sick. Your body is prepared to fight the pathogen.

Vaccines

You might have heard someone say, "I'll never get mumps because I got a mumps shot." This person probably has immunity to mumps. Mumps is a childhood disease that causes fever and swelling under the jaw. When you get a "shot," your body makes antibodies against a specific pathogen. The shot is called a **vaccine**. It has a material that causes your body to make the antibodies. Getting a vaccine is another way you can develop immunity to a disease.

Did You Know?
Have you ever had a flu shot? A flu shot is a vaccine that fights the virus that causes influenza, or the flu. Unfortunately, the influenza virus changes into a new form after some time. Therefore, a person must get a different flu shot each year.

CROSS-CURRICULAR CONNECTION

Health/Math
Have students research the number of cases of polio in 1950, 1960, 1970, etc. They may find information online through the CDC site (www.CDC.gov/) or in encyclopedias. Students can construct a graph showing what happened once polio vaccine became available in 1955.

Did You Know?

Invite volunteers to tell what they know about flu shots given in your community. They might mention where and when the shots are given and what they do. After students read Did You Know? on page 244, suggest that interested students research flu shots given in the past two years and learn how and why they differed.

Technology Note

Some of today's vaccines, such as the vaccine for measles, are made from severely weakened viruses. Other vaccines, such as vaccines for influenza, are made from dead, whole viruses. Still other vaccines are made with parts or products of a pathogen.

In 1796, an English doctor named Edward Jenner made the first vaccine. At that time, people realized that once you had smallpox, you would never get the disease again. In other words, you were immune to the disease. Jenner noticed that people who had had cowpox did not get smallpox. Cowpox is a disease that causes sores on the hands. However, cowpox is not dangerous to people.

Jenner tried using material from a cowpox sore as a vaccine against smallpox. He put the cowpox material on the scratched arm of a young boy. The boy became sick with cowpox, but then he became well. Next, Jenner gave the boy smallpox pathogens. The boy did not get smallpox. The cowpox material had given the boy immunity against smallpox.

Vaccines have also made people immune to chicken pox, measles, polio, and other infectious diseases. Because specific pathogens cause certain diseases, a different vaccine must be made for each disease.

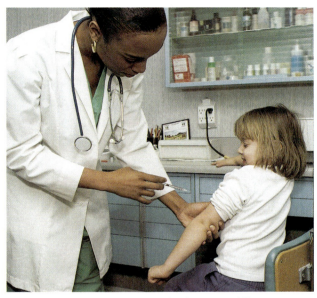

People get vaccines to protect themselves from many different infectious diseases.

LEARNING STYLES

Logical/Mathematical
Write the terms *pathogens, infectious disease, viruses, bacteria, immune system, antibodies, vaccines,* and *immunity* on the board. Have pairs of students devise a concept map that includes all the terms and shows how they are related to each other.

AT HOME

Have students check their medical records at home to see what immunizations they received and when. They can make a chart showing the diseases from which the vaccines protected them. Encourage students to research several of these diseases and find out how serious they are. They might find information in brochures at a medical clinic, encyclopedias, or online.

SCIENCE JOURNAL

Have students write their memory of what it was like to be sick with a childhood disease, a cold, or the flu. Encourage them to include descriptions of their symptoms as well as facts about how long the illness lasted, how it was treated, and what they had to do to recuperate.

CROSS-CURRICULAR CONNECTION

Health
Assign several students the task of learning about sanitation practices at a local supermarket, restaurant, and hospital. Suggest that a student interview a nurse or other health-care worker. Another can interview a butcher, and a third can interview a restaurant worker such as a cook or waiter.

LEARNING STYLES

Auditory/Verbal
Ask students to prepare a one-minute talk about sanitation measures they take at home. Speakers can explain, for example, how to sanitize kitchen counters or wash hands so that they are germ-free. They should demonstrate the procedures and use props as needed.

ONLINE CONNECTION

The United States Department of Health and Human Services Web site, www.hhs.gov/, has an extensive menu including such categories as Diseases and Conditions, Safety and Wellness, Drug and Food Information. For each category, there is a list of topics covered. This site links the reader to many articles about chapter topics, from eating right, to exercise, fitness, and specific diseases.

Sanitation
The practice of keeping things clean to prevent infectious diseases

Sanitation

When you want a drink of water, you just turn on the faucet. The water usually is clean and safe to drink. In the past, this was not always the case. People drank water that was piped in from rivers that had sewage. Sometimes people drank water from town pumps that contained many pathogens. These situations caused plagues. Thousands of people got sick and died from drinking water that was not clean.

Years ago, people did not know about pathogens. They did not realize that pathogens in water or food could harm them. Even doctors did not know that pathogens could pass from one patient to another.

Today, we know that pathogens cause many diseases. **Sanitation** is the practice of keeping things clean to prevent infectious diseases. For example, restaurants and supermarkets follow sanitation practices to keep foods safe to eat. Water treatment plants test water supplies for pathogens. Chemicals may be added to the water to kill pathogens. Hospitals sanitize linens and equipment so that diseases do not spread among the patients. Sanitation is also part of what you do every day. You practice sanitation when you wash dirty dishes, take out the garbage, and do the laundry. Washing your hands regularly is one of the best ways you can practice sanitation.

Lesson 1 REVIEW

Write your answers to these questions on a separate sheet of paper. Write complete sentences.

1. What are infectious diseases?
2. What is the body's first line of defense against pathogens?
3. How do antibodies fight pathogens?
4. How do vaccines prevent diseases?
5. Why is sanitation important?

Achievements in Science

Pasteurization Discovered

When people buy milk at a store, they can be assured that the milk does not contain pathogens, which could make them sick. Milk must be treated to kill pathogens before it can be sold. This treatment is known as pasteurization.

Pasteurization is named after the scientist Louis Pasteur. In the 1850s, he was asked to help solve a problem that local industry was having in Lille, France. During production, wine, beer, and vinegar would sometimes spoil. Pasteur discovered that microorganisms caused the spoilage and could cause disease. He also found that microorganisms could be killed by gentle heating. This process of gentle heating is known as pasteurization. It is still used today to make milk, orange juice, and many other products safe to eat and drink.

Lesson 1 Review Answers

1. diseases that can pass from one person to another **2.** The skin and lining of the nose, mouth, and throat **3.** Antibodies bind with pathogens so that they are unable to function in the body. **4.** Vaccines cause the body to make antibodies against pathogens that cause specific diseases. **5.** Sanitation helps keep food, water, and other things clean and pathogen-free.

Portfolio Assessment

Sample items include:
- Sentences using pairs of terms from Teaching the Lesson
- Models from Teaching the Lesson
- Lesson 1 Review answers

Achievements in Science

Inform students that they drink pasteurized milk. Ask them if they know what this means. Then have them read the Achievements in Science feature on page 247. After reading, ask several students to research and report briefly on Louis Pasteur and the pasteurization process.

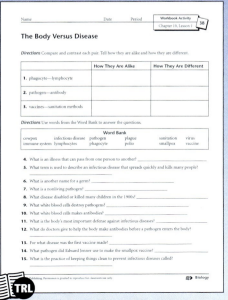

Workbook Activity 38

Lesson at a Glance

Chapter 10 Lesson 2

Overview In this lesson, students learn the six nutrients, where they are found, and how the body uses them. Students also learn how to use the Food Guide Pyramid.

Objectives
- To list the six types of nutrients your body needs every day
- To describe how your body uses those nutrients
- To use the Food Guide Pyramid to make healthy food choices

Student Pages 248–255

Teacher's Resource Library TRL

- **Workbook Activity** 39
- **Alternative Activity** 39
- **Resource File** 20
- **Lab Manual** 29

Vocabulary
nutrition
calorie
Food Guide Pyramid

Science Background
Nutrition and Nutrients

The Food Guide Pyramid was developed by the U.S. Department of Agriculture. The pyramid shows the kinds and amounts of foods a person should eat daily to get the six essential nutrients. Carbohydrates, proteins, and fats provide the body with energy. Vitamins, minerals, and water help the body use other nutrients and carry out life activities. All the nutrients except water and minerals are organic. Water makes up about 60 percent of the body's weight. Water lost through perspiration, respiration, and urination must be replaced each day.

Lesson 2 Good Nutrition

Objectives

After reading this lesson, you should be able to

- list the six types of nutrients your body needs every day.
- describe how your body uses those nutrients.
- use the Food Guide Pyramid to make healthy food choices.

Nutrition
The types and amounts of foods a person eats

Calorie
A unit used to measure the amount of energy a food contains

People often say, "Good **nutrition** is important if you want to stay healthy." Nutrition is the types and amounts of foods a person eats. Good nutrition involves eating the foods that your body needs to live and grow. The amount of food a person eats is often measured in **calories**. A calorie is a unit used to measure the energy that a food contains. The number of calories a person needs depends on several things. Some of these are a person's age, sex, body weight, activity level, and body efficiency. In this lesson, you will learn how to make sure you have good nutrition.

Nutrients

Nutrients are the parts of food that the body can use. Your body's cells use nutrients for energy, growth, and other life activities. Your cells work properly when they get the nutrients they need. There are six kinds of nutrients that your body should get every day. These are carbohydrates, fats, proteins, water, vitamins, and minerals.

Your cells use each kind of nutrient in a different way. Sugars and starches are carbohydrates. They are your body's main fuel. That's why you have so much energy after you eat foods such as pasta, rice, or potatoes. Fats and oils also store energy. Although proteins can give your body energy, they are not the best nutrients to use for energy. Cells use proteins mostly to repair themselves and to reproduce, or make copies of themselves. The photos show foods that are rich in carbohydrates, fats and oils, or proteins.

Carbohydrates

Fats and oils

Proteins

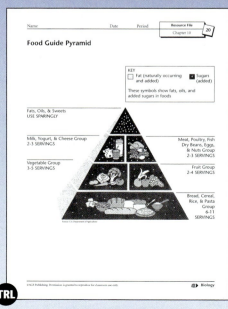

Resource File 20

The amount of energy needed to raise the temperature of 1 gram of water by 1°C is a calorie. Fats contain about 9 calories in every gram.

Did You Know?

Although sodium is a mineral your body needs, you only need a small amount. Too much sodium is not good for your health. For some people, sodium raises blood pressure and can increase the chances of disease.

Your body needs a great deal of water. Water makes up about 70 percent of your body. Your cells use water for many of their activities. However, your body loses a great deal of water each day. You lose water when you sweat. You breathe out water in the form of a gas called water vapor. You also lose water when your body releases wastes. You need to replace all the water that your body loses. For that reason, you need to make sure you get plenty of water each day.

Vitamins and Minerals

Unlike carbohydrates, proteins, and fats, vitamins and minerals do not give your body energy. However, vitamins and minerals are important to good health. They help the body use other nutrients and carry out life activities. The body needs small amounts of many different kinds of vitamins and minerals.

The vitamin chart on page 250 lists some of the jobs of different vitamins. It also shows which foods contain each vitamin. Tiny amounts of vitamins are found in foods. Because different foods contain different vitamins, it's important to eat various kinds of foods every day. That way, you can be sure to get all the vitamins your body needs.

Your body can store some vitamins, such as vitamin A. Other vitamins, such as vitamin C, pass through your body if they are not used. This means that you have to eat foods that contain vitamin C every day.

The foods you eat contain many different minerals. Calcium and phosphorus are important minerals for strong bones and teeth. Sodium and potassium help cells send messages to one another. Like vitamins, minerals are found in very small amounts in foods. Milk and cheese are two foods that have calcium and phosphorus. You can get potassium from bananas and meat. Salty foods, such as ham and crackers, have had sodium added to them. You should eat various kinds of foods every day to make sure you get all the minerals you need.

 Warm-Up Activity

Bring to class take-out menus from different types of restaurants. Menus that include a description of each dish work best. Have students analyze the menus in small groups and predict which dishes are most healthful. Use the menus to introduce the different types of nutrients found in different foods. Allow students to make their own categories of nutrients. Record these and have students revise them as they read the lesson.

 Teaching the Lesson

Have students list five foods they eat regularly. As they read, have them list the nutrients found in each food on their list. You may want to make nutrition references available so that students can find out vitamin and mineral content of certain foods.

Pass around an empty bottle that contained a multivitamin supplement. Have students read the label and compare the vitamins listed with those in the chart on page 250.

Display a Food Guide Pyramid shape on an overhead projector. As you hold up different food items, such as canned vegetables, rice, cheese, nuts, and corn oil, have students identify the region on the pyramid in which it fits. Ask them to tell whether it is a food they should eat rarely, moderately, or often, according to the size of the space for its group on the guide.

Did You Know?

Ask a volunteer to read aloud Did You Know? on page 249. Have on hand several packages of snack foods. Have students find the amount of sodium in each food, as mg and as a percentage of daily recommended amount. Discuss reasons why Americans tend to consume more salt than is healthy.

TEACHER ALERT

Students may think that, since vitamins are essential for health, taking large doses is best. Point out that vitamins are needed in certain amounts. Taking too much of certain vitamins can be unhealthy. Vitamins that are stored in the body may build up to unhealthy levels.

3 Reinforce and Extend.

TEACHER ALERT

Students may be unaware that there are different kinds of vegetarians. Ask students to find out the differences among vegans (avoid meat and foods with animal derivatives), lacto-vegetarians (avoid meat and eggs but eat dairy products), and lacto-ovo-vegetarians (avoid meat but eat eggs and dairy products).

SCIENCE INTEGRATION

Physical Science

Explain that calories measure energy in food. To lose weight, a person must burn more calories than he or she takes in. Have students use the following table to analyze the calories they burn on these exercises in a normal day.

Exercise	Calories burned per hour
sleeping	60
reading, watching TV	60
playing computer game	100
playing musical instrument	120
dancing	210
swimming	220
softball	230
walking	240
bicycling	250
basketball	348
soccer	480

Vitamin Chart		
Vitamin	Job in the Body	Foods with this Vitamin
Vitamin A	Keeps the skin, hair, and eyes healthy	Milk, egg yolk, liver carrots, spinach
B Vitamins		
Niacin	Protects the skin and nerves	Meat, cereal, whole wheat, milk, fish, legumes
Thiamin	Protects the nervous system; aids digestion	Pork, whole grains, green beans, peanut butter
Riboflavin	Protects the body from disease	Milk, eggs, bread, meat (especially liver), green vegetables
Vitamin C	Helps form bones and teeth and fights infectious diseases	Citrus fruits, tomatoes, potatoes
Vitamin D	Helps form strong bones and teeth	Milk, fish, oils
Vitamin E	Helps prevent cell damage	Vegetable oils, margarine
Vitamin K	Helps blood clot	Cabbage, spinach, soybeans, oats, egg yolks

CROSS-CURRICULAR CONNECTION

Math

Ask students to bring in nutrient labels or clean, empty containers from several of their favorite foods. Have them make a table showing what percentage of the recommended daily allowance of each vitamin each food contains. Then have them calculate how many servings of the food they would have to eat to get a day's supply of each vitamin.

Food Guide Pyramid

A guide for good nutrition

The Food Guide Pyramid

Your body needs many different kinds of nutrients in different amounts. How do you know if you are getting all the nutrients your body needs? One simple guide to good nutrition is the **Food Guide Pyramid**, which is shown below. The Food Guide Pyramid can help you choose the right foods. It can also help you eat the right amounts of different foods. That way, you are more likely to get the nutrients you need every day.

Some researchers have suggested changes in the Food Guide Pyramid. They think that certain fats, such as olive oil, deserve a lower position in the pyramid. Notice in the diagram that the lower a food is on the pyramid, the more it should be eaten. In addition, some researchers think that breads and refined starches should be higher in the pyramid. Nuts and beans may be better than fish and eggs as sources of protein. Red meat may belong at the top of the pyramid. In the coming years, the Food Guide Pyramid may change as more is learned about what the body needs for good health.

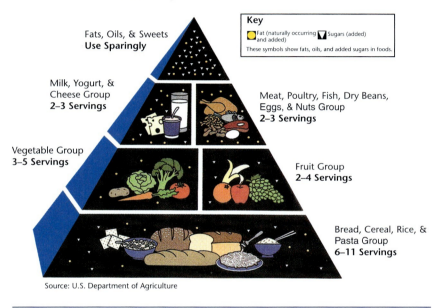

Source: U.S. Department of Agriculture

LEARNING STYLES

LEP/ESL

Students who are learning English will benefit by learning names of foods as they learn the nutrient categories. They can clip pictures of foods or bring in clean, empty food containers. Provide boxes labeled with names of nutrients. Have students write labels naming each food and then work in small groups to place each food in the box naming its chief nutrient.

IN THE COMMUNITY

Using the take-out restaurant menus from the Warm-Up Activity, have students compare the restaurant meals with the Food Guide Pyramid. Have students "order" meals from one menu that would provide the body's daily dietary needs.

LEARNING STYLES

Auditory/Verbal

Assign one area on the Food Guide Pyramid to each of six small groups of students. Group members are responsible for teaching classmates about their area of the pyramid. Suggest that they research and explain what essential nutrients each group provides. Students might also discuss why their food group has been assigned its number of servings. Finally, encourage students to list as many foods as they can that belong in that group.

LEARNING STYLES

Body/Kinesthetic
Provide students with packages of food (dry cereal works well) and measuring cups. Explain to students that they are not to eat the foods they study. Have them pour out an amount they consider a normal serving and measure it. Then have them measure the recommended serving size shown on the food label. Discuss any differences between the two measurements.

GLOBAL CONNECTION

Explain that in poor countries with large populations, the lack of food and clean water can cause many deaths. Have students research one such country's agriculture and climate in order to plan ways its people might raise food. They should summarize how these foods would provide balanced nutrition at low cost.

SCIENCE INTEGRATION

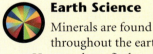

Earth Science
Minerals are found throughout the earth's crust. Have students find out the sources of minerals that are found in certain foods and in supplements. Ask them to trace these minerals back to their origin in the earth. Students can explain how certain minerals come to be in certain plants and animals, or what natural sources are used to make supplements. (For example, calcium may be derived from oyster shells.)

Serving Sizes

Eating the right foods is not enough. You also need to eat them in the right amounts. The Food Guide Pyramid gives the number of servings of each kind of food you should eat every day. How much food is one serving? In general, one serving is about the amount of a food you can hold in the palm of your hand. Below are some examples of what makes up one serving for different foods.

Grains
- 1 slice of bread
- 125 mL ($\frac{1}{2}$ cup) of rice or pasta
- 125 mL ($\frac{1}{2}$ cup) of cooked cereal

Meat and alternatives
- 1 egg
- 30 mL (2 tablespoons) of peanut butter
- 50–100 g (2–3 ounces) of meat
- 125–250 mL ($\frac{1}{2}$–1 cup) of cooked beans
- 100 g ($\frac{1}{3}$ cup) tofu

Vegetables
- 125 mL ($\frac{1}{2}$ cup) of vegetables
- 125 mL ($\frac{1}{2}$ cup) of vegetable juice

Milk, yogurt, cheese
- 250 mL (1 cup) of milk
- 235 g (1 cup) of yogurt
- 40 g (1.5 ounces) of cheese

Fruits
- 1 apple, banana, or orange
- 125 mL ($\frac{1}{2}$ cup) of fruit
- 175 mL ($\frac{3}{4}$ cup) of fruit juice

Lesson 2 REVIEW

Write your answers to these questions on a separate sheet of paper. Write complete sentences.

1. What are nutrients?
2. Which nutrients give your body energy?
3. Which nutrients do your body cells use to repair themselves?
4. List three vitamins that your body needs and the jobs they perform in the body.
5. What is the Food Guide Pyramid?

Science at Work

Dietetic Technician, Registered (DTR)

A dietetic technician needs good communication skills and a strong knowledge of food and nutrition. At least a two-year associate's degree at an approved college or university is needed. DTRs need to complete a dietetic technician program that includes 450 hours of supervised practice. Dietetic technicians must pass a national written exam to become registered.

DTRs know the foods people need to eat to stay healthy and the foods people need when they have a disease. They might work independently or as a part of a team with registered dietitians. Many DTRs work in hospitals or nursing homes, making sure patients and residents get good nutrition. They might work in schools or restaurants. DTRs buy the right foods, prepare the foods, and manage other employees. DTRs also work in health agencies or health clubs where they teach people about good nutrition. DTRs are an important part of health-care and food-service industries.

Staying Healthy Chapter 10 253

Investigation 10

This investigation will take approximately 45 minutes to complete. Students will use these process and thinking skills: communicating, collecting and interpreting data, inferring; comparing and contrasting, drawing conclusions

Preparation

- Remind students one to two weeks in advance that they need to begin collecting food labels.
- Have extra food labels available for groups that collect fewer than eight. Snack foods and breakfast cereals are good sources of food labels for this investigation.
- Students may use Lab Manual 29 to record their data and answer the questions.

Procedure

- Set up groups of four students to work cooperatively on completing the investigation. Each student can read and record the information from two food labels. All students read all labels to check each other's work.
- Before beginning the investigation, review the information on a food label. Show students where to find the different types of information requested. Using an overhead transparency of the data table, model how to record the information.
- Be sure students can distinguish the units of measure required on the table. Point out that both mass and percentage units are shown for fats, proteins, and carbohydrates.
- Students will need to extend or subdivide the vitamins and minerals column of the table to include the various nutrients in each food.

SAFETY ALERT

◆ Emphasize the need for students to bring only food labels, not packages of food. Remove any food remaining in packages so that students will not be tempted to eat it.

INVESTIGATION 10

Materials
- 8 food labels

Reading Food Labels

Purpose

How do the nutrients in different foods compare? In this investigation, you will determine the kinds and amounts of nutrients in different packaged foods.

Procedure

1. Collect food labels from eight different packaged foods. You might choose different kinds of the same food, such as different brands of cereal. That way, you can compare different brands.

2. Copy the table below on your paper.

Food	Serving Size	Carbohydrates (in Grams)	Proteins (in Grams)	Fats (in Grams)	Vitamins and Minerals (Percent)
1					
2					
3					
4					
5					
6					
7					
8					

Lab Manual 29, pages 1–2

3. Choose a food label. Write the name of the packaged food in your table.

4. Look for the size of a serving of that food. Write this information in your table.

5. Look for the kinds and amounts of carbohydrates, proteins, and fats in one serving. Notice that they are given in grams and in percents. In your table, write the number of grams of each nutrient found in one serving.

6. Are any vitamins or minerals in the food? Add to your table the names and percentages of any vitamins and minerals listed on the food label.

7. Repeat steps 3–6 for each food label you collected.

Questions and Conclusions

1. Compare the data in your table. Which food has the most carbohydrates?

2. Which food has the most proteins?

3. Which food has the most fats?

4. Which food has the most vitamins and minerals?

5. Compare the serving sizes of the foods. Which food has the smallest serving size? Which has the largest serving size?

Explore Further

1. Look at all the data you collected. Which food do you think is better for you to eat? Give reasons for your choice.

2. Does the information in your table change your ideas about eating certain foods? Why or why not?

Results
Results will vary according to the food labels students analyze. They may be surprised to find that serving sizes on high-fat snack foods are small compared to the amounts people tend to eat.

Questions and Conclusions Answers

1.–4. Answers will vary. Check to be sure answers are supported by the data.

5. Answers will vary. Snack foods and foods high in fat usually have smaller serving sizes than foods low in fat. Make sure that answers are supported by the data.

Explore Further Answers

1. Answers will vary. Make sure that answers are supported by information in the lesson.

2. Answers will vary. Make sure students give reasons for their answers.

Assessment
Check students' data tables against the food labels they interpreted to ensure that they have transferred the data accurately. Include the following items from this investigation in student portfolios:
- Data table
- Answers to Questions and Conclusions and Explore Further

Lesson at a Glance

Chapter 10 Lesson 3

Overview In this lesson, students learn about the benefits of exercise, the dangers of drug abuse, and the reasons for stress. They also explore several ways of coping with stress.

Objectives

- To explain why exercise is important and how much exercise you need
- To explain why avoiding drug abuse is important
- To describe what stress is, some signs of stress, and ways to cope with stress

Student Pages 256–259

Teacher's Resource Library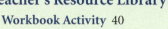

Workbook Activity 40

Alternative Activity 40

Lab Manual 30

Vocabulary

habit
drug
drug abuse

Science Background
Exercise

The U.S. Department of Health and Human Services emphasizes the importance of regular physical activity. Over 55 percent of the adult U.S. population is overweight, and poor physical fitness is an increasing problem for young people. At least 30 minutes of exercise daily is recommended. It can be spread over a number of routine or recreational activities, such as walking, bike riding, walking up stairs, mowing, swimming, playing sports, dancing, or taking part in an exercise program. Benefits from exercise include increased fitness; stronger and healthier bones, muscles, and joints; weight management; lowered risk of disease; and greater well-being and self-esteem.

256 Chapter 10 Staying Healthy

Lesson 3 Healthy Habits

Objectives

After reading this lesson, you should be able to

- explain why exercise is important and how much exercise you need.
- explain why avoiding drug abuse is important.
- describe what stress is, some signs of stress, and ways to cope with stress.

Habit
Something you usually do, often automatically

You have learned about the foods you need to keep your body healthy. You need more than just food and water. Exercise, avoiding drugs, and dealing with stress are **habits** that help keep you healthy. A habit is something you usually do. Often you do it automatically and don't even think about it. You probably have heard of bad habits. You can also have good habits.

Exercise Is Important

One of the leading health concerns in the United States is obesity. People who suffer from obesity have too much body fat and, as a result, are extremely overweight. Many more people are overweight to a lesser degree. Being overweight increases a person's chances of many different diseases, such as high blood pressure and diabetes. Sometimes, people weigh too much because of a medical condition. However, most people become overweight because they eat too many calories and don't get enough exercise.

Exercise has more benefits than helping maintain a healthy body weight. Even thin people need exercise to stay healthy. Regular exercise helps strengthen bones, muscles, and joints. It also helps reduce the risk of heart disease, stroke, diabetes, colon cancer, and high blood pressure. It helps reduce stress and depression. Regular exercise makes your heart stronger and more efficient, helps you do more things with less effort, and makes you feel better.

Having fun while exercising will help make it a habit.

Fitness experts recommend exercising at least 30 minutes on most days of the week. Recently, some experts began recommending 60 minutes on most days of the week. The amount of time a person needs to spend exercising depends on different factors, such as a person's age and weight.

256 Chapter 10 Staying Healthy

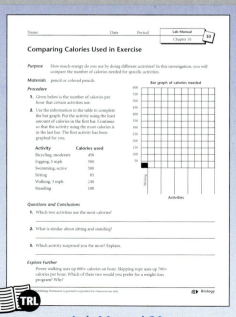

Lab Manual 30

Drug
A substance that acts on the body and changes the way the body works

Drug abuse
Using a drug when there is not a health reason for taking it

The people in the photo on page 256 are having fun with other people as they use machines in an exercise class. You don't have to use machines or take a class to get the exercise you need. Choose an activity you enjoy doing, such as playing basketball for fun or walking in the mall. Even little habits, such as riding a bicycle instead of riding in a car, add more exercise to your day.

Avoid Drug Abuse

One of the best habits you can have for your health is avoiding drug abuse. A chemical substance that acts on the body and changes the way the body works is a **drug**. Medicines your doctor prescribes and over-the-counter medicines are drugs. Always follow the directions for taking medicines and take them only for health reasons. Using a drug when there is not a health reason for taking it is **drug abuse**, or substance abuse.

Some people abuse drugs without knowing that's what they are doing. Taking drugs to improve athletic performance, lose weight without a doctor's advice, or relax can damage the body. Drugs are in forms other than pills that a person swallows. Alcohol in beer, nicotine in tobacco, chemicals in marijuana, and fumes in common substances such as gasoline are drugs.

All drugs have side effects, or unwanted effects. Even when you take a medicine, it has unwanted side effects. If you take medicine for a cold, you might notice side effects such as a dry mouth. You don't notice other side effects such as increased blood pressure.

Some of the side effects of drug abuse are dangerous immediately. Other harmful side effects of drugs may not be noticeable at first but can cause serious health problems over time. For example, smoking cigarettes increases your chances of heart attack, stroke, and lung cancer. Poisonous chemicals in tobacco smoke increase your chances of having many other kinds of cancer and diseases. Choosing not to smoke is one of the best things you can do for your body. In addition to the physical effects, drug abuse is linked to problems such as failure in school and poor judgment. When you make a habit of avoiding drugs except for health reasons, you are greatly reducing your risks of health problems.

Staying Healthy Chapter 10 **257**

 Warm-Up Activity

Invite volunteers to demonstrate forms of exercise they do often. Brainstorm a list of kinds of exercise students are likely to do. Ask students to describe what causes them stress. Discuss ways they think exercise might help them cope with stress. Explain that in this lesson, they will learn about exercise, drug abuse, and coping with stress.

2 Teaching the Lesson

Have students study the lesson objectives to write questions they will answer by reading. They should leave several blank lines after each question. As they read the lesson, they can supply answers in a different color.

Have students work in small groups to brainstorm a list of ways to add exercise to their day without changing their schedules. For example, they might use the stairs instead of elevators, take a brisk walk during lunchtime, or exercise while watching TV.

Ask students to list different drugs with which they are familiar. List these in categories on the board: over-the-counter medicines, prescription drugs, alcohol, tobacco products, athletic performance drugs, and illegal drugs. Stress the fact that all drugs affect the body and that any drug can be abused if it is taken improperly and frequently.

Have students alternate tensing muscles with relaxing all over and deep breathing in a darkened, quiet room. Discuss the sensations that signal stress versus those that indicate relaxation. Explain that learning to recognize when they are stressed is important to students as well as learning methods for relieving stress.

3 Reinforce and Extend

LEARNING STYLES

Body/Kinesthetic
Assign a small group of students to learn about different activities that provide different exercise benefits. Ask them to identify and demonstrate exercises that build muscle endurance, muscle strength, and flexibility. Discuss the benefits of each of these elements.

CAREER CONNECTION

Assign interested students to research the career of an athletic trainer or a personal trainer. They might interview a trainer who works for the school district, check online career sources, or check with a guidance counselor for references. Suggest that students find out how the trainer uses knowledge of exercise, nutrition, and human physiology.

PRONUNCIATION GUIDE

Use this list to help students pronounce difficult words in this lesson. Refer to the pronunciation key on the Chapter Planning Guide for the sounds of these symbols.

obesity [ō bē´ sə tē]
diabetes [dī ə bē´ tēz]
susceptible [sə sep´ tə bəl]

Online Connection

 Students can find out more about stress and ways to decrease it by visiting, www.aacap.org/publications/factsfam/66.htm.

Learning Styles

 Auditory/Verbal Have students write a public service announcement (PSA) for radio about the importance of exercise. Encourage students to be creative and funny while pointing out the many benefits of exercise. Have students record their PSAs on audiotape or videotape. They can then present their tapes to the class.

Cross-Curricular Connection

 Drama In small groups, have students role-play situations in which they deal with saying "no" to pressure to use drugs. In a whole-class discussion, explore students responses to these role-plays.

Learning Styles

 Interpersonal/Group Learning Form small groups of students. Have each group design and carry out an experiment to observe how laughter affects their emotions and sense of well-being. They should plan an activity that will cause them to laugh for a set time each day and note how they feel before and after. After a week, have the groups report their findings.

Coping with Stress

Everyone experiences stress. It's the reaction a person has to something that places demands on the body. A situation that causes stress for one person might not bother another person at all. For example, some people enjoy talking in front of a group of peers. Other people get sweaty palms and can feel their hearts beating in their chests. Some common causes of stress are demands in school and problems with family members or friends.

The immediate changes in your body caused by stress are the same changes that help protect you in emergencies. This feeling is known as the fight-or-flight response. It prepares your body to fight a dangerous situation or run to get away from it. The response includes a faster heart rate and breathing rate. There is also an increase in the flow of blood to the muscles in your arms and legs, and stickier platelets in your blood to help form clots if you are injured. You may also have cold, clammy hands and an upset stomach.

If the situation ends or you cope with the stress, your body returns to normal. If stress continues, you might notice headaches, grinding teeth, depression, or many other symptoms. Some people change their behaviors and overeat, become angry, abuse drugs, or withdraw from people. These behaviors are not healthy ways to deal with stress and should be avoided. Over a period of time, stress affects the immune system and makes a person more susceptible to illnesses.

You can avoid the unpleasant effects of stress by learning ways to cope with stress in your life. First identify the cause of a problem and decide on a way to solve it. Then carry out a plan to solve it. Also, exercise is a good way to cope with stress. Eating nutritious foods at regular meals and getting enough sleep are important. Another way of coping is sharing your feelings with someone you trust. Finally, do something that makes you laugh. Some studies show that laughing helps the immune system and lowers stress hormones in the body.

Lesson 3   REVIEW

Write your answers to these questions on a separate sheet of paper. Write complete sentences.

1. What is a habit?
2. What are the two main reasons people become overweight?
3. How much exercise do experts recommend?
4. Why is avoiding drug abuse important?
5. List some ways to cope with stress.

Science Myth

You only benefit from exercise if you keep exercising after your muscles start hurting.

Fact: Your muscles might be sore the day after you do a new exercise, but you do not need to be in pain while you are exercising. If your muscles hurt, stop exercising.

Staying Healthy   Chapter 10   259

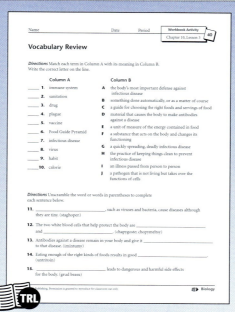

Workbook Activity 40

SCIENCE JOURNAL

Have students write a note to a friend who is stressed out. They might offer advice about ways the friend can cope with the stress in healthy ways.

Lesson 3 Review Answers

1. an activity one does routinely or automatically, without having to think about it **2.** They eat too many calories and don't get enough exercise. **3.** at least 30 minutes daily, up to 60 minutes most days **4.** Side effects of drug abuse are dangerous; other results cause serious health problems over time; taking drugs is linked to failure in school and poor judgment. **5.** Identify problems and solve them. Exercise, eat right, and get enough rest. Share your feelings with a trusted friend and find ways to laugh.

Portfolio Assessment

Sample items include:

- Questions and answers from Teaching the Lesson
- Lesson 3 Review answers

Science Myth

Read the myth portion (first sentence) of the feature on page 259 aloud. Then have a volunteer read the facts that follow. Explain that muscle soreness occurring the day after exercise is the result of muscle fiber damage. This damage occurs because you exercise hard enough to make your muscles burn. To improve any muscle takes both stress and recovery. This fact should be used as a guide to training. Exercise right up to the burn and then back off. Continue with this pattern until your muscles feel stiff and then stop. On the day after exercise, if muscles are sore, take the day off from exercise or proceed slowly. Do not train for muscle burning until soreness has gone away completely. (Most athletes take a very hard workout one day, and follow with one to seven days of easy workouts.)

Staying Healthy Chapter 10   259

Science in Your Life

Have students read Science in Your Life on page 260. Ask students to present the school's drug-related policy to the class. Discuss reasons for the policy.

Interested students might research information about the Supreme Court's decision in *Veronia School District 47J v. Acton*. Suggest that they present the case and decision to the class. Allow students to voice their opinions about the case.

Science in Your Life

What is drug testing?

Drug testing or the results of drug tests are often in the news. Reports of car accidents might mention that alcohol was involved and give the person's blood-alcohol level. An airplane pilot might be removed from a plane before takeoff if someone reports seeing the pilot drinking alcohol before boarding the plane. Bus drivers might be randomly tested. Athletes at the Olympic games or other events are suspended if they test positive for a drug. High school students might be tested in their schools or even at home if parents suspect that their child is abusing drugs.

Athletes who abuse drugs to try to improve their performance may have an unfair advantage over their opponents. At the same time, these athletes are endangering their health. Many companies also test their employees because employees who abuse drugs cost money in sick days and poor job performance. Schools sometimes conduct random blood tests to discourage drug abuse.

The drugs a person takes enter the person's blood. Some drug tests determine the presence and amount of a drug in the blood. For example, people who are involved in accidents or are driving dangerously are often tested to determine their blood-alcohol level. The amount of alcohol in the blood is an indication of how impaired the person is. Some of the alcohol in the blood is excreted from the body in urine. A blood or urine test can be used to measure alcohol in the blood. Since these tests are not practical at the side of the road, another test is used. Some alcohol passes from the blood into the lungs and is exhaled. A breath-alcohol testing device can be used to determine the blood-alcohol level from air a person exhales.

Drug tests are also used to test for past drug use. After drugs are in the blood, the body works to eliminate them. During this process, a person's body breaks down drugs into other chemicals. The presence of these chemicals indicates recent drug use. Urine tests, blood tests, and hair tests can be used to test for these chemicals.

Chapter 10 SUMMARY

- Infectious diseases can spread from one person to another. Pathogens cause infectious diseases.

- The body's first line of defense against pathogens includes the skin, nose, and throat.

- When pathogens enter the body, phagocytes may surround the pathogens and destroy them.

- The immune system is the body's most important defense against pathogens. Lymphocytes of the immune system make antibodies that fight specific pathogens.

- Suppose the body has antibodies that fight a specific pathogen. The person is unlikely to become ill the next time that pathogen enters the body. This condition is called immunity.

- A vaccine causes a person's body to make antibodies against the pathogen that causes the disease.

- Nutrients are the parts of food that the body can use. Nutrients are necessary for cell repair, growth, and other life activities.

- The body needs six nutrients: proteins, carbohydrates, fats, water, vitamins, and minerals.

- Carbohydrates, proteins, and fats and oils give the body energy. Cells use water for many of their life activities.

- Vitamins and minerals help the body use other nutrients and carry out life activities.

- The Food Guide Pyramid suggests the kinds and amounts of foods needed for good nutrition.

- Exercising, avoiding drugs, and dealing with stress are habits that help keep people healthy.

Science Words

calorie, 248	habit, 256	infectious disease, 242	phagocyte, 243
drug, 257	immune system, 243	lymphocyte, 244	plague, 242
drug abuse, 257	immunity, 244	nutrition, 248	sanitation, 246
Food Guide Pyramid, 251		pathogen, 242	vaccine, 244
			virus, 242

Chapter 10 Review

Use the Chapter Review to prepare students for tests and to reteach content from the chapter.

Chapter 10 Mastery Test

The Teacher's Resource Library includes two parallel forms of the Chapter 10 Mastery Test. The difficulty level of the two forms is equivalent. You may wish to use one form as a pretest and the other form as a posttest.

Review Answers

Vocabulary Review

1. immunity 2. Food Guide Pyramid
3. drug abuse 4. drug 5. Vaccines
6. viruses 7. Pathogens 8. sanitation
9. phagocyte 10. nutrition 11. plague
12. calorie 13. immune system
14. infectious disease

Concept Review

15. A 16. D 17. B 18. C

Chapter 10 REVIEW

Word Bank
calorie
drug
drug abuse
Food Guide Pyramid
habit
immunity
immune system
infectious disease
nutrition
pathogens
plague
sanitation
vaccines
viruses

Vocabulary Review

Choose the word or words from the Word Bank that best complete each sentence. Write the answer on a sheet of paper.

1. When antibodies remain in your body after the disease is gone, you have _____ against that disease.
2. The _____ can help you determine how much and what kinds of foods to eat.
3. Taking a drug when there is not a health reason for taking it is _____.
4. A(n) _____ is a substance that changes the way the body works.
5. _____ cause the body to produce antibodies against specific pathogens.
6. Unlike bacteria, _____ are not living organisms.
7. _____ cause all infectious diseases.
8. Keeping drinking water clean is an example of _____.
9. When you do something over and over again, it becomes a(n) _____.
10. The type and amount of food you eat is _____.
11. An infectious disease that spreads quickly through a population and kills many people is a(n) _____.
12. A(n) _____ is a unit used to measure the amount of energy a food contains.
13. Your _____ helps your body fight disease.
14. A(n) _____ can pass from person to person.

262 Chapter 10 Staying Healthy

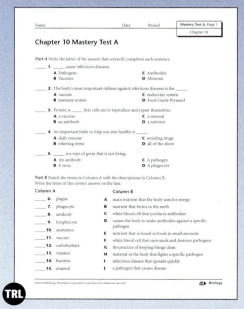

Chapter 10 Mastery Test A

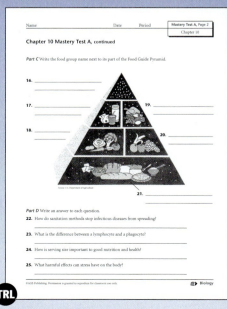

Chapter 10 Mastery Test A

262 Chapter 10 Staying Healthy

Concept Review

Choose the answer that best completes each sentence. Write the letter of the answer on your paper.

15. Your body's first line of defense against pathogens is your _____.
    - A skin
    - B stomach
    - C antibodies
    - D phagocytes

16. Nicotine and alcohol are examples of _____.
    - A carbohydrates
    - B proteins
    - C medicines
    - D drugs

17. White blood cells called _____ produce antibodies to fight disease.
    - A phagocytes
    - B lymphocytes
    - C protists
    - D platelets

18. Cells use _____ to repair themselves and to reproduce.
    - A carbohydrates
    - B fats
    - C proteins
    - D sugars

Critical Thinking

19. Give one reason why early doctors and scientists did not realize that pathogens cause infectious diseases.

20. Do you eat nutritious foods in the right amounts? Explain your answer using the Food Guide Pyramid.

Test-Taking Tip Effective studying takes place over a period of time. Spend time studying new material for several days or weeks before a test. Then spend the night before a test reviewing the material.

Staying Healthy Chapter 10 263

Critical Thinking

19. Most pathogens are microscopic, so early doctors and scientists did not know they existed. **20.** Answers will depend upon students' eating habits. They should reflect understanding of the Food Guide Pyramid groups and recommended servings.

Alternative Assessment

Alternative Assessment items correlate to the student Goals for Learning at the beginning of this chapter.

- Ask students to create a concept map with *infectious disease* in the middle. In circles around this, they can write a definition and give examples.
- Have students draw two columns. Label the first "Defense" and the second "Function." Ask students to write each of the body's defenses and what it does.
- Ask students to draw pictures or write sentences explaining three ways we protect ourselves from infectious disease.
- Ask students to list the nutrients the body needs to stay healthy. For each nutrient, have them list three foods containing it. Then ask students to make up a balanced, healthy menu choosing from these foods.
- Write descriptions of situations on slips of paper. Situations could include being overweight, being stressed out, and being pressured to take drugs. Have students draw a slip and act out a scene in which the person with the problem decides what to do about it.

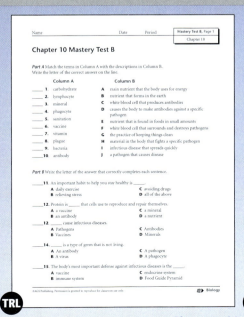
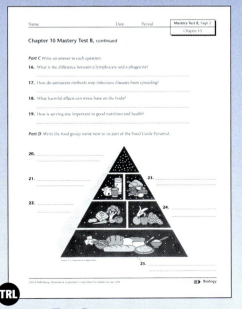

Chapter 10 Mastery Test B

Chapter 11

Planning Guide
Genetics

	Student Pages	Vocabulary	Lesson Review
Lesson 1 Heredity	266–271	✔	✔
Lesson 2 Chromosomes	272–278	✔	✔
Lesson 3 How Heredity Is Studied in Humans	279–287	✔	✔
Lesson 4 Applied Genetics	288–290	✔	✔

Student Text Lesson

Chapter Activities

Student Text
Science Center

Teacher's Resource Library
Community Connection 11:
  Applied Genetics in Your Community

Assessment Options

Student Text
Chapter 11 Review

Teacher's Resource Library
Chapter 11 Mastery Tests A and B

264A

Achievements in Science	Science at Work	Science in Your Life	Investigation	Science Myth	Note	Technology Note	Did You Know?	Science Integration	Science Journal	Cross-Curricular Connection	Online Connection	Teacher Alert	Applications (Home, Career, Community, Global, Environment)	Auditory/Verbal	Body/Kinesthetic	Interpersonal/Group Learning	Logical/Mathematical	Visual/Spatial	LEP/ESL	Workbook Activities	Alternative Workbook Activities	Lab Manual	Resource File	Self-Study Guide
						✓	267	269		268				269	269		268, 271	270		41	41	31	21, 22	✓
				278			275		277	274, 276, 277	276	273	275			277		274		42	42	32		✓
	285	284	286		✓		283	280, 284	283	282			281, 282, 283			281			280	43	43	33		✓
290										289	289	288	289	289		289				44	44			✓

Pronunciation Key

a	hat	e	let	ī	ice	ô	order	ù	put	sh she
ā	age	ē	equal	o	hot	oi	oil	ü	rule	th thin
ä	far	ėr	term	ō	open	ou	out	ch	child	ŦH then
â	care	i	it	ȯ	saw	u	cup	ng	long	zh measure

ə { a in about, e in taken, i in pencil, o in lemon, u in circus }

Alternative Workbook Activities

The Teacher's Resource Library (TRL) contains a set of lower-level worksheets called Alternative Workbook Activities. These worksheets cover the same content as the regular Workbook Activities but are written at a second-grade reading level.

Skill Track Software

Use the Skill Track Software for Biology for additional reinforcement of this chapter. The software program allows students using AGS textbooks to be assessed for mastery of each chapter and lesson of the textbook. Students access the software on an individual basis and are assessed with multiple-choice items.

Chapter at a Glance

Chapter 11: Genetics
pages 264–293

Lessons
1. Heredity pages 266–271
2. Chromosomes pages 272–278
3. How Heredity Is Studied in Humans pages 279–287

Investigation 11 pages 286–287

4. Applied Genetics pages 288–290

Chapter 11 Summary page 291
Chapter 11 Review pages 292–293

Skill Track Software for Biology

Teacher's Resource Library

Workbook Activities 41–44

Alternative Workbook Activities 41–44

Lab Manual 31–33

Community Connection 11

Resource File 21–22

Chapter 11 Self-Study Guide

Chapter 11 Mastery Tests A and B

(Answer Keys for the Teacher's Resource Library begin on page 420 of the Teacher's Edition. A list of supplies required for Lab Manual Activities in this chapter begins on page 449.)

Science Center

Use students' photographs from Introducing the Chapter to make a science center display. Have students label their descriptions of the selected characteristics on each image before adding it to the display. Later in the chapter, when the concepts of phenotype and genotype are introduced, you may want to point out that the labeled descriptions represent each individual's phenotype.

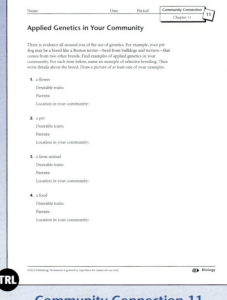

Community Connection 11

Chapter 11: Genetics

Garden peas, like the ones in the photo, are a common sight in many gardens. When peas are planted, grow, and reproduce, their characteristics are passed on to their offspring. Many important ideas about genetics were discovered over 100 years ago from studies of ordinary garden peas. In this chapter, you will learn how genes and chromosomes are passed on to offspring and affect an organism's appearance. You also will learn about twins, genetic diseases, and how knowledge of genetics can be useful.

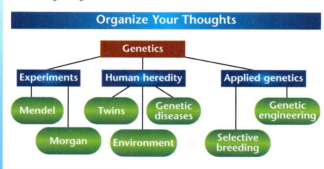

Goals for Learning

- ◆ To explain how genes pass traits from parents to offspring
- ◆ To describe the role of chromosomes in heredity
- ◆ To identify patterns of heredity in humans
- ◆ To explain how scientists are able to use the principles of genetics to affect the traits of organisms

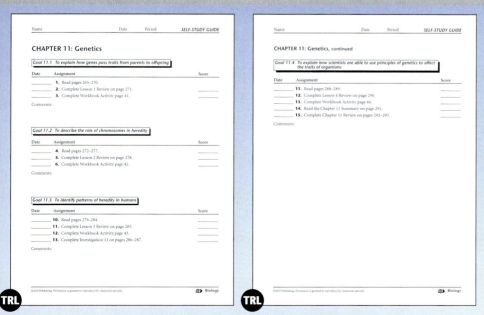

Chapter 11 Self-Study Guide

Introducing the Chapter

Divide the class into small groups. Provide each group with five photographs of the same type of organism (five people, five dogs, five pea plants, and so on). Tell students that their task will be to select ten characteristics and to describe those characteristics in each individual pictured. Begin by doing one example together as a class. For example, the class may select hair color. Individuals in the photographs may have blond hair, brown hair, black hair, or red hair. Once groups have described the characteristics of the individuals in their photos, have a representative from each group display the group's photos and share the characteristics that the group selected and described. Lead students to recognize that there are variations in the characteristics of animals, plants, and other organisms.

Notes and Technology Notes

Ask volunteers to read the notes that appear in the margins throughout the chapter. Then discuss them with the class.

TEACHER'S RESOURCE

The AGS Teaching Strategies in Science Transparencies may be used with this chapter. The transparencies add an interactive dimension to expand and enhance the *Biology* program content.

CAREER INTEREST INVENTORY

The AGS Harrington-O'Shea Career Decision-Making System-Revised (CDM) may be used with this chapter. Students can use the CDM to explore their interests and identify careers. The CDM defines career areas that are indicated by students' responses on the inventory.

Lesson at a Glance

Chapter 11 Lesson 1

Overview In this lesson, students learn about Mendel's experiments and how to use a Punnett square to predict the outcome of a genetic cross.

Objectives

- To describe Mendel's studies of pea plants
- To explain the difference between a dominant gene and a recessive gene
- To draw a Punnett square
- To explain Mendel's conclusions about heredity

Student Pages 266–271

Teacher's Resource Library

Workbook Activity 41

Alternative Workbook Activity 41

Lab Manual 31

Resource File 21, 22

Vocabulary

heredity	gene
genetics	recessive gene
self-pollination	dominant gene
P generation	genotype
cross-pollination	phenotype
F_1 generation	F_2 generation
Punnett square	factors

Science Background
Mendel's Experiment

Modern genetics began with the work of Gregor Mendel, an Austrian monk who performed controlled experiments with pea plants in the mid-1800s. Mendel performed the experiments specifically to learn about heredity. Pea plants were a good choice of plants to use because Mendel was able to control the pollination of the plants and thus be sure of the parents of the seeds that were produced. To cross-pollinate the pea plants, Mendel removed the male organs from the flowers of a plant before they produced pollen.

Lesson 1 Heredity

Objectives

After reading this lesson, you should be able to

- describe Mendel's studies of pea plants.
- explain the difference between a dominant gene and a recessive gene.
- draw a Punnett square.
- explain Mendel's conclusions about heredity.

Heredity
The passing of traits from parents to offspring

Genetics
The study of heredity

Self-pollination
The movement of pollen from the male sex organs to the female sex organs of flowers on the same plant

P generation
The pure plants that Mendel produced by self-pollination

Family members have many of the same characteristics, or traits. The passing of traits from parents to their children is called **heredity**. All organisms pass information about traits to their children, or offspring. That is why bluebirds produce bluebird chicks, and lions produce lion cubs.

Mendel's Studies

Genetics is the study of heredity. More than 100 years ago, a scientist named Gregor Mendel made important discoveries about heredity. Mendel took seeds from tall pea plants and planted them. Most of the plants that grew from the seeds were tall. However, some of the plants were short. Mendel wondered why the short pea plants were unlike their tall parents.

Mendel decided to study pea plants and their seeds. He began by trying to grow plants that would produce only tall plants from their seeds. The flowers of pea plants have both male sex organs and female sex organs. Mendel moved pollen from the male sex organs to the female sex organs on the same plant. In this way, he carried out **self-pollination** of each plant. After self-pollination occurred, each flower produced seeds. Mendel planted the seeds, and then self-pollinated the plants that grew from those seeds. He did this again and again. Finally, Mendel had seeds that would produce only tall pea plants. These were "pure tall" seeds.

Using the same method of self-pollination, Mendel grew plants that were pure for the trait of shortness. He called the "pure tall" and the "pure short" plants the **P generation**. *P* stands for parent generation. To understand what a generation is, think of your grandparents as one generation. Your parents are part of a second generation, and you are part of a third generation.

266 Chapter 11 Genetics

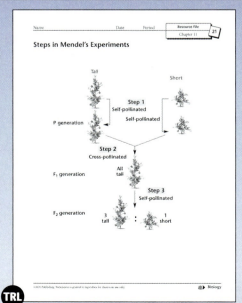

Resource File 21

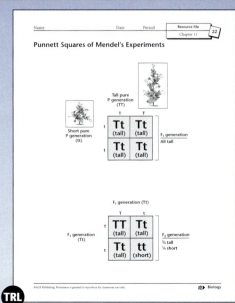

Resource File 22

Cross-pollination
The movement of pollen from the male sex organs to the female sex organs of flowers on different plants

F₁ generation
The plants that resulted when Mendel cross-pollinated two different kinds of pure plants

Did You Know?
Mendel studied seven characteristics of pea plants. Besides plant height, he studied flower position, flower color, seed color, seed shape, pod color, and pod shape. During his studies, Mendel studied almost 20,000 pea plants.

The F₁ Generation

Mendel wondered what type of offspring two *different* pure parent plants would produce. He used the method of **cross-pollination** to find out. Mendel moved the pollen from pure tall pea plants to the female sex organs of pure short pea plants. He also placed pollen from pure short plants on the female sex organs of pure tall plants.

When the cross-pollinated flowers produced seeds, Mendel took the seeds and planted them. He called these plants the **F₁ generation**. The *F* stands for filial, which means son or daughter. The *1* stands for the first filial generation.

Mendel's studies of pea plants had three main steps. The diagram below shows the results of his studies. To his surprise, Mendel found that all the F₁ generation plants were tall. Mendel crossed hundreds of pure short and pure tall pea plants. He observed the offspring plants carefully. Each time, the results were similar. All the offspring were tall. What had happened to shortness? Why had this trait disappeared?

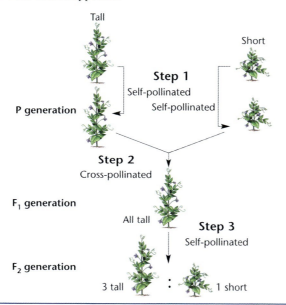

Genetics Chapter 11 **267**

Lab Manual 31, pages 1–2

Did You Know?

Ask a volunteer to read the feature on page 267 aloud. Ask students whether Mendel's experiments would have been less valid if he had tested fewer plants, helping them recognize that using a large quantity of plants ensured that his findings were consistent and not based on one or two unusual results. Explain to students that although Mendel published the results of his experiments in 1866, they were ignored until 1900. Then three European scientists rediscovered and publicized the reports.

Then Mendel used a paintbrush to transfer pollen from another plant onto the female parts of the flowers that lacked the male organs. Mendel studied the inheritance of seven characteristics in pea plants: flower color, flower position, seed color, seed shape, pod shape, pod color, and stem length.

Based on his observations, Mendel concluded that organisms pass information about their characteristics to their offspring. He called this information *factors*. Today, we know that Mendel's factors are genes.

1 Warm-Up Activity

Invite students to bring to class photographs of members of their family. Display the photos and have the class determine which students are members of which families. Discuss traits and the fact that traits are inherited from parents. If students are adopted or are uncomfortable bringing in family photos, allow them to bring in photos of a friend's family, a celebrity's family, or a pet's family instead.

2 Teaching the Lesson

Have students write sentences that link the following pairs of terms: *recessive gene/dominant gene, P generation/F1 generation, F1 generation/F2 generation, heredity/genes, self-pollination/cross-pollination*.

Bring to class two flowers of the same species. Have students use the diagram on page 157 to identify the reproductive parts of each flower. Then, use the flowers to demonstrate the processes of self-pollination and cross-pollination. Good flowers to use are gladioli or lilies.

On the board, write several pairs of letters using the capital form and/or the lowercase form of the letter. For example, you might write *Rr, ZZ, cc, Gg*, and so on. Discuss which letters stand for dominant genes and which stand for recessive genes. Have students assign traits to each letter and discuss the genotypes and phenotypes represented by each letter pair. For example, R may stand for red eye color and is dominant to r; Rr is the genotype; and Rr represents an individual with a phenotype of red eyes.

Genetics Chapter 11 **267**

3 Reinforce and Extend

LEARNING STYLES

Logical/Mathematical

On the board, draw a Punnett square. Above the top two squares, write R and r. To the left of the lefthand squares, write r and r. On the two lefthand squares, write Rr. On the two righthand squares, write rr. Ask students the following questions: Write a ratio showing the chances the offspring will be rr ($\frac{2}{4}$). What percentage does this ratio represent? (50%) Write a ratio to show the chances the offspring will be Rr ($\frac{2}{4}$). What percentage does this ratio represent? (50%).

CROSS-CURRICULAR CONNECTION

Literature

Have students look at the illustration in *The Cartoon Guide to Genetics* by Larry Gonick and Mark Wheelis. This book includes wacky and interesting cartoons that explain Mendelian genetics and other topics in genetics. Have students choose one cartoon and relate it to the information in this lesson. Or, have students make up their own cartoons to represent a concept in this lesson.

Punnett Squares

Punnett square
A model used to represent crosses between organisms

Gene
The information about a trait that a parent passes to its offspring

Recessive gene
A gene that is hidden by a dominant gene

Dominant gene
A gene that shows up in an organism

A model called a **Punnett square** can be used to explain crosses like those Mendel did. The Punnett square below shows Mendel's cross of pure tall pea plants with pure short pea plants. *TT* stands for a pure tall parent plant, and *tt* stands for a pure short parent plant. The capital *T* represents the **gene** for tallness. The lowercase *t* represents the gene for shortness. A gene is the information for a trait that a parent passes to its offspring. Therefore, a pea plant receives, or inherits, two genes for height. One gene is inherited from each parent.

Notice that although all of the F_1 pea plants were tall, they also had the gene for shortness. A gene that is hidden when it is combined with another gene is called a **recessive gene**. The gene that shows up is called a **dominant gene**. A capital letter is used to represent a dominant gene. A lowercase letter is used to represent a recessive gene.

In the case of pea plants, the gene for tallness is dominant. The gene for shortness is recessive. That explains why all the pea plants in the F_1 generation were tall.

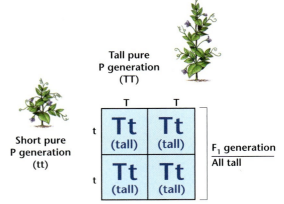

A Punnett Square of Mendel's Second Step

Genotype
An organism's combination of genes for a trait

Phenotype
An organism's appearance as a result of its combination of genes

F_2 generation
The plants that resulted when Mendel self-pollinated plants from the F_1 generation

Genotypes and Phenotypes

An organism's combination of genes for a trait is called its **genotype**. For example, the genotype of the F_1 pea plants was Tt. What an organism looks like as a result of its genes is its **phenotype**. The phenotype of the F_1 pea plants was "tall." An organism has both a genotype and a phenotype for all its traits.

The F_2 Generation

Review the diagram of Mendel's studies on page 267. Notice that Mendel produced yet another generation of pea plants by self-pollinating the F_1 generation plants. He called this third group of plants the **F_2 generation**. Mendel found that short pea plants reappeared in the F_2 generation. In fact, in the F_2 generation, about $\frac{3}{4}$ of the plants were tall, and about $\frac{1}{4}$ were short. Mendel repeated his experiments many times. He got similar results each time. The F_2 generation always included short pea plants.

The Punnett square below explains Mendel's results. Short plants reappeared in the F_2 generation because they inherited two recessive genes for shortness, one from each tall parent. The recessive genes were hidden in the tall parent plants.

The F_2 generation pea plants showed two different phenotypes: tall and short. However, notice in the Punnett square that the F_2 generation had three different genotypes. All the short plants had the genotype tt. But the tall plants had either the genotype TT or the genotype Tt.

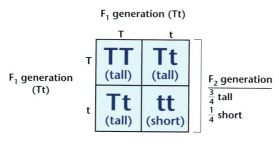

A Punnett Square of Mendel's Third Step

LEARNING STYLES

Auditory/Verbal
Students may have difficulty distinguishing *genotype* and *phenotype* because the meanings of the terms are similar and they sound much alike. Encourage students to create a mnemonic device that they can say to themselves as they use the terms. You might suggest the following mnemonic device: A **gen**otype is a trait's set of **gen**es; a **phe**notype is the inherited **fe**ature that is seen. Remind students that *ph* is pronounced *f* in phenotype.

SCIENCE JOURNAL

Have students write a journal entry about genetics that Mendel might have written as he conducted his experiments with pea plants.

LEARNING STYLES

Body/Kinesthetic
Use masking tape or chalk to outline a large Punnett square on the floor. Have students make index cards with letters for dominant and recessive genes. Two students with cards selected by you should stand in each of the four blocks of the square. The pairs tell whether they represent the inheritance of dominant or recessive traits based on the letters they are holding. Repeat the activity several times so that students recognize that to exhibit a recessive trait, an organism must inherit the recessive gene for that trait from both parents.

LEARNING STYLES

Visual/Spatial

Have students draw a two-column chart. In the first column, they should write what Gregor Mendel did in each phase of his study. In the second column, they should list the results of his work in each phase. Students can use Punnett squares in their charts if they wish. Suggest that they keep their chart as a study tool.

Factors
The name that Mendel gave to information about traits that parents pass to offspring

Mendel's Conclusions

Mendel also studied other traits of pea plants. Each trait was inherited the same way as plant height. For example, Mendel obtained only round seeds when he crossed plants pure for round seeds with plants pure for wrinkled seeds. He obtained only green pea pods when he crossed plants pure for green pea pods with those pure for yellow pea pods.

Mendel concluded that there was information in a plant that caused it to have certain traits. He called this information **factors**. He reasoned that since traits came in pairs, the factors did, too. Mendel also thought that one factor of the pair was more powerful than the other factor. The more powerful factor hid the appearance of the weaker factor. Today, we know that Mendel's factors are genes. Dominant genes hide the appearance of recessive genes.

Mendel also thought that the paired factors separated during sexual reproduction. That way, the offspring received half the factors from one parent and half from the other parent. In Lesson 2, you will learn how genes are inherited during sexual reproduction.

Genes determine whether a pea plant is short or tall. Genes also determine your traits. Genes determine whether your eyes are brown or blue. They determine whether you get freckles when you go in the sun. In all organisms, genes carry the information about traits from parents to their offspring.

Lesson 1 REVIEW

Write your answers to these questions on a separate sheet of paper. Write complete sentences.

1. What is heredity?
2. Describe the three steps in Mendel's studies of pea plants.
3. Contrast the terms *dominant gene* and *recessive gene*.
4. What is the difference between a genotype and a phenotype?
5. Draw a Punnett square to represent Mendel's cross between pure tall plants and pure short plants.

Technology Note

The Human Genome Project (HGP) was a joint international research program to map the different genes in the human body. A genome is all the DNA in an organism.

The project began in 1990 and was completed in April of 2003. Scientists have put in order the 3.1 billion units of DNA that make up the human genome.

Genes carry various hereditary conditions, ranging from cystic fibrosis to Alzheimer's. Now it's possible to test whether people might have certain genetic disorders. Rapid technological advances in genetics testing has greatly improved our ability to make the detection of genetic conditions earlier, simpler, and more precise.

Genetics Chapter 11 271

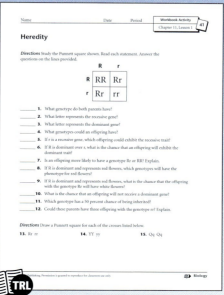

Workbook Activity 41

Lesson 1 Review Answers

1. Heredity is the passing of genes for traits from parents to offspring. 2. First, Mendel produced plants that were pure for a certain trait, such as tallness or shortness. Then, he cross-pollinated the pure plants. Finally, he self-pollinated the offspring from that cross. 3. A dominant gene shows up in an organism. A recessive gene is hidden by a dominant gene. 4. Genotype is an organism's combination of genes for a trait. Phenotype is the appearance of an organism as a result of its genotype. 5. See the Punnett square on page 268.

Portfolio Assessment

Sample items include:
- Sentences from Teaching the Lesson
- Lesson 1 Review answers

LEARNING STYLES

Logical/Mathematical Have student pairs make three sets of cards for these dominant and recessive traits of pea plants: tall plants/short plants—TT, Tt, Tt, and tt; round seeds/wrinkled seeds—RR, Rr, Rr, and rr; and green pod/yellow pod—GG, Gg, Gg, and gg. Students should mix the cards in three separate piles and take a card from each pile. They then identify what characteristics a F_2 generation would exhibit based on their cards.

Genetics Chapter 11 271

Lesson at a Glance

Chapter 11 Lesson 2

Overview In this lesson, students learn about how the processes of mitosis and meiosis relate to asexual and sexual reproduction. The lesson concludes with information about sex chromosomes and sex-linked traits.

Objectives

- To explain what a chromosome is
- To compare mitosis with meiosis
- To explain how sex is determined in humans
- To explain what a sex-linked trait is

Student Pages 272–278

Teacher's Resource Library

Workbook Activity 42
Alternative Activity 42
Lab Manual 32

Vocabulary

sex chromosome
sex-linked trait
carrier

Science Background
Basic Genetics

Genes are found in an organism's DNA molecules, which include four chemicals (adenine, thymine, guanine, and cytosine) called bases. A gene is a specific sequence of the bases in a DNA molecule. The sequence may be thousands of bases long. The base sequence in a gene provides a code that determines a certain characteristic of an organism. Organisms that reproduce sexually inherit two genes for a characteristic, one from each parent. A gene may be dominant or recessive. A dominant gene for a characteristic is always expressed in an organism. A recessive gene for a characteristic is not expressed in the presence of a dominant gene for that characteristic. An organism must inherit two recessive genes for a characteristic to be expressed.

Lesson 2: Chromosomes

Objectives

After reading this lesson, you should be able to

- explain what a chromosome is.
- compare mitosis with meiosis.
- explain how sex is determined in humans.
- explain what a sex-linked trait is.

Chromosomes are rod-shaped bodies located in the nucleus of a cell. Chromosomes are made of proteins and a chemical called DNA. Sections of DNA make up an organism's genes, which determine all the traits of an organism. A chromosome may contain hundreds of genes.

Mitosis and Cell Division

One-celled organisms, such as amoebas, reproduce by splitting in half. Before doing so, the amoeba's chromosomes make a copy of themselves. Then, the amoeba's nuclear membrane dissolves. The two sets of chromosomes separate, and a nucleus forms around each set. As you learned in Chapter 9, the division of the nucleus into two new nuclei is called mitosis.

Following mitosis, the entire cell divides. When the amoeba becomes two amoebas, each new amoeba gets one nucleus with a complete set of chromosomes. Each set of chromosomes is identical to the parent amoeba's chromosomes. Thus, each new amoeba is identical to the parent amoeba.

Just as amoebas go through mitosis and cell division, so do the cells that make up the human body. The photo shows a human cell dividing. Body cells divide by mitosis and cell division as the body grows and repairs itself. However, to reproduce itself, the body uses a different kind of process.

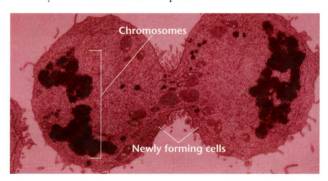

272 Chapter 11 Genetics

Sexual Reproduction

Humans and most other many-celled organisms reproduce sexually. In sexual reproduction, two cells called gametes join to form one complete cell. Males produce gametes called sperm cells. Females produce gametes called egg cells. Each gamete has only one-half of the chromosomes found in the organism's body cells. When gametes join, they form a cell that has a complete set of chromosomes.

Humans have 46 chromosomes in their body cells. Notice in the diagram below that there are only 23 chromosomes in human sperm cells. Human egg cells also have only 23 chromosomes. When humans reproduce, a sperm and an egg join to form a cell called a zygote. A zygote has 46 chromosomes and eventually develops into an adult.

Together, the 46 human chromosomes contain approximately 30,000 genes. The mix of chromosomes from both parents during sexual reproduction produces an offspring that is different from either parent.

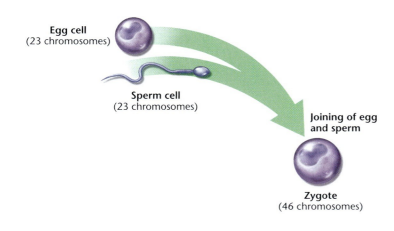

Cell Reproducing

An organism's DNA molecules are found in structures called chromosomes, which consist of DNA and proteins. In most organisms, the chromosomes are found in the nucleus of a cell. Before a cell can reproduce, its chromosomes must duplicate. Mitosis is the process by which a cell reproduces to form two cells. The body cells of a multicellular organism reproduce by mitosis. Mitosis also occurs in organisms that reproduce asexually. During mitosis, the duplicated chromosomes in the cell separate and the nucleus divides in two. Then the cell divides to form two new cells, with each cell having the same number of chromosomes found in the original cell.

Meiosis is the process by which organisms that reproduce sexually form sex cells. As in mitosis, the nucleus of a cell divides during meiosis. However, during meiosis, cell division occurs twice, producing four sex cells. Each sex cell has one-half the number of chromosomes found in the original cell.

1 Warm-Up Activity

Show students a picture of a chromosome. Explain that before a cell divides, DNA winds itself up so tightly that it has a rod shape. The rod-shaped structure is a chromosome.

2 Teaching the Lesson

After students read about mitosis and meiosis, have them make a graphic organizer to compare the two processes.

Review Morgan's work described on pages 276–277 by asking students the following questions: What is the difference between the sex chromosomes and other pairs of chromosomes? Write the symbols for the female and male sex chromosomes in fruit flies. (*Sex chromosomes determine an individual's sex. Female—XX; Male—XY*) What is the genotype of a red-eyed female fruit fly? (*$X^R X^R$ or $X^R X^r$*)

TEACHER ALERT

Tell students that in the past, the woman was frequently blamed when a couple failed to produce a child of the sex the couple wanted. Point out that the female egg does not determine the sex of the child a woman bears. Female gametes contain only an X chromosome. The male gametes contain either an X or a Y chromosome. Thus, the male's gamete actually determines the sex of the child.

3 Reinforce and Extend.

LEARNING STYLES

Visual/Spatial

Show students the video "Generations: Mitosis and Meiosis" Part 6 in the *Cycles of Life: Exploring Biology* program by The Annenberg/CPB Project. This 30-minute video includes information about cell reproduction and the passing of traits to offspring. After viewing the video, have students summarize the differences between mitosis and meiosis.

CROSS-CURRICULAR CONNECTION

Language Arts

Explain that the word *meiosis* comes from a Greek word meaning "diminish" or "reduce." Have students explain how knowing this etymology can help them remember the meaning of meiosis.

Meiosis

As you know, mitosis results in new cells that have a complete set of chromosomes. Gametes form by a different kind of process. They form by a division of the nucleus called meiosis.

The diagram below compares the process of meiosis with that of mitosis. As in mitosis, meiosis begins with the copying of the chromosomes in a parent cell. Then, the cell divides into two new cells. However, in meiosis, each new cell divides again. Thus, one parent cell results in four new sex cells. Each sex cell contains one-half the number of chromosomes of the parent cell.

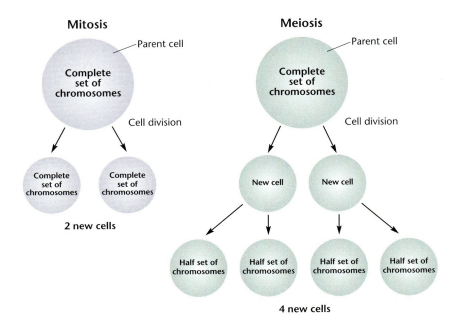

Sex chromosome
A chromosome that determines the sex of an organism

Did You Know?
Some human females and males have an extra X chromosome. Some human males have an extra Y chromosome. The extra chromosome resulted from an error that occurred during meiosis in one of their parents.

Sex Chromosomes

Humans have 46 chromosomes. The chromosomes consist of 23 pairs. Each chromosome that makes up a pair comes from a different parent. For 22 of the pairs, the two chromosomes look much alike. However, the chromosomes that make up the 23rd pair look different from each other. These two chromosomes are called **sex chromosomes** because they determine a person's sex. There are two kinds of human sex chromosomes: an X chromosome and a Y chromosome.

A human female has two X chromosomes. A human male has one X chromosome and one Y chromosome. Parents pass one of their sex chromosomes to their offspring. A male can pass an X chromosome or a Y chromosome to its offspring. A female can pass only an X chromosome.

The Punnett square below shows how human sex chromosomes determine the sex of an offspring. An offspring that inherits an X chromosome from both parents is a female (XX). An offspring that inherits an X chromosome from its mother and a Y chromosome from its father is a male (XY). Notice that the chance of producing a female offspring is 50 percent. The chance of producing a male offspring is also 50 percent.

	Female parent	
	X	X
Male parent X	XX (female)	XX (female)
Y	XY (male)	XY (male)

Offspring
50% chance of being female (XX)
50% chance of being male (XY)

Genetics Chapter 11 **275**

Career Connection

Have students interview professionals in various fields, including agriculture, biotechnology, health care, and environmental science, to discover the importance of genetics to different professions. Encourage students to seek out people in the local community, such as those working at a local farmer's market, nursery, community college or university, health-care center, and so on. In addition, many Web sites offer an "ask a professional" column that may be a good source of information for students. Have students report the results of their research to the class. You may want to use the information students find to make a concept map of the many professions that are related to genetics.

Did You Know?

Have a student read aloud Did You Know? on page 275. Explain that the Y chromosome determines the sex of a person. A male with two X chromosomes and a Y chromosome is still a male. However, most people born with this condition are unable to have children as adults.

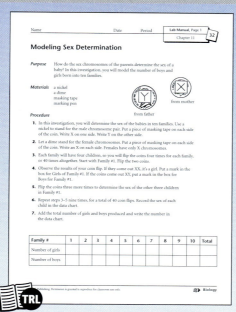

Lab Manual 32, pages 1–2

Genetics Chapter 11 **275**

CROSS-CURRICULAR CONNECTION

Math

Every species has a set number of chromosomes, which are paired, in each body cell. List the following organisms and their number of chromosomes. Ask students to identify how many pairs of chromosomes each organism has in its body cells.

Organism	Number of Chromosomes
Fruit flies	8
Petunia	14
Corn	20
Tomato	24
Yeast	32
Cat	38
Human	46
Potato	48
Horse	64
Dog	78
King crab	208

ONLINE CONNECTION

Students can learn more about chromosomes and their relationship to genes by visiting the Waksman Student Scholars Web site. The site is located at avery.rutgers.edu/WSSP/StudentScholars/Session8/Session8.html.

Sex-linked trait
A trait that is determined by an organism's sex chromosomes

Experiments with Fruit Flies

Recall that Mendel used pea plants in his study of heredity. About 50 years later, another scientist, Thomas Morgan, used fruit flies to learn about chromosomes and genes. Fruit flies are useful to study for a number of reasons. Their cells have only four pairs of chromosomes. The chromosomes are large and easy to see under a microscope. Fruit flies reproduce quickly. In addition, it is easy to tell the male fruit fly from the female.

Sex-Linked Traits

Fruit flies usually have red eyes. However, Morgan noticed that one of the male fruit flies had white eyes. Morgan mated the white-eyed male with a red-eyed female. He called the offspring the F_1 generation. All the offspring in the F_1 generation had red eyes. Morgan concluded that the gene for red eyes was dominant in fruit flies.

Next, Morgan mated the flies from the F_1 generation to produce an F_2 generation. This time, some of the offspring had red eyes and some had white eyes. However, all the white-eyed flies were males. None of the females had white eyes. Morgan concluded that white eye color in fruit flies is linked to the sex of the fly. Traits that are linked to the sex of an organism are said to be **sex-linked traits**.

As in humans, a female fruit fly has two X chromosomes. A male has one X chromosome and one Y chromosome. Morgan found that the gene for eye color in fruit flies is on the X chromosome. There is no gene for eye color on the Y chromosome. That explains why eye color in fruit flies is a sex-linked trait.

Carrier
An organism that carries a gene but does not show the effects of the gene

The Punnett squares below show Morgan's experiments. Notice that only the X chromosome carries a gene for eye color. The genotypes $X^R X^R$ and $X^R X^r$ represent a red-eyed female fly. The genotype $X^R Y$ represents a red-eyed male. White-eyed males have the genotype $X^r Y$.

Notice that a female fly with the genotype $X^R X^r$ has red eyes even though it carries a gene for white eyes. An organism that carries a gene but does not show the gene's effects is called a **carrier**.

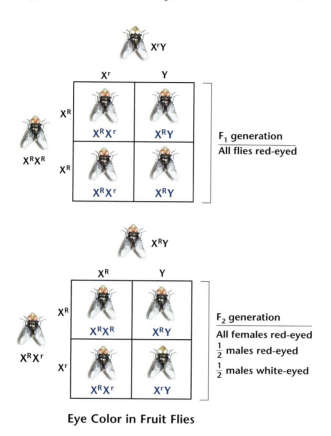

Eye Color in Fruit Flies

CROSS-CURRICULAR CONNECTION

History
Thomas Morgan won the 1933 Nobel Prize for Physiology and Medicine for his work in genetics. He was the first American-born scientist to win the award. Many of his students were involved in his study of fruit flies and their traits. One of them, Hermann Joseph Muller, also won the Nobel Prize for Physiology and Medicine. In 1946, he won the award for his work in genetic mutations.

SCIENCE JOURNAL

Ask students to predict the color of eyes of a male and female in the F_2 generation of fruit flies. The parents of the male and female have these genotypes: Female parent, $X^R X^R$, and Male parent, $X^r Y$. Students should state reasons for their predictions.

LEARNING STYLES

Interpersonal/Group Learning
Have students conduct a survey of friends and relatives to identify recessive and dominant traits. Suggest they use these traits: ability to roll tongue (dominant trait) and inability to roll tongue (recessive trait) and detached earlobes (dominant trait) and attached earlobes (recessive trait). Have students tally the results and prepare graphs. They may wish to conduct the survey so that they can tell whether the same individuals exhibit two recessive traits, two dominant traits, or a recessive and a dominant trait.

Lesson 2 Review Answers

1. Chromosomes are rod-shaped structures inside the nucleus of a cell that carry genes. **2.** Mitosis is the division of the nucleus that results in two new cells that are identical to the parent cell. Mitosis occurs in the human body during growth and repair. **3.** Meiosis is the division of the nucleus that occurs during the formation of sex cells. Meiosis results in four sex cells. **4.** human female: XX; human male: XY **5.** The gene for eye color is found only on the X chromosome, not on the Y chromosome.

Portfolio Assessment

Sample items include:
- Graphic organizer from Teaching the Lesson
- Lesson 2 Review answers

Science Myth

Have one student read the myth and one the fact portion of the Science Myth on page 278. Explain that brown eyes are a dominant trait and blue eyes are a recessive trait. Therefore, a person must inherit two recessive genes for blue eyes in order to have blue eyes.

Lesson 2 REVIEW

Write your answers to these questions on a separate sheet of paper. Write complete sentences.

1. What are chromosomes?
2. Describe mitosis.
3. Describe meiosis.
4. What are the sex chromosomes of a human female? What are the sex chromosomes of a human male?
5. Why is eye color a sex-linked trait in fruit flies?

Science Myth

Eye color is determined by one pair of genes with one gene from the father and one gene from the mother.

Fact: Your eye color is an example of a trait that is determined by many different pairs of genes. Each pair is inherited independently from the other pairs. That explains why blue eyes are not all the same blue.

Workbook Activity 42

Lesson 3: How Heredity Is Studied in Humans

Objectives

After reading this lesson, you should be able to
- explain some ways that scientists study heredity in humans.
- describe DNA and how it replicates.
- explain what a mutation is.
- give examples of genetic diseases.

Identical twins
Twins that have identical genes

Fraternal twins
Twins that do not have identical genes

Human genetics is the study of how humans inherit traits. Human genetics is more complicated than the genetics of fruit flies for a number of reasons. First, humans have more chromosomes than fruit flies do. Second, humans do not reproduce as quickly as flies. And third, humans cannot be used in laboratory experiments. Scientists study heredity in humans by studying identical twins. They also look at families to learn how genetic diseases are passed from one generation to the next.

Different Kinds of Twins

There are two types of twins: **identical twins** and **fraternal twins**. As you can see below, they form in different ways. Identical twins form from the same zygote and therefore have identical genes. First, a sperm cell joins with an egg cell to form a zygote. Then, the zygote divides into two cells that separate completely. Each cell develops into an offspring, resulting in identical twins.

Fraternal twins do not have identical genes. They are just like other brothers and sisters. Fraternal twins form from two different zygotes. The zygotes develop into offspring with different sets of genes.

A woman has an increased chance of having fraternal twins if she has a mother, a sister, or an aunt who had fraternal twins. The father does not seem to influence the chance of his offspring being twins.

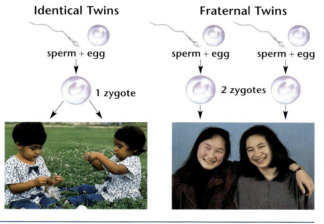

Genetics Chapter 11 279

Teaching the Lesson

Have students write sentences that explain how scientists study heredity in humans.

Tell students that the two strands in DNA are called complementary strands. Explain that complementary means "completing." Point out that two strands join together to form a complete DNA molecule. You may want to encourage students to suggest examples of things that are complementary from other areas, such as complementary angles in geometry. Distinguish *complementary* from *complimentary*.

Write the following words on the chalkboard: *bag, bad, bat*. Replace each letter *a* with an *e*. Discuss how changing one letter can change the meaning of a word. Explain that analogous changes in the base sequence of an individual's DNA can change the trait that is inherited.

 Reinforce and Extend

LEARNING STYLES

 LEP/ESL
Review the steps in DNA replication using the illustration on page 281. Ensure that students can identify the original strands of DNA and the new strands. Have students compare the two new DNA molecules with the original DNA molecule.

SCIENCE INTEGRATION

 Physical Science
After students have read about DNA replication, point out that the bases are always paired—T (thymine) is always paired with A (adenine) and G (guanine) is always paired with C (cytosine). In DNA, these bases are arranged like steps on a spiral staircase. Suggest that students identify the chemical composition of each base.

Environment
An organism's surroundings

Mutation
A change in a gene

Heredity and Environment

You are born with certain genes. That is your heredity. Your genes determine your skin color, eye color, body shape, and other characteristics. But, your **environment** also may affect your characteristics. Your family, the air you breathe, and everything else in your surroundings make up your environment. To find out how environment affects a person's characteristics, scientists study identical twins who have been separated since birth. Both twins have the same genes, but they grew up in different environments. Scientists look for differences in the characteristics of the twins. If they find any different characteristics, they know the environment caused those differences.

The Influence of Environment

Food, sunlight, air, and other parts of the environment can affect a person's characteristics. For example, a person who doesn't have good nutrition may not grow tall, even though he has a gene for tallness. A person who avoids sunlight may not form freckles, even though she has a gene for freckles. The environment also affects a developing baby. Studies show that babies born to women who smoke cigarettes weigh less than babies born to women who are nonsmokers.

The environment can directly affect a person's genes. For example, X rays and some types of chemicals cause changes in genes. These changes are called **mutations**. Mutations can cause problems in humans and other organisms. How do mutations occur? Before you can understand mutations better, you need to know more about DNA.

DNA

DNA in chromosomes is the material that contains an organism's genes. Each DNA molecule in a cell makes one chromosome. DNA passes the genes from one cell to another during cell division. All the information needed to carry out life activities is in DNA. All the information that makes a duck a duck is in the duck's DNA. All the information that makes you a human is in your DNA.

Base
A molecule found in DNA that is used to code information

You can see below that DNA is a large molecule shaped like a twisted ladder. The rungs of the ladder are made of four different kinds of molecules called **bases**. The letters *T*, *A*, *C*, and *G* are used as abbreviations for the names of the four bases. The order of the bases in the DNA molecules of a cell provides a code for all the information that the cell needs to live.

The DNA molecules of different organisms have different orders of bases. The greater the difference is between the organisms, the greater is the difference in the order of their bases. Thus, the order of bases in a frog's DNA is very different from the order of bases in your DNA. The difference in the order of bases between your DNA and your friend's DNA is not as great.

Recall that a gene is a section of a DNA molecule. Each DNA molecule has thousands of genes. A gene is made up of a certain order of bases. Different genes have a different order of bases. This difference allows genes to provide different kinds of information. For example, the order of bases in a gene for hair color determines whether the hair will be black, brown, red, or blonde.

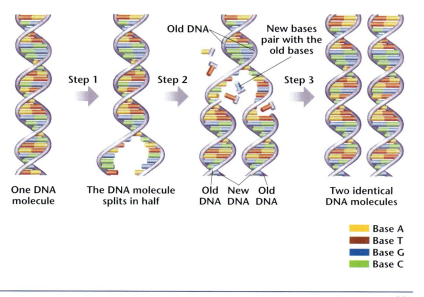

LEARNING STYLES

Interpersonal/Group Learning

Body weight is one area in which the effects of environment on a person's physical appearance are evident. Researchers have found that more Americans are obese than people from most other nations. Have students find more facts about this phenomenon and conduct research to find reasons for it. For example, experts have theorized that fast foods and processed foods contributed to Americans' obesity; and that the whole grain breads and rice eaten by people in other areas of the world are less fattening. Have students make their own hypotheses, conduct studies, and report on their findings.

IN THE ENVIRONMENT

Point out that certain substances that are released into our environment can be mutagens, or agents that cause gene mutations. For example, mustard gas, used for chemical warfare in World War I, is a mutagen. Some pesticides and chemicals released into the environment by industry are also mutagens. Encourage interested students to find out some specific chemicals that are proven mutagens.

CROSS-CURRICULAR CONNECTION

Health

Discuss the fact that the sun's ultraviolet rays can cause changes to DNA that can cause skin cancer. Explain that ultraviolet rays are a form of radiation. Excessive amounts of other types of radiation, such as X rays, can also cause changes in genes that can cause disease. Have students discuss the importance of using sunscreen and avoiding excessive sunbathing and tanning beds.

GLOBAL CONNECTION

Explain that Huntington's disease, discussed in Investigation 11, is a genetic disorder. It is more common in South Africa than in other parts of the world. Researchers determined that one Dutch settler, who moved to South Africa in 1658, had the disease. All South Africans who suffer from the disease are directly or indirectly descended from this settler. This is a phenomenon known as the founder effect: one or several people with a genetic abnormality create a new population of people with the abnormality. The founder effect is more likely to occur in a place that is isolated and has a small population. For example, other genetic disorders caused by the founder effect have been found among the Amish in Lancaster, Pennsylvania, and on Martha's Vineyard in the 1700s, when it was isolated from the American mainland.

Replicate
To make a copy of

Genetic disease
A disease that is caused by a mutated gene

Diabetes
A genetic disease in which a person has too much sugar in the blood

An important feature of DNA is that it can **replicate**, or copy, itself. A DNA molecule is held together at its rungs. Notice in the diagram on page 281 that each rung is made of a pair of bases. The bases pair in certain ways. Base A pairs with base T. Base C pairs with base G. When DNA replicates, it first splits down the middle of its rungs. The paired bases separate. Then, new bases pair with the separated bases on each half of the DNA molecule. The result is two identical copies of the original DNA molecule.

DNA replication occurs every time a cell divides normally. The new cells receive copies of the DNA molecules. In this way, genetic information is passed on from cell to cell.

Mutations

Sometimes there is a change in the order of bases in a DNA molecule. This change is called a mutation. Recall that parts of the environment, such as X rays and chemicals, can cause mutations.

Mutations cause changes in genes. Since genes determine traits, mutations may affect the traits of an organism. For example, white eye color in fruit flies is the result of a mutation. Mutations may be harmful or helpful to organisms or may have no effect at all. Only gene mutations in sex cells can be passed on to offspring.

Genetic Diseases

A **genetic disease** is a disease that results from the genes a person inherits. Recessive genes cause most genetic diseases. That means a person must inherit the recessive gene for the disease from both parents to have the disease. Dominant genes cause other genetic diseases. Genetic diseases result from mutations.

A recessive gene causes a form of **diabetes**. The gene prevents the body from making insulin. Insulin is a protein that controls the amount of sugar in the bloodstream. People who have diabetes have too much sugar in their blood. People with this form of diabetes must take insulin, either by mouth or by injection, to lower the amount of blood sugar.

Sickle-cell anemia
A genetic disease in which a person's red blood cells have a sickle shape

Gene pool
The genes found within a population

Inbreeding
Sexual reproduction between organisms within a small gene pool

Hemophilia
A genetic disease in which a person's blood fails to clot

Did You Know?

Queen Victoria of England might have been the first carrier of hemophilia within the royal families of Europe. A mutation may have been present in her mother's or father's sex cell. Although Queen Victoria did not have hemophilia, she passed the gene for it to her daughters. The daughters then passed the gene to their sons.

A recessive gene also causes **sickle-cell anemia**. You can see in the photos that the disease causes a person's red blood cells to have a sickle shape. The sickle cells clog blood vessels. People with sickle-cell anemia have weakness and an irregular heart beat.

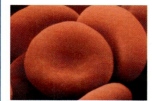
Normal red blood cells

Sickled red blood cell

Inbreeding

With the billions of people who live in the world, there is a great variety of human genes. A **gene pool** is all the genes that are found within a certain population. The larger the population is, the larger is the gene pool. In a large population, a person who is a carrier of a genetic disease is unlikely to marry someone who also is a carrier. Their children will probably not have the disease. However, people living in a small population often marry each other. Their gene pool is small. After several generations, there is a greater likelihood of couples having children with genetic diseases. The same is true when people marry close relatives. Sexual reproduction among people who are part of a small gene pool is called **inbreeding**.

Inbreeding caused a disease called **hemophilia** to occur frequently within the royal families of Europe. In hemophilia, a gene that causes blood to clump, or clot, is not normal. As a result, the blood does not have a protein it needs to clot. People who have hemophilia bleed a great deal when they are just slightly injured. They need to receive the protein from normal blood to help their blood clot.

Hemophilia is sometimes called the "Royal Disease." People from royal families generally married other people from royal families, including cousins. The royal gene pool was very small. This increased the likelihood of hemophilia showing up in the family members.

SCIENCE JOURNAL

Have students describe some of their most striking physical traits, such as skin color, eye color, and body shape, and trace the trait through their parents, grandparents, aunts and uncles, to themselves.

IN THE COMMUNITY

Have students seek out information about services and support groups for people who have different genetic diseases. You may want to have a counselor or representative of a group speak to the class about a specific genetic disease and people who have that disorder. Help students understand that many people who have a genetic disease can lead full lives. Encourage students to be accepting of others who may have different abilities. Be sensitive to the feelings of students in the class who may have a genetic disease.

Did You Know?

Have a student read aloud Did You Know? on page 283. Students may be surprised to learn that hemophilia may have had a great effect on history. The hemophilia gene passed from Queen Victoria to her granddaughter, Alexandra, who married the Russian czar Nicholas II. Their son Alexis was born with the disease. Historians speculate that Alexandra's obsession with her son's health problems caused her to rely on the advice of Rasputin, a monk who said he could cure Alexis. Having gained the trust of the royal family, Rasputin soon exerted enormous and damaging control over the king. The Russian people revolted against Nicholas in the Revolution of 1917.

Science Integration

Technology

Medical technology has enabled doctors to test fetuses for genetic diseases. This is done through amniocentesis, in which a small amount of amniotic fluid is taken from a pregnant woman's uterus. When done in the fourth month of pregnancy, this enables doctors to test cells for many genetic diseases, such as Down's syndrome and Tay-Sachs disease. The test can also be done in the last three months of pregnancy to check a baby's stage of development. Interested students can research how amniocentesis is done and what specific information it can provide.

Science in Your Life

Have volunteers read aloud the feature on page 284. Point out that DNA profiling has been used in recent years to exonerate individuals who had previously been convicted of certain crimes. Have students describe cases they recall or research one or more well-known cases.

Human Sex-Linked Traits

In Lesson 2, you learned that eye color in fruit flies is sex-linked. A number of human traits also are sex-linked. For example, the gene for color blindness is a recessive gene found on the X chromosome. People who are color-blind have trouble telling certain colors apart. For a female to be color-blind, she must inherit two genes for color blindness, one on each X chromosome. However, a male who inherits a single gene for color blindness on his X chromosome is color-blind. For that reason, color blindness is more common in males than in females. Hemophilia and muscular dystrophy are two other sex-linked traits found in humans. They are both caused by a recessive gene found on the X chromosome. Therefore, they also are more common in males than in females.

Science in Your Life

What is DNA fingerprinting?

You know what fingerprints are. Sometimes you can see them when your hands are dirty and you leave a fingerprint on a glass or other object. Even if your hands are clean, you leave your fingerprints on just about everything you touch. You usually can't see them. Unless you have an identical twin, your fingerprints are unique. They can be used to identify you.

You have probably heard about DNA fingerprints in the news. They are used to place a criminal at the scene of a crime, prove a person innocent, identify victims, and prove or disprove that two people are related. What is a DNA fingerprint?

Most of the DNA in human cells contains information that makes a person a human instead of another kind of organism. Only one-tenth of one percent of a person's DNA is unique to that person. This amount may seem small, but it includes about 3 million bases. Samples of blood, bone, hair, semen, or other body tissues and fluids contain a person's DNA. By analyzing samples for regions of the DNA that vary from person to person, scientists can create a DNA fingerprint, or profile, for the person.

There is a very small chance that two people have the same DNA profile for a particular set of regions or markers. The more regions that match, the more likely the DNA profile is unique to an individual. For example, when matching a sample from a crime scene to a suspect, four or five matching regions are usually required. It is rare for two individuals to match in five regions. If you are ever on a jury and DNA fingerprint evidence is being given, pay attention to how many regions or markers match.

Lesson 3 REVIEW

Write your answers to these questions on a separate sheet of paper. Write complete sentences.

1. What is the difference between identical twins and fraternal twins?
2. Give an example of how both heredity and environment may affect a certain human characteristic.
3. What is DNA?
4. What is a mutation?
5. Define the term *genetic disease* and give an example of a genetic disease.

Science at Work

Genetic Counselor

A genetic counselor must be good at solving problems, interpreting medical information, and analyzing patterns and risks. A college degree in a field such as biology, genetics, nursing, or psychology is needed. A graduate degree in genetic counseling is also needed. Certification as a genetic counselor requires clinical experience and passing the examination of the American Board of Genetic Counseling (ABGC).

Genetic counselors counsel families who might have offspring with genetic diseases or birth defects. Genetic counselors identify possible risks, investigate the problem in the family's history, and explain information to the family. Most genetic counselors work in medical centers or hospitals. Some work in research laboratories that diagnose genetic problems, drug companies, private practice, or public-health agencies.

Genetics Chapter 11 285

Workbook Activity 43

Lesson 3 Review Answers

1. Identical twins form from the same zygote and have identical genes. Fraternal twins form from different zygotes and do not have identical genes. 2. Answers may vary. Students may describe the effect of heredity and environment on height, freckles, or birth weight of newborns. 3. DNA is a large molecule that contains an organism's genes. It stores all the information that an organism needs to carry out its life activities. 4. a change in the order of bases in a DNA molecule 5. A genetic disease is a disease that is caused by a mutated gene. Examples include diabetes, sickle-cell anemia, hemophilia, and muscular dystrophy.

Portfolio Assessment

Sample items include:
- Sentences from Teaching the Lesson
- Lesson 3 Review answers

Science at Work

Have volunteers read aloud Science at Work on page 285. Hold a class discussion about whether they think all couples should participate in genetic counseling before having children.

Genetics Chapter 11 285

Investigation 9

This investigation will take approximately 45 minutes. Students will use these process and thinking skills: observing, communicating, inferring, interpreting charts, making and using models.

Preparation

- You may wish to prepare a transparency for the pedigree shown on page 286. Use the transparency to discuss the pedigree with the class.

Procedure

- To help students orient themselves to the pedigree, review the key with them before allowing them to begin the investigation.
- Students should use the pedigree given and its key as the basis for the pedigree that they draw. Their pedigree should show three generations: John, his parents, and his grandparents.
- Students may use Lab Manual 33 to record their data and answer the questions.

INVESTIGATION 11

Materials
- paper
- pencil

Tracing a Genetic Disease

Purpose

How can a genetic disease in a family be traced from generation to generation? In this investigation, you will trace a genetic disease by using a family history and a diagram called a pedigree.

Procedure

1. The diagram below is called a pedigree. Pedigrees are used to trace traits from generation to generation. This pedigree shows the inheritance of a trait called albinism. A person who is an albino lacks coloring. As a result, the person has milky-white skin and hair and pink eyes.

2. Compare each set of parents in the pedigree with their children. Based on the pedigree, determine whether a recessive gene or a dominant gene causes albinism. (*Hint:* If a recessive gene causes albinism, then two normal parents are able to produce a child with albinism. If a dominant gene causes albinism, then every albino child will have at least one albino parent.)

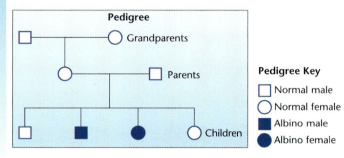

286 *Chapter 11 Genetics*

3. Read the following family history.

   Huntington's disease is a genetic disorder that affects the nervous system. John has Huntington's disease, but his sister, Nancy, does not. John's father also had Huntington's disease, but his mother did not. John's father's mother (John's grandmother) had Huntington's disease. John's father's father (John's grandfather) did not.

4. On a sheet of paper, draw a pedigree to show the history of Huntington's disease in John's family. Make a pedigree key similar to the one shown on page 286.

Questions and Conclusions

1. Based on the pedigree shown on page 286, is albinism caused by a recessive gene or by a dominant gene? Explain.
2. Look at the pedigree you drew in step 4. How many generations does your pedigree show?
3. Does the pedigree you drew show a genetic disease that is caused by a recessive gene or by a dominant gene? How do you know?
4. From which parent did John inherit Huntington's disease?
5. Do you think that Nancy is a carrier for Huntington's disease? Why or why not?

Explore Further

1. Based on your answer to question 3 above, give John's genotype for Huntington's disease. Explain your answer.
2. Assume that John marries a woman who does not have Huntington's disease. Draw a Punnett square to show the chance that a child of theirs will inherit the disease.

Results

Students' pedigrees should show John's grandmother as a female with Huntington's disease and the grandfather as a male without the disease; John's mother as a female without the disease and his father as a male with the disease; and John as a male with the disease and his sister as a female without the disease.

Questions and Conclusions Answers

1. A recessive gene causes albinism. Two normal parents can produce an albino child.
2. The pedigree should show three generations: John, his parents, and his grandparents.
3. Huntington's disease is inherited by a dominant gene. Every person with Huntington's disease has a parent who also has the disease.
4. John probably inherited Huntington's disease from his father.
5. Nancy is not a carrier. If she had inherited a gene for Huntington's disease, she would also have the disease because the gene is dominant.

Explore Further Answers

1. John's genotype is probably Hh. He received a dominant gene for Huntington's disease from his father. He received a normal gene from his mother.
2. The Punnett square should show a cross between the genotypes Hh and hh. John's child has a 50 percent chance of inheriting Huntington's disease.

Assessment

Check students' answers against the pedigree they drew. You might include the following items from this investigation in student portfolios:

- Pedigree
- Answers to Questions and Conclusions and Explore Further sections
- Punnett square from Explore Further

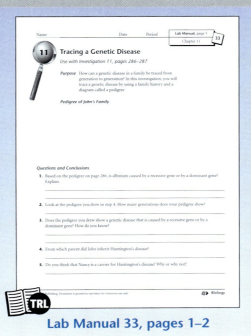

Lab Manual 33, pages 1–2

Genetic engineering *The process of transferring genes from one organism to another*

Breeding Racehorses

Selective breeding is used to produce great racehorses. The horse named Secretariat won the three most important horse races in the United States. To breed a winner, Secretariat's owner selected parents with desirable genes. Secretariat's father was known for fathering horses that won short races. From his mother's side, Secretariat inherited strength and the ability to go long distances. The result was an oversized horse with excellent speed and a stride of about 25 feet. Breeders select horses that have a good temperament as well. An animal's genes also affect its behavior.

Genetic Engineering

In the 1970s, scientists began introducing new genes into organisms. This process is called **genetic engineering**. At first, scientists transferred genes from one species of bacterium to another species of bacterium. Today, scientists are able to transfer genes between entirely different organisms. For example, scientists have transferred human genes into bacteria. The human genes function in the bacteria as they do in human cells. When scientists transferred the gene for human insulin into bacteria, the bacteria produced human insulin. The insulin has been used to treat people with diabetes.

Scientists also have used genetic engineering to improve crop plants. For example, they have transferred a bacterial gene into plants that makes the plants resistant to insect attacks.

Some scientists would like to use genetic engineering to cure genetic diseases. So far, they have not been able to permanently replace disease-causing genes in humans with normal genes. However, scientists have found ways to temporarily introduce normal genes into people who have certain genetic diseases. As a result, the people are free of the diseases for a while.

Genetics Chapter 11 **289**

3 Reinforce and Extend

LEARNING STYLES

Auditory/Verbal
Show students the video "DNA: Blueprint of Life," Part 9 in the *Cycles of Life: Exploring Biology* program by The Annenberg/CPB Project. This 30-minute video discusses how scientists use genetic engineering to design new plants and animals. The video asks the question: "How far should we go with this technology?" After viewing, lead a discussion about that question.

CROSS-CURRICULAR CONNECTION

Health
Explain that in addition to insulin, another product of genetic engineering is interferon. Interferon is produced in human cells when they are attacked by viruses and tumors. Through genetic engineering, scientists have been able to engineer E. coli into interferons. These have been tested in the treatment of many diseases including various cancers and AIDS. Have interested students research interferons and their uses.

ONLINE CONNECTION

Students can review main ideas from the chapter, find genetics-related activities, and read current genetics news at gslc.genetics.utah.edu/.

LEARNING STYLES

Interpersonal/Group Learning
Have students work in four small groups. Assign each group a specific plant or animal, such as dogs, cats, flowers, and cattle. Have each group research one or more examples of selective breeding in the assigned field. For example, students might find that Brangus cattle have been bred from Brahman and Black Angus. Have each group illustrate their findings by showing each breed and its traits as well as the resulting breed and its traits.

AT HOME

Students may not realize how many things they use or come in contact with that are genetically engineered or have been produced using the techniques of genetic engineering. Have students make a list of such products. Encourage them to find out which fruits or vegetables in the local supermarket are genetically engineered, which medicines and vaccines are the result of genetic engineering, and whether any plants at a local nursery, in the local park, or in their own yard are genetically engineered.

Lesson 4 Review Answers

1. the process of using knowledge of genetics to affect heredity **2.** Answers will vary. Students might say that white fur is a helpful mutation for an animal that lives in a snowy environment. **3.** using breeding techniques to produce plants and animals with desirable traits **4.** the process of moving genes from one organism to another **5.** Answers will vary. Answers include producing insulin, producing crops that are resistant to insects, and introducing normal genes into humans who have genetic diseases.

Portfolio Assessment

Sample items include:
- Word lists and definitions from Teaching the Lesson
- Lesson 4 Review answers

Achievements in Science

Read Achievements in Science on page 290. Point out that Burbank is described not as a scientist but as an inventor. He did not have formal academic training as a scientist; he did not use pure pollens and keep careful records. Instead, he was a working horticulturist who used his instincts for successful plant breeding. In addition to breeding the fruits described in the feature, Burbank is known for breeding the Shasta daisy from four chrysanthemum species.

Lesson 4 REVIEW

Write your answers to these questions on a separate sheet of paper. Write complete sentences.

1. What is applied genetics?
2. Give an example of a mutation that is helpful to an organism.
3. What is selective breeding?
4. Define the term *genetic engineering*.
5. Describe one way that scientists have used genetic engineering to help humans.

Achievements in Science

Breeding Leads to New Plants

Luther Burbank was a famous inventor who devoted his life to breeding plants for certain characteristics. By using selective breeding and crossbreeding plants, he produced better plants and new varieties of plants. In 1871, he developed what is known as the Burbank potato. This potato helped people in Ireland because it was resistant to a plant disease that destroyed their potato crops.

Luther Burbank spent over 55 years doing research and experimenting on plants. Among his other discoveries, he created 113 new kinds of plums and prunes, ten different kinds of apples, five kinds of nectarines, and eight kinds of peaches. In 1986, Luther Burbank was inducted into the National Inventors Hall of Fame.

Workbook Activity 44

Chapter 11 SUMMARY

- Heredity is the passing of traits from parents to their offspring.
- Gregor Mendel, while working with pea plants, discovered that traits in organisms are due to paired factors.
- Mendel's factors are now called genes. Genes are located on chromosomes, which are found in the nucleus of a cell.
- Chromosomes are able to make copies of themselves. During mitosis and cell division, each new cell gets an exact copy of the parent cell's chromosomes.
- Meiosis is the process by which sex cells form. Each sex cell contains one-half the number of chromosomes in the parent cell.
- There are two kinds of sex chromosomes in humans: an X chromosome and a Y chromosome. A human female has two X chromosomes. A human male has one X chromosome and one Y chromosome.
- Chromosomes contain DNA. The order of bases in DNA provides a code for information about all the traits of an organism.
- Mutations are changes in DNA. Harmful mutations cause genetic diseases, such as diabetes and hemophilia.
- In selective breeding, the helpful results of mutations produce better varieties of plants and animals.
- The transfer of genes from one organism to another is called genetic engineering.

Science Words

applied genetics, 288	F_1 generation, 267	genotype, 269	replicate, 282
base, 281	F_2 generation, 269	hemophilia, 283	selective breeding, 288
carrier, 277	factors, 270	heredity, 266	self-pollination, 266
cross-pollination, 267	fraternal twins, 279	identical twins, 279	sex chromosome, 275
diabetes, 282	gene, 268	inbreeding, 283	sex-linked trait, 276
dominant gene, 268	gene pool, 283	mutation, 280	sickle-cell anemia, 283
environment, 280	genetic disease, 282	P generation, 266	
	genetic engineering, 289	phenotype, 269	
	genetics, 266	Punnett square, 268	
		recessive gene, 268	

Chapter 11 Review

Use the Chapter Review to prepare students for tests and to reteach content from the chapter.

Chapter 11 Mastery Test

The Teacher's Resource Library includes two parallel forms of the Chapter 11 Mastery Test. The difficulty level of the two forms is equivalent. You may wish to use one form as a pretest and the other form as a posttest.

Review Answers

Vocabulary Review

1. P generation 2. genotype 3. recessive
4. heredity 5. dominant 6. Punnett square 7. environment 8. mutations
9. sex-linked 10. genetic engineering
11. genes 12. selective breeding
13. identical 14. replicate

Concept Review

15. B 16. D 17. A 18. C

TEACHER ALERT

In the Chapter Review, the Vocabulary Review activity includes a sample of the chapter's vocabulary terms. The activity will help determine students' understanding of key vocabulary terms and concepts presented in the chapter. Other vocabulary terms used in the chapter are listed below:

applied genetics	gene pool
base	genetics
carrier	genetic disease
cross-pollination	hemophilia
diabetes	inbreeding
F_1 generation	phenotype
F_2 generation	self-pollination
factors	sex chromosome
fraternal twins	sickle-cell anemia

Chapter 11 REVIEW

Vocabulary Review

Choose the word or words from the Word Bank that best complete each sentence. Write the answer on a sheet of paper.

Word Bank
- dominant
- environment
- genes
- genetic engineering
- genotype
- heredity
- identical
- mutations
- P generation
- Punnett square
- recessive
- replicate
- selective breeding
- sex-linked

1. The pure breeding parent pea plants that Mendel studied were called the _____.
2. _____ is an organism's combination of genes for a trait.
3. A gene that is hidden when combined with another gene is said to be _____.
4. The passing of traits from parents to offspring is _____.
5. _____ genes are usually represented by capital letters.
6. The crossing of two organisms' traits can be shown in a model called a (an) _____.
7. Everything in your surroundings make up your _____.
8. Changes in an organism's DNA are _____.
9. A gene located on a sex chromosome is said to be _____.
10. The process of moving genes from one organism to another is called _____.
11. Parents pass traits to their offspring through their _____.
12. Breeding dogs with desirable traits is an example of _____.
13. _____ twins have identical genes.
14. Before a cell can divide normally, its DNA molecules have to _____.

292 Chapter 11 Genetics

Chapter 11 Mastery Test A

Concept Review

Choose the answer that best completes each sentence. Write the letter of the answer on your paper.

15. After _____ and cell division, each new cell has the same number of chromosomes as the parent cell.
 A meiosis   C reproduction
 B mitosis   D genetics

16. When Mendel crossed pure tall pea plants with pure short pea plants, all the offspring were tall because the gene for tallness is _____.
 A recessive   C a factor
 B sex-linked   D dominant

17. _____ is a genetic disease in which a person has too much sugar in the bloodstream.
 A Diabetes   C Insulin
 B Hemophilia   D Sickle-cell anemia

18. A human female passes one _____ chromosome to her offspring.
 A Y   B sex-linked   C X   D body

Critical Thinking

Write the answer to each of the following questions.

19. When Mendel studied flower color in pea plants, the F_1 generation had only purple flowers. The F_2 generation had three purple flowers for every white flower. Which gene do you think is probably dominant: the gene for purple flowers or the gene for white flowers?

20. In humans, does the mother or the father of an offspring determine the sex of the offspring? Explain.

Test-Taking Tip Read the directions carefully. Don't assume that you know what you're supposed to do.

Critical Thinking

19. Following the reasoning used in Lesson 1 to infer the dominance of the gene for tallness, students should conclude that the gene for purple flower color is dominant. **20.** The father determines the sex of the offspring because, unlike the mother, he can pass on either an X chromosome or a Y chromosome.

ALTERNATIVE ASSESSMENT

Alternative Assessment items correlate to the student Goals for Learning at the beginning of this chapter.

- Have student pairs make up an inherited trait and demonstrate its inheritance using at least two Punnett squares. Ask students to identify the genotypes and phenotypes of the offspring.

- Label slips of paper with the genotypes of male fruit flies and female fruit flies with different eye colors. Place the male slips and female slips in separate containers. Then, have students draw a slip and make a Punnett square to predict the genotypes and phenotypes of the parents' offspring.

- Ask students to make a pedigree for a trait they exhibit. The pedigree should show how the trait could be passed from grandparent to parent to student.

- Provide students with pieces of paper with the letter *A*, *C*, *G*, or *T* written on each piece. Have students use the pieces to show how the bases are paired in a DNA molecule. (*Base A pairs with base T. Base C pairs with base G.*) Then ask students to use the pieces to show a mutation in DNA.

- Have students design a genetic engineering project. Projects should have a clear goal and a plausible process. Students can illustrate, model, or write about their project and the benefits it would bring to society.

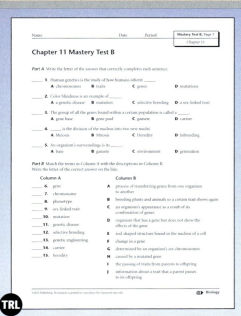

Chapter 11 Mastery Test B

Chapter 12

Planning Guide
Ecology

	Student Pages	Vocabulary	Lesson Review
Lesson 1 Living Things and Nonliving Things	296–305	✔	✔
Lesson 2 Food Chains and Food Webs	306–311	✔	✔
Lesson 3 How Energy Flows Through Ecosystems	312–316	✔	✔
Lesson 4 How Materials Cycle Through Ecosystems	317–322	✔	✔

Chapter Activities

Student Text
Science Center

Teacher's Resource Library
Community Connection 12:
   An Ecosystem in the Community

Assessment Options

Student Text
Chapter 12 Review

Teacher's Resource Library
Chapter 12 Mastery Tests A and B

294A

Student Text Features								Teaching Strategies						Learning Styles						Teacher's Resource Library				
Achievements in Science	Science at Work	Science in Your Life	Investigation	Science Myth	Note	Technology Note	Did You Know?	Science Integration	Science Journal	Cross-Curricular Connection	Online Connection	Teacher Alert	Applications (Home, Career, Community, Global, Environment)	Auditory/Verbal	Body/Kinesthetic	Interpersonal/Group Learning	Logical/Mathematical	Visual/Spatial	LEP/ESL	Workbook Activities	Alternative Workbook Activities	Lab Manual	Resource File	Self-Study Guide
	303		304	302					300	299, 301, 302	301	297	298, 300	298		299			301	45	45	34, 35	23	✓
309		311							308			307	309, 311		310		308	308	309	46	46	36	24	✓
							313, 315	314		314, 315		313								47	47			✓
					✓	✓		319	318	320	321		321			319		320		48	48			✓

Pronunciation Key

a	hat	e	let	ī	ice	ô	order	ù	put	sh	she	ə { a	in about
ā	age	ē	equal	o	hot	oi	oil	ü	rule	th	thin	e	in taken
ä	far	ėr	term	ō	open	ou	out	ch	child	ᵺ	then	i	in pencil
â	care	i	it	ò	saw	u	cup	ng	long	zh	measure	o	in lemon
												u	in circus

Alternative Workbook Activities

The Teacher's Resource Library (TRL) contains a set of lower-level worksheets called Alternative Workbook Activities. These worksheets cover the same content as the regular Workbook Activities but are written at a second-grade reading level.

Skill Track Software

Use the Skill Track Software for Biology for additional reinforcement of this chapter. The software program allows students using AGS textbooks to be assessed for mastery of each chapter and lesson of the textbook. Students access the software on an individual basis and are assessed with multiple-choice items.

Chapter at a Glance

Chapter 12: Ecology
pages 294–325

Lessons

1. **Living Things and Nonliving Things** pages 296–305

   Investigation 12 pages 304–305

2. **Food Chains and Food Webs** pages 306–311

3. **How Energy Flows Through Ecosystems** pages 312–316

4. **How Materials Cycle Through Ecosystems** pages 317–322

Chapter 12 Summary page 323

Chapter 12 Review pages 324–325

Skill Track Software for Biology

Teacher's Resource Library

- Workbook Activities 45–48
- Alternative Workbook Activities 45–48
- Lab Manual 34–36
- Community Connection 12
- Resource File 23–24
- Chapter 12 Self-Study Guide
- Chapter 12 Mastery Tests A and B

(Answer Keys for the Teacher's Resource Library begin on page 420 of the Teacher's Edition. A list of supplies required for Lab Manual Activities in this chapter begins on page 448.)

Science Center

Provide students with materials to use to build models of ecosystems. Possible materials include boxes, white paper, construction paper, toy animals, sticks, moss, stones, markers, glue, modeling clay, tape, cotton (for clouds), and foil or cellophane (for water). At the end of the chapter, have students use their models to address the goals on page 295.

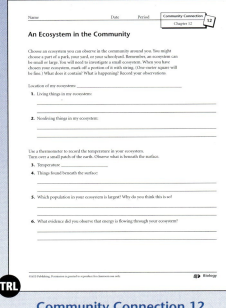

Community Connection 12

Chapter 12: Ecology

The cattle in the photo interact with living and nonliving parts of the environment. They do this as they eat grass, drink water, and breathe air. They get food and energy from the grass. Then, they pass this food and energy on to other organisms. The cattle also have a role in cycles. In this chapter, you will learn about the levels of organization of living and nonliving things. You will learn how food and energy flow through the environment. You also will learn about the cycle of different materials through the environment.

Organize Your Thoughts

Biosphere — Ecosystems
Organisms — Living Things — Energy — Non-living Things
Populations — Communities — Materials

Goals for Learning

◆ To identify ways in which living things interact with one another and with nonliving things
◆ To describe feeding relationships among the organisms in a community
◆ To explain how energy flows through ecosystems
◆ To identify materials that cycle through ecosystems
◆ To understand that human activities have an impact on ecosystems

295

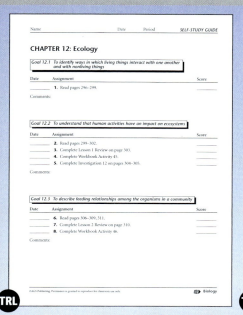

Chapter 12 Self-Study Guide

Introducing the Chapter

Direct students' attention to the photograph on page 294. Ask them to identify living and nonliving things in the picture, including things they do not see directly, such as sunlight and air. Have students list some ways the cattle are interacting with other things in the picture. Invite discussion of what would happen if one of the things were missing from the picture. Lead students to conclude that everything in an environment is connected to everything else.

Read the introductory paragraph together, then have students study the graphic organizer and Goals for Learning in order to ask questions they would like to answer as they read.

Notes and Technology Notes

Ask volunteers to read the notes that appear in the margins throughout the chapter. Then discuss them with the class.

TEACHER'S RESOURCE

The AGS Teaching Strategies in Science Transparencies may be used with this chapter. The transparencies add an interactive dimension to expand and enhance the *Biology* program content.

CAREER INTEREST INVENTORY

The AGS Harrington-O'Shea Career Decision-Making System-Revised (CDM) may be used with this chapter. Students can use the CDM to explore their interests and identify careers. The CDM defines career areas that are indicated by students' responses on the inventory.

Ecology Chapter 12 295

Lesson at a Glance

Chapter 12 Lesson 1

Overview In this lesson, students learn how living things can be organized into levels based on types of interactions. They explore how communities change over time, how pollution damages ecosystems, and how biomes and resources are related.

Objectives

- To explain the relationships among organisms, populations, communities, and ecosystems
- To describe the process of succession
- To describe how pollution affects ecosystems
- To understand how human activities affect the environment
- To list examples of biomes
- To describe two types of natural resources

Student Pages 296–305

Teacher's Resource Library

Workbook Activity 45
Alternative Workbook Activity 45
Lab Manual 34, 35
Resource File 23

Vocabulary

interact	threatened
ecology	endangered
biotic	extinct
abiotic	biome
habitat	resource
population	renewable
community	resources
ecosystem	nonrenewable
succession	resources
climax community	fossil fuels
pollution	biosphere
acid rain	

296 Chapter 12 Ecology

Lesson 1 Living Things and Nonliving Things

Objectives

After reading this lesson, you should be able to

- explain the relationships among organisms, populations, communities, and ecosystems.
- describe the process of succession.
- describe how pollution affects ecosystems.
- understand how human activities affect the environment.
- list examples of biomes.
- describe two types of natural resources.

Interact
To act upon or influence something

Ecology
The study of the interactions among living things and the nonliving things in their environment

Organisms act upon, or **interact** with, one another and with nonliving things in their environment. For example, you interact with the air when you inhale oxygen and exhale carbon dioxide. You interact with plants when you eat fruits and vegetables. **Ecology** is the study of the interactions among living things and the nonliving things in their environment. Living things are the **biotic** factors in the environment. Nonliving things, such as light, temperature, water, and air are **abiotic** factors.

Levels of Organization

Organisms interact at different levels. For example, organisms of the same species interact with one another. Organisms of different species also interact. The diagram shows the organization of living things into different levels. The higher the level, the more interactions there are.

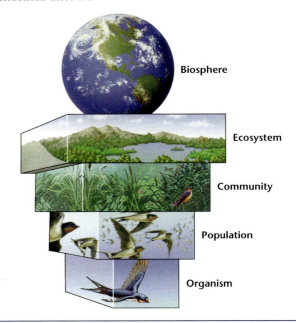

296 Chapter 12 Ecology

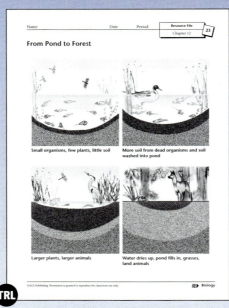

Resource File 23

Biotic
Living

Abiotic
Nonliving

Habitat
The place where an organism lives

Population
A group of organisms of the same species that lives in the same area

Community
A group of different populations that live in the same area

Ecosystem
The interactions among the populations of a community and the nonliving things in their environment

Notice that the lowest level of organization is the individual organism. The place where an organism lives is its **habitat**. Each organism is adapted to live in its habitat. For example, tuna use their fins to swim through the ocean. A spider monkey uses its long tail to hang from trees.

Populations

The next level of organization is a **population**. A group of organisms of the same species that lives in the same area form a population. The grizzly bears in Yellowstone National Park make up a population. All the people in the United States make up the country's human population.

The individual members of a population interact with one another. For example, the males and females in a population interact when they mate. The individuals also interact when they compete for food, water, and space.

Communities

A **community** is the third level of organization of living things. Populations of different species that live in the same area make up a community. Bears, rabbits, pine trees, and grass are different populations of organisms, but they all may live together in the same forest community.

The populations in a community interact with one another in many ways. In a forest, large trees determine how much light gets through to the shrubs and grass on the forest floor. The trees also provide shelter and food for animals, such as squirrels. Some of the nuts that squirrels bury grow to become new trees. A hawk may eat squirrels and use a tree as a nest.

Ecosystems

All the interactions among the populations of a community and the nonliving things in their environment make up an **ecosystem**. Organisms interact with nonliving things when they breathe air, drink water, or grow from the soil. Ecosystems occur on land, in water, and in air. The interactions that occur in a fish tank make up an ecosystem. So do the interactions in the ocean.

Science Background
The Biosphere

The biosphere extends from the deepest ocean trenches (about 11 km below the sea) to about 12 km into the atmosphere. Ecologists divide the biosphere into ecosystems. An ecosystem includes a community of organisms, all the nonliving things that affect it, and the interactions between the living and nonliving things. The size of an ecosystem depends in part on the interactions being studied. For example, the base of a tree can be an ecosystem if a scientist is studying the interactions between the moss, fungi, insects, and other populations at the base of that tree. On the other hand, if the study concerns the feeding patterns of foxes, owls, and other larger animals, the entire forest may be the ecosystem.

  Warm-Up Activity

Ask students to name some living and nonliving things with which they have interacted today. Encourage them to think beyond the obvious. For example, they may have provided food for a mosquito or their presence may have prevented a bird or squirrel from feeding. Have them describe other effects of those occurrences. For example, the cars and buses on which many of them rode to school produced gases that polluted the air and affected many living things. A farmer grew the grain or fruit they ate for breakfast; the crop fields used habitats where many wild animals once lived.

  Teaching the Lesson

Review the terms *organism*, *environment*, and *species*. As they read the lesson, ask students to write an original sentence for each vocabulary term.

Ask: What parts of Earth are not part of the biosphere? (*areas where life does not exist*)

Discuss the concept of population with students. Have them investigate to estimate the population of humans, dogs, and cats in your community and of ants in the school playground.

Discuss local examples of succession. For example a vacant lot or abandoned field may show succession as grasses give way to shrubs, which give way to conifers.

TEACHER ALERT

Students may think that succession occurs over several years. In fact, it may take hundreds, even thousands, of years. Once a climax community is established, it tends to remain stable. However, disturbances such as a volcanic eruption, severe weather, or certain human activities may destroy a climax community.

3 Reinforce and Extend

LEARNING STYLES

Auditory/Verbal
Sounds from animals, wind, and water vary in different environments. Play sound recordings of environments such as rain forests, streams, oceans, and the plains. Invite students to imagine the ecosystem and describe the living and nonliving things in it.

IN THE ENVIRONMENT

Ask students who have visited a regional, state, or national park to bring in pictures, brochures, and pamphlets that will help them explain the type of ecosystem stage it illustrates. Have them describe various communities within the park, such as pond, marsh, forest, and try to identify any communities that are climax communities.

Succession
The process by which a community changes over time

Changes in Ecosystems

As the community of organisms and the nonliving things of an ecosystem interact, they may cause changes. These changes may result in the community changing into a different type of community. For example, a pond community may change into a forest community. The changes that occur over time in a community are called **succession**.

The diagrams below show the succession of a pond community into a forest community. Notice that small organisms and few plants live in a young pond. The bottom of the pond has little soil. When these organisms die, their bodies sink to the bottom and decay. The dead matter helps to form a layer of soil on the bottom of the pond. Soil is also washed into the pond from the surrounding land. Soon, larger plants are able to grow from the soil in the pond. Larger animals that feed on those plants move into the pond. Over the years, the pond fills up with even more soil. Grasses and other small land plants begin to grow on the dry edges of the pond. Small land animals, such as mice and rabbits, move into the area. Changes continue. The pond completely fills in. Bushes shade out the grass. Then trees overgrow the bushes. Squirrels and deer move into the area. The area that was once a pond becomes a forest.

Succession of Pond to Forest

298 Chapter 12 Ecology

Lab Manual 34, pages 1–2

Climax community
A community that changes little over time

Pollution
Anything added to the environment that is harmful to living things

Acid rain
Rain that is caused by pollution and is harmful to organisms because it is acidic

Eventually, a community reaches a point at which it changes little over time. A community that is stable is called a **climax community**. A climax community, such as an oak-hickory forest, may stay nearly the same for hundreds of years. A climax community usually has a great diversity of organisms. However, a volcano, a forest fire, or an earthquake can destroy large parts of a climax community within a short time. When that happens, the community goes through succession once again.

Human Impact on Ecosystems

People produce a variety of wastes and waste products that affect ecosystems. **Pollution** is anything added to the environment that is harmful to living things. Pollution is most often caused by human activities. For example, the burning of coal, oil, or gasoline releases a colorless, poisonous gas called sulfur dioxide. This gas poisons organisms that breathe it. Sulfur dioxide in the air also makes rainwater more acidic. This **acid rain** decreases the growth of plants and harms their leaves. Acid rain that falls into lakes and streams can harm or kill organisms living in the water.

Succession of Pond to Forest Continues

LEARNING STYLES

Interpersonal/Group Learning
Have students work in small groups to research the following ways people affect the biosphere: smog, acid rain, land pollution, and water pollution. Suggest that students use magazine articles, encyclopedia yearbooks, and Web sites as sources of information. Encourage cooperative work as groups create cause-and-effect charts and illustrations about the types of pollution they explored. After groups present their reports, invite them to work together to form a plan for reducing waste and pollution at your school.

CROSS-CURRICULAR CONNECTION

History/Social Studies
Have several interested students read about the industrial pollution of the Great Lakes in the twentieth century. They can make a timeline summarizing how the lake water changed for the worse and then for the better as laws were passed and pollution lessened. Students might illustrate sections of their timelines, showing the variety of organisms in one of the lakes when it was most polluted and then when its waters became less polluted.

GLOBAL CONNECTION

Identify the biome that you and your students live in. Then have students research the people, animals, and plants of the same biome in another part of the world. Compare and contrast the organisms in the two biomes.

SCIENCE JOURNAL

Display pictures of various endangered and threatened species of animals. Ask students to select one and write a letter telling why they believe it should be saved from extinction.

Threatened
There are fewer of a species of animal than there used to be

Endangered
There are almost no animals left of a certain species

Extinct
All the members of a species are dead

Biome
An ecosystem found over a large geographic area

Other types of pollution affect lakes and other bodies of water as well. Topsoil is washed off the land because of construction and bad farming practices. It fills up streams and lakes. Fertilizer washes off the land into bodies of water. This pollutes the water with chemicals. Factories dump chemicals into lakes and streams. The chemicals kill plants and animals in the water.

Another way that human activities affect wildlife is by causing the loss of habitat. In order to build roads and shopping centers, animal habitats are sometimes destroyed. Birds and other animals that live nearby can no longer build nests or eat the plants in the developed area. It becomes harder for those animals to survive. If habitat destruction becomes widespread, there are fewer animals to reproduce.

Over time, this can result in a species being classified as **threatened**. This means that there are fewer of these animals than there used to be. If there are almost no animals left of a certain species, that species is **endangered**. When all the members of a species are dead, that species has become **extinct**. In Chapter 14, you will learn about dinosaurs, which became extinct millions of years ago. But there are many other animals that have become extinct more recently. They will never return. Pollution and habitat destruction are two reasons animals become extinct.

Biomes

Some ecosystems are found over large geographic areas. These ecosystems are called **biomes**. Some biomes, such as deserts, pine forests, and grasslands, are on land. Water biomes include the oceans, lakes, and rivers.

The photos on page 301 show two completely different biomes. They are a tundra and a tropical rain forest. Different biomes are found in different climates. Temperature, sunlight, and rainfall are all part of a biome's climate. For example, tropical rain forests get plenty of rainfall and are hot. Tundras are found in areas that are dry and cold. A desert is the driest of all biomes. Some deserts get as little as 2 centimeters of rain in a year.

Resource
A thing that an organism uses to live

Renewable resources
Resources that are replaced by nature

Nonrenewable resources
Resources that cannot be replaced

The types of organisms found in a particular biome depend on the **resources** available to the organisms. Resources are things that organisms use to live. Resources include water, air, sunlight, and soil. Fish are not found in a desert biome because a desert has little water. Most cactuses do not grow in a tropical rain forest because the soil in the forest is too wet.

Tundra

Tropical rain forest

In order to be good citizens of Earth, we should use our natural resources wisely. Some resources, such as water, air, and sunlight, are **renewable resources**. This means that they are resources that are replaced constantly by nature. Other resources, such as coal, minerals, and natural gas, are **nonrenewable resources**. These materials cannot be replaced once they are used up.

LEARNING STYLES

LEP/ESL
Students may need help learning the extensive vocabulary in this lesson. Have them write each vocabulary word on a large index card and draw a picture that illustrates the word's meaning. Have pairs of students organize the cards into groups of related terms, for example *pollution, acid rain, threatened, endangered,* and *extinct*. Ask partners to place the groups of cards into graphic organizers to show how they are related.

CROSS-CURRICULAR CONNECTION

Music
Play a version of the song "Big Yellow Taxi" with the lyrics "They paved paradise and put up a parking lot." Discuss the message of the lyrics. Then invite students to write a song (or write new lyrics for a familiar melody) about how people are polluting earth and affecting ecosystems. After students share their songs, collect them in a class songbook.

ONLINE CONNECTION

For information, activities, and fun facts on ecology, students can visit library.thinkquest.org/11353/ecosystems.htm and http://www.fi.edu/tfi/units/life/habitat/habitat.html.

Science Myth

Ask students where they would expect acid rain to be the biggest problem. After they have predicted the location, read aloud the first sentence of the Science Myth feature on page 302. Have a volunteer read aloud the fact portion of the feature. Discuss the role of climate and weather (e.g., prevailing winds) on air quality.

CROSS-CURRICULAR CONNECTION

Art

Study of the biosphere presents a great opportunity for a classroom mural project. Divide the class into groups and assign a biome to each group. Have the groups design a visual panel for their biome that details living and nonliving things. Once they have their design, they should then draw and color or paint it on butcher paper. The groups can hang their completed panels to form a classroom mural of the biosphere.

Fossil fuels
Fuels formed millions of years ago from the remains of plants and animals

Biosphere
The part of earth where living things can exist

Science Myth

Acid rain falls only in cities, which are sources of air pollution.

Fact: Winds carry air pollution from cities to ecosystems away from cities. Acid rain falls on forests, in ponds, or wherever the winds carry it. In the United States, the Northeast has the most acid rain. Much of this acid rain is cause by pollution from cities in the Northeast. Some of the acid rain is from the Midwest.

Both kinds of resources are important. One thing you can do to reduce the use of nonrenewable resources is to recycle, or reuse, some materials. Recycling newspapers means that fewer trees need to be cut down to make paper. Recycling aluminum cans means that the minerals used to make these products can be conserved, or saved. Another way to save resources is to use alternative energy resources. This kind of energy does not use **fossil fuels**, or fuels formed millions of years ago from the remains of plants and animals. You can learn more about different kinds of energy in Appendix D at the end of this book.

The Biosphere

Look back at the diagram on page 296. Notice that the highest level of organization of life is the **biosphere**. The desert biome, the ocean biome, and all the other biomes on Earth together form the biosphere. The biosphere is the part of Earth where living things can exist.

Think of Earth as an orange or an onion. The biosphere is like the peel of the orange or the skin of the onion. It is the thin layer on a large sphere. The biosphere includes the organisms living on Earth's surface, in water, underground, and in the air. The biosphere also includes nonliving things, such as water, minerals, and air.

The biosphere is a tiny part of Earth. This thin surface layer can easily be damaged. Thus, humans need to be aware of how they can protect the biosphere. One way is to avoid polluting it. The survival of living things depends on the conditions of the nonliving parts of the biosphere.

Lesson 1 REVIEW

Write your answers to these questions on a separate sheet of paper. Write complete sentences.

1. What is the difference between a population and a community?
2. What kinds of interactions make up an ecosystem?
3. Describe the difference between renewable and nonrenewable resources.
4. How does acid rain affect plants and animals?
5. Define *biome* and list three or more examples of biomes.

Science at Work

Ecologist

An ecologist needs to be able to identify and solve problems and must communicate well. Ecologists usually need a master's degree in science or a Ph.D. in ecology or environmental science. Jobs are also available for ecologists with a bachelor's degree in biology, ecology, or a related field. There are some jobs in ecology that require a two-year associate of science degree.

Ecologists are scientists who study the relationships between organisms and their environments. Ecologists study organisms in ecosystems as different as a city, a desert, a tropical rain forest, or the ocean. Some ecologists work for universities, parks, museums, and government agencies. As a part of their jobs, ecologists do research, help identify and solve environmental problems, and manage resources. Ecologists also provide advice to different agencies and communicate what they have learned to others.

Ecology Chapter 12 303

Investigation 12

This investigation will take approximately 30 minutes to complete. Students will use these process and thinking skills: measuring, collecting, and interpreting data; comparing and contrasting; and drawing conclusions

Preparation

- Locate a position for the trash cans where rain will not hit trees or buildings on the way down and water from puddles will not splash into the cans.
- Cut strips of pH paper ahead of time so students will not use too much.
- Have plastic cups available into which students can pour the distilled water for testing.

Procedure

- Work groups may include 3 to 6 students. Two students can be assigned the task of finding appropriate places for the trash cans. Have students take turns testing water samples. Another student can calculate the average. Students could reverse roles when repeating the activity on other rainy days.
- Have different students independently determine the pH of different samples. If the readings on the same samples are different, have students compare the two strips of pH paper. How different are they? Discuss how this is a subjective reading.
- Use pH paper with a range of pH 0–7, in increments of 0.25.
- Position a can where the rainwater will have touched a building or plant and test whether the pH is different. Test the pH of the rainwater that collects in puddles on a parking lot.

SAFETY ALERT

- Do not go out in the rain during a storm with lightning.
- Have students clean up spilled water immediately.
- Provide protective gloves for students handling the rain water.

INVESTIGATION 12

Materials
- 3 small trash cans
- 3 new plastic trash bags
- pH paper
- pH scale
- distilled water

Testing the pH of Rain

Purpose
How can you tell if rain is acid rain? In this investigation, you will find out if the rain in your area is acid rain.

Procedure

1. Copy the data table below on a sheet of paper.

Date	pH of Sample 1	pH of Sample 2	pH of Sample 3	pH of Sample 4	Average pH of Samples
pH of distilled water:					

2. On a rainy day, place open plastic bags inside the trash cans.

3. If there is no thunder and lightning, place the containers outside in the rain. Make sure that the containers collect rainwater that has not touched anything on the way down, such as a roof or the leaves of a tree.

4. When a small amount of rainwater has been collected, touch the edge of the pH paper to the rainwater in one of the containers.

5. Notice the change in color of the pH paper. Compare the color of the pH paper to the colors on the pH scale. The matching color on the scale indicates the pH of the sample of rainwater. In your data table under sample 1, record the pH value of the rainwater.

6. Repeat steps 4 and 5 for the other two samples of rainwater. Record the pH of the samples in the correct columns.

7. If the pH values that you recorded for the three samples are not the same, compute the average pH of the samples. To do this, add all of the pH values and divide by 3. Record this number in the last column in your data table.

8. Use the pH paper to determine the pH of distilled water. Record the pH in your data table.

9. Repeat this investigation on one or two more rainy days. Record in your data table the dates on which you collected the rainwater.

Questions and Conclusions

1. How does the pH of distilled water compare with the pH of the rainwater you tested?

2. When water has a pH lower than 7, it is acidic. Normal rain is always slightly acidic and has a pH between 4.9 and 6.5. When rain has a pH of less than 4.9, it is called acid rain. Were any of the samples of rainwater you tested acid rain?

3. Was the pH of the rainwater the same every day you collected samples? What are some reasons the pH of rainwater could vary from day to day?

Explore Further

Test the pH of the water in a local pond, lake, or stream. **Safety Alert: Wear protective gloves when you collect the water samples.** Is the pH of the body of water the same as that of rainwater? Why might the values be different?

Ecology Chapter 12 305

Results

Results will vary according to the condition of the local atmosphere.

Questions and Conclusions Answers

1. The pH of distilled water is 7.0. The pH of rainwater will vary. It should be lower than that of distilled water.

2. Answers will vary. Any samples with pH less than 4.9 indicate acid rain.

3. The pH of rainwater may vary on different days because the level of pollution in the air may vary. In addition, rain comes from different areas. If it is blown in from a large city or highly industrialized area, it may be more acidic.

Explore Further Answers

1. The pH of a pond, lake, or stream is likely to be different from that of rainwater. Streams and lakes fed by streams get water from other locations; they are not filled only by local rain. Natural materials, such as limestone, and the activities of organisms that live in the water also affect the pH of a body of water.

Assessment

Check students' data tables to be sure they completely filled out the table and correctly computed the average pH. Include the following items from this investigation in student portfolios:

- Data table
- Answers to Questions and Conclusions and Explore Further

Lab Manual 35

Ecology Chapter 12 305

Lesson at a Glance

Chapter 12 Lesson 2

Overview In this lesson, students learn how living things are organized into food chains and food webs, about levels and numbers of producers and consumers, and about the importance of decomposers to a community.

Objectives

- To distinguish between producers and consumers
- To trace a food chain from producer through three levels of consumers
- To explain how a food chain is related to a food web
- To understand the importance of decomposers in communities

Student Pages 306–311

Teacher's Resource Library (TRL)

Workbook Activity 46
Alternative Activity 46
Lab Manual 36
Resource File 24

Vocabulary

food chain   pyramid of
producer   numbers
consumer   food web
omnivore

Science Background
Producers and Consumers

Living things are autotrophs or heterotrophs. Autotrophs are organisms that make food from nonliving materials (plants, some bacteria, and algae). Most autotrophs use the energy from the Sun to make food. They are called producers because they produce food for other organisms.

Unable to make their own food, heterotrophs require organic matter to live. They are consumers in the ecosystem. Consumers include animals, fungi, some protists, and some bacteria.

Lesson 2: Food Chains and Food Webs

Objectives

After reading this lesson, you should be able to
- distinguish between producers and consumers.
- trace a food chain from producer through three levels of consumers.
- explain how a food chain is related to a food web.
- understand the importance of decomposers in communities.

A water plant captures the energy from sunlight. It uses the energy to make sugars and other molecules. A small fish eats the plant. A bigger fish eats the small fish. A bird eats the big fish. This feeding order is called a **food chain**. Almost all food chains begin with plants or other organisms that capture the energy of the sun.

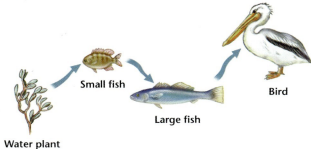

A Food Chain

Food chain
The feeding order of organisms in a community

Producer
An organism that makes its own food

Consumer
An organism that feeds on other organisms

Producers

Plants, some protists, and some bacteria make their own food. Organisms that make, or produce, their own food are called **producers**. Every food chain begins with a producer. Most producers use the energy of sunlight to make food by the process of photosynthesis.

Consumers

Organisms that cannot make their own food must get food from outside their bodies. These organisms get their food from other organisms. **Consumers** are organisms that feed on, or consume, other organisms. All animals and fungi and some protists and bacteria are consumers.

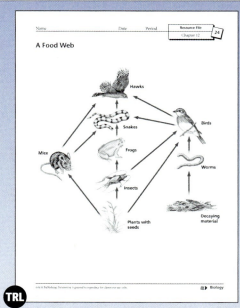

Resource File 24

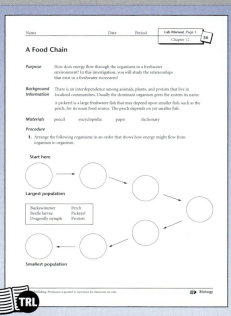

Lab Manual 36, pages 1–2

Omnivore
A consumer that eats both plants and animals

Pyramid of numbers
A diagram that compares the sizes of populations at different levels of a food chain

Consumers may eat plants or other consumers. Consumers, such as rabbits, that eat only plants are called herbivores. Consumers, such as lions, that eat only animals are carnivores. Some consumers, such as bears, are **omnivores**. They eat both plants and animals.

The consumers in a food chain are classified into different feeding levels, called orders, depending on what they consume. First-order consumers eat plants. Rabbits are first-order consumers. Second-order consumers eat animals that eat plants. A snake eats rabbits. Thus, a snake is a second-order consumer. A hawk that eats the snake is a third-order consumer.

Numbers of Producers and Consumers

The African plains are covered with billions of grass plants. Herds of antelope feed on the grasses. However, the number of antelope is far less than the number of grass plants. In the same area, there may be only a few dozen lions. Lions need a wide range in which to hunt antelope and other animals.

You might think of a food chain as a pyramid with the highest-level consumers at the top. Notice in the **pyramid of numbers** that a food chain begins with a large number of producers. There are more producers in a community than there are first-order consumers feeding on the producers. The sizes of the populations decrease at each higher level of a food chain.

A Pyramid of Numbers

Decomposers are heterotrophs that break down organic matter into inorganic matter (fungi and some protists and bacteria). In this way, chemicals are returned to the nonliving parts of the environment.

 Warm-Up Activity

Show the class pictures of organisms that could be links in a food chain. Examples might include plants, a rabbit, and a hawk. Ask which organism eats which organism. Arrange the pictures on a board and connect them with string. Explain that this is a food chain. Then add other organisms and ask where these organisms fit as you form a food web.

 Teaching the Lesson

Have students write the following sets of terms several lines apart on a sheet of paper: *producers/consumers/decomposers* and *food chain/food web*. As they read the lesson, have them compare and contrast the terms. They can write sentences that clarify the relationships among the sets of terms.

After having students trace the food chain on page 306, invite them to give other examples of food chains.

Explain that the same organism can be a different-level consumer in different food chains. For example, when a crow eats grain, it is a first-order consumer. When it eats insects, it is a second- or third-order consumer, depending on whether the insect ate plants or other insects.

Discuss reasons why populations of second- and third-order consumers are small compared to those of producers and first-order consumers. (*Food for first-order consumers is plentiful; producers are limited only by amount of space, rain, and sunlight. Higher order consumers would starve if their populations grew too large because there would not be enough food to go around.*)

Point out that decomposers play a vital role because all plants and animals die. Have students study a portion of a rotting log with magnifying glasses. Discuss the life-forms students observe and classify them as decomposers or consumers.

 TEACHER ALERT

Some students might confuse the pyramid of numbers with the Food Guide Pyramid. Explain that the latter is a guide to help people choose a healthful diet. Compare the pyramids on pages 251 and 307.

3 Reinforce and Extend

LEARNING STYLES

Logical/Mathematical
Emphasize how the number of organisms decreases as you go up the pyramid of numbers. Give students mathematical problems such as the following. Suppose 150 grain plants support 4 rats, and 20 rats support 1 bird. How many grain plants does it take to support one bird? ($\frac{20}{4} \times 150 = 5 \times 150 = 750$)

SCIENCE JOURNAL

Ask students to write what they eat in a day and identify which order of consumer they are for each food.

LEARNING STYLES

Visual/Spatial
Give students a list of organisms, such as owl, leaf, beetle, and mouse. Have them draw a food chain, picturing each animal and showing how energy flows from producer to consumers.

Food web
All the food chains in a community that are linked to one another

Food Webs

Few consumers eat only one kind of food. The American crow is an example of a bird that eats a wide range of foods. It eats grains, seeds, berries, insects, dead animals, and the eggs of other birds. Eating a variety of foods helps to ensure that the consumer has a sufficient food supply.

The frogs in a pond eat a variety of foods, including insects and worms. The frogs in turn may be eaten by snakes or birds. Thus, frogs are part of more than one food chain. The diagram shows some food chains in a community. Trace the different food chains. Notice that the food chains are linked to one another at certain points. Together, the food chains form a **food web**.

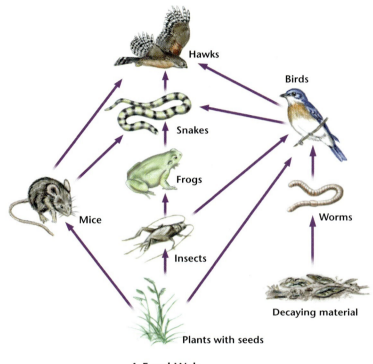

A Food Web

308 Chapter 12 Ecology

Decomposers

Suppose that a high-level consumer, such as a lion, dies but is not eaten by another animal. Does the food chain stop there? No, because decomposers continue the food chain by feeding on the dead animal. Recall that decomposers are certain bacteria, fungi, and protists that feed on dead organisms. Decomposers feed on dead organisms at each level of a food chain. They feed on producers and consumers.

Decomposers get food by breaking down complex chemicals in dead organisms into simple chemicals. The simple chemicals become part of the soil. Plants take in these chemicals through their roots and use them to grow. Over time, the chemicals are used again and again. They are taken up by plants, transferred to herbivores and carnivores, and returned to the soil by decomposers. The total amount of the chemicals stays the same even though their form and location change.

Achievements in Science

Crop Rotation Recommended

By the late 1800s, the soil in the southern states of the United States could not produce as much cotton as it once did. Decades of growing crops, such as cotton, had taken nitrogen out of the soil without replacing it. Farmers did not know that plants need nitrogen. Even if the farmers did know this, fertilizers were not available to add nitrogen to the soil.

George Washington Carver was one of the United States' greatest plant researchers. He found that crops such as cotton take nitrates, a form of nitrogen needed for plant growth, out of the soil. Rotating these crops with other crops such as peanuts put nitrates back into the soil. The soil was improved, and crops grew better.

Carver shared his research and helped improve the productivity of farmers. Even today, crop rotation is used. It is better for the soil and the environment than using fertilizers.

Lesson 2 Review Answers

1. A producer makes its own food. A consumer gets food by feeding on other organisms. 2. A herbivore is a consumer that eats only plants. A carnivore is a consumer that eats only animals. An omnivore is a consumer that eats both plants and animals. 3. Food chain diagrams will vary. The highest level of a food chain has the smallest population. 4. A food web consists of all the food chains within a given community. 5. Decomposers break down the chemicals in dead organisms into simpler chemicals that plants use for growth.

Portfolio Assessment

Sample items include:
- Sentences of comparison and contrast from Teaching the Lesson
- Lesson 2 Review answers

LEARNING STYLES

Body/Kinesthetic
Assign an organism to each student, using a name tag. Include a variety of producers, herbivores, carnivores, omnivores, and decomposers. Have students stand in a circle. Then give a ball of yarn to a "producer." The "producer" holds onto the end of the yarn and tosses the ball to an "herbivore" or "omnivore." Have students continue in this manner until an intricate food web is formed with the yarn. Then have one student drop the yarn to represent a sudden decrease in population of that organism. Discuss how that decrease affects the community.

Lesson 2 REVIEW

Write your answers to these questions on a separate sheet of paper. Write complete sentences.

1. What is the difference between a producer and a consumer?
2. Define the terms *herbivore, carnivore,* and *omnivore.*
3. Diagram a food chain that includes three levels of consumers. Which level of consumers has the smallest population size?
4. What is the relationship between food chains and food webs?
5. What is the role of decomposers in a community?

310   Chapter 12   Ecology

Workbook Activity 46

Science in Your Life

What kind of consumer are you?

Like all consumers, you cannot make your own food. You must eat, or consume, food. Are you a first-order, second-order, or third-order consumer? When you eat plants or parts of plants, you are a first-order consumer. For example, if you eat an apple or a peanut, you are a first-order consumer.

When you eat the meat or products of animals that feed on plants, you are a second-order consumer. For example, if you eat a hamburger or drink a glass of milk, you are a second-order consumer. Milk and hamburger come from cows, which feed on plants. When you eat the meat or products of animals that feed on other animals, you are a third-order consumer. If you eat swordfish or lobster for dinner, you are a third-order consumer.

You probably eat many different kinds of food. Depending on what you eat, you are part of different food chains. The diagram shows the food chains that you are a part of when you eat a chicken sandwich. Trace the different food chains. What kind of consumer are you when you eat a chicken sandwich?

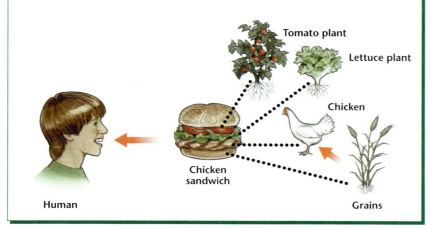

Science in Your Life

Ask students to recall what is meant by a first-order, second-order, and third-order consumer. Then read aloud the bold question at the top of the Science in Your Life feature on page 311. After students respond to the question, have volunteers read the paragraphs aloud. Call attention to the diagram and have students explain what orders of consumption are represented. (*first-order and second-order consumers*)

List other foods suggested by students and establish food chains for them on the board. Ask students to practice tracing the food chains and determining the order of consumer they are when eating each food.

AT HOME

Students can plan and help prepare a meal to include foods that make them first-, second-, and third-order consumers. Suggest that students draw a picture of the meal in the center of a sheet of paper. Around this, they can diagram the food chains that show the orders of consumers. (For example, the salmon you eat fed on smaller fish and crustaceans; the crustaceans fed on small marine animals, which in turn fed on marine plants. When you eat salmon, you are a third-order consumer.)

Lesson at a Glance

Chapter 12 Lesson 3

Overview In this lesson, students explore how energy flows from the Sun to plants and then to other organisms in an ecosystem. They also learn how the amount of energy changes as it flows through the food chain.

Objectives

- To explain why organisms need energy
- To describe how energy flows through a food chain
- To compare the amount of energy available at different levels of a food chain
- To explain how the amount of available energy affects the sizes of populations

Student Pages 312–316

Teacher's Resource Library
Workbook Activity 47
Alternative Activity 47

Vocabulary
energy pyramid

Science Background
Energy in Ecosystems

Food chains are feeding patterns that show the transfer of energy through ecosystems, or simply put, who eats whom. Most of the energy for living things comes from the Sun. Plants capture this energy to make food. Animals get their energy by eating plants or other animals. Because organisms use some of this energy to survive and give off more as heat, energy is lost at each step of consumption.

 Warm-Up Activity

Ask students where they get their energy. (from food) Remind them that one of the laws of physics is that energy is neither created nor destroyed. Then ask them if they know where the energy in food comes from.

312 Chapter 12 Ecology

Lesson 3 — How Energy Flows Through Ecosystems

Objectives

After reading this lesson, you should be able to
- explain why organisms need energy.
- describe how energy flows through a food chain.
- compare the amount of energy available at different levels of a food chain.
- explain how the amount of available energy affects the sizes of populations.

Plants use energy from the Sun to make food. You get energy from the food you eat. What is energy? Energy is the ability to do work, to move things, or to change things. Energy comes in many forms. For example, light from the Sun is energy. Heat from the Sun also is energy. Batteries store chemical energy. A moving bicycle has mechanical energy.

You and all organisms need energy to live. Your muscles use energy to contract. Your heart uses energy to pump blood. Your brain uses energy when you think. Your cells use energy when they make new molecules.

You probably get tired when you work hard. You might even say that you have "run out of energy" to describe how you feel. Like the boy in the photo, you take a break and eat some lunch to "get your energy back." Food contains chemical energy.

Food gives us the energy we need to do work.

312 Chapter 12 Ecology

Did You Know?

Only about one-tenth of the available energy at each level of a food chain is passed on to the next level.

Energy in Food

Recall from Chapter 7 that plants absorb energy from the Sun. A plant's chlorophyll and other pigments absorb some of the light energy. By the process of photosynthesis, the plant uses the absorbed energy to make sugar molecules. Photosynthesis is a series of chemical reactions. During these reactions, light energy is changed into chemical energy. The chemical energy is stored in the sugar molecules.

Plants use the sugar to make other food molecules, such as starches, fats, and proteins. All these nutrients store chemical energy. When plants need energy, they release the energy stored in the nutrients. Plants need energy to grow, reproduce, make new molecules, and perform other life processes. To use the energy stored in nutrients, plant cells break down the molecules into simpler molecules. As the molecules are broken down, they release the stored energy.

Plants also use the nutrients they make to produce tissues in their leaves, roots, and stems. The nutrients' chemical energy is stored in the tissues. When you eat potatoes, asparagus, or other plant parts, you are taking in the plants' stored chemical energy.

Flow of Energy Through Food Chains

Animals and other consumers are unable to make their own food. They must eat plants or other organisms for food. As organisms feed on one another, the energy stored in the organisms moves from one level of the food chain to the next.

The flow of energy in a food chain begins with the producers, such as plants. As you know, plants absorb the Sun's energy to make food. They use some of the energy in the food for life processes. As plants use this energy, some is changed to heat. The heat becomes part of the environment. The rest of the energy is stored as chemical energy in the plants' tissues.

The energy stored in plants is passed on to the organisms that eat the plants. These first-order consumers use some of the food energy and lose some energy as heat. The rest of the energy is stored as chemical energy in the nutrients in their body.

2 Teaching the Lesson

Review the terms *molecule, chlorophyll, pigment, starch, fat, protein, nutrient,* and *tissue* with students.

As they read, students should write the main idea of each section. Then have them write one or two supporting sentences for each main idea.

Relate the energy pyramid to the pyramid of numbers discussed in the previous lesson. As you move up the food chain, available energy from the food source decreases. As a result, each successive level of consumer must have a smaller population.

Refer to some of the sample food chains discussed in Lesson 2. On the board, have students draw energy pyramids for those food chains.

TEACHER ALERT

Many students will not know of organisms that do not get energy from the Sun. Some bacteria near hydrothermal vents on the ocean floor get their energy by breaking down chemicals that contain sulfur.

3 Reinforce and Extend

Did You Know?

Have a volunteer read aloud the Did You Know? feature on page 313. Ask students to list reasons why such a small portion of the available energy is passed on to the next level. (*Not all organisms are consumed; some organisms die of natural causes; energy is used for other purposes by the organisms.*)

CROSS-CURRICULAR CONNECTION

Health

The amount of energy foods contain is directly proportional to their calorie content. Have students study food labels to compare the amounts of energy in different foods. Ask students what happens when they eat food that has more energy than they need. (*The energy is stored as fat.*)

SCIENCE INTEGRATION

Physical Science

Discuss with students that energy can neither be created nor destroyed. However, it can be changed from one form to another. Pair students and ask partners to diagram the changes in energy illustrated by photosynthesis, consumption of plants by animals, and animals performing life activities. (*Photosynthesis changes solar energy to chemical energy. When animals eat plants and move around, they change some chemical energy into kinetic energy and some into heat energy, though some is stored as chemical energy.*)

Energy pyramid
A diagram that compares the amounts of energy available to the populations at different levels of a food chain

The energy stored in the first-order consumers is passed on to the second-order consumers. Then, energy stored in the second-order consumers is passed on to the third-order consumers. At each level of the food chain, some energy is used for life processes, some is lost as heat, and the rest is stored in the organisms.

Energy Pyramid

The **energy pyramid** below compares the amounts of energy available to the populations at different levels of a food chain. The most energy is available to the producers. They get energy directly from the Sun. Less energy is available to the insects, the first-order consumers that feed on the producers. That is because the producers have used some of the Sun's energy for their own needs. Also, some of the energy was lost as heat. Only the energy that is stored in the producers is passed on to the insects.

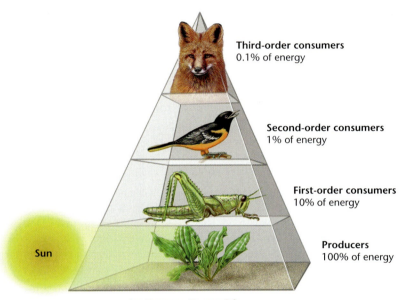

An Energy Pyramid

314 Chapter 12 Ecology

Did You Know?

Communities that do not use energy from the Sun live deep in the ocean near hot water vents. The producers in these communities are a kind of bacteria that use chemicals for energy. Other organisms in these communities include tube worms, clams, and sea urchins.

The insects use some of the energy they take in for their own needs. Again, some energy is lost as heat. Thus, the amount of energy available to the birds, the second-order consumers, is less than the amount of energy available to the insects. The amount of available energy decreases at each higher level of a food chain. The foxes have the least amount of energy available. They are at the highest level of the food chain.

The amounts of energy available to the different populations of a food chain affect the sizes of the populations. Look back to the pyramid of numbers shown on page 307. Recall that the size of the population decreases at each higher level of a food chain. That is because less energy is available to the population at each higher level.

Importance of the Sun

Without the Sun, there would be no life on Earth. Plants, the animals that eat plants, and most other organisms depend on energy from the Sun. Energy flows in one direction from the Sun to producers, and then to consumers. Communities lose energy as it flows through food chains. The Sun continuously replaces the lost energy.

Did You Know?

Have students read the Did You Know? feature on page 315. Show pictures of tube worms, clams, and sea urchins. Ask students to suggest the energy pyramid that might exist in this community and place the pictures into the pyramid. (*Producer: bacteria; first-order consumers: tube worms, clams; second-order consumer: sea urchin*)

CROSS-CURRICULAR CONNECTION

Math

Give pairs of students a square meter area to analyze. They can count the number of producers their area contains and then the number of visible consumers they observe. Ask partners to create a population "map" of their ecosystem and a graph comparing the numbers of producers and consumers.

Lesson 3 Review Answers

1. for various life processes, such as reproducing, growing, and making new molecules **2.** producers, usually plants **3.** First, producers absorb energy and use it to make food. Some is stored as chemical energy in the producers' tissues. Energy flows through the food chain as consumers feed on one another and take in chemical energy stored in organisms' tissues. **4.** Organisms use some of their available food energy for life processes and lose some as heat. The rest is stored in body tissues. Only this energy is available for the next higher level. **5.** Moving up the food chain, less and less energy is available to the consumers; therefore, it supports a smaller and smaller population of consumers.

Portfolio Assessment

Sample items include:

- Main idea and supporting detail sentences from Teaching the Lesson
- Lesson 3 Review answers

Lesson 3   REVIEW

Write your answers to these questions on a separate sheet of paper. Write complete sentences.

1. Why do organisms need energy?
2. Which organisms in a community get energy directly from the Sun?
3. How does energy flow through a food chain?
4. Explain why less energy is available at each higher level of a food chain.
5. How does the amount of energy available to a population affect the size of the population?

Workbook Activity 47

Lesson 4: How Materials Cycle Through Ecosystems

Objectives

After reading this lesson, you should be able to
- describe how water cycles through ecosystems.
- explain the roles of photosynthesis and cellular respiration in the carbon, oxygen, and water cycles.
- describe how nitrogen cycles through ecosystems.
- explain how cycles in ecosystems are linked to one another.

Groundwater
The water under Earth's surface

The planet Earth is sometimes compared to a spaceship. Like a spaceship, Earth is isolated in space. All the materials we use to build homes, to make tools, and to eat come from the biosphere. If a material is in short supply, there is no way to get more of it. Materials in the biosphere must be used over and over again. For example, chemicals continuously cycle between organisms and the nonliving parts of Earth. Some chemicals important for life are water, carbon, oxygen, and nitrogen.

The Water Cycle

The diagram shows the water cycle between the living and nonliving parts of an ecosystem. The most noticeable water in ecosystems is the liquid water in lakes, rivers, and the ocean. In addition, **groundwater** exists beneath the surface of the land.

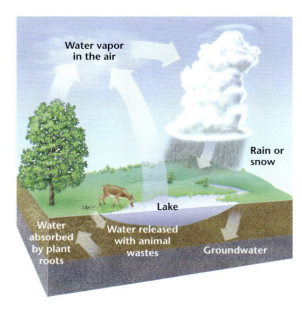

Ecology Chapter 12 317

Warm-Up Activity

Write the words *circle* and *cycle* on the board. Ask students how they are similar in meaning. (*In a cycle, events occur in a series, leading back to the starting point. In this respect, a cycle forms a circle.*) Then ask: Why is there still oxygen in the air if animals have been breathing it in for millions of years? Challenge students to relate their explanations to a cycle.

Teaching the Lesson

Review the concepts of photosynthesis and cellular respiration with students.

As they read the lesson, have students write a main idea sentence for each paragraph. After they finish reading, discuss students' sentences.

Discuss the processes of evaporation and condensation and how they relate to the water cycle.

Demonstrate condensation by adding water to a shiny metal can. Add ice until beads of water appear on the outside of the can. Ask a volunteer to explain where the water came from. (*water vapor in the air near the outside of the can*)

Use overhead transparencies of each cycle or draw each cycle on the board as you discuss it. Ask students where each cycle begins. They should recognize that, once a cycle is begun, it has no beginning, just as a wheel has no beginning.

Reinforce and Extend

SCIENCE JOURNAL

Ask students to write a description or story about what it would be like if your area suddenly became low on water. They should identify why this has happened and how more water could enter your area's supply.

318 *Chapter 12 Ecology*

Evaporate
To change from a liquid to a gas

Condense
To change from a gas to a liquid

What happens when a puddle of water dries up? As the puddle dries, the liquid water changes into a gas, or **evaporates**. This gaseous water is called water vapor. Water from the ocean, lakes, and rivers evaporates and becomes part of the air. Water vapor comes from other places, too. Organisms produce water when they get energy from food during the process of cellular respiration. Plants release water vapor through their leaves. Animals release water vapor with their breath. They also release liquid water with their wastes.

Water vapor is always in the air, but you cannot see it. Have you ever noticed that the outside of a glass of ice water becomes wet? Water vapor from the air **condenses**, or changes into a liquid, on the outside of the glass. The ice water cools the air next to the glass, causing the water vapor to condense.

Water vapor in the air may cool and condense into water droplets in a cloud. When enough water gathers in the cloud, rain or snow may fall. That water may be used by organisms. Organisms need water for various life processes. For example, plants need water to make food during photosynthesis. Plants take in water from the soil through their roots. Animals may drink water from ponds or streams. Animals also get water from the food they eat.

The Carbon Cycle

All living things are made up of chemicals that include carbon. Carbohydrates, fats, and proteins contain carbon. Carbon also is found in the nonliving parts of the environment. For example, carbon dioxide gas is in the air and in bodies of water. Carbon is found in fossil fuels, such as coal and oil.

The diagram on page 319 shows how carbon cycles through an ecosystem. Plants and other organisms that undergo photosynthesis take in carbon dioxide and use it to make food. Animals take in carbon-containing chemicals when they eat plants or other animals.

318 *Chapter 12 Ecology*

During cellular respiration, plants, animals, and other organisms produce carbon dioxide. Plants release carbon dioxide through their leaves and other plant parts. Animals release carbon dioxide when they exhale. Decomposers release carbon dioxide as they break down dead organisms. The carbon dioxide that is released by organisms may become part of the air or a body of water. People also produce carbon dioxide when they burn fossil fuels. In these ways, carbon continues to cycle through ecosystems.

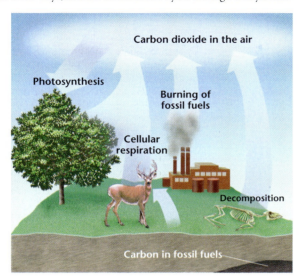

The Oxygen Cycle

Oxygen is a gas that is essential to almost every form of life. Oxygen is found in air and in bodies of water. Organisms use oxygen for cellular respiration. The oxygen is used to release energy that is stored in food.

Most of the oxygen that organisms use comes from producers, such as plants. Producers release oxygen as a waste product during photosynthesis. Producers use some of the oxygen for cellular respiration. Consumers take in some of the oxygen and use it themselves. Thus, oxygen continuously cycles between producers and consumers in an ecosystem.

LEARNING STYLES

Interpersonal/Group Learning
Have students work together to plan and carry out a demonstration that models the water cycle. Groups should develop a set of procedural steps, including safety precautions, and identify needed materials before asking your approval for their investigation. Have groups perform their approved demonstrations for the class.

SCIENCE INTEGRATION

Technology
Have students read articles about global warming and report on how the increased use of fossil fuels and the loss of forests have increased the amount of carbon dioxide in the atmosphere. Discuss with students how this imbalance of the carbon cycle may have created problems for future generations.

CROSS-CURRICULAR CONNECTION

Home Economics

Have students investigate the minerals present in various foods: e.g., dairy products are rich in calcium, bananas in potassium, meats contain proteins, which are rich in nitrogen, oxygen, and carbon. Ask students to diagram how one of the minerals got into the food and where it will pass next in its cycle.

LEARNING STYLES

Visual/Spatial

Review photosynthesis, pages 148–150, and cellular respiration, pages 152–154, with students. Ask some students to draw a diagram that incorporates photosynthesis and the water, oxygen, or carbon cycle. Have other students draw a diagram that incorporates cellular respiration and the water, oxygen, or carbon cycle. Ask students to present their diagrams to the class, and then display them on a bulletin board.

Nitrogen fixation
The process by which certain bacteria change nitrogen gas from the air into ammonia

The Nitrogen Cycle

Nitrogen is a gas that makes up about 78 percent of the air. Many chemicals important to living things, such as proteins and DNA, contain nitrogen. However, the nitrogen in the air is not in a form that organisms can use. Certain bacteria are able to change nitrogen gas into a chemical, called ammonia, that plants can use. These bacteria live in the soil and in the roots of some plants. The process by which the bacteria change nitrogen gas into ammonia is called **nitrogen fixation**.

The diagram shows that plants take in the ammonia through their roots. Some of the ammonia, however, is changed by certain bacteria into chemicals called nitrates. Plants use both ammonia and nitrates to make proteins and other chemicals they need.

Not all of the nitrates are used by plants. Notice in the diagram below that bacteria change some of the nitrates back into nitrogen gas. The return of nitrogen gas to the air allows the nitrogen cycle to continue.

Animals get the nitrogen they need by feeding on plants or on animals that eat plants. When organisms die, decomposers change the nitrogen-containing chemicals in the organisms into ammonia. The ammonia may then be used by plants or may be changed into nitrates by bacteria.

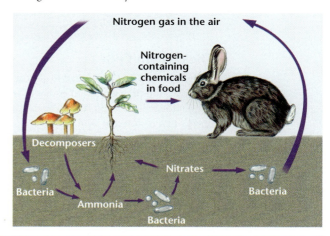

It is estimated that the population of people on Earth will grow from just over 6 billion in 2003 to over 7 billion in 2013 and to over 9 billion by 2050.

System of Cycles

The different cycles in an ecosystem are linked to one another. For example, the carbon cycle, oxygen cycle, and water cycle are linked by photosynthesis and cellular respiration. Plants take in carbon dioxide and water for photosynthesis, and release oxygen. Animals, plants, and other organisms use the oxygen for cellular respiration and release carbon dioxide and water.

When plants make nitrogen-containing chemicals from ammonia and nitrates, they use carbon and oxygen. Many other materials cycle through ecosystems. Iron, calcium, phosphorus, and other chemicals used by living organisms are cycled.

Scientists may study one cycle at a time to make it easier to understand the cycle. However, each cycle is only a small part of a system of cycles that interact with one another.

ONLINE CONNECTION

For more information about biogeochemical cycles, students can contact library.thinkquest.org/11353/nitrogen.htm and pages.nyu.edu/~pet205/biogeochem1.html, which includes detailed diagrams.

CAREER CONNECTION

Explain that ecologists take an active role in solving environmental problems, restoring and re-establishing natural ecosystems, advising policy makers, educating the public about ecosystems, conducting research, and managing natural resources. Ask students who think they would like to work in this field to explore one or more of the following resources to find out what they can do today:

- Join an after-school ecology club (or start one).
- Volunteer at a park, nature center, wildlife refuge, zoo, aquarium, or lab where ecological work is being done.
- Contact the Ecological Society of America (1707 H St., NW, Suite 400, Washington, DC 20006, www.esa.org/) for information.
- Get a list of internships and volunteer opportunities from The Environmental Careers Organization (www.eco.org/).

Lesson 4 Review Answers

1. A plant gives off water vapor as it respires. Water vapor condenses in the atmosphere to form a cloud. When rain falls from it, water enters the soil and can be taken in again by the plant. **2.** The complex molecules that living things are made of, such as carbohydrates, proteins, and fats, contain carbon. **3.** Producers take in carbon dioxide for photosynthesis. Producers and consumers release carbon dioxide as a by-product of cellular respiration. **4.** Oxygen is released by producers during photosynthesis. Producers and consumers use oxygen for cellular respiration. **5.** Certain bacteria change nitrogen gas into ammonia in the soil; other bacteria change ammonia into nitrates. Producers take in these usable forms of nitrogen.

Portfolio Assessment

Sample items include:
- Main idea sentences from Teaching the Lesson
- Lesson 4 Review answers

Lesson 4 REVIEW

Write your answers to these questions on a separate sheet of paper. Write complete sentences.

1. Describe how water may move from a plant to a cloud and back to a plant.
2. Why is carbon important to living things?
3. How are photosynthesis and cellular respiration part of the carbon cycle?
4. How are photosynthesis and cellular respiration part of the oxygen cycle?
5. How is nitrogen gas from the air changed into a form that plants can use?

Technology Note

Farmers need to add nitrogen to soil to help plants grow. Adding too much nitrogen, though, causes pollution of groundwater, lakes, and streams. Different kinds of technology can help farmers apply just the amount the crop needs. As farmers drive tractors across fields, sensors can collect information about the soil and the plants. Computers on the tractors use this information. They calculate how much nitrogen is needed. The right amount of nitrogen is then applied to different parts of the field.

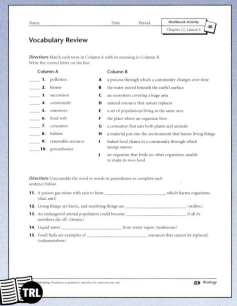

Workbook Activity 48

Chapter 12 SUMMARY

- A population is a group of organisms of the same species that lives in the same area.
- A community is a group of different populations that live in the same area.
- Communities may change over time by a process called succession.
- All the interactions among populations and the nonliving things in their environment make up an ecosystem.
- Human activities can cause pollution and loss of habitat.
- Natural resources can be renewable or nonrenewable.
- The feeding order of organisms is called a food chain. Every food chain begins with producers, which can make their own food. Consumers must take in food.
- Energy flows through food chains. All the food chains in a community that are linked make up a food web.
- Water, carbon, oxygen, and nitrogen cycle through ecosystems.
- Water evaporates and condenses as it cycles between organisms and the environment.
- Oxygen and carbon dioxide are cycled through the environment by the processes of respiration and photosynthesis.
- In the nitrogen cycle, bacteria change nitrogen gas from the air into ammonia. Other bacteria change nitrates back to nitrogen gas.

Science Words

abiotic, 297	ecology, 296	groundwater, 317	producer, 306
acid rain, 299	ecosystem, 297	habitat, 297	pyramid of numbers, 307
biome, 300	endangered, 300	interact, 296	resource, 301
biosphere, 302	energy pyramid, 314	nitrogen fixation, 320	renewable resources, 301
biotic, 297	evaporate, 318	nonrenewable resources, 301	succession, 298
climax community, 299	extinct, 300	omnivore, 307	threatened, 300
community, 297	food chain, 306	pollution, 299	
condense, 318	food web, 308	population, 297	
consumer, 306	fossil fuels, 302		

Chapter 12 Review

Use the Chapter Review to prepare students for tests and to reteach content from the chapter.

Chapter 12 Mastery Test

The Teacher's Resource Library includes two parallel forms of the Chapter 12 Mastery Test. The difficulty level of the two forms is equivalent. You may wish to use one form as a pretest and the other form as a posttest.

Review Answers

Vocabulary Review

1. Producers 2. ecology 3. omnivores
4. biomes 5. pollution 6. habitat
7. biosphere 8. Consumers 9. food chains 10. Ecosystems 11. succession
12. climax community 13. population
14. nitrogen fixation

TEACHER ALERT

In the Chapter Review, the Vocabulary Review activity includes a sample of the chapter's vocabulary terms. The activity will help determine students' understanding of key vocabulary terms and concepts presented in the chapter. Other vocabulary terms used in the chapter are listed below:

abiotic	groundwater
acid rain	interact
biotic	nonrenewable
community	resources
condense	pyramid of
endangered	numbers
energy pyramid	renewable
evaporate	resources
extinct	resource
food web	threatened
fossil fuels	

Chapter 12 REVIEW

Vocabulary Review

Choose the word or words from the Word Bank that best complete each sentence. Write your answer on your paper.

Word Bank
biomes
biosphere
climax community
consumers
ecology
ecosystems
food chains
habitat
nitrogen fixation
omnivores
pollution
population
producers
succession

1. _____ absorb the energy of the Sun.

2. The study of how living things interact with one another and with nonliving things in their environment is _____.

3. Consumers that eat both plants and animals are _____.

4. Deserts, grasslands, and the ocean are all examples of _____.

5. Anything added to the environment that is harmful to living things is _____.

6. The place where an organism lives is its _____.

7. All the biomes on Earth together form the _____.

8. _____ feed on other organisms.

9. All the _____ in a community that are linked to one another make up a food web.

10. _____ consist of all the interactions among populations and the nonliving things in their environment.

11. The process by which a community changes over time is called _____.

12. A community that changes very little over time is called a(n) _____.

13. All the deer of the same species living in a forest make up a(n) _____.

14. Certain bacteria change nitrogen gas from the air into ammonia by the process of _____.

324 Chapter 12 Ecology

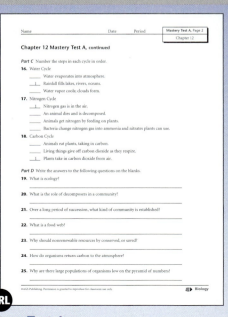

Chapter 12 Mastery Test A

Concept Review

Choose the answer that best completes each sentence. Write the letter of the answer on your paper.

15. Water, air, and sunlight are examples of _____.
  A organisms
  B renewable resources
  C nonrenewable resources
  D pollution

16. Photosynthesis and cellular respiration help to cycle _____ through ecosystems.
  A carbon, oxygen, and water
  B nitrogen
  C only carbon
  D only oxygen

17. When water in a pond _____, it changes into water vapor.
  A condenses
  B falls as rain
  C evaporates
  D becomes groundwater

18. Plants store _____ energy in their tissues.
  A mechanical
  B heat
  C light
  D chemical

Critical Thinking

Write the answer to each of the following questions.

19. What would Earth be like if there were no decomposers?

20. Identify the producers and the first-order, second-order, and third-order consumers in the food web shown here. Remember that organisms may be part of more than one food chain.

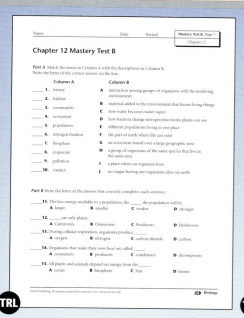

Test-Taking Tip When choosing answers from a word bank, answer all of the questions you know first. Then, study the remaining words to choose the answers for the questions you are not sure about.

Ecology Chapter 12 325

Concept Review
15. B **16.** A **17.** C **18.** D

Critical Thinking
19. The bodies of dead animals would not decompose, or break into chemicals usable by plants. After a time, plants would be unable to live. Then organisms that depend on plants would die.

20. producers: plants; first-order consumers: mice, insects, birds; second-order consumers: frogs, birds, snakes, hawks; third-order consumers: snakes, hawks

ALTERNATIVE ASSESSMENT

Alternative Assessment items correlate to the student Goals for Learning at the beginning of this chapter.

- Display a picture of an ecosystem, such as a pond or forest scene with plants and animals. Have students name living and nonliving things in the system and explain ways they interact, or depend on one another.

- Have students cut out pictures of plants and animals from old magazines and arrange them into food chains or a food web. Ask each student to describe who eats whom in one food chain.

- Have students use their food chains from the preceding activity to explain the source of energy for each organism and to tell how energy is changed as it flows up the chain.

- Give students unlabeled, untitled diagrams that illustrate the water, carbon, oxygen, and nitrogen cycles. Have them identify the material that is being cycled through the ecosystem in each illustration and explain how it moves and changes.

- Ask students to draw a poster illustrating several types of pollution caused by humans. Students can use their posters to explain how ecosystems are harmed.

Ecology Chapter 12 325

Chapter 13

Planning Guide
The Behavior of Organisms

	Student Pages	Vocabulary	Lesson Review
Lesson 1 Innate Behavior	328–334	✔	✔
Lesson 2 Learned Behavior	335–340	✔	✔
Lesson 3 How Animals Communicate	341–346	✔	✔

Student Text Lesson

Chapter Activities

Student Text
Science Center

Teacher's Resource Library
Community Connection 13: Dog Watching

Assessment Options

Student Text
Chapter 13 Review

Teacher's Resource Library
Chapter 13 Mastery Tests A and B

Student Text Features								Teaching Strategies						Learning Styles						Teacher's Resource Library				
Achievements in Science	Science at Work	Science in Your Life	Investigation	Science Myth	Note	Technology Note	Did You Know?	Science Integration	Science Journal	Cross-Curricular Connection	Online Connection	Teacher Alert	Applications (Home, Career, Community, Global, Environment)	Auditory/Verbal	Body/Kinesthetic	Interpersonal/Group Learning	Logical/Mathematical	Visual/Spatial	LEP/ESL	Workbook Activities	Alternative Workbook Activities	Lab Manual	Resource File	Self-Study Guide
				329		✓	331	331, 333	330	333	334	329	330, 333	330	332	330		331	332	49	49	37, 38		✓
	338		339				336		336	337	337		337				337	336, 337		50	50	39		✓
345		346			✓	✓	344	342		344		342	343, 346	344	343			343		51	51		25, 26	✓

Pronunciation Key

a	hat	e	let	ī	ice	ô	order	ů	put
ā	age	ē	equal	o	hot	oi	oil	ü	rule
ä	far	ėr	term	ō	open	ou	out	ch	child
â	care	i	it	ȯ	saw	u	cup	ng	long

sh	she	a	in about
th	thin	e	in taken
ŦH	then	ə { i	in pencil
zh	measure	o	in lemon
		u	in circus

Alternative Workbook Activities

The Teacher's Resource Library (TRL) contains a set of lower-level worksheets called Alternative Workbook Activities. These worksheets cover the same content as the regular Workbook Activities but are written at a second-grade reading level.

Skill Track Software

Use the Skill Track Software for Biology for additional reinforcement of this chapter. The software program allows students using AGS textbooks to be assessed for mastery of each chapter and lesson of the textbook. Students access the software on an individual basis and are assessed with multiple-choice items.

Chapter at a Glance

Chapter 13: The Behavior of Organisms
pages 326–349

Lessons
1. Innate Behavior pages 328–334
2. Learned Behavior pages 335–340

   Investigation 13 pages 339–340
3. How Animals Communicate pages 341–346

Chapter 13 Summary page 347
Chapter 13 Review pages 348–349
Skill Track Software for Biology
Teacher's Resource Library
   Workbook Activities 49–51

   Alternative Workbook Activities 49–51

   Lab Manual 37–39

   Community Connection 13

   Resource File 25–26

   Chapter 13 Self-Study Guide

   Chapter 13 Mastery Tests A and B

(Answer Keys for the Teacher's Resource Library begin on page 420 of the Teacher's Edition. A list of supplies required for Lab Manual Activities in this chapter begins on page 449.)

Science Center

Provide students with discarded nature magazines. Have students cut out pictures that show the behavior of different types of organisms. Display them for the duration of the chapter. After students have studied Lesson 1, have them identify and label the stimulus and response in each picture.

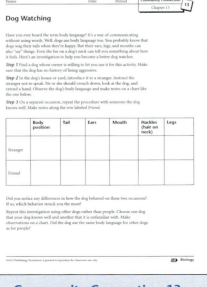

Community Connection 13

Chapter 13
The Behavior of Organisms

The brown bear cubs in the photo are watching the fish. The cubs are learning from their mother how to hunt and catch food. The cubs will learn many behaviors from their mother that will help them survive. The cubs also were born with many behaviors. In this chapter, you will learn about behaviors that animals are born with and behaviors that they learn. You also will learn how animals communicate.

Organize Your Thoughts

Goals for Learning

◆ To recognize that behaviors are responses to stimuli
◆ To identify behavior as innate or learned
◆ To identify ways that animals communicate
◆ To recognize that behaviors help organisms survive

Introducing the Chapter

Have students look closely at the bears pictured on page 326. Point out how curious the cubs appear. Ask students what information the cubs might learn from their mother in a situation like this. Invite them to compose questions the cubs might be "asking." (Examples: Are those animals good to eat? Are they dangerous?) List students' questions on the board and discuss how the mother bear might supply answers.

Point out that the chapter distinguishes between behaviors animals learn and behaviors they are born with. Invite students to provide examples of behaviors that they have observed in animals. Discuss with the class whether their examples might be learned or innate. List their ideas on the board and return to them as students read the chapter.

Notes and Technology Notes

Ask volunteers to read the notes that appear in the margins throughout the chapter. Then discuss them with the class.

TEACHER'S RESOURCE

The AGS Teaching Strategies in Science Transparencies may be used with this chapter. The transparencies add an interactive dimension to expand and enhance the *Biology* program content.

CAREER INTEREST INVENTORY

The AGS Harrington-O'Shea Career Decision-Making System-Revised (CDM) may be used with this chapter. Students can use the CDM to explore their interests and identify careers. The CDM defines career areas that are indicated by students' responses on the inventory.

The Behavior of Organisms Chapter 13

Lesson at a Glance

Chapter 13 Lesson 1

Overview Students learn about innate behaviors and how these behaviors help animals survive. They also learn how plants respond to different stimuli.

Objectives

- To define *innate behavior* and give two examples
- To list examples of courtship behavior, nest-building behavior, and territorial behavior
- To describe plant responses to light, gravity, and touch

Student Pages 328–334

Teacher's Resource Library

Workbook Activity 49
Alternative Workbook Activity 49
Lab Manual 37, 38

Vocabulary

behavior
stimulus
innate behavior
reflex
instinct
courtship behavior
territorial behavior
phototropism
gravitropism

Science Background
Innate Behaviors

Behavior has both a genetic (inborn) and an experiential component. Behaviors that are inborn, preprogrammed, genetically determined, or inherited are called innate behaviors, or instincts. Behaviors that are environmentally determined or based on experience are said to be learned behaviors. Most behaviors have both components.

Though they are inborn, innate behaviors require environmental input. An innate behavior is one that is performed perfectly or nearly perfectly the first time an organism of an appropriate maturity attempts it. For example, once their bodies are able to fly, birds fly well the first time they try.

Lesson 1: Innate Behavior

Objectives

After reading this lesson, you should be able to
- define *innate behavior* and give two examples.
- list examples of courtship behavior, nest-building behavior, and territorial behavior.
- describe plant responses to light, gravity, and touch.

Behavior is an interaction of heredity and experience. It is the way an organism acts. When organisms behave, they are reacting to something. Anything to which an organism reacts is called a **stimulus**. There are two main types of behavior: innate and learned.

A behavior that is present at birth is called an **innate behavior**. The behavior is inherited. It does not have to be learned. An animal usually performs the behavior correctly on the first try. Ask a classmate to pass a hand close to your eyes. Chances are you will blink. Blinking is a **reflex** that protects the eyes from harmful objects. A reflex is an innate behavior.

Unlike a reflex, an **instinct** is a pattern of behavior. Many parts of the body work together to produce actions in a certain order. For example, the eyes, nerves, and brain of a deer help the deer to see a wolf. The brain, nerves, and muscles work together to cause the deer to stand still. Then, the wolf may not see the deer in the trees. The deer did not learn this behavior. It is instinctive. Instinctive behaviors help animals to survive.

Behavior
The way an organism acts

Stimulus
Anything to which an organism reacts

Innate behavior
A behavior that is present at birth

Reflex
An automatic response

Instinct
A pattern of innate behavior

Nest-Building Behavior

Nest building is one example of an instinct. Each species of bird builds its own kind of nest. Birds build nests of all sizes and in different places. They use materials they find in their habitats. A habitat is the part of an ecosystem where each species lives.

Woodpeckers drill a hole in the side of a tree to make a nest. The nest helps protect their eggs and young.

328 Chapter 13 The Behavior of Organisms

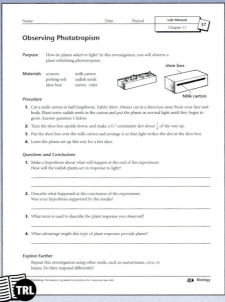

Lab Manual 37

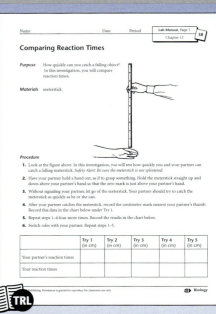

Lab Manual 38, pages 1–2

Courtship behavior
Behavior that helps attract a mate

Science Myth

All birds build nests and care for their young.

Fact: Some species of birds, such as the brown-headed cowbird, do not build nests. A female cowbird watches until a bird of another species, such as a cardinal, leaves its nest. The cowbird then lays its egg in the other bird's nest. The cardinal keeps the egg warm and takes care of the baby cowbird when it hatches.

Gulls use seaweed and plants to build a large nest on the ground. Terns make a hole in the sand where they lay their eggs. Nests provide safe places for eggs and young birds. The instinct of nest building helps each species survive.

Flying Behavior

You may have heard that birds learn to fly. Birds do improve their flight skills as they grow older. However, flying is an innate behavior. In one experiment, young birds were prevented from flapping their wings. The birds grew to the age when they normally would fly. The birds were released. They were able to fly on their first try. They had never practiced. The birds did not need to learn how to fly. Their bodies had developed to the point where they could fly. This shows that flying is innate.

Courtship Behavior

Courtship behaviors are behaviors to attract a mate. Animals must make sure they mate with the same species. Bluejays must mate with bluejays. Swallowtail butterflies must mate with swallowtail butterflies. Two different species may be able to produce offspring. However, the offspring will not be able to produce more offspring. Courtship behaviors help the species to survive. Each species behaves in certain ways to attract a mate. One animal starts the courtship behavior. The other animal responds in a certain way. These behaviors help the animals recognize their species.

Courtship behaviors are different for different animals. Birds may attract a mate with a certain song. Some birds, such as peacocks, show off their feathers to get a mate. Frogs have special mating sounds. Think of several species of birds and frogs that live in the same field or pond. Each species needs a way to identify its kind. Special songs or sounds help animals recognize their kind.

The Behavior of Organisms Chapter 13

Teacher Alert

Students may think that because innate behaviors are inborn and genetically determined, they do not require environmental input. This is not true. Almost all innate behaviors require some environmental input. The innate part of the behavior is the readiness of the organism to respond to certain stimuli in a certain, almost perfect way, the first time exposed.

 Warm-Up Activity

Have students make a paper fan, using notebook paper. Ask partners to wear their safety glasses. Partners take turns waving a fan increasingly close to the other's face. Have student discuss the experience, focusing on the stimulus and the response.

 Teaching the Lesson

Have students scan the lesson and write down the headings. Under each heading, they should write one question that they expect to be answered in the lesson. As students read the text, they should write the answers. Allow students to share their work with partners. Discuss unanswered questions as a class.

Help students distinguish between a stimulus and a response. On the board, write these or similar sentences: *The bird flew away when the cat came toward it. The birds flew south when winter approached. The dog barked when the stranger walked by.* Then have students identify the stimulus and the response in each scenario. Have students develop their own sentences and trade them with a partner, who identifies each stimulus and response.

Science Myth

Read Science Myth on page 329 with students, and then write the word *parasite* on the board. Ask students what they know about parasites. Explain that a parasite is an organism that benefits from or lives off another organism without contributing anything in return. Point out that the cowbird is known as a *brood parasite*. Ask students to think of other parasites in the natural world. (*Examples include fleas, ticks, and tapeworms.*)

3 Reinforce and Extend

SCIENCE JOURNAL

Have students observe animals in an aquarium, at a zoo, at a local park, or on videotape. Have students record a description of each behavior. They should try to identify stimuli and responses and name the different behaviors they observe.

LEARNING STYLES

Interpersonal/Group Learning
Point out that human beings frequently engage in territorial behavior. Have students form small groups and ask them to brainstorm a list of specific examples of territorial behavior that they have observed or heard about. Encourage students to consider the behavior of individuals ("Keep Out" sign on a bedroom door) and of large groups (barbed wire at border crossings). Suggest that students consider the following categories: home, school, neighborhood, town or city, and country. Invite groups to share their examples in a class discussion about territorial behavior. List their ideas on the board.

IN THE ENVIRONMENT

Take students on a nature walk to find spider webs. Have students compare the different types of webs they find. They may wish to sketch different webs. Caution students not to disturb the webs or the spiders.

LEARNING STYLES

Auditory/Verbal
Play for the class a recording of bird songs. As they listen, help students list the species and make notes identifying each of the songs. Suggest that they use descriptive phrases (*a high whistle*) or syllables reproducing the sounds (*too-wee*). Then assign partners a bird species. Play the recording again, this time with the songs in a different order. Ask students to imagine that they are birds listening for the songs of their own species. Have them stand up when they hear their song.

Territorial behavior
Behavior that claims and defends an area

Territorial Behavior

A territory is an area that is defended by an animal. Usually, the animal defends the territory from other animals of the same species. Behaviors that claim and defend an area are called **territorial behaviors**. These behaviors help animals survive. All animals need a place to find food, shelter, and a mate. The survival of a species is tied to its ability to reproduce. In their territories, animals can court, mate, produce offspring, and raise their young.

Birds often claim their territories with their songs. Mockingbirds find the tallest trees on which to sing their territorial song. In neighboring territories, other male mockingbirds sing to claim their own space. Animals may mark an area with their scent. The scent warns others that the area is claimed.

An animal that claims a territory may be challenged to give it up. Other animals may try to fight the animal. The animal that already has the territory usually wins. The animal fights harder to keep the territory. It also may be older and more experienced in defending the area.

Behavior in Spiders and Bees

A young spider knows at birth how to build a web. The ability to weave a web is inherited. The parents do not take care of newly hatched young. The young do not have time to learn how to weave a web. They need the web to trap prey. Without this instinct, young spiders would not survive long. The same is true for bees. Bees are born with the ability to build a hive and find food. They also know how to care for the queen bee and protect the hive.

---

330 Chapter 13 The Behavior of Organisms

Phototropism
The response of a plant to light

Did You Know?
Other things respond to the time of day just as plants do. Your pulse rate, blood pressure, and temperature change with the time of day.

Responses of Plants

Plants do not behave in the same ways as animals. However, they do respond to their environment. Plants respond to light, gravity, water, temperature, and other factors. Parts of plants turn toward light. Roots grow downward. Some flowers open and close according to the time of day.

Response to Light

The response of plants to light is called **phototropism**. *Photo* means "light," and *tropic* means "turning." Geraniums placed near a window will lean toward the sunlight. Suppose an indoor plant is turned around so that it faces away from a window. In just hours, the leaves turn toward the light. This happens because of chemical changes inside the plant. The cells on the leaves opposite the light get longer. This causes the plant to bend toward the light. Sunflowers in a field will follow the sun from sunrise to sunset. Phototropism helps a plant survive by getting sunlight to make food.

These plants are leaning toward the sunlight. This response is called phototropism.

The Behavior of Organisms Chapter 13 **331**

Did You Know?

After students read Did You Know? on page 331, write the term *jet lag* on the board. Explain that the pace of modern life can sometimes throw the body from its natural rhythm. Give the example of a passenger who leaves New York at 8 P.M. and arrives in London, England, at 7 A.M. the next morning. Explain that the flight took only six hours. Ask students to calculate what time the passenger's body feels it is. Discuss with the class how jet lag can affect the way a traveler feels.

SCIENCE INTEGRATION

Technology
Remind students that the Sun is the energy source that sustains most life on Earth. Explain that people have only just begun to explore ways of tapping the Sun's energy for purposes such as electricity and transport. Have students research the subject of solar energy from print or Internet sources. Invite them to share their findings in small groups. Discuss the potential of solar energy with the class.

LEARNING STYLES

Visual/Spatial
Have students make a flip book that shows a sunflower turning toward the Sun. To make the book, students draw a sunflower on a sheet of paper and then make other drawings of the flower in slightly different positions on separate sheets of paper. They may want to include the Sun in their drawings. To see the sunflower moving, they should staple the pages together and then flip through them.

The Behavior of Organisms Chapter 13 **331**

Learning Styles

LEP/ESL

Write the word *response* on the board. Circle the prefix *re-* and point out that many words in English begin with these letters. Point out that sometimes *re-* can mean "again" or "anew"; it can also mean "backward" or "back." (*Response* is formed from *re-*, meaning "back" plus the root *spondere*, meaning "to promise.") Write the following words on the board: *reflect, report, renew, rethink, resist, refresh*. Then ask students to use a dictionary to identify three words in which the prefix *re-* is used to mean "again" (*renew, rethink, refresh*) and the three in which *re-* means "back" (*reflect, report, resist*).

Pronunciation Guide

Use this list to help students pronounce difficult words in this lesson. Refer to the pronunciation key on the Chapter Planning Guide for the sounds of these symbols.

geraniums	(jə rā′ nē əms)
mimosa	(mə mō′ sə)

Learning Styles

Body/Kinesthetic

Provide students with seeds for pole beans. Have them plant two or three beans each in two flowerpots. In one pot, students should insert a dowel or straight stick at least 30 cm long. Instruct students to place the pots on a bright windowsill or outside in a warm, sunny spot. When the bean shoots appear above the ground, have students draw a simple sketch of the growing plants every five days. Challenge students to draw conclusions about the responses of pole beans.

Gravitropism
The response of a plant to gravity

Response to Gravity

If you plant seeds, the roots grow downward and the sprouts grow upward. What would happen if you planted the seeds sideways or upside down? The seedling still would grow in the same way. This is a response to gravity called **gravitropism**. The stem grows upward to get sunlight. The root grows downward to get nutrients and water from the soil.

Response to Touch

Picture an insect landing on the Venus's-flytrap in the photo. The pressure of the insect stimulates sensitive hairs that line the surface of the leaves. The leaves close and capture the insect. All this happens within seconds of the fly touching the leaf. The response of the Venus's-flytrap to touch helps this plant to get food.

Another touch response occurs in mimosa plants. The leaflets of the mimosa plant usually are spread apart. When it is windy, the leaflets move closer together. The closing of its leaves helps prevent evaporation. This behavior helps the mimosa to conserve water. It can also protect the plant from animals. Thorns on the stems stand out when the leaves fold.

Sensitive hairs line the leaves of the Venus's-flytrap. When an insect touches them, the leaves close together and trap the insect.

Response to touch helps some plants to grow. Morning glories coil around an object that they touch. Vines wrap around tree trunks, fences, and posts. They grow in response to the object they are touching. These plants need support to grow upright. This helps them get more sunlight.

Other Plant Responses

Have you ever seen a plant that raises its leaves during the day and lowers them at night? These are called sleep movements. They are an example of plant responses to light. Other plant responses follow the seasons. Flowers bloom at certain times of the year. Some trees lose their leaves in autumn. These are examples of plant responses to length of daylight and temperature.

Technology Note

Scientists at NASA are exploring ways to grow plants in spacecraft. Plants would provide fresh food for space travelers, and purify water and air. One problem is that there is no gravity in space. On Earth, roots grow downward to get nutrients from the soil. The stem grows upward, reaching for sunlight. Without gravity, the roots of the plant don't grow correctly. This is just one problem scientists are trying to solve. If they do, gardens in space may someday be a reality.

AT HOME

Have students select a houseplant in their homes or a plant growing outside in their yards or neighborhoods. Ask them to make notes on the changes this plant undergoes for one week. Have students suggest the stimuli that may cause the changes they observe. Invite volunteers to present their notes to the class.

CROSS-CURRICULAR CONNECTION

Health

People, like plants, are sometimes affected by the amount of daylight. The limited hours of daylight during the winter months results in a condition known as seasonal affective disorder (SAD) in some people. Sufferers feel sad, tired, and unable to handle many daily tasks. Treatments include several hours of exposure to intense light that simulates sunlight. Suggest that students research information about SAD and the effect of light on humans.

SCIENCE INTEGRATION

Technology

Ask students what problems weightlessness might create for astronauts at mealtimes. (*food flies around*) Point out that salt and pepper are provided in liquid form so that the grains can't float away and get into machinery or the astronauts' eyes, ears, or lungs. Challenge students to come up with reasons why astronauts eat tortillas instead of bread. (*According to NASA's official Web site, bread takes up too much room; tortillas don't crumble; Tex-Mex food tastes better; and tortillas make great space frisbees!*)

Online Connection

For more information on gravitropism, have students visit www.geocities.com/CapeCanaveral/5229/n_grvtro.htm.

Lesson 1 Review Answers

1. behavior that is present at birth **2.** Courtship behaviors help animals recognize other animals of the same species for mating. **3.** Territorial behaviors help animals defend an area in which to find food, shelter, and a mate and to raise their young. **4.** Responses to light help plants get sunlight to make food. Responses to gravity help plants get needed nutrients. Responses to touch help plants get food, protect them from animals, conserve water, and grow in certain ways. **5.** Plants do not behave as animals do. However, plants do respond to stimuli, such as light, gravity, and touch.

Portfolio Assessment

Sample items include:
- Questions and answers from Teaching the Lesson
- Entries from Science Journal
- Lesson 1 Review answers

Lesson 1 REVIEW

Write your answers to these questions on a separate sheet of paper. Write complete sentences.

1. What is innate behavior?
2. Why are courtship behaviors important?
3. How do territorial behaviors help an animal to survive?
4. How do plant responses help plants to survive?
5. Do plants have behaviors? Explain your answer.

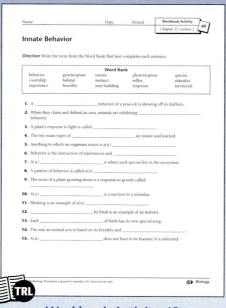

Workbook Activity 49

Lesson 2: Learned Behavior

Objectives

After reading this lesson, you should be able to

- tell the difference between innate and learned behaviors.
- define and give an example of each of the different types of learned behaviors.
- explain how learned behavior helps animals survive.

Learned behavior
Behavior that results from experience

Imprinting
Learning in which an animal bonds with the first object it sees

Observational learning
Learning by watching or listening to the behavior of others

In Lesson 1, you learned about innate behaviors. Unlike innate behaviors, **learned behaviors** are not present at birth. They are behaviors that are the result of experience. Learned behaviors can change over the lifetime of an organism. Five types of learned behaviors are imprinting, observational learning, trial-and-error, conditioning, and insight.

Imprinting

Baby geese bond with the first object they see. This behavior is called **imprinting**. In nature, the first object is the mother goose. A scientist named Konrad Lorenz removed goose eggs from their nest. He hatched them in a lab. The baby geese saw Lorenz first. They learned that Lorenz was their "mother" and followed him wherever he went.

Young geese will follow their "first sight" leader. It can be the mother goose, a person, or even a ticking clock. Lorenz found that once imprinting is set, it does not change. His geese followed him everywhere. They even preferred him to their mother or other geese.

In nature, imprinting helps animals survive. It keeps young geese in the family circle. The young stay close to the mother, who protects and feeds them.

Observational Learning

Birdsongs are both innate and learned. The behavior is innate because birds are born knowing how to sing. A bird that is raised alone will sing. However, the song is different from songs by the same species. Birds must learn their songs by hearing other adult birds sing. This type of behavior is called **observational learning**. An animal learns by watching or listening to another animal.

Warm-Up Activity

Lead a discussion about training animals. Allow students who have trained a pet to discuss their strategies and the behaviors their pets have learned. Students who do not have pets may want to discuss animals they have seen at theme parks or on television.

Teaching the Lesson

From a college biology text or reference source, show students a picture of Lorenz's geese following him as a result of imprinting.

Tell students that they can think of trial-and-error learning as learning by doing and correcting mistakes. Ask students to describe situations in which they used trial and error to learn. Examples include learning a new sport or game, learning to use a computer or a new computer application, and learning the best route to take to school.

Have students outline the lesson, using the lesson title and headings as the main outline headings.

Reinforce and Extend

LEARNING STYLES

Visual/Spatial
Students may enjoy watching *Fly Away Home*, a 1996 movie in which a young girl raises geese and helps them migrate to a winter home. Ask students to write a review of the behaviors the geese exhibit in this movie.

SCIENCE JOURNAL

Have students find out more about what happened to the birds that imprinted on Konrad Lorenz. A number of humorous anecdotes describe the animals trying to feed Lorenz, for example. Then have students write about what it would be like to have a bird imprint on them.

336 Chapter 13 *The Behavior of Organisms*

Trial-and-error learning
Learning in which an animal connects a behavior with a reward or a punishment

Conditioning
Learning in which an animal connects one stimulus with another stimulus

Did You Know?
Salmon travel from the stream where they were hatched to the ocean. After many years, they may return to their stream to mate. The journey may be long. How do they know the location of the stream? Newly hatched salmon use a form of imprinting in which odors are stored in their memory. They use this information to find their way back.

Trial-and-Error Learning

B. F. Skinner, an American scientist, studied behavior in rats. He designed a box with a lever inside. A food pellet dropped into the box when the lever was pressed. At first, rats pressed the lever by accident as they moved. Soon they learned that pressing the lever resulted in food. This type of learning is called **trial-and-error learning**. Animals connect a behavior with a reward or punishment. In this case, the food pellet was the reward.

In this Skinner box, a rat presses a lever and is rewarded with food.

This type of learning is used to train pets and performing animals. In nature, trial-and-error learning helps animals survive. Animals learn which foods taste good or bad. They learn where to find food.

Conditioning

The keeper of an aquarium feeds the fish twice a day. At feeding time, the keeper taps lightly on the glass. The fish swim to the corner near the keeper. Every day, the actions are the same. The keeper taps before sprinkling the food. The fish have learned that they will get food when the keeper taps on the glass.

What if the keeper approached the aquarium but did not tap the glass? The fish would not move to the corner. They have learned to connect the tap with food. The tap is the first stimulus. The food is the second stimulus. This type of behavior is called **conditioning**. Conditioning is a type of learning in which an animal connects one stimulus with another stimulus.

336 Chapter 13 The Behavior of Organisms

Did You Know?

Read aloud Did You Know? on page 336 and discuss how many different kinds of animals migrate—or move from one point to another and back again. Point out that much information has been collected about migrating animals, but mysteries still remain, especially regarding how these creatures navigate. Encourage interested students to find out more about migrating species such as monarch butterflies, arctic terns, salmon, wildebeest, gray whales, and hummingbirds.

Insight
The ability to solve a new problem based on experience

The work of Ivan Pavlov is a famous example of conditioning. Pavlov was a Russian scientist who studied digestion and feeding behavior in dogs. Dogs naturally make saliva at the sight and smell of meat. Pavlov showed the dogs meat. He collected and measured their saliva. Then he began ringing a bell before he gave the dogs meat. After a while, he discovered that the dogs made saliva when he rang the bell. He did not have to show the meat. The dogs had learned to connect the bell with food. The dogs' response was conditioned.

This raven solved the problem of getting the food on the string. It pulled on the string with one foot. It held the pulled string with the other foot.

Insight

Imagine a chimpanzee in a room with a banana hanging from the ceiling. Boxes are scattered about the room. The chimpanzee stacks the boxes. It climbs on them to reach the banana. The chimpanzee has never done this before. It gets the banana on the first try. However, this behavior is not innate. The chimpanzee was not born knowing how to get a banana from a ceiling. It solved the problem by using previous experience. For example, the chimpanzee might have climbed rocks or seen an animal climb in the past. Based on this and other experiences, the chimpanzee solved this problem. This type of learning is called **insight**.

Insight learning is the highest type of learning. Insight is most common in apes, monkeys, and humans. However, other animals also show insight.

LEARNING STYLES

Visual/Spatial
Demonstrate how experience affects behavior. Provide each student with a sheet of paper on which letters of the alphabet are randomly scattered. Give students 15 seconds to consecutively number the letters. For example, students should write the numeral 1 beside the letter *A*, the numeral 2 beside the letter *B*, and so on. Have students record the last letter they numbered in the allotted time. Repeat the activity twice. Have students share any improvements they made over the three trials. Link the improvements to their learning the worksheet.

CROSS-CURRICULAR CONNECTION

Literature
Suggest that students interested in animal intelligence and learning read *Alex and Friends: Animal Talk, Animal Thinking* by Dorothy Hinshaw Patent. The book explores how Irene Pepperberg taught Alex, a parrot, to understand the meaning of the words he uses. Ask students to write a book review for the class.

ONLINE CONNECTION

Refer students interested in dog training to *Dr. P's Dog Training*, maintained by Dr. Mark Plonsky, professor of psychology at University of Wisconsin, Steven's Point. This Web site refers visitors to accessible information on all aspects of managing dogs from house training to catching disks. The address is www.uwsp.edu/psych/dog/dog.htm.

LEARNING STYLES

Logical/Mathematical
Point out that human beings frequently use insight to solve problems. Have students form pairs, and ask them to think of specific problems that require past experience or knowledge to solve. Invite partners to describe the situations to the class, explaining how insight was the key to solving the problem.

GLOBAL CONNECTION

Discuss with students how scientists from around the world have contributed to the understanding of animals and their behaviors. Draw students' attention to the three great scientists referred to in the text: Konrad Lorenz, B. F. Skinner, and Ivan Pavlov. Ask students to research information about the life and work of one of these or other scientists who studied animal behavior. Their reports should identify where the scientist lived and worked. When students present their reports, have them label a map or globe to show where the scientists worked.

Lesson 2 Review Answers

1. Learned behaviors may change over an organism's lifetime with experience. Innate behaviors do not change over a lifetime, but they can change over generations. **2.** because an animal bonds with the first object it sees after hatching **3.** Birds use innate and learned behaviors. A bird is born knowing how to sing. However, birds must hear adult birds sing to learn the song of their species. **4.** Conditioning is a type of learning in which an organism learns to connect one stimulus with another. **5.** Insight is a result of experience. The animal solves the problem mentally before performing it. An animal acting on instinct performs correctly the first time but does not have to think about how to perform it. The animal that acts on instinct is not solving a problem.

Portfolio Assessment

Sample items include:
- Outlines from Teaching the Lesson
- Entries from Science Journal
- Lesson 2 Review answers

Science at Work

Read Science at Work on page 338 together with students. Encourage students who have dogs as pets to share training successes and failures. Point out that trial-and-error learning is central to dog training. Invite pairs of students to write simple instructions for teaching a dog to sit, lie down, or stay.

338 Chapter 13 The Behavior of Organisms

Lesson 2 REVIEW

Write your answers to these questions on a separate sheet of paper. Write complete sentences.

1. How is learned behavior different from innate behavior?
2. Why is imprinting a learned behavior?
3. Explain how some birds learn their songs.
4. What is conditioning?
5. Animals use insight to do something correctly on the first try. Why is this not instinct?

Science at Work

Dog Trainer

Dog trainers need to understand dogs and know the best way to train them. They also should enjoy spending time with dogs. They need patience and the ability to identify and solve problems. Dog trainers learn their job by reading and attending seminars and workshops. They also can work with an experienced trainer or attend a school for dog training.

Dog trainers should be able to recognize behavior problems and correct them.

Dog trainers often teach classes, such as obedience training, for pet owners and their dogs. Dog trainers get dogs ready for dog shows, and teach tracking and retrieving. Dog trainers can be self-employed or can work for someone who has a business that trains dogs.

338 Chapter 13 The Behavior of Organisms

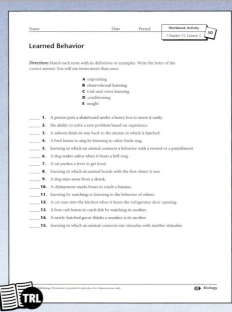

Workbook Activity 50

INVESTIGATION 13

Observing Learning Patterns

Purpose
Can you learn to do a task faster by practicing it? In this investigation, you will compare the learning rates for completing a maze.

Materials
- large sheet of unlined paper
- drinking straws
- safety glasses
- scissors
- transparent tape
- blindfold
- stopwatch or watch with a second hand

Procedure

1. Look at the maze below. You will build a maze but make it different from the one shown. As you work on your maze, do not show it to anyone else.

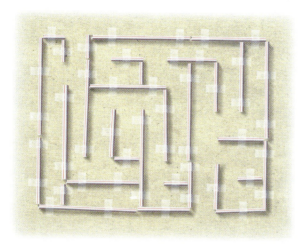

2. Draw an outline of your maze on a sheet of unlined paper. Make sure your maze has a start and a finish. Make only one path between the start and the finish.

The Behavior of Organisms Chapter 13

Investigation 13

This investigation will take 45 minutes to complete. Students will use these process and thinking skills: observing, communicating, inferring, organizing information, and making generalizations.

Preparation
- A local business may be willing to donate drinking straws for this investigation.
- Students may use Lab Manual 39 to record their data and answer the questions.

Procedure
- Students should work in pairs. Each student should design a maze.
- Tell students to allow a student who is not their partner to check their maze for accuracy before timing their partner.
- Have students practice starting and stopping the stopwatch before they begin timing their partner.
- Tell students to make the data table before they begin to time their partner.

SAFETY ALERT
- Have students wear safety glasses when working with sharp objects, such as scissors.
- Model correct cutting before having students cut their straws. Show students how to cut in a direction away from their body.

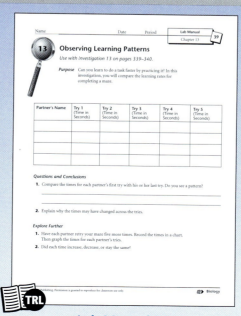

Lab Manual 39

Results

Students should find that times improve with practice. However, times may begin to decrease as the person becomes fatigued.

Questions and Conclusions Answers

1. Answers may vary and should reflect the data collected. Times should improve with practice.
2. Answers may vary and should reflect the data collected. Students may suggest that times improved with practice because learning occurred with each try.

Explore Further Answers

1. Students may choose to use different colors to graph all partners on one graph. Graphs must reflect the data collected.
2. Answers will vary. Students may find that partners are able to complete the maze in record time because they have learned the maze well. Students may find that times begin to increase because of fatigue.

Assessment

Compare students' answers with their data tables. You might include the following items from this investigation in student portfolios:

- Data table
- Answers to Questions and Conclusions and Explore Further

3. Cut and tape straws to make the walls. Place the straws lengthwise over the lines you drew. Leave openings for the start and finish. **Safety Alert: Wear safety glasses. Cut in a direction away from your body. Keep scissors and straws away from your eyes.**

4. Copy the data table below on a sheet of paper.

Partner's Name	Try 1 (Time in Seconds)	Try 2 (Time in Seconds)	Try 3 (Time in Seconds)	Try 4 (Time in Seconds)	Try 5 (Time in Seconds)

5. Blindfold your partner. Place his or her finger at the start of your maze. Have your partner try to find the finish. Your partner must not cross any straws or lift his or her finger.

6. Time how long it takes your partner to finish. Record the time in seconds on your chart.

7. Have your partner complete your maze five times. Have other classmates try your maze five times. Record all the times in your chart.

Questions and Conclusions

1. Compare the times for each partner's first try with his or her last try. Do you see a pattern?
2. Explain why the times may have changed across the tries.

Explore Further

1. Have each partner retry your maze five more times. Record the times in a chart. Then, graph the times for each partner's tries.
2. Did each time increase, decrease, or stay the same?

Lesson 3: How Animals Communicate

Objectives

After reading this lesson, you should be able to
- define *communication*.
- identify different channels of communication.
- give examples of different kinds of animal communication.
- explain how human language helps humans survive.

When a frog croaks, it is communicating. When a bee dances, it is communicating. When a mockingbird sings, it is also communicating. When you speak or nod your head, you are communicating. In this lesson, you will learn about different ways that animals communicate.

What Is Communication?

Communication is sending information. You may think only of language and speech when you think of communication. There are many other ways, or **channels**, to communicate. Animals use actions, such as showing their teeth. They also use smells, electricity, and sounds.

Communication has a purpose. Frogs croak to find a mate. Mockingbirds sing to protect their territory. Bees dance to show the location of food. You nod your head to mean "yes" or you speak to welcome a friend. Nodding also is a type of behavior.

Chemical Signals

Some animals use odors to send information. These are chemical signals. They are common among mammals and insects. Dogs and cats mark their territory with urine. Female silkworm moths give off a chemical that attracts males. The chemical can travel several kilometers.

Some ants release an alarm scent when other ants invade their territory. The scent calls other ants to help defend the area. You have probably seen a trail of ants. They are following a path that other ants made with chemicals. The path leads to food.

Communication
Sending information

Channel
A way of communicating

Lesson at a Glance

Chapter 13 Lesson 3

Overview In this lesson, students learn how animals communicate and how language helps humans survive.

Objectives
- To define *communication*
- To identify different channels of communication
- To give examples of different kinds of animal communication
- To explain how human language helps humans survive

Student Pages 341–346

Teacher's Resource Library
- Workbook Activity 51
- Alternative Activity 51
- Resource File 25, 26

Vocabulary
communication
channel

Science Background
Animal Communication

Animal communication is much broader than language. Communication is the sending and receiving of messages among organisms. Animal communication systems are an evolutionary adaptation that benefits both the sender and the receiver of the message. Animals use many different channels in their communication systems.

Human language is one form of communication. Unlike other forms of animal communication, language is rule based. Languages include thousands of different spoken languages and some sign languages. Human infants are receptive to speech discrimination at birth. They begin to speak as they mature. Humans acquire the language that is spoken in their environment. Studying a language later in life often results in an accent and other imperfections in language acquisition.

1 Warm-Up Activity

Allow students who speak or study a language other than English to share their experiences. Ask: What is the difference between picking up a language naturally and studying a second language later in life? Are there certain sounds in a second language that are harder to learn after one is older?

2 Teaching the Lesson

Have students work in pairs to prepare a graphic organizer showing the different communication channels animals use.

Show students "Language Development" part 6 in the *Discovering Psychology* program by the Annenberg/CPB Project. This 30-minute video includes information on acquiring language. Have students relate the information in this video to the process of chimpanzees trying to learn human language.

TEACHER ALERT

Make sure students understand that animal communication is not a form of speech. The bee's dance or the firefly's light pattern cannot be put into words. Point out that human speech is only one way of sending information—one aspect of the larger field of communication.

3 Reinforce and Extend

SCIENCE INTEGRATION

Earth Science

Point out that geologists receive important visual signals about life on earth from the rocks they study. Ask students what messages ancient rocks might provide. Elicit that fossils of plants and animals give important evidence about life on earth. Have students research the importance of fossils. Encourage them to share their findings with the class.

Visual Signals

Visual signals are signals that other organisms can see. The diagram below shows honeybees performing a dance that tells bees where to find food. The dance pattern shows other bees the direction and distance of flowers from their hive. If food is close to the hive, a bee might dance in circles.

Birds use many visual signals. Baby birds open their mouths to be fed. Male birds get into fighting postures with other males. They gain or keep their place as head of the flock by such behavior. Many male birds show off their beautiful feathers to attract a mate.

Round dance   **Waggle dance**

Fireflies make flashes of light to find a mate. Usually, a male firefly makes a certain pattern of flashes as he flies. A female flashes back in response. Different species of fireflies use different patterns of flashes. Trace the flight paths of the seven different firefly species shown in the diagram. Can you tell their courtship signals from one another?

LEARNING STYLES

Visual/Spatial
Point out that many animals communicate their desire to mate by means of elaborate courtship behaviors. Have students find pictures of animals attempting to attract a mate. By keying in the words *courtship display* and *pictures* on a search engine, students will find photographs of a range of animals communicating visually—often in spectacular fashion. Ask students to print out favorite images and share them with their classmates.

LEARNING STYLES

Body/Kinesthetic
Have students practice and perform the bee dance or to use flashlights to recreate the patterns of flashes by fireflies. Discuss the importance of both forms of communication to the survival of the insects involved.

CAREER CONNECTION

Invite a sign language interpreter or a speech-language pathologist to class. The interpreter can demonstrate American Sign Language (ASL), teach students a few words, and discuss his or her profession. The speech-language pathologist can discuss the types of work he or she does. Speech-language pathologists usually specialize in one or two areas such as speech impediments, language acquisition problems, accent reduction, voice disorders, and neurological problems.

The Behavior of Organisms Chapter 13 **343**

Resource File 25

Resource File 26

The Behavior of Organisms Chapter 13 **343**

Did You Know?

After they read Did You Know? on page 344, encourage students to share examples of language development that they have observed in younger siblings or other relatives. Help them compile a chronological list of words and phrases that young children typically learn. Discuss with students the types of information human beings need to communicate at an early age.

CROSS-CURRICULAR CONNECTION

Music
Point out that people's ability to communicate by sound is not accomplished entirely by means of words. Suggest to students that music can communicate love, anger, excitement, sorrow, mystery, beauty, and other powerful human feelings. Invite students to select and play for the class examples of music that they feel express universal emotions.

LEARNING STYLES

Auditory/Verbal
Many animal sounds are available on CD and a number can be downloaded online. If possible, expose students to a recorded sampling of wolves, whales, loons, and other dramatic natural performers. Discuss with the class the purpose of these calls and their effect on human beings. Encourage students interested in animal sounds to learn more about this science on one of the many Web sites devoted to bioacoustics.

Human babies automatically learn languages that are spoken around them. Children who are deaf will learn sign language that is signed around them.

While human communication is the most developed communication, it also can be the most difficult. Humans interpret information they receive in different ways.

Sound Signals

Many animals produce sounds to attract mates and keep their territory. Birds sing. Crickets rub their forewings together. Frogs croak. Female mosquitoes buzz. Female elephants send out a sound that can be heard only by male elephants.

Human Language

Human language is the most highly developed communication. Humans can speak, write, and use sign language. More than 3,000 different human languages have been studied. Humans are able to speak many different languages.

Languages allow us to talk about things that happened in the past or that will happen in the future. They also allow us to talk about ideas or about things that are not present.

Human language is a behavior that helps humans survive. Humans can put ideas into words. They can tell other humans how they changed their behavior when the environment changed. They can pass their knowledge from one generation to the next. New generations are helped by knowing what has worked. They use this knowledge to solve new problems. With new digital communication, people from anywhere on Earth can easily communicate. The Internet is changing the way people live, work, play, and learn.

Technology Note

Bioacoustics is the study of sound and living things, including how animals communicate. Special equipment for recording, analyzing, and playing back sounds is used. This technology can record and analyze infrasounds, which are sounds that are too low for humans to hear. It can record and analyze ultrasounds, which are too high for humans to hear. Giraffes, lions, rhinos, and whales make infrasounds. Dolphins and bats make ultrasounds.

Lesson 3 REVIEW

Write your answers to these questions on a separate sheet of paper. Write complete sentences.

1. Why do animals communicate?
2. List three types of channels that animals use to communicate.
3. How do bees communicate about food with other bees?
4. Give two examples of how animals use chemical signals.
5. How does language help humans to survive?

Achievements in Science

Chimpanzee Behavior Observed

Before 1960, people believed that humans were the only organisms that used tools. People also believed that animals had little in common with humans. One scientist challenged these beliefs when she observed chimpanzees making and using tools to catch termites to eat.

Imagine living among wild chimpanzees in the forest of East Africa. What if you got up early every morning to spend the day observing chimpanzees and didn't go home until after dark? Jane Goodall did just that. In 1960, she moved to the Gombe Stream National Park in Tanzania, East Africa, to begin decades of study of wild chimpanzees.

In addition to discovering that chimpanzees use tools, Goodall learned many other things about chimpanzees. She saw that chimpanzees hunt for meat and have a variety of personalities. Because of her research and discoveries, scientists now know that chimpanzees are very similar to humans.

Lesson 3 Review Answers

1. to send and receive information
2. chemical signals, visual signals, and sounds
3. Bees do a dance pattern that shows other bees the direction and the distance of food from their hive.
4. Dogs and cats mark a territory; ants mark a path.
5. Humans can pass their knowledge to others to help solve problems.

Portfolio Assessment

Sample items include:
- Graphic organizers from Teaching the Lesson
- Lesson 3 Review answers

Achievements in Science

Students can learn more about Jane Goodall, her work with chimpanzees, and her efforts to conserve the world's vanishing wild places by visiting the Jane Goodall Institute at www.janegoodall.org. Goodall also tells her own story in *My Life with the Chimpanzees*, written for readers aged 9–12.

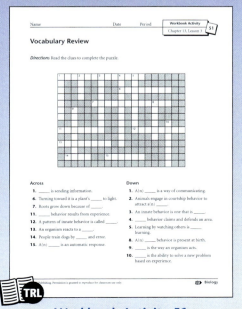

Workbook Activity 51

Science in Your Life

Read Science in Your Life on page 346 with students. If possible, invite a visually-impaired dog handler to visit your classroom. Give students time beforehand to prepare questions about communication between animal and owner. Encourage interested students to learn more about service animals and their training.

IN THE COMMUNITY

Point out that loss of vision seriously reduces a person's ability to receive information. Introduce students to the Braille alphabet—made from combinations of six raised dots—and explain that it substitutes touch for sight and permits the visually impaired to read. Ask students where they might find samples of Braille in their community (*banks, elevators, public offices*). Encourage students to copy and translate examples. Invite volunteers to share their translations with the class.

Science in Your Life

How do dog guides and their handlers communicate?

You may have seen a dog leading a person who is blind across a busy street. A special communication between the dog guide and its handler makes going places safer. Dog guides are bred, raised, and trained to do this job.

The training process begins with puppies. They are evaluated to see which dogs are most likely to be good dog guides. These puppies are given to qualified families. The families provide the puppies with good homes. They also provide obedience training and expose the puppies to new situations. The puppies stay with the families for about a year. Then they leave the families to attend a dog guide school, which may last four or five months.

Unlike regular dog training, which rewards behaviors with treats, dog guides are not given food as a reward. They have to learn not to be distracted by food when they are working. Praise is used to reinforce desired behavior.

Dog guides learn to walk in a straight line a little to the left and in front of the handler. They have to learn many skills. They learn to stop at curbs and stairs and recognize obstacles. They also have to learn many commands. Dog guides have to learn when *not* to obey a command. For example, suppose a dog guide stops at a curb and his handler gives a command to cross the street. If the dog sees any danger, the dog must not obey the command.

After dog guides are trained, they are matched with a person who needs a dog guide. The dog and its new handler spend time working as a team. The handler needs to learn all the commands and what the dog's movements mean.

The next time you see a dog guide, remember that it is working. It needs to pay attention to its surroundings and handler. Do not talk to the dog or try to pet it.

Chapter 13 SUMMARY

- Behavior is the way an organism acts. The two main types of behavior are innate behavior and learned behavior. Behaviors help animals to survive.

- Innate behavior is behavior that is present at birth. It is inherited. An instinct is a pattern of innate behavior.

- Nest building, courtship behaviors, and flying are innate behaviors in birds. Web weaving is innate in spiders. In bees, building and protecting a hive are innate behaviors.

- Plants do not behave the way animals do. Plants do respond to their environment. They respond to light, gravity, touch, and other factors.

- Behavior that can change with experience is called learned behavior.

- Some types of learning are observational learning, imprinting, trial-and-error learning, conditioning, and insight.

- Communication is sending information. Animals communicate through chemical signals, sounds, and visual signals.

- Communication has a purpose. Animals attract mates, defend territory, and give the location of food.

- Humans can communicate by using language. In this way, people can store information and pass it from one generation to the next. People use this information to help solve new problems.

Science Words

behavior, 328
channel, 341
communication, 341
conditioning, 336
courtship behavior, 329
gravitropism, 332
imprinting, 335
innate behavior, 328
insight, 337
instinct, 328
learned behavior, 335
observational learning, 335
phototropism, 331
stimulus, 328
reflex, 328
territorial behavior, 330
trial-and-error learning, 336

Chapter 13 Review

Use the Chapter Review to prepare students for tests and to reteach content from the chapter.

Chapter 13 Mastery Test

The Teacher's Resource Library includes two parallel forms of the Chapter 13 Mastery Test. The difficulty level of the two forms is equivalent. You may wish to use one form as a pretest and the other form as a posttest.

Review Answers

Vocabulary Review

1. behavior 2. innate 3. learned
4. communication 5. instinct
6. phototropism 7. gravitropism
8. stimulus 9. trial-and-error
10. insight 11. conditioning
12. channel

Concept Review

13. B 14. D 15. C 16. A 17. B 18. A

Critical Thinking

19. This is a form of courtship behavior. The communication channel is electrical.
20. Students may say yes because nonhuman primates and other animals have not been able to speak in the ways that humans have. Students may say no because other animals have been taught to vocalize. They might also say that humans do not yet fully understand the brains and communication systems of other animals enough to judge.

TEACHER ALERT

In the Chapter Review, the Vocabulary Review activity includes a sample of the chapter's vocabulary terms. The activity will help determine students' understanding of key vocabulary terms and concepts presented in the chapter. Other vocabulary terms used in the chapter are listed below:

  courtship behavior
  imprinting
  observational behavior
  reflex
  territorial behavior

Chapter 13 REVIEW

Word Bank
behavior
channel
communication
conditioning
gravitropism
innate
insight
instinct
learned
phototropism
stimulus
trial-and-error

Vocabulary Review

Choose the word or words from the Word Bank that best complete each sentence. Write your answer on a sheet of paper.

1. The way an organism acts is its _____.
2. Inherited behavior is _____ behavior.
3. _____ behavior changes over an organism's lifetime.
4. Sending information is called _____.
5. An innate behavior that follows a pattern is called a(n) _____.
6. A plant's response to light is called _____.
7. Roots growing downward is an example of _____.
8. Anything that causes an organism to react is called a(n) _____.
9. Most animals are trained using _____ learning.
10. When an animal solves a new problem based on experience, this is called _____.
11. A response to one stimulus that is connected with another stimulus is called _____.
12. A visual signal is an example of a(n) _____ that an organism may use to communicate.

Concept Review

Choose the answer that best completes each sentence. Write the letter of the answer on your paper.

13. Fighting is an example of _____ behavior.
    A courtship   C observational
    B territorial   D imprinting

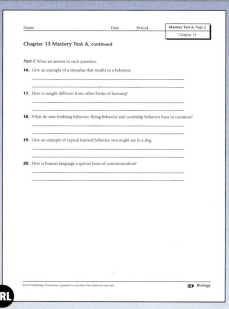

Chapter 13 Mastery Test A

14. A bird learns to sing the song of its species by _____.
    A imprinting   C trial-and-error learning
    B conditioning   D observational learning

15. Learning that keeps geese close to their mother is called _____.
    A insight   C imprinting
    B conditioning   D territorial learning

16. A dog that was stung has learned by _____ to leave bees alone.
    A trial-and-error   C insight
    B conditioning   D communication

17. Flying is _____ behavior in birds.
    A territorial   C courtship
    B innate   D learned

18. Chemical signals can be _____.
    A odors   C bee dances
    B sounds   D colors

Critical Thinking

19. Certain electric female fish give off an electric pulse during breeding season. Males of the same species respond with their own electric signals. What kind of behavior is this? Which channel is being used?

20. The famous biologist John Maynard Smith once said, "The one thing that really separates us from other animals is our ability to talk." Does the information in this chapter support this statement? Explain your answer.

Test-Taking Tip When studying for a test, work with a partner to write your own test items. Then complete each other's test. Double-check your answers.

Alternative Assessment

Alternative Assessment items correlate to the student Goals for Learning at the beginning of this chapter.

- Have groups of students create a two-column puzzle. Column 1 should consist of sentences describing animal or plant behavior (examples: The dog growled; The Venus's flytrap snapped shut). Column 2 should describe stimuli relating to the behaviors in column 1 (examples: It felt a fly. A stranger approached.) Groups exchange completed puzzles and draw lines connecting stimuli and responses.

- Remind students that learned behaviors may change over an organism's lifetime while innate behaviors do not change. Then have pairs of students write a list of behaviors in dogs that are either learned or innate. Invite pairs to write their ideas in a chart on the board with columns labeled "Innate" and "Learned."

- Secretly assign student pairs one of the types of learned behavior. Have each pair perform a charade that demonstrates the learned behavior. The rest of the class should guess the type of behavior.

- Ask students to give an example of animal communication using each of the following channels: visual cues, sounds, and smells.

- Have groups of students identify ways that behaviors help plants and animals survive. Invite groups to name a behavior and to challenge other students to explain how this behavior aids survival. For example one group might say, "Bees dance." The answer would be, "They need food to survive."

Chapter 14

Planning Guide
Evolution

	Student Text Lesson		
	Student Pages	Vocabulary	Lesson Review
Lesson 1 — Change over Time	352–357	✔	✔
Lesson 2 — What Fossils Show	358–368	✔	✔
Lesson 3 — The Theory of Evolution	369–375	✔	✔
Lesson 4 — What Humanlike Fossils Show	376–380	✔	✔

Chapter Activities

Student Text
Science Center

Teacher's Resource Library
Community Connection 14: Adaptation of an Organism in Your Community

Assessment Options

Student Text
Chapter 14 Review

Teacher's Resource Library
Chapter 14 Mastery Tests A and B

Student Text Features								Teaching Strategies						Learning Styles						Teacher's Resource Library				
Achievements in Science	Science at Work	Science in Your Life	Investigation	Science Myth	Note	Technology Note	Did You Know?	Science Integration	Science Journal	Cross-Curricular Connection	Online Connection	Teacher Alert	Applications (Home, Career, Community, Global, Environment)	Auditory/Verbal	Body/Kinesthetic	Interpersonal/Group Learning	Logical/Mathematical	Visual/Spatial	LEP/ESL	Workbook Activities	Alternative Workbook Activities	Lab Manual	Resource File	Self-Study Guide
							355	355	355	354	356	353	354	355				354, 356		52	52			✓
	365	368	366	361	✓	✓			363	361, 362	363	364	360, 364	360	361	362		368	359	53	53	40	27	✓
								372	371	371, 373		370	372		374	373		371, 374		54	54	41, 42		✓
379							377	378		379	378		379				378			55	55		28	✓

Pronunciation Key

a	hat	e	let	ī	ice	ô	order	ů	put	sh	she
ā	age	ē	equal	o	hot	oi	oil	ü	rule	th	thin
ä	far	ėr	term	ō	open	ou	out	ch	child	ŦH	then
â	care	i	it	ȯ	saw	u	cup	ng	long	zh	measure

ə { a in about, e in taken, i in pencil, o in lemon, u in circus }

Alternative Workbook Activities

The Teacher's Resource Library (TRL) contains a set of lower-level worksheets called Alternative Workbook Activities. These worksheets cover the same content as the regular Workbook Activities but are written at a second-grade reading level.

Skill Track Software

Use the Skill Track Software for Biology for additional reinforcement of this chapter. The software program allows students using AGS textbooks to be assessed for mastery of each chapter and lesson of the textbook. Students access the software on an individual basis and are assessed with multiple-choice items.

Chapter at a Glance

Chapter 14: Evolution
pages 350–383

Lessons
1. Change over Time
   pages 352–357
2. What Fossils Show
   pages 358–368

   Investigation 14 pages 366–367
3. The Theory of Evolution
   pages 369–375
4. What Humanlike Fossils Show
   pages 376–380

Chapter 14 Summary page 381

Chapter 14 Review
pages 382–383

Skill Track Software for Biology

Teacher's Resource Library
   Workbook Activities 52–55
   Alternative Workbook Activities 52–55
   Lab Manual 40–42
   Community Connection 14
   Resource File 27–28
   Chapter 14 Self-Study Guide
   Chapter 14 Mastery Tests A and B

(Answer Keys for the Teacher's Resource Library begin on page 420 of the Teacher's Edition. A list of supplies required for Lab Manual Activities in this chapter begins on page 449.)

Science Center

Ask students to design and make a display titled "They Lived Long Ago." Students can search reference books such as *The Encyclopedia of Dinosaurs*, science magazines such as *National Geographic* or *Natural History*, and the Internet for information about organisms that lived in the past. Make craft materials available. As they complete the chapter, have students add to and reorganize their display to show when organisms lived or how they are connected to organisms living today.

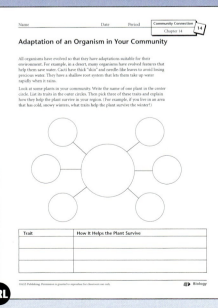

Community Connection 14

Chapter 14: Evolution

Cave drawings like the one in the photo provide clues about living things on Earth many thousands of years ago. Scientists study the traces and remains of living things to help determine what life was like long ago. They study differences in populations today to find out how living things change over time. In this chapter, you will learn how living things change. You will find out what scientists have learned from fossils. You also will learn about the theory of evolution.

Organize Your Thoughts

- Evolution
  - Natural selection
  - Changes in populations
  - Evidence
    - Fossil record
    - Embryo development
    - Vestigial structures
    - Homologous structures

Goals for Learning

- To recognize patterns of change in populations over time
- To give examples of evidence that supports evolution
- To explain the theory of evolution by natural selection
- To describe the fossils of early human ancestors

Introducing the Chapter

Have students draw a timeline of their life history. Tell them to list important events in their life, including dates, in the correct order. Have students organize the events into units of time such as baby, preschool, elementary school, and middle school. Compare students' life-history diagrams to the geologic time scale on page 360. Point out to students that in this chapter, they will learn what scientists have learned about the history of life on Earth.

After they have read the introductory paragraph on page 351, invite students to list clues to the past that they think scientists get from cave paintings and fossils.

Notes and Technology Notes

Ask volunteers to read the notes that appear in the margins throughout the chapter. Then discuss them with the class.

TEACHER'S RESOURCE

The AGS Teaching Strategies in Science Transparencies may be used with this chapter. The transparencies add an interactive dimension to expand and enhance the *Biology* program content.

CAREER INTEREST INVENTORY

The AGS Harrington-O'Shea Career Decision-Making System-Revised (CDM) may be used with this chapter. Students can use the CDM to explore their interests and identify careers. The CDM defines career areas that are indicated by students' responses on the inventory.

Lesson at a Glance

Chapter 14 Lesson 1

Overview This lesson presents an overview of the concept of evolution and explains the genetic and ecological concepts that relate to evolution.

Objectives

- To define *evolution*
- To relate genes and mutations to the process of evolution
- To describe how new species can form

Student Pages 352–357

Teacher's Resource Library

Workbook Activity 52

Alternative Workbook Activity 52

Vocabulary

evolution
lethal mutation
geographic isolation

Science Background
Evolution

The process of evolution accounts for the many different kinds of organisms alive today. It also explains the differences between ancient organisms and modern organisms and how they are related.

The theory of evolution by natural selection states that evolution occurs as the gene pool of a population changes. Natural selection is not purposeful. Organisms and populations are not able to choose which traits they will acquire or will pass on to future generations. Furthermore, only genetically inherited traits contribute to evolution. Traits that an organism acquires during its life are not passed on to progeny and do not contribute to evolution.

Lesson 1 Change over Time

Objectives

After reading this lesson, you should be able to

◆ define *evolution*.
◆ relate genes and mutations to the process of evolution.
◆ describe how new species can form.

Evolution
The changes in a population over time

You have probably noticed that there are many different kinds of organisms alive today. Many of these organisms are quite different in their structure and behavior from organisms that lived in the past. For example, modern reptiles are smaller than dinosaurs and other ancient reptiles. Modern reptiles also eat different foods. The diversity of organisms today and in the past is the result of a process called **evolution**, or change over time.

Organisms, Populations, and Change

An organism changes as it grows and develops. However, these changes are not evolution. Individual organisms do not evolve. Evolution is the changes that occur in a population of organisms over time. As you learned in Chapter 12, a population is made up of individuals of the same species that live in the same place. A species is made up of individuals of the same kind that are able to interbreed and reproduce. What species makes up the population shown in the picture?

Evolution occurs within populations.

352 Chapter 14 Evolution

Most populations of organisms exist over long periods of time. A population exists much longer than any individual in the population. For example, as individual penguins die and leave the population, other penguins are born into it. The population of penguins continues even though individuals die.

The process of evolution takes place over many generations. For example, it may take millions of years for a population of short-necked animals to change, or evolve, into a population of long-necked animals. Evolution involves changes in populations over extremely long periods of time. Evolution has occurred on Earth over many millions of years.

Changes in Genes

As you know, genes determine the traits of all living things. Thus, genes determine the characteristics of a population. Evolution involves changes in a population's gene pool. As you learned in Chapter 11, radiation and chemicals in the environment can cause genes to change. Genes also change if they are copied incorrectly during DNA replication.

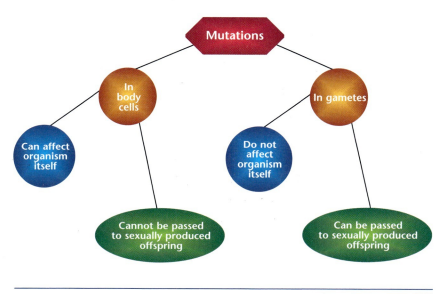

1 Warm-Up Activity

Display pictures of dinosaurs and modern-day reptiles. Ask students to point out ways the two types of organisms are alike and different. Explain that in this lesson they will learn how a population of organisms can evolve, or take on new characteristics gradually, over a long period of time.

2 Teaching the Lesson

Review the terms *population, environment, gene, DNA, mutation, gamete, sexual reproduction,* and *species.*

Have students write a question from each objective on page 352. As they read, they can answer their questions. Allow students to share their questions and answers to check their reading.

Remind students that a population includes members of one species in a given location. This is the smallest unit in which evolution occurs.

Have students consider these questions as you discuss the lesson:

- Why does a change in the genes of some members of a species living on an island not find its way into the genes of the same species living on the mainland? (*Because they are separated geographically, the two populations do not breed and so cannot pass on genetic traits to one another.*)

- How does climate change figure into the process of evolution? (*When a climate becomes hotter, colder, wetter, or drier, organisms need different traits to survive. Genetic changes that favor survival will be passed on to more offspring over time.*)

TEACHER ALERT

Students may think that all mutations affect an organism's phenotype. Point out that some mutations have no effect; they are neither helpful nor harmful. Such mutations would not improve an organism's chances for survival and thus would not lead to the evolution of a population.

3 Reinforce and Extend

LEARNING STYLES

Visual/Spatial
Have students work in pairs to make a diagram showing how genetic mutations are passed on to offspring. They might draw gametes and body cells for an organism, highlighting the mutated chromosome portion in red. Ask students to use their diagrams to explain why mutations in gametes can lead to evolution but body cell mutations cannot.

IN THE COMMUNITY

Have students observe animals in an aquarium, at a zoo or local park, or on a televised nature program and record their observations. They should try to identify ways that each organism is adapted to its natural environment. Encourage students to think of mutations that might be lethal for these organisms.

CROSS-CURRICULAR CONNECTION

Literature
For insights about how a population changes over time, have students read *SuperCroc and the Origin of Crocodiles* by Christopher Sloan. In the book, ancient crocodiles are compared to crocodiles of today.

Lethal mutation
A mutation that results in the death of an organism

Recall that a change in a gene is a mutation. Notice in the diagram on page 353 that mutations may occur in two different kinds of cells. Some mutations occur in an organism's body cells. These mutations can cause cancers and other changes in the organism. In organisms that reproduce sexually, body-cell mutations are not passed on to the offspring. For this reason, body-cell mutations are not involved in the evolution of most plants and animals.

Mutations can also occur in an organism's gametes. If the genes in an organism's gametes change, these mutations usually do not affect the organism itself. However, the mutations are passed on to the organism's offspring. The mutations can affect the traits of the offspring and of future generations.

Effects of Mutations over Time

Often, mutations cause harmful changes in traits. A **lethal mutation** is one that results in an organism's death. Organisms that inherit a lethal mutation usually do not live long enough to reproduce. As a result, the mutation is not passed on to offspring.

Sometimes a mutation results in a trait that improves an organism's chances for survival. An organism that survives is more likely to reproduce. The favorable mutation is then passed on to the offspring. As the mutation is passed on to future generations, it becomes more and more common within the population. Over time, all of the members of the population may have the mutation.

Mutations usually occur at a slow rate in populations. For example, the mutation rate in one generation of a population of fruit flies is about 0.93. This means that less than one fly per generation is likely to have a mutation. Because mutations are rare, populations usually evolve slowly.

Geographic isolation

The separation of a population into two populations that have no contact with each other, caused by a change in the environment

Did You Know?

Horses and donkeys are different species of animals, but mules are not a species. A mule is the offspring of a female horse and a male donkey. Mules are infertile and cannot have offspring.

Changes in a Population's Environment

If a population's environment changes, members of the population that have certain traits may be more likely to survive. For example, if the environment turns cold and snowy, an animal with white fur may be more likely to survive. The white fur helps to hide the animal against the white background of snow.

Organisms that survive and reproduce pass their genes on to following generations of the population. In this way, traits that help animals survive in the environment become more common within the population. Over time, the population evolves.

Environmental changes can affect populations in other ways. A change in the environment may split a population into two isolated groups. For example, a barrier, such as a large river or canyon, may form and divide a population in two. The single population becomes two populations that have no contact with each other. This is called **geographic isolation**. Over time, different mutations may occur in each isolated population. As a result, the populations have different gene pools. Eventually, the populations may evolve into different species. Look at the squirrels in the photos below. They live on opposite sides of the Grand Canyon. Scientists think the differences between the squirrels are the result of geographic isolation as the canyon formed over millions of years.

Kaibab squirrels on the north side of the Grand Canyon are isolated from the Abert's squirrels on the south side.

Evolution Chapter 14 **355**

Did You Know?

Display a picture of a horse, a donkey, and a mule. Ask students to predict how they are related. Discuss and list differences that might show that the organisms are separate species. Ask a volunteer to read aloud Did You Know? on page 355. Ask students why they think a mule is infertile. (*Species are not usually able to interbreed successfully, so it is logical that the offspring of these two species would be unable to reproduce.*)

SCIENCE INTEGRATION

 Earth Science

According to the theory of continental drift, Earth's continents were once a solid landmass called Pangaea. As a result of sea-floor spreading, the continents have moved apart gradually over time. Have students study the coastlines of Africa and South America to see how they may once have fit together. Discuss what they would expect to find in species on these two continents and explain why. (*Similar species because animals that were one species when the continents were connected became isolated as an ocean came between. Over time, each population changed, but fossils show similarities between them.*)

SCIENCE JOURNAL

 Have students look at the pictures on page 355 and then research information to write paragraphs comparing and contrasting the Kaibab and Abert's squirrels of the Grand Canyon.

LEARNING STYLES

 Body/Kinesthetic

Briefly discuss the theories of Pangaea and continental drift with students. Then give them an outline map of Earth and have them cut out the continents. Ask them to try to fit the continents together much like a jigsaw puzzle. Once they have connected the continents have them move them slowly back into their present-day positions. This activity can help students understand geographic isolation and the development of different organisms on the continents.

LEARNING STYLES

Visual/Spatial
Use Punnett squares to review with students how dominant and recessive traits exhibit themselves. Ask students to use Punnett squares to show how a mutated gene might begin to exhibit itself in a population.

ONLINE CONNECTION

 As part of the National Health Museum's Access Excellence, the activities exchange offers a number of investigations and concept maps on evolution: www.accessexcellence.org/AE/AEPC/WWC/1995/.

A population may also become divided into groups when members of the population move out of the area. For example, individuals from a population may travel from a mainland to a neighboring island. On the island, they interbreed and reproduce for generations. Over time, the population's genes change. If enough change occurs, the island population may evolve into a different species.

Large changes in the gene pool of a population can lead to the formation of new species. The chart below shows what might happen for a population of one species to evolve into two species.

There are variations in organisms within a species. Because of these variations, some organisms might survive changed environmental conditions. If these organisms survive, they reproduce and pass the genes for the variation on to following generations. A great diversity of organisms increases the chances that some living things will survive major changes in the environment.

One Species Evolves into Two Species
Groups become physically separated.
Groups move to different habitats.
Groups reproduce at different times.
Groups are not attracted to one another.
Mating results in infertile offspring.
Differences in body structures prevent mating.

Lesson 1 REVIEW

Write your answers to these questions on a separate sheet of paper. Write complete sentences.

1. What is evolution?
2. Do individual organisms evolve? Explain your answer.
3. In organisms that reproduce sexually, through what type of cells are mutations passed to offspring?
4. Relate mutations to a population's ability to evolve.
5. Describe one way in which one species might become two species.

Lesson 1 Review Answers

1. Evolution is the changes in populations of organisms over many generations.
2. Individual changes in an organism are not evolution. They may affect the organism, but they are not passed on to offspring, so they do not affect the population. 3. sex cells or gametes
4. Some mutations result in traits that improve or diminish an organism's chances for survival. An organism that does not survive to reproduce will not pass the new trait on to future generations. An organism that survives is more likely to reproduce and pass the new trait on to the next generation. If it is passed on over many generations, the population will slowly change, or evolve.
5. See the chart on page 356. Possible answer: One part of the population has a mutation the other part lacks. If an organism with the mutation tries to mate with an organism lacking the mutation, their offspring are infertile.

Portfolio Assessment

Sample items include:

- Questions and answers based on objectives from Teaching the Lesson
- Lesson 1 Review answers

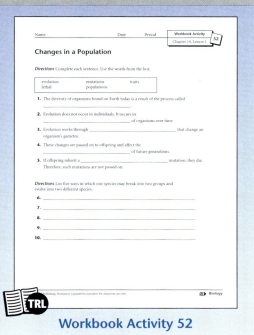

Workbook Activity 52

Lesson at a Glance

Chapter 14 Lesson 2

Overview In this lesson, students learn about fossils, how scientists date fossils, and what the fossil record shows about the history of life on Earth.

Objectives
- To explain how some fossils form
- To give examples of information learned from fossils
- To discuss how scientists determine the age of fossils

Student Pages 358–368

Teacher's Resource Library TRL

- **Workbook Activity** 53
- **Alternative Activity** 53
- **Lab Manual** 40
- **Resource File** 27

Vocabulary

fossil	geologic time
sediment	scale
mold	radioactive
cast	mineral
fossil record	half-life
paleontologist	mass extinction

Science Background
Fossils

Fossils usually occur in rock, but other materials may also preserve evidence about ancient organisms. Besides the fossilization described in the text, they may also form through mummification (when an organism dries out quickly and its soft parts don't decay), freezing (an organism freezes and is imbedded in a glacier, remaining preserved until the glacier thaws), and preservation in tar (an organism sinks in a tar pit and is preserved).

Lesson 2 What Fossils Show

Objectives
After reading this lesson, you should be able to
- explain how some fossils form.
- give examples of information learned from fossils.
- discuss how scientists determine the age of fossils.

Fossils are the remains or traces of organisms that lived in the past. Fossils are found in rock, in tree sap, and in other materials. The word *fossil* comes from the Latin word *fossilis*, which means "dug up." People have dug up different kinds of fossils all over the world.

Types of Fossils

Dinosaur skeletons are fossils. When an organism dies, its soft parts, such as skin and hair, usually decay rapidly. But its hard parts, such as bones and teeth, may remain as fossils for millions of years.

Sometimes when organisms die, they become buried in mud and bits of rock, called **sediment**, at the bottom of bodies of water. Over time, layers of sediment harden into rock. When the trapped remains of the organisms decay, they leave a type of fossil called a **mold**. The mold is an empty space inside the rock. Minerals and other particles dissolved in water may fill the mold and harden into a **cast**. The cast resembles the organism that formed the mold. What organism is shown in the cast in the photo?

Fossil
The remains or traces of an organism that lived in the past

Sediment
The bits of rock and mud that settle to the bottom of a body of water

Mold
A type of fossil that is formed when a dead organism decays and leaves an empty space in rock

Cast
A type of fossil that is formed when a mold in rock is filled with minerals that harden

Trilobites lived at the bottom of oceans about 500 million years ago.

Fossil record
The history of life on Earth, based on fossils that have been discovered

Paleontologist
A scientist who studies life in the past

Geologic time scale
A chart that divides Earth's history into time periods

Organisms may become trapped in certain materials in which their bodies do not decay. For example, insects have been found trapped in hardened tree sap. Large mammals have been found frozen in ice.

The preserved solid wastes of animals also have been found. These fossils tell the types of food eaten by animals in the past. Animal tracks and imprints of plants in rock are other types of fossils.

The many layers of rock provide evidence for the history of Earth. As scientists study the fossils found in the rock layers, they learn about the history of changing life-forms on Earth. More recent rock layers have fossils that more closely resemble existing species.

The Fossil Record

Paleontologists, or scientists who study life in the past, look for and study fossils. Their findings have formed a **fossil record** that tells about the organisms that have lived on Earth. The fossil record also shows how organisms have changed over time.

The history of life on Earth is a long one, going back about 3.5 billion years. Using fossils, paleontologists have put together the **geologic time scale** shown on page 360. The geologic time scale is a chart that divides Earth's history into different time periods. It shows the kinds of organisms that first appeared during each time period.

Technology Note

The fossils of dinosaur skeletons can deteriorate in museums. Using computer technology and scanners, digital copies of the dinosaur skeleton are stored in electronic files. Some fossil bones are used to make molds and casts. Then, the fossils are preserved and stored. The electronic files can be used to make molds of bones that are missing. A computer-guided laser can form the missing bones from a vat of liquid plastic.

Evolution   Chapter 14   **359**

1 Warm-Up Activity

Bring in fossils and encourage students to bring any fossils they may have from home to share with the class. Have students discuss which kinds of organisms are most likely to form fossils. (*those with hard parts; those that are entrapped in a substance, such as amber, that permeates and preserves body parts*)

2 Teaching the Lesson

After students read the first section of the lesson, have them make a graphic organizer that compares the different types of fossils.

When they finish reading the lesson, ask students to write a paragraph summarizing how the fossil record enabled scientists to create a geologic time scale.

Use an ice tray and a gelatin mold to point out the difference between a mold and a cast. (*The tray and mold are molds. The ice cubes and hardened gelatin are casts.*)

Ask students to infer some different ways fossils are useful to natural history museums. (*They can be used to reconstruct ancient life-forms for display; they show form, size, and habits of ancient organisms; they can be scanned, digitally copied, and used to create molds of rare or damaged bones.*)

3 Reinforce and Extend

LEARNING STYLES

LEP/ESL
Remind students that many words in English are derived from other languages, and some words are homographs, or words that have the same spelling but different meanings or derivations. Use *mold* as an example. In Chapter 5, students were introduced to the term *mold* as a fungus. The term *mold* as a fungus is derived from an Old Norse word. The term *mold* as a type of fossil comes from a Latin word. Tell students that they must use context when identifying the meaning of a homograph.

Evolution   Chapter 14   **359**

CAREER CONNECTION

Paleontologists study life of the ancient geological times mainly through fossils. Jack Horner of Montana is an example of a paleontologist with hands-on experience. He and his crew were the first to find dinosaur eggs in North America. Suggest that students find out more about Jack Horner and his work or about paleontology as a career. They might use the Internet to begin their research by typing the key words *Jack Horner, paleontologist* on a search engine.

LEARNING STYLES

Auditory/Verbal
Assign each student one time division of the geologic time scale represented in the chart on page 360. Ask students to prepare a short oral report on an epoch, a period, or an era. Students can research more facts about their time period and make pictures showing the appearance of Earth at the time. Have students present information in order, from most recent age to most distant age, and remain standing in a timeline.

PRONUNCIATION GUIDE

Use this list to help students pronounce difficult words in this lesson. Refer to the pronunciation key on the Chapter Planning Guide for the sounds of these symbols.

Quaternary	(kwä′ tər ner′ē)
Tertiary	(tėr′ shē er′ ē)
Cretaceous	(kri tā′ shəs)
Jurassic	(jủ ras′ ik)
Silurian	(sə lủr′ē ən)
Ordovician	(ôr′ də vish′ ən)
Pleistocene	(plīs′ tə sēn)
Pliocene	(plī′ ə sēn)

The Geologic Time Scale

Era	Period	Epoch	Years Before the Present (approximate) Began	Ended	Life-forms	Physical Events
Cenozoic	Quaternary	Recent	11,000		Humans dominant	West Coast uplift continues in U.S.; Great Lakes form
		Pleistocene	2,000,000	11,000	Primitive humans appear	Ice Age
	Tertiary	Pliocene	7,000,000	2,000,000	Modern horse, camel, elephant develop	North America joined to South America
		Miocene	23,000,000	7,000,000	Grasses, grazing animals thrive	North America joined to Asia; Columbia Plateau
		Oligocene	38,000,000	23,000,000	Mammals progress; elephants in Africa	Himalayas start forming; Alps continue rising
		Eocene	53,000,000	38,000,000	Ancestors of modern horse, other mammals	Coal forming in western U.S.
		Paleocene	65,000,000	53,000,000	Many new mammals appear	Uplift in western U.S. continues; Alps rising
Mesozoic	Cretaceous		145,000,000	65,000,000	Dinosaurs die out; flowering plants	Uplift of Rockies and Colorado Plateau begins
	Jurassic		208,000,000	145,000,000	First birds appear; giant dinosaurs	Rise of Sierra Nevadas and Coast Ranges
	Triassic		245,000,000	208,000,000	First dinosaurs and mammals appear	Palisades of Hudson River form
Paleozoic	Permian		280,000,000	245,000,000	Trilobites die out	Ice age in South America; deserts in western U.S.
	Pennsylvanian		310,000,000	280,000,000	First reptiles, giant insects; ferns, conifers	Coal-forming swamps in North America and Europe
	Mississippian		345,000,000	310,000,000		
	Devonian		395,000,000	345,000,000	First amphibians appear	Mountain building in New England
	Silurian		435,000,000	395,000,000	First land animals (spiders, scorpions)	Deserts in eastern U.S.
	Ordovician		500,000,000	435,000,000	First vertebrates (fish)	Half of North America submerged
	Cambrian		570,000,000	500,000,000	Trilobites, snails; seaweed	Extensive deposition of sediments in inland seas
Precambrian			4,600,000,000	570,000,000	First jellyfish, bacteria, algae	Great volcanic activity, lava flows, metamorphism of rocks. Evolution of crust, mantle, core

Radioactive mineral
A mineral that gives off energy as it changes to another substance over time

Half-life
The amount of time required for one-half of a radioactive mineral to decay

Mass extinction
The dying out of large numbers of species within a short period of time

Dating Fossils

The geologic time scale is based, in part, on the ages of fossils. Scientists can determine the relative ages of fossils by comparing their locations in rock. Relative age shows whether a fossil is older or younger than another fossil. Fossils in lower layers of rock are older than fossils in upper layers of rock if the layers are undisturbed. That is because the lower rock layers were formed first.

Scientists also can determine the actual ages of fossils. Actual age tells the number of years ago the fossil formed. Scientists use **radioactive minerals** to date fossils. Radioactive minerals give off energy as they decay into another substance over time. For example, uranium-238 changes to lead very slowly. The rate of change of uranium-238 to lead is like a clock ticking off time. The "radioactive clock" is measured in a unit called a **half-life**. A half-life is the amount of time it takes for one-half of a radioactive mineral to decay. For example, the half-life of uranium-238 is 4.5 billion years. This means that every 4.5 billion years, one-half of a given sample of uranium-238 becomes lead.

To determine the ages of fossils, scientists compare the amount of uranium-238 to the amount of lead in the rocks in which the fossils are found. Carbon-14, thorium-230, and potassium-40 are other radioactive minerals that scientists use to date fossils.

Extinctions

The fossil record shows that different species of organisms lived and then disappeared during Earth's history. When a species disappears, it is said to be extinct. As species become extinct, new species appear.

During certain periods of Earth's history, large numbers of species became extinct. These **mass extinctions** were probably caused by major changes in the environment, such as the cooling of the climate. For example, dinosaurs and some other reptiles became extinct about 65 million years ago.

Science Myth

Dinosaurs are the only extinct animals.

Fact: Many animal species have become extinct. Some scientists estimate that, at the current rate, half of all species now on Earth will be extinct within 100 years. Most extinctions are caused by human activities that result in habitat loss.

Cross-Curricular Connection

Language Arts

Tell students that *equus* is the Latin word for *horse*. Ask students to use this fact and their understanding of English suffixes to suggest definitions for the terms *equestrian* and *equine*. Have students check their definitions against dictionary definitions. Discuss how knowing the Latin term *equus* helps students understand the English terms.

Learning Styles

Interpersonal/Group Learning

Encourage students to join groups that combine talents and skills—design, computer, writing, and speaking. Ask them to develop a video that follows the evolution of the horse. For example, as some students write and record a script describing the evolution of the horse, other students might prepare computer graphics that show eohippus morphing into today's saddle horse. Students can then combine the audio and visuals for a video presentation to the class.

Fossils and Evolution

Fossils show how groups of organisms have changed over time. The diagram shows the evolution of the horse as indicated by the fossil record. You can see that about 50 million years ago, the earliest horses had four toes on each front foot. These horses were less than 48 centimeters high, about the size of a dog. As the horse evolved, it became larger. Its four toes became a single hoof.

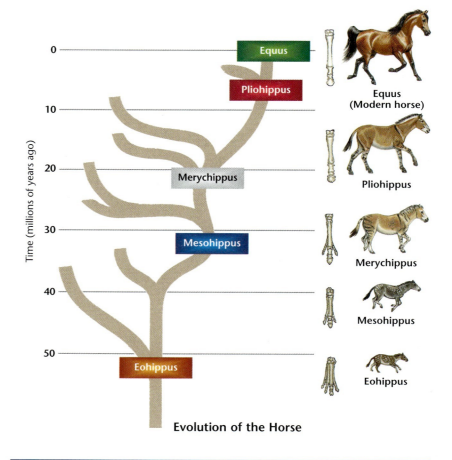

Evolution of the Horse

362   Chapter 14   Evolution

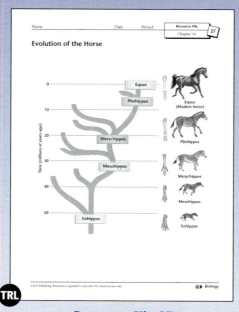

Resource File 27

Some dinosaurs may have been more like birds than like reptiles. They may have been warm-blooded, laid eggs in nests, and kept their eggs warm. They also might have fed and protected their young.

Fossils and Living Organisms

Paleontologists learn more about fossil organisms by comparing them to organisms that are alive today. They look for similarities in body structures, such as teeth and bones. Paleontologists also compare the organisms' environments.

The study of fossils has helped paleontologists discover the relationships among different groups of organisms. For example, the fossil record indicates that certain fishes evolved into amphibians. Some reptiles evolved into mammals. Other reptiles evolved into birds. The photo shows a fossil of an extinct organism called *Archaeopteryx*. The organism lived about 150 million years ago. *Archaeopteryx* had features of both reptiles and birds. Like a bird, *Archaeopteryx* had wings, feathers, and a beak. Like a reptile, it had a tail and teeth.

This fossil of Archaeopteryx *shows that the organism had characteristics of both birds and reptiles.*

SCIENCE JOURNAL

 Ask students to imagine they are the scientists who discovered *Archaeopteryx*. Have them write a letter to a colleague describing the fossil and explaining what they think it shows.

ONLINE CONNECTION

 Students can find out more about *Archaeopteryx* by visiting www.ucmp.berkeley.edu/diapsids/birds/archaeopteryx.html.

Evolution Chapter 14

Hunting for Fossils Today

Paleontologists continue to search for and study fossils. New fossils are found all the time. As fossils are discovered, scientists try to fit the information they provide into the history of life on Earth.

New techniques are helping scientists learn more information from fossils. Scientists are now able to compare the DNA of modern organisms. They can also compare the DNA of modern organisms with the DNA found in fossils. For example, in 1994, a scientist named Mary Schweitzer found DNA in a dinosaur bone. By comparing the DNA of organisms, scientists can study changes that occurred in genes over time. This can help scientists piece together the history of life on Earth.

More and more evidence is becoming available to show the relationships between organisms that are alive today and those that appear in the fossil record. These relationships are used to explain how evolution occurs. In Lesson 3, you will learn more about the process of evolution.

GLOBAL CONNECTION

The compilation of information about dinosaurs has resulted from the efforts of amateur and professional scientists from around the world. The American father and son team Luis and Walter Alvarez developed the theory that an asteroid striking Earth led to the mass extinction of dinosaurs and other animals. American Sue Hendrickson, Argentinian José F. Bonaparte, Australian Thomas Rich, Canadian Philip Currie, Chinese Dong Zhiming, German Werner Janensch, and Mongolian Rinchen Barsbold are just a few of the scientists who found or named dinosaurs. Challenge students to find out about these and other "dinosaur hunters" and to locate their country on a map. Then ask students to suggest reasons why dinosaur hunting has worldwide appeal.

TEACHER ALERT

Some students may benefit from a review of information about DNA. Refer them to Chapter 11, pages 280–282.

Lesson 2 REVIEW

Write your answers to these questions on a separate sheet of paper. Write complete sentences.

1. What are fossils?
2. Name two kinds of fossils.
3. What does the fossil record show about the evolution of the horse?
4. How are radioactive minerals used to date fossils?
5. How might studying DNA help scientists piece together the history of life on Earth?

Science at Work

Preparator

Preparators need to have a good knowledge of fossils and how to prepare them for storage. A preparator can gain experience as a volunteer for a museum or an archaeological dig. Entry-level jobs require a high school diploma and experience. Some jobs require a college degree.

A preparator cleans and prepares fossils. Most fossils are embedded in rock when they are found. A preparator uses tools, such as dental drills, brushes, and a microscope, to remove the rock. The fossil is cleaned, coated with a substance to protect it, then identified and stored. Preparators work for museums, universities, or research centers. Some preparators are self-employed and work on fossils for private collectors. Much of the work is done in laboratories. Preparators may also work in the field, where they record data and get fossils ready to transport.

Lesson 2 Review Answers

1. the remains or traces of organisms that lived in the past 2. Answers may include fossil bones and teeth, stone or mineral molds and casts of bodies or body parts, animal tracks, plant imprints, preserved solid wastes, and organisms trapped in tree sap or ice. 3. Over a period of 50 million years, the horse became larger and the structure of its front feet changed from four toes to a single hoof. 4. The rate at which radioactive elements decay into another substance is constant. By measuring and comparing the amount of a radioactive element to the amount of substance it becomes, scientists can determine how long ago the stone or fossil formed. 5. Scientists compare the DNA of modern organisms with that in fossils to observe changes that have occurred at the molecular level and similarities that suggest relationships between today's organisms and those of the ancient past.

Portfolio Assessment

Sample items include:
- Fossil graphic organizer from Teaching the Lesson
- Paragraph from Teaching the Lesson
- Lesson 2 Review answers

Science at Work

Have students read the title and look at the photograph for the Science at Work feature on page 365. Ask them to predict what a preparator does. (*Point out the root* prepar *in the term and have students explain how something is prepared.*) Have volunteers read the feature aloud. Discuss ways students might gain experience in this work before graduating from high school. Interested students might talk to a guidance counselor to find out more about the career and any local or regional opportunities.

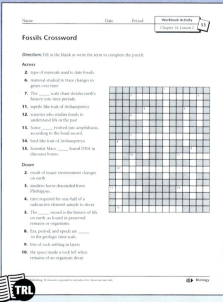

Workbook Activity 53

Investigation 14

This investigation will take about 45 minutes the first day and 15 to 20 minutes on the second day. Students will use these process and thinking skills: observing, communicating, inferring; organizing information, and making generalizations.

Preparation

- Purchase a commercially prepared plaster mixture or prepare the mixture yourself in advance of class. Do not allow students to prepare the mixture.
- Ask students to bring in disposable aluminum or plastic containers that can be used for this investigation.

Procedure

- Have students work cooperatively in pairs, but have each student make one mold.
- Set aside an area where students' molds can dry overnight. Make sure students label their molds.
- After the molds have hardened, have students exchange molds with other groups and identify the specimens that were used to make the molds.
- Students may use Lab Manual 40 to record their data and answer the questions.

SAFETY ALERT

- Have students wear disposable gloves and safety glasses when working with the wet plaster mixture.
- Instruct students to immediately wash any skin that comes in contact with the plaster mixture.
- Make sure that your room is well ventilated during the time that students use the plaster mixture.
- The sharp edges of shells can cut skin. Caution students to handle the shells carefully.
- Plant juices and parts may irritate skin. Caution students to keep their hands away from their face and to wash their hands after handling plant parts. Students should not taste any plant parts.

INVESTIGATION 14

Materials
- seashell
- leaf with thick veins and a petiole
- petroleum jelly
- disposable gloves
- safety glasses
- plaster mixture
- small containers, such as aluminum pie pans or disposable plastic bowls

Making Molds

Purpose

How are fossils formed? In this investigation, you will learn how to make molds of plant and animal remains. This will help you understand the process of fossil formation.

Procedure

1. Copy the data table below on your paper.

	Observations
Mold of shell	
Mold of leaf	

2. Choose a shell and a leaf from the collection provided by your teacher. **Safety Alert: The sharp edges of shells can cut your skin. Handle them carefully.**

3. Choose a disposable container that fits the shape and size of the shell. **Safety Alert: Wear safety glasses and disposable gloves when handling the wet plaster mixture. Keep the plaster mixture away from your face and mouth.** Fill the container with the plaster mixture.

4. Choose the side of the shell that you would like to make a mold of. Cover this side with a thin layer of petroleum jelly. Then, set the shell on top of the plaster mixture with the side you want to make a mold of facing down.

5. Press the shell into the plaster mixture. Do not push the shell below the surface of the mixture.

6. Set the container in a place where it will not be disturbed. Allow the plaster mixture to harden.

7. Repeat steps 3–6, using the leaf you selected.

8. After the plaster mixtures have hardened, carefully remove the shell and leaf. Record your observations of the two molds.

Questions and Conclusions

1. How are the shell and leaf similar to their molds?
2. How are the shell and leaf different from their molds?
3. Based on your observations in this investigation, what kinds of remains would make the best fossil molds?

Explore Further

Cover each mold with a thin layer of petroleum jelly. Then use more plaster to make casts.

Results

Each group should produce at least two molds of specimens.

Questions and Conclusions Answers

1. Students may notice that the specimens and molds have similar sizes, shapes, and textures.
2. Students may point out that the specimens' color is missing from the mold. In addition, the mold does not show all sides of the specimen.
3. Students should conclude that remains with greater depth, such as shells or bones, would produce a better mold than would flat remains, such as a leaf or feather.

Explore Further Answers

The casts that students make should resemble the specimens that were used to make the molds.

Assessment

Compare students' molds with their specimens to evaluate how carefully students worked. If students completed the Explore Further, also compare the casts with the molds and specimens. Include the following items from this investigation in student portfolios:

- Data table
- Mold
- Answers to Questions and Conclusions

Science in Your Life

As students complete their model of the geologic time scale from the Science in Your Life activity on page 368, help them realize that they cannot locate their birth date precisely on their scale. One mm represents one million years; one year or even 100 years represents too tiny a fraction of the scale to be differentiated on this scale.

LEARNING STYLES

Visual/Spatial
Extend the timeline activity by having students break out and extend the final million years. You may wish to show them how to use leaders and extensions to expand their timelines to show important events from the past 1,000 years and then to show important events from the last 100 years. Have students display their extended timelines and explain their choice of events.

Science in Your Life

How long is 4.6 billion years?

Many scientists think that Earth is about 4.6 billion years old. That is a long time compared to the time spans we deal with every day. One way to understand how long 4.6 billion years is would be to make a model of the geologic time scale.

You will need to gather colored pencils, a meterstick, tape, scissors, and sheets of unlined paper. Then, follow the steps below.

1. Using the meterstick and scissors, tape the sheets of paper together end-to-end until you reach 4.6 meters. On the paper, 1 meter will represent 1 billion years.
2. Find an area of the floor where you can spread out the sheets of paper. Then, tape the paper to the floor. This will be your model.
3. Mark one end of your model "Origin of Earth." Mark the opposite end of your model "Today."
4. Refer to the geologic time scale on page 360. Using the chart below, map each event in the geologic time scale on your model.
5. Where is the year of your birth on the time scale? Is it possible to find it?

Geological Time Scale	
Length	Number of Years
1 meter =	1 billion years
10 centimeters =	100 million years
1 centimeter =	10 million years
1 millimeter =	1 million years

368 Chapter 14 Evolution

Lesson 3: The Theory of Evolution

Objectives

After reading this lesson, you should be able to
- define the term *scientific theory*.
- state the two theories that come from Darwin's work.
- give two types of evidence that support the theory of evolution.

Scientific theory
A generally accepted and well-tested scientific explanation

Thousands of years ago, the early Greeks believed that the diversity of organisms on Earth resulted from evolution. However, it was not until the mid-1800s that people could explain how evolution occurs. These ideas came from two British scientists, Charles Darwin and Alfred Wallace. In this lesson, you will learn about the current theory of evolution.

Scientific Theory

In science, the term *theory* is used in a way that is different from its use in common language. In everyday language, you might say that you have a theory about why a friend did not go to a party. That is, you might have a hunch. But in science, the term *theory* indicates more than a hunch or a guess. A **scientific theory** is an explanation that has undergone many tests. Many different kinds of evidence support a scientific theory. For an explanation to be a scientific theory, no evidence can contradict, or disagree with, the explanation.

The theory of evolution is not the only theory in science. For example, the cell theory states that all living things are made of cells. The germ theory states that germs cause disease. Theories explain the basic ideas of science. As scientists find new evidence, they compare the evidence to the theory. If the evidence contradicts the theory, the theory is changed.

Darwin's Travels and Observations

Charles Darwin spent five years traveling around the world, from 1831 to 1836. During his travels, Darwin studied fossils from rock formations that were known to be very old. He compared those fossils with fossils from younger rock formations. Darwin found similar organisms in the different rock formations. He noted that the similar organisms had undergone change. Darwin was trying to answer the question "Do species ever change?" The fossil evidence showed that species did change over time.

Evolution Chapter 14   **369**

Lesson at a Glance

Chapter 14 Lesson 3

Overview This lesson provides an overview of Charles Darwin's work and the theories that resulted from it.

Objectives
- To define the term *scientific theory*
- To state the two theories that come from Darwin's work
- To give two types of evidence that support the theory of evolution

Student Pages 369–375

Teacher's Resource Library
  Workbook Activity 54
  Alternative Activity 54
  Lab Manual 41, 42

Vocabulary

scientific theory
hypothesis
descent with modification
natural selection
adaptive advantage
vestigial structure
homologous structure

Science Background
Science Theory

A scientific theory is an explanation of a scientific phenomenon that is supported by many lines of evidence and has no counter evidence. As new evidence is found, it is compared with the theory. If evidence contradicts the theory, the theory is modified or discarded and replaced with a new theory that can accommodate all the data.

The theory of evolution by natural selection is the primary scientific theory that explains how populations change over time. It describes the mechanism by which evolution is thought to occur. It is considered the core theme of biology. Theories are a part of all branches of science. Examples of important theories are the cell theory that living things are made up of cells and the germ theory that germs cause disease. The atomic theory states that all matter is made up of atoms.

Lab Manual 41, pages 1-2   Lab Manual 42

Evolution Chapter 14   **369**

Warm-Up Activity

Locate the Galapagos Islands on a globe or map. Show pictures or a video of the area. Tell students that many of Darwin's ideas about evolution came from the observations he made on the Galapagos Islands.

Teaching the Lesson

As they read the first two pages of the lesson, ask students to write a definition for the scientific meaning of *theory* and tell how a hypothesis is different. When students have finished reading the lesson, ask them to write a paragraph explaining the theory of descent with modification or the theory of natural selection.

Check students' understanding of natural selection with these questions: Could the environment select for a trait that is not determined by genetics? If so, would the selection of the trait contribute to the evolution of the species? Why or why not? (*Yes, such a trait could be selected for. However, since it would not be passed to future generations, it would not contribute to the evolution of the species.*)

Teacher Alert

Students may think that evolution can occur only over extremely long periods of time. Point out that the amount of time required for evolution to take place depends on how long it takes an organism to produce a new generation. Ask students what kinds of organisms are likely to evolve more quickly based on their generation time. (*Bacteria, some of which can produce a generation in an hour, evolve more quickly than organisms that require years to produce a new generation.*)

Hypothesis
A testable explanation of a question or problem (plural is hypotheses)

Darwin also collected and made sketches of plants and animals during his travels. The map below shows the route that Darwin took aboard the ship H.M.S. *Beagle*. Based on his observations and collections, Darwin came up with ideas of how evolution occurs. Darwin shared his ideas with other scientists but did not publish them right away.

Later, from 1848 to 1852, Alfred Wallace traveled in South America. Then, from 1854 to 1862, he traveled in the East Indies. As he collected information about organisms, he also formed an explanation of how evolution occurs. Wallace's explanation was similar to Darwin's. When Darwin learned of this, he hurried to write his ideas in a book. The book, *The Origin of Species*, was published in 1859. Darwin became famous for his ideas about evolution.

Darwin's Theories

When Darwin first proposed his ideas about evolution, the ideas were **hypotheses**. A hypothesis is a testable explanation of a question or problem. Darwin's hypotheses have been tested. Many kinds of evidence have been found to support them. Today, Darwin's hypotheses about evolution are stated as two theories.

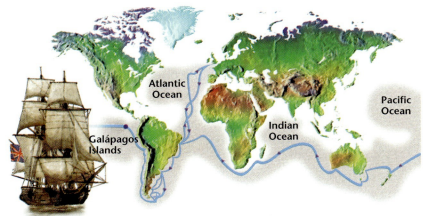

Darwin's Route

Descent with modification
The theory that more recent species of organisms are changed descendants of earlier species

Natural selection
The process by which organisms best suited to the environment survive, reproduce, and pass their genes to the next generation

The first theory is called **descent with modification**. It states that more recent species found in the fossil record are changed descendants of earlier species. In other words, present organisms are related to past organisms. In fact, all organisms have descended from one or a few original life-forms. Descent with modification basically says that evolution occurs in nature.

The second theory is called **natural selection**. This theory explains how evolution occurs. Recall from Lesson 1 that environmental conditions influence which organisms in a population survive. According to the theory of natural selection, organisms that are best suited to the environment are more likely to survive and reproduce. Those organisms will pass on their genes to their offspring. In this way, the organisms in a population with traits that are best suited to the environment increase in number. The population adapts to its environment as it gains genes that are suited to the environment. This process results in the evolution of a new species over a long period of time.

The following four points summarize Darwin's theory of natural selection to explain how evolution occurs.

1. Organisms tend to produce more offspring than can survive. For example, fish lay thousands of eggs, but only a few live to be adult fish.

2. Individuals in a population have slight variations. For example, fish in a population may differ slightly in color, length, fin size, or speed.

3. Individuals struggle to survive. Individuals that have variations best suited to the environment are more likely to survive.

4. Survivors pass on their genes to their offspring. Gradually, the population changes.

Evolution Chapter 14 371

3 Reinforce and Extend

LEARNING STYLES

Visual/Spatial
Use a globe to trace the route of the H.M.S. *Beagle*. Point out that the ship set sail at the end of 1831 and returned to its home in October 1836. Darwin served as a naturalist with the expedition.

CROSS-CURRICULAR CONNECTION

Language Arts
Refer students to the explanation for the term *fossil* on SE page 358. (It comes from the Latin word *fossilis*, meaning "dug up.") Explain that many terms used in science are built from Latin roots. Have students use a dictionary to explore the origins of *hypothesis* (from *hypo*, meaning "under" and *tithenai*, meaning "to put"—"to put under, suppose"). Invite students to explain how the modern meaning of *hypothesis* fits with the original meanings of the word parts.

SCIENCE JOURNAL

Have students write a journal entry that Darwin might have written as he traveled aboard H.M.S. *Beagle*. The entry should include thoughts that Darwin may have had about evolution.

Evolution Chapter 14 371

SCIENCE INTEGRATION

Physical Science

Refer students to the two photos on page 372. Have them describe the difference in the beaks of the warbler finch and the medium ground finch and make sketches of how the birds would use their beaks. Discuss the advantage of each beak as a tool for its owner. For example, the medium ground finch needs a powerful vise to hold and crack the covers off seeds; the warbler finch needs a pair of tweezers to extract insects from hiding places. You may wish to display a vise and a tweezers as you compare the beaks to these tools.

AT HOME

Have students search at home for antibacterial soaps and cleaners. Explain that the widespread use of antibacterial products like these is increasing the number of resistant strains of bacteria. This means that the bacteria are not killed by the antibacterial products. Ask students to explain how this phenomenon is related to the theory of natural selection.

Descent with Modification

Darwin observed descent with modification when he visited the Galápagos Islands off the coast of South America. He was amazed by the variety of living things he saw. Darwin found a variety of birds called finches. He observed 14 different species of island finches. The finches differed in the shapes and sizes of their beaks.

Notice the differences in the beaks in the photos below. The beaks are related to differences in the birds' diets. For example, large, thick beaks are used to crack seeds. Long, pointed beaks are used to eat insects.

Darwin noted that the island finches resembled the finch species on the nearby coast of South America. He reasoned that the island species were probably descended from the mainland species. Isolated by the water barrier for a long time, the finch populations on the different islands had undergone different genetic changes. These genetic changes resulted in the evolution of the 14 finch species.

The beaks of different finch species on the Galápagos Islands are adapted for eating different kinds of foods. The warbler finch, on the left, eats insects. The medium ground finch, on the right, eats seeds.

Adaptive advantage
The greater likelihood that an organism will survive, due to characteristics that allow it to be more successful than other organisms

Vestigial structure
A body part that appears to be useless to an organism but was probably useful to the organism's ancestors

Natural Selection

Natural selection affects all populations of organisms. An example is the natural selection of snakes that have a specialized upper tooth. Young snakes use the tooth to cut their way out of their shell. Snakes that are able to leave their shell with ease are more likely to survive and reproduce. The presence and use of the specialized tooth give the young snakes an **adaptive advantage**. They have a greater likelihood of surviving because of their characteristics. The environment selects individuals in a population that have an adaptive advantage. Young snakes that do not have the specialized tooth do not live to reproduce. Their genes are lost from the population.

Evidence of Evolution

If populations really do change slowly over time, then there should be evidence of the change. As you know, the fossil record shows recent species that are slightly different from earlier species. This is one type of evidence that supports the theory of evolution.

Other evidence is that the embryos of some kinds of organisms go through similar stages of development. For example, an early human embryo has a tail and gill pouches, just as an early fish embryo has. Similarities in the development of vertebrate embryos are an indication that all vertebrates descended from a common ancestor.

Another indication that certain organisms are related to one another is the presence of **vestigial structures**. A vestigial structure is a body part that appears to be useless to an organism. For example, snakes and whales have the remnants of leg bones and pelvic bones. These vestigial structures are not used to help snakes and whales move. They are probably "leftover" structures from ancestors of snakes and whales that had legs for walking on land.

LEARNING STYLES

Interpersonal/Group Learning
Ask small groups of students to research information about the peppered moth of England. Students should explore the environmental changes in industrialized England and the hypothesis that the peppered moth evolved to blend in with the changes in the environment.

CROSS-CURRICULAR CONNECTION

Health
Explain that many scientists consider the appendix to be vestigial in humans. Invite students to tell about their own or family members' experiences with the appendix. Have a volunteer research its purpose in other mammals. (*The appendix appears to be a vestigial remnant of the cecum, a blind sac located where the large intestine begins. In mice, for example, the cecum is large and is used for storage of bulk cellulose.*) Discuss medical problems caused by the appendix in some humans. (*Appendicitis is an inflammation of the appendix requiring its removal. If the appendix bursts, material in the digestive tract is introduced into the body cavity lining. Lethal infection can result.*)

LEARNING STYLES

Body/Kinesthetic

Have students analyze the limbs of the four animals in the illustration on page 374. They can feel the bones of their own arms to locate the pictured parts and demonstrate their types and range of movements. Discuss the differences in limb bones of the other animals. For what different jobs are these limbs used by the animals? Ask students to try to model these uses using their arms and tell why they cannot do so successfully.

LEARNING STYLES

Visual/Spatial

Tell students that some animals have structures that have the same function but very different constructions. Examples include the butterfly and the bird. Both have wings for flying, but the construction of the wings is very different. Encourage students to show the similarities and differences on a comparison-and-contrast chart. Ask why the wings of butterflies and birds are not homologous. (*The body parts are not similar in structure; the animals are not related.*)

Homologous structures
Body parts that are similar in related organisms

Look at the front limbs of the vertebrates in the diagram below. Notice how similar they are. The limbs are **homologous structures**. Homologous structures are body parts that are similar in related organisms. Homologous structures are thought to have first appeared in an ancestor that is common to all the organisms that have the homologous structures. Thus, vertebrates probably share a common ancestor that had front limbs like those you see in the diagram.

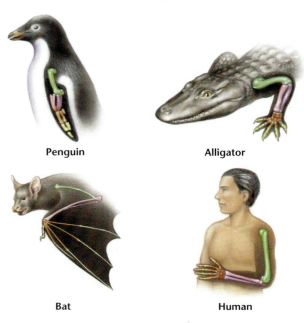

Homologous Structures in Vertebrates

374 Chapter 14 Evolution

374 Chapter 14 Evolution

Lesson 3 REVIEW

Write your answers to these questions on a separate sheet of paper. Write complete sentences.

1. Distinguish between the meaning of the term *scientific theory* and the everyday meaning of the term *theory*.
2. What is meant by descent with modification?
3. Explain the process of natural selection.
4. Explain the process that led to the evolution of the finch species that Darwin observed on the Galápagos Islands.
5. Give two types of evidence other than fossils that support the theory of evolution.

Lesson 3 Review Answers

1. A scientific theory is an explanation that has undergone many tests and is supported by many different kinds of evidence. In everyday language, a theory is usually a hunch or a guess. 2. Descent with modification means that present organisms are related to past organisms; they are changed descendants of past organisms. 3. Organisms produce many offspring. They have slight variations. Those with variations best suited to their environment are most likely to survive. Survivors pass on their traits to offspring. 4. The finch populations on the various islands were isolated from one another by water. Therefore, different populations bred only with each other and underwent different genetic changes. 5. parallel embryo development, presence of vestigial structures and homologous structures in vertebrates

Portfolio Assessment

Sample items include:

- Theory definition, hypothesis comparison, and paragraph on a theory from Teaching the Lesson
- Lesson 3 Review answers

Workbook Activity 54

Lesson at a Glance

Chapter 14 Lesson 4

Overview This lesson provides an overview of hominid fossil finds. Students learn about the characteristics of different hominids.

Objectives

- To list characteristics that all primates share
- To trace the evolution of hominids, based on fossil evidence
- To describe how hominids changed over time

Student Pages 376–380

Teacher's Resource Library

- Workbook Activity 55
- Alternative Activity 55
- Resource File 28

Vocabulary

primate	Neanderthals
hominid	Cro-Magnons
Homo sapiens	

Science Background
Hominids

Scientists now think that the ancestor of humans beings probably evolved from an apelike common ancestor of apes and human beings between 5 and 10 million years ago. That split marked the beginning of development of hominids (early humanlike ancestors and human beings). Ability to walk on two legs is a unique feature of hominids. Scientists look for characteristic changes in bones and muscles of the hips and legs in deciding whether to classify a fossil as a hominid.

Hominids are believed to have originated in Africa. That is where the oldest hominid fossils have been found, and where chimpanzees and gorillas (the closest living relatives to human beings) originated.

376 Chapter 14 Evolution

Lesson 4 — What Humanlike Fossils Show

Objectives

After reading this lesson, you should be able to

- list characteristics that all primates share.
- trace the evolution of hominids, based on fossil evidence.
- describe how hominids changed over time.

Primate
The group of mammals that includes humans, apes, monkeys, and similar animals

Hominid
The group that includes humans and humanlike primates

Homo sapiens
The species to which humans belong

Humans belong to a group of mammals called **primates**. Besides humans, primates include monkeys, apes, and other similar animals. Primates share several characteristics. They can see color. They have fingers that can move and grasp. They have nails instead of claws.

Hominids

Humans and their humanlike relatives make up a group of primates called **hominids**. The hominid species to which humans belong is **Homo sapiens**, which means "wise man." *Homo sapiens* is the only living hominid species. All other hominid species are extinct.

Most of what is known about the history of humans is based on fossil evidence. As you can see below, fossils of a number of different hominid species have been found. Hominid species that belong to the genus *Homo* are more similar to humans than are those that belong to the genus *Australopithecus*.

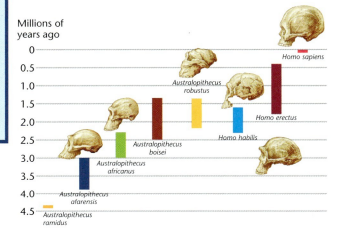

Timeline of the Hominid Species

376 Chapter 14 Evolution

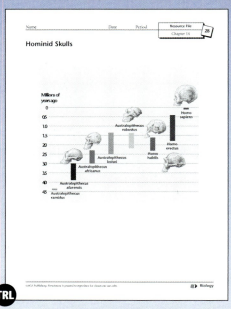

Resource File 28

Did You Know?

Lucy was only 1 meter tall, and her head was the size of a softball.

Homo sapiens is the only primate living today that walks upright on two legs. All other hominid species walked on two legs as well. Hominids are thought to have started walking on two legs early in their history. One of the earliest fossils of a primate that walked upright is 3.2 million years old. This fossil, which scientists named Lucy, was found in 1974. Lucy's scientific name is *Australopithecus afarensis*. The photo shows parts of Lucy's skeleton.

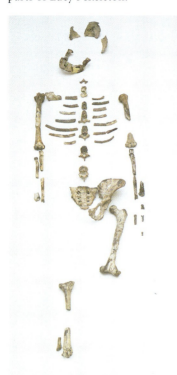

The bottom of Lucy's skull has an opening for the spinal cord. This suggests that Lucy walked upright. Lucy's hip and leg bones show that they could support her body while walking upright on two legs. Scientists estimate that Lucy's brain was only about one-third the size of the brain of a modern human.

Since the 1970s, many hominid fossils have been found. In addition to fossilized bones, fossilized footprints have shown that early hominids walked upright.

In 1974, Donald Johanson found this hominid skeleton, now called Lucy, in eastern Africa. The fossil is about 3.2 million years old.

 Warm-Up Activity

Show students photographs of chimpanzees and humans doing various activities, such as walking and eating. Tell students that both chimpanzees and humans, as well as monkeys, gorillas, and some other animals, are primates. Ask students to describe similarities and differences between chimpanzees and humans.

2 Teaching the Lesson

Review the method of deriving a scientific name (genus + species) and the meanings of *genus* and *extinct* before students read the lesson.

Help students scan the lesson heads and graphics and write questions they want to answer as they read.

After they have finished reading, ask students to define the vocabulary and make a graphic organizer showing the relationships among these terms. For example, you might ask them to rank the terms to show which includes the greatest number or type of organisms. (*Hominids are a subgroup of primates; the genus* Homo *is a subgroup of hominids; and the species* Homo sapiens *is a subgroup of the genus* Homo.)

PRONUNCIATION GUIDE

Use this list to help students pronounce difficult words in this lesson. Refer to the pronunciation key on the Chapter Planning Guide for the sounds of these symbols.

Australopithecus
(ȯ strā′ lō pi′ thə kəs)

Refer students to the diagram on page 376. Point out the species names and ask students to describe how they are all similar. (*All are two words, written in italics, and begin with one of two genus names*—Australopithecus *or* Homo.) Remind students that species belonging to the same genus are more closely related to one another than area species belonging to different genera.

Did You Know?

Have a volunteer read Did You Know? on page 377. Display a meterstick and a softball to help students visualize Lucy's size. Have students find *Australopithecus afarensis* on the chart on page 376 to help place Lucy in context of the hominid timeline.

3 Reinforce and Extend

LEARNING STYLES

Logical/Mathematical
Have students interpret the diagram on page 376 to figure out how long each hominid species lived. Tell them that the top of the bar shows the end of the species' existence, and the bottom shows the beginning of its existence. Challenge them to estimate their answers to the nearest 100,000 years.

ONLINE CONNECTION

To learn more about hominids, students can visit www.bbc.co.uk/science/cavemen/index.shtml. This site describes hominids and primates. Students can find out about the age of Nanjing man by visiting exn.ca/hominids/home.cfm.

SCIENCE INTEGRATION

Technology
With the development of tool-making ability, humans made many advances and improved their standard of living. Have several students research the differences in tools of Neanderthals and Cro-Magnons, both of which lived during the Stone Age. They can then compare the differences in the way these two types of humans are believed to have lived. Also suggest that they contrast these lives with those of people of the Bronze Age (beginning about 5,000 years ago).

The Genus *Homo*

In the 1950s, the fossil hunters Mary and Louis Leakey discovered fossils of a hominid species in Africa. They found stone tools near the fossils, suggesting that the hominid used tools. They named the hominid species *Homo habilis*, which means "handy man." You can see a skull of this species on page 376. The *Homo habilis* brain was larger than the brains of species belonging to the genus *Australopithecus*. Fossils of *Homo habilis* are between 1.6 and 2.4 million years old.

Homo erectus is another hominid species belonging to the genus *Homo*. This species lived between about 400,000 and 1.9 million years ago. *Homo erectus* means "upright man." Compare the skull of *Homo erectus* with the skull of *Homo habilis* on page 376. Notice that the skull of *Homo erectus* could hold a larger brain. *Homo erectus* is thought to be the first group of hominids to leave Africa. Burned bones found near the fossils of *Homo erectus* suggest that this hominid used fire for cooking and staying warm.

Mary and Louis Leakey discovered fossils of the hominid species *Homo habilis*. Here, Mary Leaky uncovers another fossil find.

Neanderthals
Homo sapiens who lived between about 35,000 and 150,000 years ago but are not thought to be direct ancestors of humans living today

Cro-Magnons
Homo sapiens who lived about 35,000 years ago and are direct ancestors of humans living today

Homo sapiens

Most scientists think *Homo erectus* evolved into *Homo sapiens*. **Neanderthals** are *Homo sapiens* who lived between about 35,000 and 150,000 years ago. Neanderthals are thought to have lived in caves and used tools. Some of the tools suggest that Neanderthals wore clothes made from animal hides. The Neanderthal brain was larger than that of modern humans. Although Neanderthals were *Homo sapiens*, some evidence suggests that the Neanderthals are not direct ancestors of humans living today.

Cro-Magnons are *Homo sapiens* who lived about 35,000 years ago in Europe. Scientists think Cro-Magnons were an early form of modern humans. Cro-Magnons made their own shelters. They made beautiful cave drawings and carvings of animals. Their brain was about the same size as that of humans living today. Cro-Magnons are our direct ancestors.

Achievements in Science

First Ichthyosaur and Plesiosaur Discovered

Even today, fossils that are found are new discoveries. These fossils provide clues about life on Earth in the past. It was only 200 years ago that people didn't know that dinosaurs used to roam Earth.

One of the first people to collect fossils was Mary Anning. She was born in Great Britain about 200 years ago. She is credited with discovering the first complete *Ichthyosaurus* skeleton when she was only 12 years old. Most sources say that her brother found the first fossil. Anning was the one who found most of the ichthyosaur's remaining fossils. She also found the skeletons of other ichthyosaurs. One of her most important finds was the discovery of a plesiosaur.

Anning sold the fossils she found to scientists, museums, and wealthy people with private collections. In addition to collecting fossils, she had a good understanding of fossils and how they helped determine the history of life on Earth. She made great contributions to the science of paleontology when it was just beginning.

Evolution Chapter 14 **379**

CROSS-CURRICULAR CONNECTION

Art
Display photographs of the art of hominids, such as that found in the Lascaux caves in France. Ask students to describe these paintings. Discuss what the appearance of art (as recently as 30,000 years ago) suggests about the evolution of hominids. (*For example, making ornaments, carving sculptures, and making engravings and paintings suggests improved tools, more sophisticated methods of food getting, and a more specialized and community-oriented society.*) Ask students to list things scientists could learn about early people from their artwork. (*kinds of tools they used, kinds of animals they hunted, plants they used for dyes, and how they dressed, for example*) Encourage students to learn more about one type of prehistoric art.

IN THE ENVIRONMENT

Explain that the interaction of a population with the environment is key to its evolution and survival. During the ice ages of the Pleistocene Epoch (from about 2 million to 11,000 years ago) humans could not have survived in Europe during times when the region was covered by glaciers. Fossils have been found, however, for times in this period when glaciers were known to have retreated. Remind students of current developments in Earth's environment, such as global warming and pollution. Discuss what students think might happen to human evolution if such trends continue.

Achievements in Science

Have volunteers read aloud the Achievements in Science feature on page 379. Display illustrations of Icthyosaurs and plesiosaurs. Discuss how scientists might have concluded what these organisms look like by studying fossils. Interested students might read further about the life of Mary Anning and create a poster or give a report on her accomplishments.

Lesson 4 Review Answers

1. Primates can see colors, have grasping fingers, and have nails instead of claws.
2. Hominids are those primates which have humanlike characteristics, including humans. Students may mention any of the species of genus *Australopithecus* or *Homo*, listed on page 376. **3.** The bottom of Lucy's skull has an opening for the spinal cord. Her hip and leg bones show they would support her while walking upright on two legs. **4.** The hominid brain increased in size over time although the Neanderthal brain was larger than that of modern humans. **5.** Neanderthals, Cro-Magnons, and humans living today are all members of the species *Homo sapiens*, which probably descended directly from *Homo erectus*. However, Cro-Magnons are believed to be our direct ancestors while Neanderthals are not thought to be.

Portfolio Assessment

Sample items include:
- Questions and answers from Teaching the Lesson
- Definitions and graphic organizers from Teaching the Lesson
- Lesson 4 Review answers

Lesson 4 REVIEW

Write your answers to these questions on a separate sheet of paper. Write complete sentences.

1. List three primate characteristics.
2. What are hominids? Give the species name of one hominid, other than *Homo sapiens*.
3. What evidence suggests that Lucy walked upright?
4. How did the hominid brain change over time?
5. How are *Homo erectus*, Neanderthals, Cro-Magnons, and humans living today related to one another?

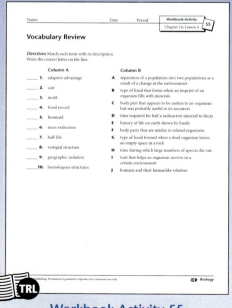

Workbook Activity 55

Chapter 14 SUMMARY

- Evolution is the result of changes in the gene pools of populations over long periods of time. Populations are the smallest units in which evolution occurs. Mutations and changes in a population's environment may cause the population's gene pool to change.

- Fossils are the remains or traces of organisms that lived in the past. Fossils are preserved in Earth's rock, in tree sap, and in other materials.

- The history of life on Earth has been pieced together using the fossil record. The geologic time scale divides Earth's history into time periods. It also shows when certain kinds of organisms first appeared on Earth.

- The geologic time scale is based, in part, on the ages of fossils. Scientists determine the relative ages of fossils by comparing the fossils' locations in rock. Scientists use radioactive minerals to determine the actual ages of fossils.

- A scientific theory is an explanation that has undergone many tests and is supported by different kinds of evidence. Today, Darwin's ideas about evolution are stated as two theories. The first theory is called descent with modification. Darwin's second theory is called natural selection.

- Humans and their humanlike ancestors make up a group of primates called hominids. *Homo sapiens* is the only hominid species living today.

Science Words

adaptive advantage, 373
cast, 358
Cro-Magnons, 379
descent with modification, 371
evolution, 352
fossil, 358
fossil record, 359
geographic isolation, 355
geologic time scale, 359
half-life, 361
hominid, 376
homologous structures, 374
Homo sapiens, 376
hypothesis, 370
lethal mutation, 354
mass extinction, 361
mold, 358
natural selection, 371
Neanderthals, 379
paleontologist, 359
primate, 376
radioactive mineral, 361
scientific theory, 369
sediment, 358
vestigial structure, 373

Chapter 14 Review

Use the Chapter Review to prepare students for tests and to reteach content from the chapter.

Chapter 14 Mastery Test

The Teacher's Resource Library includes two parallel forms of the Chapter 14 Mastery Test. The difficulty level of the two forms is equivalent. You may wish to use one form as a pretest and the other form as a posttest.

Review Answers

Vocabulary Review

1. evolution 2. Lethal mutations
3. fossils 4. geologic time scale
5. scientific theories 6. fossil record
7. extinct 8. natural selection
9. hominids 10. *Homo sapiens*
11. paleontologists 12. radioactive minerals 13. hypothesis 14. mold

Concept Review

15. C 16. B 17. C 18. D

TEACHER ALERT

In the Chapter Review, the Vocabulary Review activity includes a sample of the chapter's vocabulary terms. The activity will help determine students' understanding of key vocabulary terms and concepts presented in the chapter. Other vocabulary terms used in the chapter are listed below:

adaptive advantage
cast
Cro-Magnons
descent with modification
geographic isolation
half-life
homologous structure
mass extinction
Neanderthals
primates
sediment
vestigial structure

Chapter 14 REVIEW

Vocabulary Review

Choose the word or words from the Word Bank that best complete each sentence. Write the answer on your paper.

Word Bank
evolution
extinct
fossil record
fossils
geologic time scale
hominids
Homo sapiens
hypothesis
lethal mutations
mold
natural selection
paleontologists
radioactive minerals
scientific theories

1. The process called _____ is the changes in a population over time.
2. _____ result in the death of organisms.
3. The remains or other traces of organisms that lived in the past are _____.
4. The _____ divides Earth's history into time periods.
5. Explanations that are supported by many different kinds of evidence are _____.
6. The _____ is made up of fossils that show the history of life on Earth.
7. A species that is _____ no longer exists.
8. The theory of _____ states that organisms that are best suited to live in a certain environment are more likely to reproduce.
9. Humans and their humanlike ancestors make up a group of primates called _____.
10. Humans belong to the species _____.
11. Scientists who study life in the past are called _____.
12. Scientists use _____ to determine the actual ages of fossils.
13. An explanation of a question or a problem that can be tested is a(n) _____.
14. When an organism that has been trapped in sediment decays, it may leave a fossil called a(n) _____.

382 Chapter 14 Evolution

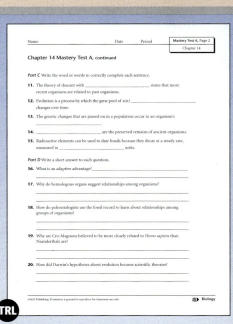

Chapter 14 Mastery Test A

Concept Review

Choose the answer that best completes each sentence. Write the letter of the answer on your paper.

15. Being able to see color, having fingers that can grasp, and having nails instead of claws are characteristics shared by _____.
  A only *Homo sapiens*
  B only hominids
  C all primates
  D all mammals

16. Evolution involves changes in the gene pool of a(n) _____.
  A individual
  B population
  C body cell
  D gamete

17. Suppose a scientist found equal amounts of uranium-238 and lead in a rock. The rock is about _____ years old.
  A 2.25 billion
  B 238 million
  C 4.5 billion
  D 9 billion

18. *Homo erectus* mostly likely evolved before _____.
  A *Australopithecus robustus*
  B *Homo habilis*
  C Lucy
  D Neanderthals

Critical Thinking

Write the answers to each of the following questions.

19. Using radioactive minerals to date fossils was not known during Darwin's times. How was Darwin able to estimate the ages of his fossil finds?

20. Two species of foxes live in different geographic areas. Scientists believe they once belonged to the same species. Explain how they could have become two species.

Test-Taking Tip If a question asks you to describe or explain, try to answer the question as completely as possible. Write your answer in complete sentences.

Evolution Chapter 14 383

Appendix A: Animal Kingdom

Invertebrates (No backbone)

Group	Description	Example
Porifera *Phylum*	Body wall of two cell layers; pores and canals; no tissues or organs; no symmetry; live in water; strain food from water	sponge
Cnidarian *Phylum*	Baglike body of two cell layers; one opening leading into a hollow body; tissues; usually mouth surrounded by tentacles; stinging cells; radial symmetry; live in water	jellyfish, coral, hydra
Flatworm *Phylum*	Ribbonlike body; three cell layers; organs; flat, unsegmented body; digestive system with one opening; nervous system; bilateral symmetry; most are parasitic	tapeworm, planarian, fluke
Nematode *Phylum*	Round, slender body; unsegmented body; digestive system with two openings; nervous system; bilateral symmetry; some are parasitic	hookworm, pinworm, vinegar eel
Mollusk *Phylum*	Soft body covered by a fleshy mantle; move with muscular foot; some have shells; all organ systems; bilateral symmetry	
Gastropod *Class*	One shell (slugs have no shell); head, eyes, and tentacles	snail, slug
Bivalve *Class*	Two shells; no head, eyes, or tentacles	clam, oyster, scallop
Cephalopod *Class*	No shell; head; eyes; foot divided into tentacles	squid, octopus, nautilus
Annelid *Phylum*	Round, segmented body; digestive system, nervous system, circulatory system; bilateral symmetry; most are not parasitic	earthworm, leech
Arthropod *Phylum*	Segmented body; jointed legs; most have antennae; all organ systems; external skeleton	
Arachnid *Class*	Two body segments; four pairs of legs; no antennae	spider, scorpion, tick, mite
Crustacean *Class*	Two body segments; usually five pairs of legs; two pairs of antennae; breathe with gills	crayfish, lobster, crab, shrimp, sowbug, barnacle
Chilopod *Class*	usually 15–170 body segments; one pair of legs on each segment; poison claws; flattened body	centipede
Diplopod *Class*	usually 25–100 body segments; two pairs of legs on each segment; body often rounded	millipede
Insect *Class*	Three body segments; three pairs of legs; one pair of antennae; most have two pairs of wings	fly, beetle, grasshopper, earwig, silverfish, water strider, butterfly, bee, ant
Echinoderm *Phylum*	Covered with spines; body has five parts; radial symmetry; live in ocean	sea star, sea urchin, sand dollar

Vertebrates (Backbone)

Group	Description	Examples
Chordate *Phylum*	Internal skeleton of bone or cartilage; skull; sexual reproduction; bilateral symmetry	
Jawless Fish *Class*	Skeleton of cartilage; no scales or jaw; unpaired fins; breathe with gills; live in water; cold-blooded	lamprey, hagfish
Cartilage Fish *Class*	Skeleton of cartilage; toothlike scales; jaw; paired fins; breathe with gills; live in water; cold-blooded	shark, ray
Bony Fish *Class*	Skeleton of bone; bony scales; jaw; paired fins; breathe with gills; live in water; swim bladder in most; cold-blooded	trout, salmon, swordfish, goldfish
Amphibian *Class*	Skeleton of bone; moist, smooth skin; no claws; breathe with lungs or through skin as adults; young live in water, adults live on land; four legs; eggs lack shells; cold-blooded	newt, frog, toad
Reptile *Class*	Skeleton of bone; dry, scaly skin; claws; breathe with lungs all stages; four legs except snakes; eggs have shell; cold-blooded	turtle, snake, alligator, lizard
Bird *Class*	Skeleton of bone; feathers; wings; beaks; claws; breathe with lungs all stages; eggs have shell; warm-blooded	hawk, goose, quail, robin, penguin
Mammal *Class*	Skeleton of bone; hair; mammary glands; breathe with lungs all stages; young develop within mother; warm-blooded	bat, kangaroo, mouse, dog, whale, seal, human

Appendix B: Plant Kingdom

Spore Plants

Group	Description	Examples
Bryophyte Division*	Nonvascular (no tubes for carrying materials in plant); live in moist places	liverwort, hornwort, moss
Club Moss Division	Spores in cones at end of stems; simple leaves	club moss
Horsetail Division	Spores in cones at end of stems; hollow, jointed stem	horsetail
Fern Division	Spores in sori; fronds	fern

Seed Plants

Group	Description	Examples
Palmlike Division	Naked seeds in cones; gymnosperm (nonflowering seed plant); male and female cones on different trees; palm-shaped leaves	cycad, sago palm
Ginkgo Division	Naked seeds in conelike structures; gymnosperm; male and female cones on different trees; fan-shaped leaves; only one known species	ginkgo
Conifer Division	Naked seeds in cones; gymnosperm; male and female cones; most are evergreen; needlelike or scalelike leaves	pine, fir, spruce, yew
Angiosperm Division	Produce flowers; seeds protected by ovary that ripens into a fruit; organs of both sexes often in same flower	
Monocot Class	One cotyledon; parallel veins; flower parts in multiples of three	grass, palm, corn, lily
Dicot Class	Two cotyledons; branching veins; flower parts in multiples of four or five	cactus, maple, rose, daisy

*For plants, biologists use *Division* instead of *Phylum*.

Appendix C: Body Systems

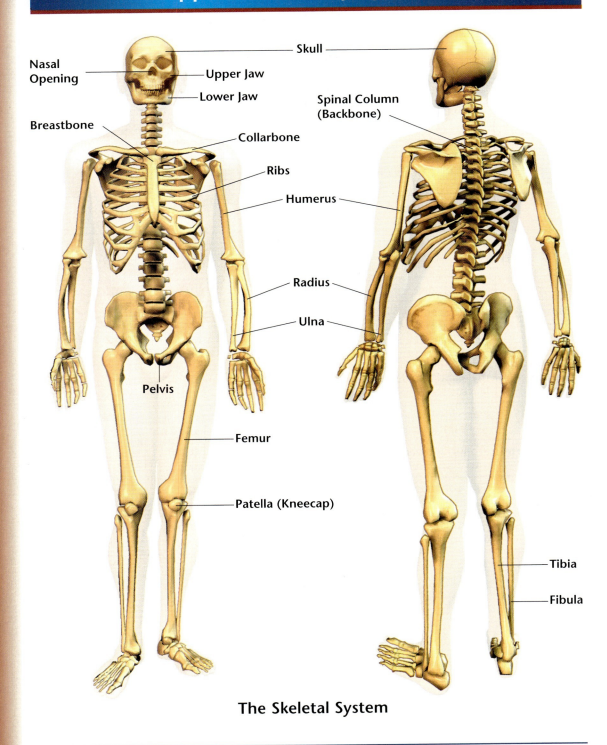

The Skeletal System

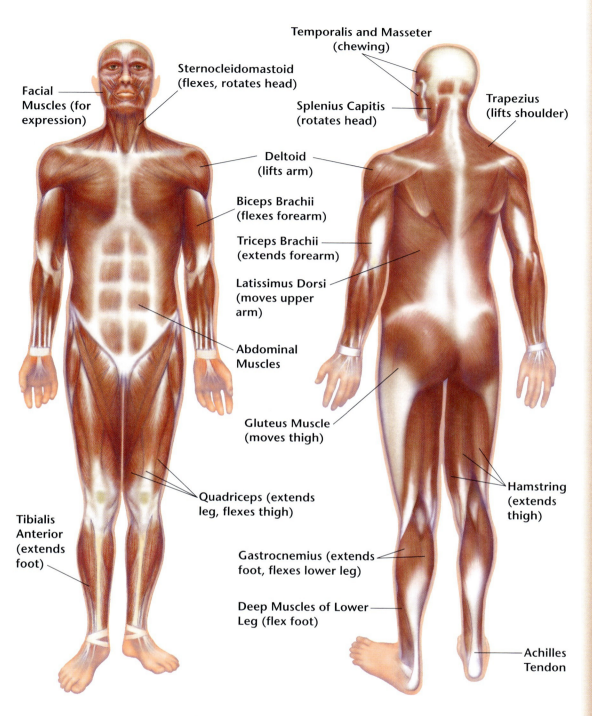

The Muscular System

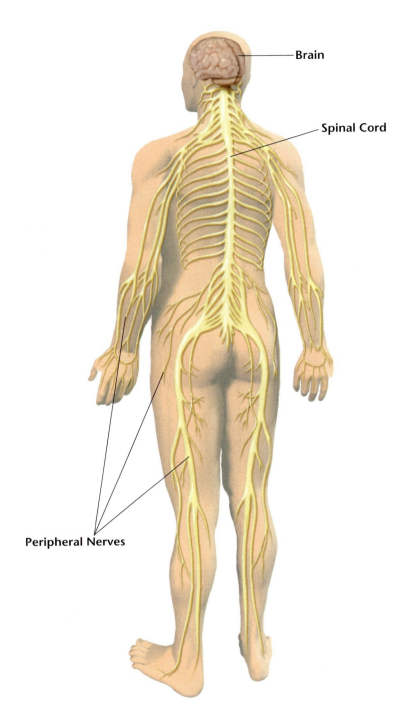

The Nervous System

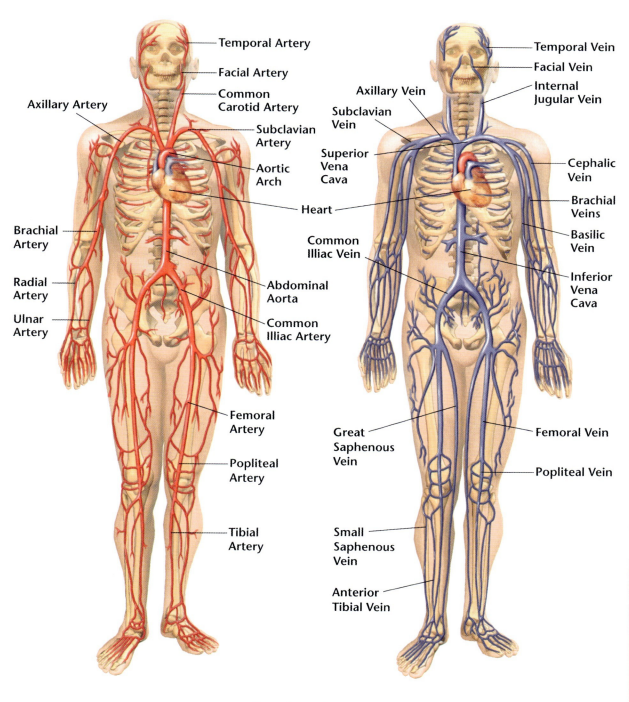

The Circulatory System

Appendix D: Measurement Conversion Factors

Metric Measures

Length
1,000 meters (m) = 1 kilometer (km)
100 centimeters (cm) = 1 m
10 decimeters (dm) = 1 m
1,000 millimeters (mm) = 1 m
10 cm = 1 decimeter (dm)
10 mm = 1 cm

Area
100 square millimeters (mm^2) = 1 square centimeter (cm^2)
10,000 cm^2 = 1 square meter (m^2)
10,000 m^2 = 1 hectare (ha)

Volume
1,000 cubic meters (m^3) = 1 cubic centimeter (cm^3)
1,000 cubic centimeters (cm^3) = 1 liter (L)
1 cubic centimeter (cm^3) = 1 milliliter (mL)
100 cm^3 = 1 cubic decimeter (dm^3)
1,000,000 cm^3 = 1 cubic meter (m^3)

Capacity
1,000 milliliters (mL) = 1 liter (L)
1,000 L = 1 kiloliter (kL)

Mass
100 grams (g) = 1 centigram (cg)
1,000 kilograms (kg) = 1 metric ton (t)
1,000 grams (g) = 1 kg
1,000 milligrams (mg) = 1 g

Temperature Degrees Celsius (°C)
0°C = freezing point of water
37°C = normal body temperature
100°C = boiling point of water

Time
60 seconds (sec) = 1 minute (min)
60 min = 1 hour (hr)
24 hr = 1 day

Customary Measures

Length
12 inches (in.) = 1 foot (ft)
3 ft = 1 yard (yd)
36 in. = 1 yd
5,280 ft = 1 mile (mi)
1,760 yd = 1 mi
6,076 feet = 1 nautical mile

Area
144 square inches (sq in.) = 1 square foot (sq ft)
9 sq ft = 1 square yard (sq yd)
43,560 sq ft = 1 acre (A)

Volume
1,728 cubic inches (cu in.) = 1 cubic foot (cu ft)
27 cu ft = 1 cubic yard (cu yard)

Capacity
8 fluid ounces (fl oz) = 1 cup (c)
2 c = 1 pint (pt)
2 pt = 1 quart (qt)
4 qt = 1 gallon (gal)

Weight
16 ounces (oz) = 1 pound (lb)
2,000 lb = 1 ton (T)

Temperature Degrees Fahrenheit (°F)
32°F = freezing point of water
98.6°F = normal body temperature
212°F = boiling point of water

To change	To	Multiply by	To change	To	Multiply by
centimeters	inches	0.3937	meters	feet	3.2808
centimeters	feet	0.03281	meters	miles	0.0006214
cubic feet	cubic meters	0.0283	meters	yards	1.0936
cubic meters	cubic feet	35.3145	metric tons	tons (long)	0.9842
cubic meters	cubic yards	1.3079	metric tons	tons (short)	1.1023
cubic yards	cubic meters	0.7646	miles	kilometers	1.6093
feet	meters	0.3048	miles	feet	5,280
feet	miles (nautical)	0.0001645	miles (statute)	miles (nautical)	0.8684
feet	miles (statute)	0.0001894	miles/hour	feet/minute	88
feet/second	miles/hour	0.6818	millimeters	inches	0.0394
gallons (U.S.)	liters	3.7853	ounces avdp	grams	28.3495
grams	ounces avdp	0.0353	ounces	pounds	0.0625
grams	pounds	0.002205	pecks	liters	8.8096
hours	days	0.04167	pints (dry)	liters	0.5506
inches	millimeters	25.4000	pints (liquid)	liters	0.4732
inches	centimeters	2.5400	pounds advp	kilograms	0.4536
kilograms	pounds avdp	2.2046	pounds	ounces	16
kilometers	miles	0.6214	quarts (dry)	liters	1.1012
liters	gallons (U.S.)	0.2642	quarts (liquid)	liters	0.9463
liters	pecks	0.1135	square feet	square meters	0.0929
liters	pints (dry)	1.8162	square meters	square feet	10.7639
liters	pints (liquid)	2.1134	square meters	square yards	1.1960
liters	quarts (dry)	0.9081	square yards	square meters	0.8361
liters	quarts (liquid)	1.0567	yards	meters	0.9144

Appendix E: Alternative Energy Sources

Fossil Fuels

We fly through the air in planes. We roll down highways in cars. On the coldest days, our homes are warm. Our stores are full of products to satisfy our needs and wants.

The power that runs our lives comes from fossil fuels. A fossil is the remains of ancient life. Fossil fuels formed from the remains of dead matter—animals and plants. Over millions of years, forests of plants died, fell, and became buried in the earth. Over time, the layers of ancient, dead matter changed. The carbon in the animals and plants turned into a material we now use as fuel. Fossil fuels include coal, oil, natural gas, and gasoline.

Fossil fuels power our lives and our society. In the United States, electricity comes mainly from power plants that burn coal. Industries use electricity to run machines. In our homes, we use electricity to power lightbulbs, TVs, and everything else electric. Heat and hot water for many homes come from natural gas or oil, or from fuels that come from oil.

Of course, cars and trucks run on gasoline, which is also made from oil. Powering our society with fossil fuels has made our lives more comfortable. Yet our need for fossil fuels has caused problems. Fossil fuels are a nonrenewable source of energy. That means that there is a limited supply of these fuels. At some point, fossil fuels will become scarce. Their cost will increase. And one day the supply of fossil fuels will run out. We need to find ways now to depend less and less on fossil fuels.

Fossil fuels cause pollution. The pollution comes from burning them. It is like the exhaust from a car. The pollution enters the air and causes disease. It harms the environment. One serious effect of burning fossil fuels is global warming. Carbon dioxide comes from the burning of fossil fuels. When a large amount of this gas enters the air, it warms the earth's climate. Scientists believe that warming of the climate will cause serious problems.

Renewable Energy

Many people believe that we should use renewable fuels as sources of energy. Renewable fuels never run out. They last forever.

What kinds of fuels last forever? The energy from the Sun. The energy in the wind. The energy in oceans and rivers. We can use these forms of energy to power our lives. Then we will never run out of fuel. We will cut down on pollution and climate warming. Using renewable energy is not a dream for the future. It is happening right now—right here—today.

Energy from the Sun

As long as the Sun keeps shining, the earth will get energy from sunlight. Energy from the Sun is called solar energy. It is the energy in light. When you lie in the Sun, your skin becomes hot. The heat comes from the energy in sunlight. Sunlight is a form of renewable energy we can use forever.

We use solar energy to make electricity. The electricity can power homes and businesses. Turning solar energy into electricity is called photovoltaics, or PV for short. Here's how PV works.

Flat solar panels are put near a building or on its roof. The panels face the direction that gets the most sunlight. The panels contain many PV cells. The cells are made from silicon—a material that absorbs light. When sunlight strikes the cells, some of the light energy is absorbed. The energy knocks some electrons loose in the silicon. The electrons begin to flow. The electron flow is controlled. An electric current is produced. Pieces of metal at the top and bottom of each cell make a path for electrons. The path leads the electric current away from the solar panel. The electric current flows through wires to a battery. The battery stores the electrical energy. The electrical wiring in a building is connected to the battery. All the electricity used in the building comes from the battery.

Today, PV use is 500 times greater than it was 20 years ago. And PV use is growing about 20 percent per year. Yet solar energy systems are still not perfect. PV cells do not absorb all the sunlight that strikes them, so some energy is lost. Solar energy systems also are not cheap. Still, every year, PV systems are improved. The cost of PV electricity has decreased. The amount of sunlight PV cells absorb has increased.

On a sunny day, every square meter of the earth receives 1,000 watts of energy from sunlight. Someday, when PV systems are able to use all this energy, our energy problems may be solved.

Energy from the Wind

Sunlight warms different parts of the earth differently. The North Pole gets little sunlight, so it is cold. Areas near the equator get lots of sunlight, so they are warm. The uneven warming of the earth by the Sun creates the wind. As the earth turns, the wind moves, or blows. The blowing wind can be used to make electricity. This is wind energy. Because the earth's winds will blow forever, the wind is a renewable source of energy.

Wind energy is not new. Hundreds of years ago, windmills created energy. The wind turned the large fins on a windmill. As the fins spun around, they turned huge stones inside the mill. The stones ground grain into flour.

Modern windmills are tall, metal towers with spinning blades, called wind turbines. Each wind turbine has three main parts. It has blades that are turned by blowing wind. The turning blades are attached to a shaft that runs the length of the tower. The turning blades spin the shaft. The spinning shaft is connected to a generator. A generator changes the energy from movement into electrical energy. It feeds the electricity into wires, which carry it to homes and factories.

Wind turbines are placed in areas where strong winds blow. A single house may have one small wind turbine near it to produce its electricity. The electricity produced by the wind turbine is stored in batteries. Many wind turbines may be linked together to produce electricity for an entire town. In these systems, the electricity moves from the generator to the electric company's wires. The wires carry the electricity to homes and businesses.

Studies show that in the United States, 34 of the 50 states have good wind conditions. These states could use wind to meet up to 20 percent of their electric power needs. Canada's wind conditions could produce up to 20 percent of its energy from wind too. Alberta already produces a lot of energy from wind, and the amount is expected to increase.

Energy from Inside the Earth

Deep inside the earth, the rocks are burning hot. Beneath them it is even hotter. There, rocks melt into liquid. The earth's inner heat rises to the surface in some places. Today, people have developed ways to use this heat to create energy. Because the inside of the earth will always be very hot, this energy is renewable. It is called geothermal energy (*geo* means "earth"; *thermal* means "heat").

Geothermal energy is used where hot water or steam from deep inside the earth moves near the surface. These areas are called "hot spots." At hot spots, we can use geothermal energy directly. Pumps raise the hot water, and pipes carry it to buildings. The water is used to heat the space in the buildings or to heat water.

Geothermal energy may also be used indirectly to make electricity. A power plant is built near a hot spot. Wells are drilled deep into the hot spot. The wells carry hot water or steam into the power plant. There, it is used to boil more water. The boiling water makes steam. The steam turns the blades of a turbine. This energy is carried to a generator, which turns the energy into electricity. The electricity moves through the electric company's wires to homes and factories.

Everywhere on the earth, several miles beneath the surface, there is hot material. Scientists are improving ways of tapping the earth's inner heat. Some day, this renewable, pollution-free source of energy may be available everywhere.

Energy from Trash

We can use the leftover products that come from plants to make electricity. For example, we can use the stalks from corn or wheat to make fuel. Many leftover products from crops and lumber can fuel power plants. Because this fuel comes from living plants, it is called bioenergy (*bio* means "life" or "living"). The plant waste itself is called biomass.

People have used bioenergy for thousands of years. Burning wood in a fireplace is a form of bioenergy. That's because wood comes from trees. Bioenergy is renewable because people will always grow crops. There will always be crop waste we can burn as fuel.

Some power plants burn biomass to heat water. The steam from the boiling water turns turbines. The turbines create electricity. In other power plants, biomass is changed into a gas. The gas is used as fuel to boil water, which turns the turbine.

Biomass can also be made into a fuel for cars and trucks. Scientists use a special process to turn biomass into fuels, such as ethanol. Car makers are designing cars that can run on these fuels. Cars that use these fuels produce far less pollution than cars that run on gas.

Bioenergy can help solve our garbage problem. Many cities are having trouble finding places to dump all their trash. There would be fewer garbage dumps if we burned more trash to make electricity.

Bioenergy is a renewable energy. But it is not a perfect solution to our energy problems. Burning biomass creates air pollution.

Energy from the Ocean

Have you ever been knocked over by a small wave while wading in the ocean? If so, you know how much power ocean water has. The motion of ocean waves can be a source of energy. So can the rise and fall of ocean tides. There are several systems that use the energy in ocean waves and tides. All of them are very new and still being developed.

In one system, ocean waves enter a funnel. The water flows into a reservoir, an area behind a dam where water is stored. When the dam opens, water flows out of the reservoir. This powers a turbine, which creates electricity. Another system uses the waves' motion to operate water pumps, which run an electric generator. There is also a system that uses the rise and fall of ocean waves. The waves compress air in a container. During high tide, large amounts of ocean water enter the container. The air in the container is under great pressure. When the high-pressure air in the container is released, it drives a turbine. This creates electricity.

Energy can also come from the rise and fall of ocean tides. A dam is built across a tidal basin. This is an area where land surrounds the sea on three sides. At high tide, ocean water is allowed to flow through the dam. The water flow turns turbines, which generate electricity. There is one serious problem with tidal energy. It damages the

Biology   Appendix E   **397**

environment of the tidal basin and can harm animals that live there.

The oceans also contain a great deal of thermal (heat) energy. The Sun heats the surface of the oceans more than it heats deep ocean water. In one day, ocean surfaces absorb solar energy equal to 250 billion barrels of oil! Deep ocean water, which gets no sunlight, is much colder than the surface.

Scientists are developing ways to use this temperature difference to create energy. The systems they are currently designing are complicated and expensive.

Energy from Rivers and Dams

Dams built across rivers also produce electricity. When the dam is open, the flowing water turns turbines, which make electricity. This is called hydroelectric power (*hydro* means water). The United States gets 7 percent of its electricity from hydroelectric power. Canada gets up to 60 percent of its electricity from hydroelectric plants built across its many rivers.

Hydroelectric power is a nonpolluting and renewable form of energy—in a way. There will always be fresh water. However, more and more people are taking water from rivers for different uses. These uses include drinking, watering crops, and supplying industry. Some rivers are becoming smaller and weaker because of the water taken from them. Also, in many places, dams built across rivers hurt the environment. The land behind the dam is "drowned." Once the dam is built, fish may not be able swim up or down the river. In northwestern states, salmon have completely disappeared from many rivers that have dams.

Energy from Hydrogen Fuel

Hydrogen is a gas that is abundant everywhere on the earth. It is in the air. It is a part of water. Because there is so much hydrogen, it is a renewable energy source. And hydrogen can produce energy without any pollution.

The most likely source of hydrogen fuel is water. Water is made up of hydrogen and oxygen. A special process separates these elements in water. The process produces oxygen gas and hydrogen gas. The hydrogen gas is changed into a liquid or solid. This hydrogen fuel is used to produce energy in a fuel cell.

Look at the diagram on page 399. Hydrogen fuel (H_2) is fed into one part of the fuel cell. It is then stripped of its electrons. The free electrons create an electric current (e). The electric current powers a lightbulb or whatever is connected to the fuel cell.

Meanwhile, oxygen (O_2) from the air enters another part of the fuel cell. The stripped hydrogen ($H+$) bonds with the oxygen, forming water (H_2O). So a car powered by a fuel cell has pure water leaving its tailpipe. There is no exhaust to pollute the air.

When a regular battery's power is used up,

the battery dies. A fuel cell never runs down as long as it gets hydrogen fuel.

A single fuel cell produces little electricity. To make more electricity, fuel cells come in "stacks" of many fuel cells packaged together. Stacked fuel cells are used to power cars and buses. Soon, they may provide electric power to homes and factories.

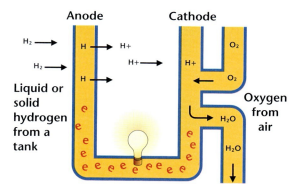
Hydrogen Fuel Cell

Hydrogen fuel shows great promise, but it still has problems. First, hydrogen fuel is difficult to store and distribute. Today's gas stations would have to be changed into hydrogen-fuel stations. Homes and factories would need safe ways to store solid hydrogen.

Second, producing hydrogen fuel by separating water is expensive. It is cheaper to make hydrogen fuel from oil. But that would create pollution and use nonrenewable resources. Scientists continue to look for solutions to these problems.

Energy from Atoms

Our Sun gets its energy—its heat and light—from fusion. Fusion is the joining together of parts of atoms. Fusion produces enormous amounts of energy. But conditions like those on the Sun are needed for fusion to occur. Fusion requires incredibly high temperatures.

In the next few decades, scientists may find ways to fuse atoms at lower temperatures. When this happens, we may be able to use fusion for energy. Fusion is a renewable form of energy because it uses hydrogen atoms. It also produces no pollution. And it produces no dangerous radiation. Using fusion to produce power is a long way off. But if the technology can be developed, fusion could provide us with renewable, clean energy.

Today's nuclear power plants produce energy by splitting atoms. This creates no air pollution. But nuclear energy has other problems. Nuclear energy is fueled by uranium, a substance we get from mines. There is only a limited amount of uranium in the earth. So it is not renewable. And uranium produces dangerous radiation, which can harm or kill living things if it escapes the power plant. Used uranium must be thrown out, even though it is radioactive and dangerous. In 1999, the United States produced nearly 41 tons of radioactive waste from nuclear power plants. However, less uranium is being mined. No new nuclear power plants have been built. The amount of energy produced from nuclear power is expected to fall. People are turning toward less harmful, renewable energy sources: the Sun, wind, underground heat, biomass, water, and hydrogen fuel.

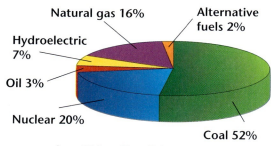
Fuel That U.S. Electric Utilities Used to Generate Electricity in 2000

Source: U.S. Dept. of Energy Hydropower Program

Glossary

A

Abiotic (ā bī ot´ik) Nonliving (p. 297)

Acid rain (as´id rān) Rain that is caused by pollution and is harmful to organisms because it is acidic (p. 299)

Adaptive advantage (ə dap´tiv ad van´tij) The greater likelihood that an organism will survive, due to characteristics that allow it to be more successful than other organisms (p. 373)

Adolescence (ad l es´ns) The teenage years of a human (p. 234)

Aflatoxin (a flə tok´ sən) A chemical that causes liver cancer and is produced by molds growing on stored crops (p. 103)

Algae (al´jē) Protists that make their own food and usually live in water (p. 32)

Alveolus (al vē´ə ləs) A tiny air sac at the end of each bronchiole that holds air (plural is *alveoli*) (p. 182)

Amino acid (ə mē´nō as´id) Molecules that make up proteins (p. 14)

Amoeba (ə mē´ bə) A protozoan that moves by pushing out parts of its cell (p. 33)

Amphibian (am fib´ē ən) A vertebrate that lives at first in water and then on land (p. 52)

Anal pore (ā´nl pôr) The opening through which undigested food leaves a paramecium (p. 95)

Angiosperm (an´jē ə spėrm) A flowering plant (p. 71)

Annelid (an´l id) A segmented worm or a worm whose body is divided into sections (p. 125)

Annual growth ring (an´yü əl grōth ring) Ring in a tree trunk formed by the growth of wood in layers (p. 145)

Antibody (an´ti bod ē) A material in the body that fights a specific pathogen; a protein in plasma that fights disease (p. 176)

Anus (ā´nəs) The opening through which material that is not digested leaves the digestive tract (p. 119)

Aorta (ā ôr´ tə) A large vessel through which the left ventricle sends blood to the body (p. 174)

Applied genetics (ə plīd´ jə net´iks) The process of using knowledge of genetics to affect heredity (p. 288)

Arachnid (ə rak´nid) A class of arthropods that includes spiders, scorpions, mites, and ticks (p. 59)

Artery (är´tər ē) A blood vessel that carries blood away from the heart (p. 175)

Arthropod (är´thrə pod) A member of the largest group of invertebrates, which includes insects (p. 58)

Asexual reproduction (ā sek´shü əl rē prə duk´shən) Reproduction that involves one parent and no egg or sperm (p. 97)

Atom (at´əm) The smallest particle of an element that still has the properties of that element (p. 8)

ATP High-energy molecules that store energy in a form the cells can easily use (p. 8)

Atrium (ā´trē əm) A heart chamber that receives blood returning to the heart (plural is *atria*) (p. 126)

Auditory nerve (o´də tôr ē nėrv) A bundle of nerves that carry impulses from the ear to the brain (p. 193)

B

Bacteria (bak tir´ē ə) The simplest single cells that carry out all basic life activities (p. 2)

Base (bās) A molecule found in DNA that is used to code information (p. 281)

Behavior (bi hā´vyər) The way an organism acts (p. 328)

Bilateral symmetry (bī lat´ər əl sim´ə trē) A body plan that consists of left and right halves that are the same (p. 57)

Bile (bīl) A substance made in the liver that breaks down fats (p. 170)

Binary fission (bī´nər ē fi´zhən) Reproduction in which a bacterial cell divides into two cells that look the same as the original cell (p. 87)

Biology (bī ol´ə jē) The study of living things (p. 31)

Biome (bī´ōm) An ecosystem found over a large geographic area (p. 300)

Biosphere (bī´ə sfir) The part of earth where living things can exist (p. 302)

Biotic (bī ot´ik) Living (p. 297)

Blood pressure (blud presh´ər) The force of blood against the walls of blood vessels (p. 176)

Brain stem (brān stem) The part of the brain that controls automatic activities and connects the brain and the spinal cord (p. 188)

Bronchiole (brong′ kē ōl) A tube that branches off the bronchus (p. 181)

Bronchus (brong′kəs) A tube that connects the trachea to a lung (plural is *bronchi*) (p. 181)

Budding (bud′ing) Reproduction in which part of an organism pinches off to form a new organism (p. 108)

C

Calorie (kal′ər ē) A unit used to measure the amount of energy a food contains (p. 248)

Capillary (kap′ə ler ē) A blood vessel with a wall one cell thick through which oxygen and food molecules pass to body cells (p. 175)

Carbohydrate (kär bō hī′drāt) A sugar or starch, which living things use for energy (p. 13)

Cardiac (kär′dē ak) Relating to the heart (p. 173)

Carnivore (kär′nə vôr) An animal that eats other animals (p. 117)

Carrier (kar′ē ər) An organism that carries a gene but does not show the effects of the gene (p. 277)

Cartilage (kär′tl ij) A soft material found in vertebrate skeletons (p. 51)

Cast (kast) A type of fossil that is formed when a mold in rock is filled with minerals that harden (p. 358)

Cell (sel) The basic unit of life (p. 2)

Cell differentiation (sel dif ə ren shē ā′shən) The process of cells taking on different jobs in the body (p. 220)

Cell membrane (sel mem′brān) A thin layer that surrounds and holds a cell together (p. 6)

Cell wall (sel wȯl) The outer part of a plant cell that provides structure to the cell (p. 7)

Cellular respiration (sel′yə lər res pə rā′shən) The process in which cells break down food to release energy (p. 152)

Central nervous system (sen′trəl nėr′vəs sis′təm) The brain and spinal cord (p. 136)

Cerebellum (ser ə bel′əm) The part of the brain that controls balance (p. 188)

Cerebrum (sə rē′brəm) The largest part of the brain that controls thought, memory, learning, feeling, and body movement (p. 137)

Channel (chan′l) A way of communicating (p. 341)

Chlorophyll (klôr′ ə fil) The green pigment in plants that absorbs light energy for photosynthesis (p. 149)

Chloroplast (klôr′ə plast) Captures the light energy from the Sun to make food (p. 7)

Chromosome (krō′mə sōm) A rod-shaped structure that contains DNA and is found in the nucleus of a cell (p. 215)

Chyme (kīm) Liquid food in the digestive tract that is partly digested (p. 170)

Cilia (sil′ē ə) Hairlike structures that help some one-celled organisms move (p. 33)

Circulatory (sėr′kyə lə tôr ē) Flowing in a circle (p. 125)

Classify (klas′ə fī) To group things based on the features they share (p. 42)

Climax community (klī′maks kə myü′nə tē) A community that changes little over time (p. 299)

Closed circulatory system (klōzd′ sėr′kyə lə tôr ē sis′təm) A system in which blood stays inside vessels at all times (p. 125)

Cnidarian (nī dar′ē ən) An invertebrate animal that includes jellyfish, corals, and hydras (p. 56)

Cochlea (kok′lē ə) The organ in the ear that sends impulses to the auditory nerve (p. 193)

Cold-blooded (kōld′ blud′ id) Having a body temperature that changes with temperature of surroundings (p. 53)

Commensalism (kə men′sə liz əm) A relationship in which one organism benefits and the other is not affected (p. 87)

Communication (kə myü nə kā′shən) Sending information (p. 341)

Community (kə myü′nə tē) A group of different populations that live in the same area (p. 297)

Complete metamorphosis (kəm plēt′ met ə môr′fə sis) Changes in form during development in which earlier stages do not look like the adult (p. 60)

Condense (kən dens′) To change from a gas to a liquid (p. 318)

a	hat	e	let	ī	ice	ȯ	order	ü	put	sh	she		a in about
ā	age	ē	equal	o	hot	oi	oil	ü	rule	th	thin	ə {	e in taken
ä	far	ėr	term	ō	open	ou	out	ch	child	ᴛʜ	then		i in pencil
â	care	i	it	ȯ	saw	u	cup	ng	long	zh	measure		o in lemon
													u in circus

Biology Glossary

Conditioning (kən dish′ə ning) Learning in which an animal connects one stimulus with another stimulus (p. 336)

Conifer (kon′ə fər) A cone-bearing gymnosperm (p. 73)

Consumer (kən sü′mər) An organism that feeds on other organisms (p. 306)

Contractile vacuole (kən trak′təl vak′yü ōl) A structure in a protist that removes water that is not needed (p. 96)

Coordinate (kō ôrd′n āt) To work together (p. 133)

Cornea (kôr′nē ə) A clear layer of the eye that light passes through (p. 192)

Cotyledon (kot l ēd′n) A structure in the seeds of angiosperms that contains food for the plant (p. 72)

Courtship behavior (kôrt′ship bi hā′vyər) Behavior that helps attract and get a mate (p. 329)

Cro-Magnons (krō mag′nənz) Homo sapiens who lived about 35,000 years ago and are direct ancestors of humans living today (p. 379)

Crop (krop) The part of the digestive tract of some animals where food is stored (p. 119)

Cross-pollination (krȯs′ pol ə nā′shən) The movement of pollen from the male sex organs to the female sex organs of flowers on different plants (p. 267)

Crustacean (krus tā′shən) A class of arthropods that includes crabs, lobsters, crayfish, and sow bugs (p. 59)

Cyclosporine (sī klə spôr′ēn) A drug that is produced from mold and that helps prevent the rejection of transplanted organs (p. 102)

Cytoplasm (sī′tə plaz əm) A gel-like substance containing chemicals needed by the cell (p. 6)

D

Decompose (dē kəm pōz′) To break down or decay matter into simpler substances (p. 34)

Dermis (dėr′mis) The thick layer of cells below the epidermis (p. 184)

Descent with modification (di sent′ wiŧH mod ə fə kā′shən) The theory that more recent species of organisms are changed descendants of earlier species (p. 371)

Development (di vel′əp mənt) The changes that occur as a living thing grows (p. 18)

Diabetes (dī ə bē′tis) A genetic disease in which a person has too much sugar in the blood (p. 282)

Diatom (dī′ə tom) Microscopic alga that has a hard shell (p. 91)

Dicot (dī′kot) An angiosperm that has two seed leaves (p. 72)

Diffusion (di fyü′zhən) The movement of materials from an area of high concentration to an area of low concentration (p. 123)

Digestion (də jes′chən) The process by which living things break down food (p. 17)

Digestive enzyme (də jes′ tiv en′zīm) A chemical that helps break down food (p. 107)

Digestive tract (də jes′tiv trakt) A tubelike digestive space with an opening at each end (p. 119)

Diversity (də vėr′sə tē) The range of differences among the individuals in a population (p. 212)

DNA The chemical inside cells that stores information about an organism (p. 6)

Dominant gene (dom′ə nənt jēn) A gene that shows up in an organism (p. 268)

Drug (drug) A substance that acts on the body and changes the way the body works (p. 257)

Drug abuse (drug ə byüz′) Using a drug when there is not a health reason for taking it (p. 257)

E

Eardrum (ir′drum) A thin tissue in the middle ear that vibrates when sound waves strike it (p. 193)

Ecology (ē kol′ə jē) The study of the interactions among living things and the nonliving things in their environment (p. 296)

Ecosystem (ē′kō sis təm) The interactions among the populations of a community and the nonliving things in their environment (p. 297)

Electron microscope (i lek′tron mī′krə skōp) An instrument that uses a beam of tiny particles called electrons to magnify things (p. 3)

Embryo (em′brē ō) A beginning plant; an early stage in the development of an organism (pp. 71, 220)

Endangered (en dān′jərd) There are almost no animals left of a certain species (p. 300)

Endoplasmic reticulum (en′də plaz′mik ri ti′kyə ləm) A system of tubes that processes and transports proteins within the cell (p. 7)

Endospores (en′dō spōrz) Bacteria that dry up and form into cells with thick walls (p. 88)

Energy pyramid (en′ər jē pir′ə mid) A diagram that compares the amounts of energy available to the populations at different levels of a food chain (p. 314)

Environment (en vī′rən mənt) An organism's surroundings (p. 280)

Enzyme (en′zīm) A substance that speeds up chemical changes (p. 118)

Epidermis (ep ə dėr′mis) The thin outer layer of skin (p. 184)

Estrogen (es′trə jən) Female sex hormone (p. 230)

Evaporate (i vap′ə rāt) To change from a liquid to a gas (p. 318)

Evolution (ev ə lü′shən) The changes in a population over time (p. 352)

Excrete (ek skrēt′) To get rid of wastes or substances that are not needed (p. 128)

Excretion (ek skrē′shən) The process by which living things get rid of wastes (p. 17)

Excretory system (ek′skrə tôr ē sis′təm) A series of organs that get rid of cell wastes in the form of urine (p. 185)

External fertilization (ek stėr′ nl fėr tl ə zā′shən) The type of fertilization that occurs outside the female's body (p. 218)

Extinct (ek stingkt′) When all the members of a species are dead (p. 300)

Eyespot (ī′spot) A structure on many protists that senses changes in the brightness of light (p. 97)

F

F_1 generation (jen ə rā′shən) The plants that resulted when Mendel cross-pollinated two different kinds of pure plants (p. 267)

F_2 generation (jen ə rā′shən) The plants that resulted when Mendel self-pollinated plants from the F1 generation (p. 269)

Factors (fak′tərsz) The name that Mendel gave to information about traits that parents pass to offspring (p. 270)

Fallopian tube (fə lō′pē ən tüb) A tube through which eggs pass from an ovary to the uterus (p. 230)

Fat (fat) A chemical that stores large amounts of energy (p. 13)

Fatty layer (fat′ē lā′ər) Protects organs, keeps in heat (p. 184)

Feces (fē′sēz) Solid waste material remaining in the large intestine after digestion (p. 171)

Fern (fėrn) A seedless vascular plant (p. 68)

Fetus (fē′təs) An embryo after eight weeks of development in the uterus (p. 232)

Filter feeding (fil′tər fēd′ ing) A way of getting food by straining it out of the water (p. 116)

Flagellum (flə jel′əm) A whiplike tail that helps some one-celled organisms move (plural is *flagella*) (p. 33)

Flame cell (flām′ sel) A cell that collects excess water in a flatworm (p. 129)

Flatworm (flat′wėrm) A simple worm that is flat and thin (p. 57)

Food chain (füd′ chān) The feeding order of organisms in a community (p. 306)

Food Guide Pyramid (füd gīd pir′ə mid) A guide for good nutrition (p. 251)

Food vacuole (füd vak′yü ōl) A bubble-like structure where food is digested inside a protozoan (p. 95)

Food web (füd′ web) All the food chains in a community that are linked to one another (p. 308)

Fossil (fos′əl) The remains or traces of an organism that lived in the past (p. 358)

Fossil fuel (fos′əl fyü′əl) Fuels formed millions of years ago from the remains of plants and animals (p. 302)

Fossil record (fos′əl rek′ərd) The history of life on earth, based on fossils that have been discovered (p. 359)

Fraternal twins (frə tėr′nl twinz) Twins that do not have identical genes (p. 279)

Frond (frond) A large feathery leaf of a fern (p. 77)

Fungus (fung′gəs) An organism that usually has many cells and decomposes material for its food (plural is *fungi*) (p. 33)

G

Gallbladder (gòl′ blad ər) The digestive organ that stores bile (p. 170)

Gamete (gam′ēt) A sex cell, such as sperm or egg (p. 216)

Gastrovascular cavity (gas trō vas′kyə lər kav′ə tē) A digestive space with a single opening (p. 118)

Gene (jēn) The information about a trait that a parent passes to its offspring (p. 268)

Gene pool (jēn pül) The genes found within a population (p. 283)

a	hat	e	let	ī	ice	ò	order	ù	put	sh	she	ə {	a in about
ā	age	ē	equal	o	hot	oi	oil	ü	rule	th	thin		e in taken
ä	far	ėr	term	ō	open	ou	out	ch	child	ᴛʜ	then		i in pencil
â	care	i	it	ô	saw	u	cup	ng	long	zh	measure		o in lemon
													u in circus

Genetic disease (jə net′ik də zēz′) A disease that is caused by a mutated gene (p. 282)

Genetic engineering (jə net′ik en jə nir′ing) The process of transferring genes from one organism to another (p. 289)

Genetics (jə net′iks) The study of heredity (p. 266)

Genotype (jen′ə tīp) An organism's combination of genes for a trait (p. 269)

Genus (jē′nəs) A group of living things that includes separate species (p. 44)

Geographic isolation (jē ə graf′ik ī sə lā′shən) The separation of a population into two populations that have no contact with each other, caused by a change in the environment (p. 355)

Geologic time scale (jē ə loj′ik tīm skāl) A chart that divides Earth's history into time periods (p. 359)

Germinate (jėr′mə nāt) Start to grow into a new plant (p. 158)

Gestation time (je stā′shən tīm) The period of development of a mammal, from fertilization until birth (p. 224)

Gill (gil) A structure used by some animals to breathe in water (p. 52)

Gizzard (giz′ərd) The part of the digestive tract of some animals that grinds food (p. 119)

Golgi body (gȯl′jē bod′ē) Packages and distributes proteins outside the cell (p. 7)

Gravitropism (grav ə trō′piz əm) The response of a plant to gravity (p. 332)

Groundwater (ground′wȯ tər) The water under earth's surface (p. 317)

Guard cell (gärd′ sel) A cell that opens and closes stomata (p. 154)

Gullet (gul′it) The opening through which a paramecium takes in food (p. 95)

Gymnosperm (jim′nə spėrm) A nonflowering seed plant (p. 73)

H

Habit (hab′it) Something you usually do, often automatically (p. 256)

Habitat (hab′ə tat) The place where an organism lives (p. 297)

Half-life (haf′līf) The amount of time required for one-half of a sample of radioactive mineral to decay (p. 361)

Hemoglobin (hē′mə glō bən) A substance in red blood cells that carries oxygen (p. 176)

Hemophilia (hē mə fil′ē ə) A genetic disease in which a person's blood fails to clot (p. 283)

Herbivore (ėr′bə vôr) An animal that eats plants (p. 117)

Heredity (hə red′ə tē) The passing of traits from parents to offspring (p. 266)

Homeostasis (hō mē ō stā′səs) The ability of organisms to maintain their internal conditions (p. 18)

Hominid (hom′ə nid) The group that includes humans and humanlike primates (p. 376)

Homo sapiens (hō′mō sā′pē enz) The species to which humans belong (p. 376)

Homologous structures (hō mo′lə gəs struk′ chərz) Body parts that are similar in related organisms (p. 374)

Hormone (hôr′mōn) A chemical signal that glands produce (p. 133)

Humus (hyü′məs) Decayed plant and animal matter that is part of the topsoil (p. 79)

Hyphae (hī′fē) Thin, tubelike threads produced by a fungus (p. 99)

Hypothesis (hī poth′ə sis) A testable explanation of a question or problem (plural is *hypotheses*) (p. 370)

I

Identical twins (ī den′tə kəl twinz) Twins that have identical genes (p. 279)

Immune system (i myün′ sis′təm) The body's most important defense against infectious diseases (p. 243)

Immunity (i myü′nə tē) The ability of the body to fight off a specific pathogen (p. 244)

Imprinting (im′print ing) Learning in which an animal bonds with the first object it sees (p. 335)

Impulse (im′puls) A message that travels along nerve cells (p. 134)

Inbreeding (in′brēd ing) Sexual reproduction between organisms within a small gene pool (p. 283)

Incomplete metamorphosis (in kəm plēt′ met ə môr′fə sis) Changes in form during development in which earlier stages look like the adult (p. 60)

Infectious disease (in fek′shəs də zēz′) An illness that can pass from person to person (p. 242)

Innate behavior (i nāt′ bi hā′vyər) A behavior that is present at birth (p. 328)

Insight (in′sīt) The ability to solve a new problem based on experience (p. 337)

Instinct (in′stingkt) A pattern of innate behavior (p. 328)

Interact (in tər akt′) To act upon or influence something (p. 296)

Internal fertilization (in tėr′nl fėr tl ə zā′shən) The type of fertilization that occurs inside the female's body (p. 218)

Invertebrate (in vėr′tə brit) An animal that does not have a backbone (p. 56)

Involuntary muscle (in vol′ən ter ē mus′əl) A muscle that a person cannot control (p. 203)

Iris (ī′ris) The part of the eye that controls the amount of light that enters (p. 192)

K

Kidney (kid′nē) An organ of excretion found in vertebrates (p. 130)

Kingdom (king′dəm) One of the five groups into which living things are classified (p. 31)

L

Larynx (lar′ingks) The voice box (p. 181)

Learned behavior (lėrnd bi hā′vyər) Behavior that results from experience (p. 335)

Lethal mutation (lē′thəl myü tā′shən) A mutation that results in the death of an organism (p. 354)

Lichen (lī′kən) An organism that is made up of a fungus and an alga or a bacterium (p. 109)

Ligament (lig′ə mənt) A tissue that connects bone to bone (p. 200)

Lymphocyte (lim′fə sīt) A white blood cell that produces antibodies (p. 244)

Lysosome (lī′sə sōm) A tiny structure in a cell containing enzymes that break down substances (p. 7)

M

Mammary gland (mam′ər ē gland) A milk-producing structure on the chest or abdomen of a mammal (p. 54)

Marsupial (mär sü′pē əl) A mammal that gives birth to young that are very undeveloped (p. 222)

Mass extinction (mas ek stingk′shən) The dying out of large numbers of species within a short period of time (p. 361)

Meiosis (mī ō′sis) The process that results in sex cells (p. 217)

Menstruation (men strü ā′shən) The process during which an unfertilized egg, blood, and pieces of the lining of the uterus exit the female body (p. 230)

Metamorphosis (met ə môr′fə sis) A major change in form that occurs as some animals develop into adults (p. 52)

Methane (meth′ān) A gas produced by bacteria from hydrogen and carbon dioxide (p. 87)

Microorganism (mī krō ôr′gə niz əm) An organism that is too small to be seen without a microscope (p. 32)

Microscope (mī′krə skōp) An instrument used to magnify things (p. 2)

Mineral (min′ər əl) A chemical found in foods that is needed by living things in small amounts (p. 15)

Mitochondrion (mī tə kon′drē ən) An organelle that uses oxygen to break down food and release energy in chemical bonds (plural is *mitochondria*) (p. 6)

Mitosis (mī tō′sis) The process that results in two cells identical to the parent cell (p. 215)

Mold (mōld) A type of fossil that is formed when a dead organism decays and leaves an empty space in rock (p. 358)

Molecule (mol′ə kyül) Two or more atoms joined together (p. 8)

Mollusk (mol′əsk) An invertebrate divided into three parts (p. 58)

Molting (mōlt′ing) The process by which an arthropod sheds its external skeleton (p. 58)

Moneran (mə nir′ən) An organism that is one-celled and does not have organelles (p. 34)

Monocot (mon′ə kot) An angiosperm that has one seed leaf (p. 72)

Moss (mȯs) A nonvascular plant that has simple parts (p. 68)

Mutation (myü tā′shən) A change in a gene (p. 280)

Mutualism (myü′chü ə liz əm) A closeness in which two organisms live together and help each other (p. 87)

Mycelium (mī sē′lē əm) A mass of hyphae (plural is *mycelia*) (p. 99)

a hat	e let	ī ice	ȯ order	ủ put	sh she	ə { a in about
ā age	ē equal	o hot	oi oil	ü rule	th thin	e in taken
ä far	ėr term	ō open	ou out	ch child	ᴛʜ then	i in pencil
â care	i it	ȯ saw	u cup	ng long	zh measure	o in lemon, u in circus

Biology Glossary **405**

Mycorrhiza (mī kə rī′zə) A mutualism between a fungus and the roots of a plant (plural is *mycorrhizae*) (p. 108)

N

Natural selection (nach′ər əl si lek′shən) The process by which organisms best suited to the environment survive, reproduce, and pass their genes to the next generation (p. 371)

Neanderthals (nē an′dər tälz) Homo sapiens who lived between about 35,000 and 150,000 years ago but are not thought to be direct ancestors of humans living today (p. 379)

Nectar (nek′tər) A sweet liquid that many kinds of flowers produce (p. 158)

Nerve net (nėrv′ net) A bunch of nerve cells that are loosely connected (p. 135)

Neuron (nùr′on) A nerve cell (p. 189)

Neurotransmitter (nùr ō trans mi′tər) A chemical signal that a nerve cell releases (p. 134)

Nitrogen fixation (nī′trə jən fik sā′shən) The process by which certain bacteria change nitrogen gas from the air into ammonia (p. 320)

Nonrenewable resources (non ri nü′ə bel ri sôrs′əz) Resources that cannot be replaced (p. 301)

Nonvascular plant (non vas′kyə lər plant) A plant that does not have tubelike cells (p. 69)

Nucleus (nü′klē əs) Information and control center of the cell (p. 6)

Nutrient (nü′trē ənt) Any chemical found in foods that is needed by living things (p. 15)

Nutrition (nü trish′ən) The types and amounts of foods a person eats (p. 248)

Nymph (nimf) A young insect that looks like the adult (p. 221)

O

Observational learning (ob zər vā′shə nəl lėr′ning) Learning by watching or listening to the behavior of others (p. 335)

Omnivore (om′nə vôr) A consumer that eats both plants and animals (p. 307)

Open circulatory system (ō′pən sėr′kyə lə tôr ē sis′təm) A system in which blood makes direct contact with cells (p. 125)

Optic nerve (op′tik nėrv) A bundle of nerves that carry impulses from the eye to the brain (p. 193)

Organ (ôr′gən) A group of different tissues that work together (p. 4)

Organelle (ôr gə nel′) A tiny structure inside a cell (p. 3)

Organism (ôr′gə niz əm) A living thing that can carry out all the basic life activities (p. 27)

Osmosis (oz mō′sis) The movement of water through a cell membrane (p. 96)

Osteoporosis (os tē ō pə rō′sis) A disease in which bones become lighter and break easily (p. 200)

Ovary (ō′vər ē) The lower part of the pistil, that contains eggs (p. 157)

Ovulation (ō vyə lā′shən) The process of releasing an egg from an ovary (p. 230)

P

Paleontologist (pā lē on tol′ə jist) A scientist who studies life in the past (p. 359)

Paramecium (par ə mē′sē əm) A protozoan that moves by using its hairlike cilia (plural is *paramecia*) (p. 92)

Parasite (par′ə sīt) An organism that absorbs food from a living organism and harms it (p. 34)

Pathogen (path′ə jən) A germ (p. 242)

Penis (pē′nis) The male organ that delivers sperm to the female body (p. 229)

Peripheral nervous system (pə rif′ər əl nėr′vəs sis′təm) The nerves that send messages between the central nervous system and other body parts (p. 136)

Peristalsis (per ə stòl′sis) The movement of digestive organs that pushes food through the digestive tract (p. 169)

Perspiration (pėr spə rā′shən) Liquid waste made of heat, water, and salt released through the skin (p. 184)

Petiole (pet′ē ōl) The stalk that attaches a leaf to a stem (p. 146)

P generation (jen ə rā′shən) The pure plants that Mendel produced by self-pollination (p. 266)

Phagocyte (fag′ə sīt) A white blood cell that surrounds and destroys pathogens (p. 243)

Pharynx (far′ingks) The passageway between the mouth and the esophagus for air and food (p. 181)

Phenotype (fē′nə tīp) An organism's appearance as a result of its combination of genes (p. 269)

Phloem (flō′em) The vascular tissue in plants that carries food from leaves to other parts of the plant (p. 145)

Photosynthesis (fō tō sin′thə sis) The process in which a plant makes food (p. 148)

Phototropism (fō to′trə piz əm) The response of a plant to light (p. 331)

Phylum (fī′ləm) Subdivision of a kingdom (plural is *phyla*) (p. 44)

Pigment (pig′mənt) A chemical that absorbs certain types of light (p. 149)

Pistil (pis′tl) The female organ of reproduction in a flower (p. 157)

Placenta (plə sen′tə) A tissue that provides the embryo with food and oxygen from its mother's body (p. 223)

Plague (plāg) An infectious disease that spreads quickly and kills many people (p. 242)

Plasma (plaz′mə) The liquid part of blood (p. 176)

Platelet (plāt′lit) A tiny piece of cell that helps form clots (p. 177)

Pollen (pol′ən) Tiny grains containing sperm (p. 157)

Pollination (pol ə nā′shən) The process by which pollen is transferred from the stamen to the pistil (p. 158)

Pollution (pə lü′shən) Anything added to the environment that is harmful to living things (p. 299)

Population (pop yə lā′shən) A group of organisms of the same species that live in the same area (p. 297)

Pregnancy (preg′nən sē) The development of a fertilized egg into a baby inside a female's body (p. 231)

Primate (prī′māt) The group of mammals that includes humans, apes, monkeys, and similar animals (p. 376)

Producer (prə dü′sər) An organism that makes its own food (p. 306)

Progesterone (prō jes′tə rōn) Female sex hormone (p. 230)

Property (prop′ər tē) A quality that describes an object (p. 26)

Prostate gland (pros′tāt gland) The gland that produces the fluid found in semen (p. 229)

Protein (prō′tēn) A chemical used by living things to build and repair body parts and regulate body activities (p. 14)

Protist (prō′tist) An organism that usually is one-celled and has plantlike or animal-like properties (p. 32)

Protozoan (prō tə zō′ən) A protist that has animal-like qualities (p. 32)

Pseudopod (sü′də pod) Part of some one-celled organisms that sticks out like a foot to move the cell along (p. 33)

Puberty (pyü′bər tē) The period of rapid growth and physical changes that occurs in males and females during early adolescence (p. 234)

Punnett square (pə′nət skwâr) A model used to represent crosses between organisms (p. 268)

Pupa (pyü′pə) A stage in the development of some insects that leads to the adult stage (p. 60)

Pupil (pyü′pəl) The black circle in the center of the iris (p. 192)

Pyramid of numbers (pir′ə mid ov num′bərz) A diagram that compares the sizes of populations at different levels of a food chain (p. 307)

R

Radial symmetry (rā′dē əl sim′ə trē) An arrangement of body parts that resembles the arrangement of spokes on a wheel (p. 56)

Radioactive mineral (rā dē ō ak′tiv min′ər əl) A mineral that gives off energy as it changes to another substance over time (p. 361)

Receptor cell (ri sep′tər sel) A cell that receives information about the environment and starts nerve impulses to send that information to the brain (p. 192)

Recessive gene (ri ses′iv jēn) A gene that is hidden by a dominant gene (p. 268)

Rectum (rek′təm) Lower part of the large intestine where feces are stored (p. 171)

Red marrow (red mar′ō) The spongy material in bones that makes blood cells (p. 200)

Reflex (rē′fleks) An automatic response (p. 328)

Renewable resources (ri nü′ə bəl ri sôrs′əz) Resources that are replaced by nature (p. 301)

Replicate (rep′lə kāt) To make a copy of (p. 282)

Reproduction (rē prə duk′shən) The process by which living things produce offspring (p. 19)

Reptile (rep′tīl) An egg-laying vertebrate that breathes with lungs (p. 53)

Resource (ri sôrs′) A thing that an organism uses to live (p. 301)

Respiration (res pə rā′shən) The process by which living things release energy from food (p. 17)

a	hat	e	let	ī	ice	ȯ	order	u̇	put	sh	she		ə	a	in about
ā	age	ē	equal	o	hot	oi	oil	ü	rule	th	thin			e	in taken
ä	far	ėr	term	ō	open	ou	out	ch	child	ᴛʜ	then			i	in pencil
â	care	i	it	ȯ	saw	u	cup	ng	long	zh	measure			o	in lemon
														u	in circus

Respire (ri spīr´) Take in oxygen and give off carbon dioxide (p. 123)
Retina (ret´n ə) The back part of the eye where light rays are formed (p. 193)
Rhizoid (rī´zoid) A tiny rootlike thread of a moss plant (p. 78)
Rhizome (rī´zōm) A plant part that has shoots above ground and roots below (p. 77)
Ribosome (rī´bə sōm) A protein builder of the cell (p. 6)
RNA A molecule that works together with DNA to make proteins (p. 8)
Roundworm (round´wėrm) A worm with a smooth, round body and pointed ends (p. 57)

S

Sanitation (san ə tā´shən) The practice of keeping things clean to prevent infectious diseases (p. 246)
Saprophyte (sap´rō fīt) An organism that decomposes dead organisms or waste matter (p. 87)
Scientific name (sī ən tif´ik nām) The name given to each species, consisting of its genus and its species label (p. 46)
Scientific theory (sī ən tif´ik thē´ər ē) A generally accepted and well-tested scientific explanation (p. 369)
Scrotum (skrō´təm) The sac that holds the testes (p. 228)
Secrete (si krēt´) Form and release, or give off (p. 118)
Sediment (sed´ə mənt) The bits of rock and mud that settle to the bottom of a body of water (p. 358)
Seed (sēd) A plant part that contains a beginning plant and stored food (p. 68)
Segmented worm (seg mən´təd wėrm) A worm whose body is divided into sections, such as earthworms or leeches (p. 57)
Selective breeding (si lek´tiv brē´ding) The process of breeding plants and animals so that certain traits repeatedly show up in future generations (p. 288)
Self-pollination (self´pol ə nā´shən) The movement of pollen from the male sex organs to the female sex organs of flowers on the same plant (p. 266)
Semen (sē´mən) A mixture of fluid and sperm cells (p. 229)

Sex chromosome (seks krō´mə sōm) A chromosome that determines the sex of an organism (p. 275)
Sex-linked trait (seks´ lingkt trāt) A trait that is determined by an organism's sex chromosomes (p. 276)
Sexual reproduction (sek´shü əl rē prə duk´shən) Reproduction that involves two parents, an egg, and sperm (p. 97)
Sickle-cell anemia (sik´əl sel ə nē´mē ə) A genetic disease in which a person's red blood cells have a sickle shape (p. 283)
Skeletal system (skel´ə təl sis´təm) The network of bones in the body (p. 199)
Solution (sə lü´shən) A mixture in which the particles are evenly mixed (p.12)
Sori (sôr´ī) Clusters of reproductive cells on the underside of a frond (p. 77)
Species (spē´shēz) A group of organisms that can breed with each other to produce offspring like themselves (p. 44)
Spontaneous generation (spon tā´nē əs jen ə rā´shən) The idea that living things can come from nonliving things (p. 210)
Spore (spôr) The reproductive cell of some organisms (p. 77)
Sporozoan (spôr ə zō´ən) A protozoan that is a parasite, and lives in blood; may cause malaria (p. 93)
Stamen (stā´mən) The male organ of reproduction in a flower that includes the anther and filament (p. 157)
Stigma (stig´mə) The upper part of the pistil, on the tip of the style (p. 157)
Stimulus (stim´yə ləs) Anything to which an organism reacts (p. 328)
Stoma (stō´mə) A small opening in a leaf that allows gases to enter and leave (plural is *stomata*) (p. 146)
Succession (sək sesh´ən) The process by which a community changes over time (p. 298)
Swim bladder (swim blad´ər) A gas-filled organ that allows a bony fish to move up and down in water (p. 52)
Synapse (sin´aps) A tiny gap between neurons (p. 189)

T

Taxonomy (tak son´ə mē) The science of classifying organisms based on the features they share (p. 31)

Tentacle (ten′tə kəl) An armlike body part in invertebrates that is used for capturing prey (p. 56)

Territorial behavior (ter ə tôr′ē əl bi hā′vyər) Behavior that claims and defends an area (p. 330)

Testis (tes′tis) The male sex organ that produces sperm cells (plural is *testes*) (p. 196)

Testosterone (te stos′tə rōn) Male sex hormone (p. 229)

Threatened (thret′nd) When there are fewer of a species of animal than there used to be (p. 300)

Tissue (tish′ü) A group of cells that are similar and work together (p. 3)

Toxin (tok′sən) A poison produced by bacteria or other organisms (p. 88)

Trachea (trā′kē ə) The tube that carries air to the bronchi (p. 181)

Trait (trāt) A characteristic of an organism (p. 211)

Trial-and-error learning (trī′əl and er′ər ler′ning) Learning in which an animal connects a behavior with a reward or a punishment (p. 336)

Trypanosome (tri pan′ə sōm) Protozoans that are parasites, and live in blood; may cause sleeping sickness (p. 93)

Tube foot (tüb fùt) A small structure used by echinoderms for movement (p. 61)

U

Umbilical cord (um bil′ə kəl kôrd) The cord that connects an embryo to the placenta (p. 231)

Ureter (yù rē′tər) A tube that carries urine from the kidney to the urinary bladder (p. 185)

Urethra (yù rē′thrə) The tube that carries urine out of the body (p. 185)

Urine (yùr′ən) Liquid waste formed in the kidneys (p. 185)

Uterus (yü′tər əs) An organ in most female mammals that holds and protects an embryo (p. 223)

V

Vaccine (vak sēn′) A material that causes the body to make antibodies against a specific pathogen before that pathogen enters the body (p. 244)

Vacuole (vak′yü ōl) Stores substances such as food, water, and waste products (p. 7)

Vagina (və jī′nə) The tubelike canal in the female body through which sperm enter the body (p. 229)

Vascular plant (vas′kyə lər plant) A plant that has tubelike cells (p. 68)

Vascular tissue (vas′kyə lər tish′ü) A group of plant cells that form tubes through which food and water move (p. 68)

Vein (vān) A blood vessel that carries blood back to the heart (p. 175)

Ventricle (ven′trə kəl) A heart chamber that pumps blood out of the heart (p. 126)

Vertebra (vėr′tə brə) One of the bones or blocks of cartilage that make up a backbone (p. 51)

Vertebrate (vėr′tə brit) An animal with a backbone (p. 51)

Vestigial structure (ve stij′ē əl struk′chər) A body part that appears to be useless to an organism but was probably useful to the organism's ancestors (p. 373)

Villi (vil′ī) Tiny fingerlike structures in the small intestine through which food molecules enter the blood (p. 171)

Virus (vī′rəs) A type of germ that is not living (p. 242)

Vitamin (vī′tə mən) A chemical found in foods that is needed by living things in small amounts (p. 15)

Voluntary muscle (vol′ən ter ē mus′əl) A muscle that a person can control (p. 203)

W

Warm-blooded (wôrm′ blud′ id) Having a body temperature that stays the same (p. 54)

X

Xylem (zī′lem) The vascular tissue in plants that carries water and minerals from roots to stems and leaves (p. 145)

Z

Zygote (zī′gōt) A fertilized cell (p. 156)

a	hat	e	let	ī	ice	ȯ	order	ù	put	sh	she	ə {	a in about, e in taken, i in pencil, o in lemon, u in circus
ā	age	ē	equal	o	hot	oi	oil	ü	rule	th	thin		
ä	far	ėr	term	ō	open	ou	out	ch	child	ŦH	then		
â	care	i	it	ò	saw	u	cup	ng	long	zh	measure		

Biology Glossary **409**

Index

A

Abiotic factors, 296
Achievements in Science
　Ancient Bacteria Discovered in Antarctic Ice, 94
　Aquariums Invented, 120
　Breeding Leads to New Plants, 290
　Chimpanzee Behavior Observed, 345
　Crop Rotation Recommended, 309
　First Ichthyosaur and Plesiosaur Discovered, 379
　Human Blood Groups Discovered, 178
　Human Digestion Observed, 172
　Mammal Egg Discovered, 219
　Microscopic Bacteria Observed, 28
　Pasteurization Discovered, 247
　Plant Cells Observed, 16
　Study of Invertebrates Begins, 62
　The Discovery of Photosynthesis, 151
　World's Oldest Flowering Plant Discovered, 74
Acid rain, 299, 302
Adaptations, 212
Adaptive advantage, 373
Adolescence, 234
Adrenal glands, 196, 197
Adrenaline, 197
Aflatoxins, 103
AIDS, 242, 243
Air pollution, 109, 299
Alcohol, pregnancy and, 236
Aldosterone, 196
Algae, 32, 90–91, 109
Alligators, 53
Alveoli, 181, 182
Alzheimer's disease, 14, 190, 271
Amino acids, 14, 15
Ammonia, 87, 131
Amoebas, 33, 92, 95, 272
Amphibians, 52–53, 124
Anal pore, 95
Angiosperms, 71, 72, 159
Animal kingdom, 31–32, 44
Animals, 31
　cells of, 6–8
　classifying, 41–62
　communication of, 341–344
　gas exchange in, 123
　getting and digesting food, 116–119
　growth and development of, 220–221
Annelids, 125
Anning, Mary, 379
Annual growth rings, 145
Anteater, 222
Antibiotics, 36, 88, 102
Antibodies, 176, 177
Anus, 119, 171
Aorta, 174, 175
Aphids, 116
Applied genetics, 288
Aqua-Lung, 127
Aquariums, 120
Arachnids, 58, 59
Archaeopteryx, 363
Aristotle, 45, 68
Arteries, 175, 176
Arthropods, 58–60, 125, 135
Asexual reproduction, 97
　advantages and disadvantages of, 214
　in fungi, 108
　in seedless plants, 156, 214–215, 216
　mitosis in, 215
Assassin bugs, 116
Association neurons, 189, 190
Athlete's foot, 33, 102
Atoms, 8
ATP, 8
Atria, 126, 174
Auditory nerve, 193
Australopithecus, 376, 378
Australopithecus afarensis, 377

B

Baby, 232
　birth of, 233
　premature, 232
Backbone, 51, 136
Bacteria, 2, 27, 34, 86–88, 89, 116, 242–243
　asexual reproduction of, 214
　harmful, 36, 88
　helpful, 36, 87–88
　life activities of, 87
　microscopic, 28
　properties of, 86
Bald eagle, 115
Ball-and-socket joint, 201
Bases, 281
Beaumont, William, 172
Bees, 60, 116
　behavior in, 330
Beetles, 117
Behavior
　innate, 328–333
　learned, 335–337
Biceps, 202
Bilateral symmetry, 57
Bile, 170
Bile duct, 170
Binary fission, 87
Bioacoustics, 344
Bioengineering, 225
Biologists, classification of animals by, 42–47
Biology, 31
Biomes, 300–302
Biometrics, 186
Biosphere, 302, 317
Biotic factors, 296
Birds, 43, 54
Black Death, 242
Blinking, 190
Blood, 173
　circulation of, 174
　parts of, 176–177
Blood-alcohol level, 260
Blood banks, 177
Blood pressure, 176, 249
Blood transfusions, 178
Blood types, 177
Blood vessels, 125, 173, 175, 176
Body odor identification, 186
Bog builders, 79
Bogs, 79
Bonds, chemical, 8
Bones, 136, 199–200
Bony fish, 52, 128, 130
Brain, 136, 187, 188, 189, 194
Brain stem, 188
Bread mold, 101
Breathing, 182
Breeding, selective, 288–289, 290
Bronchi, 181
Bronchial tubes, 181
Bronchioles, 181
Bubonic plague, 242
Budding, 108
Bulbs, 145
Burbank, Luther, 290
Butterflies, 116
B vitamins, 250

C

Calcium, 200, 249
Calories, 248, 249
Capillaries, 175, 177, 181
Carbohydrates, 13, 15, 90, 101, 118, 168, 169, 170, 248
Carbon, 87
Carbon cycle, 318–319
Carbon dioxide, 13, 101, 123, 124, 125, 126, 146, 148, 174, 175, 181, 183, 184
Cardiac muscle, 173, 203
Carnivores, 117, 307, 309
Carrier, 277
Cartilage, 51, 52, 200
Carver, George Washington, 309
Cast, 358

Caterpillars, 60
Cell differentiation, 220
Cell division, 215, 217–218, 272
Cell membranes, 6
Cells, 1, 2
    animal, 6–8, 9
    comparing plant and animal, 6–8
    functions of, 2, 8
    observing, 2–3
    plant, 6–8, 16
Cellular respiration, 152–153, 318, 319
Cell walls, 7
Centipedes, 58, 59
Central nervous system, 136, 187
Cerebellum, 188
Cerebrum, 137, 188
Chambers, 126
Channels, 341
Chemical bonds, 8
Chemical energy, 150, 313
Chemicals, importance of, for life, 12–15
Chemical signals, 341
Chicken pox, 243, 245
Chimpanzees, 337, 345
Chlorophyll, 149, 150, 313
Chloroplasts, 7, 8, 90, 99, 149
Cholera, 242
Chromosomes, 213, 215, 217, 265, 272–277
    sex, 275
Chyme, 170
Cicadas, 116
Cilia, 33, 129
Circulatory system, 125–126, 173–177, 181
    closed, 125
    open, 125
Clams, 58, 116, 124
Class, 44
Classification, 42
    levels of, 44–45
    of plants, 68–69
Climax community, 299
Closed circulatory systems, 125
Clot, 177
Club fungi, 100
Cnidarians, 56, 117, 118, 135

Coal, 79
Cocaine, 236
Cochlea, 193
Cold-blooded, 53
Color blindness, 284
Commensalism, 87
Common cold, 243
Common names, 46
Communication
    in animals, 341–344
Communities, 297, 315
Complete metamorphosis, 60
Complex carbohydrates, 13
Compound microscopes, 28
Computer models, 14
Computers in research, 48
Condensation, 318
Conditioning, 336–337
Conifers, 73, 159
Consumers, 306–307, 311, 319
Contractile vacuoles, 96
Coordinate, 133
Corals, 56
Cornea, 192
Cotyledons, 72
Coughing, 190
Courtship behavior, 329
Cousteau, Jacques, 127
Cowpox, 245
Crabs, 59
Crayfish, 59
Crickets, 60
Crocodiles, 53
Cro-Magnons, 379
Crop rotation, 309
Crop, 119
Crossbreeding, 290
Cross-pollination, 267
Crustaceans, 58, 59
Cycles
    carbon, 318
    nitrogen, 320
    oxygen, 319
    systems of, 321
    water, 317–318
Cyclosporine, 102
Cystic fibrosis, 271
Cytoplasm, 6

D

Darwin, Charles
    life and travels of, 369–370
    theories of, 370–373

Dead Sea, 86
Decompose, 34
Decomposers, 309, 319
Dermis, 184
Descent with modification, 371, 372
Development, 18, 220–221
Diabetes, 256, 282
Dialysis machine, 132
Diaphragm, 182
Diatoms, 91
Dicots, 72
Diet, importance of balanced, during pregnancy, 236
Dietetic technician, registered, 253
Diffusion, 123
Digestion, 17, 118, 168–171
Digestive enzymes, 107, 118, 119, 169
Digestive system, 196, 203
Digestive tracts, 119
Dinosaurs, 53, 300, 358, 359
Disease, role of body in fighting, 242–246
Diversity, 212, 214
DNA, 6, 8, 15, 211, 280–282, 364
    genetic engineering and, 160
    genetics and, 27, 213
    reproduction and, 211–212, 214, 215, 216, 220, 271, 272
    in viruses, 242
DNA replication, 353
Dolphins, 54
Dominant genes, 268, 270, 282
Dragonflies, 117
Drugs, 257
    abuse of, 257

E

Eardrum, 193
Ears, 188, 192
Earthworm, 57
Echinoderms, 61
Ecologist, 303
Ecology, 295–321
Ecosystems, 297, 328
    changes in, 298–299
    energy flows in, 312–315
    human impact on, 299–300
    material cycle through, 317–321
Egg, 156, 209, 216, 217, 218, 230, 231, 273, 279
Electron microscopes, 3
Elephant, 125
Embryo, 71, 209, 219, 220, 223, 230, 231, 232, 236
Embryology, 219
Emus, 54
Endangered species, 300
Endocrine glands, 196
Endocrine system, 133–134, 196–197
Endoplasmic reticulum, 7
Endospores, 88
Energy, 8
    chemical, 150, 313
    flow of, through ecosystems, 312–315
    in food, 313
Energy pyramid, 314–315
Environment
    heredity and, 280
    influence of, 280
Enzymes, 118, 196
    digestive, 107, 169
Epidermis, 184
Esophagus, 171, 203
    digestion in, 169
Estrogen, 230
Euglena gracilis, 90, 95
Euglenas, 33
Evaporation, 318
Evergreens, 73
Evolution, 351–379
    evidence of, 373
    fossils and, 362
    theory of, 369–374
Excretion, 17, 128, 130
Excretory system, 185
Exercise, 256–257, 259
External fertilization, 218
Extinctions, 361
Extinct species, 300
Eye color, 278
Eyes, 188, 192
Eyespot, 97

F

F_1 generation, 267, 276
F_2 generation, 269
Factors, 270
Fallopian tubes, 230, 231

Biology Index **411**

Family, 44
Fats, 13, 15, 118, 168, 170, 248
Fatty layer, 184
Feces, 171
Feedback loop, 197
Feeding, 116
Female reproductive system, 230
Ferns, 68, 77–78
Fertilization, 158, 218
  external, 218
  internal, 218
Fertilizers, 309
Fetus, 232, 236
Fight-or-flight response, 258
Filter feeding, 116
Fingerprint identification, 186
Fireflies, 343
Fish, 52, 54, 124, 130
Flagella, 33, 90, 93
Flame cells, 129
Flatworms, 57, 123, 129, 135
Fleas, 60
Florist, 147
Flowering plants, 71, 74
Flu shot, 244
Flying behavior, 329
Food
  digestion of, 168–171
  energy in, 313
  getting, 17
Food chain, 306–307, 315
  flow of energy through, 313–314
Food Guide Pyramid, 251, 252
Food vacuole, 95
Food webs, 308
Fossil fuels, 302, 318
Fossil record, 359
Fossils, 351
  dating, 361
  evolution and, 362
  humanlike, 376–379
  hunting for, 364
  living organisms and, 363
  types of, 358–359
Fraternal twins, 279
Frogs, 52, 124
Fronds, 77
Fruit, 71
Fungi, 31, 33–34, 99–103, 242–243
  harmful, 102–103
  helpful, 102
  living with other organisms, 108–109
  properties of, 99
  reproduction of, 108
  survival of, 107–109

G
Gagnan, Emile, 127
Gallbladder, 170
Gametes, 216–217, 273
Gas, properties of, 152
Gas exchange, 123–124
Gastrovascular cavity, 118
Gene pool, 283
Genes, 265, 268, 270, 281
  changes in, 353–354
  dominant, 268, 270, 282
  recessive, 268, 270, 282–283, 284
Genetically modified organism (GMO), 155, 225
Genetic counselor, 285
Genetic diseases, 271, 282
Genetic engineering, 155, 160, 289
Genetics, 265–289
  applied, 288
Genetics testing, 271
Genotype, 269
Genus, 44, 46, 68
Geobacter, 110
Geographic isolation, 355
Geologic time scale, 359, 360, 368
Germination, 158
Germs, 242
Gestation, 224
Gestation times, 224
Giant redwood trees, 71
Giardia lamblia, 93
Gills, 52, 100, 116, 124, 125
Ginkgo trees, 73
Gizzard, 119
Glands, 170, 196, 197
Glucose, 150, 152, 196, 197
Golgi bodies, 7
Goodall, Jane, 345
Gram, Hans Christian, 89
Gram Stain, 89
Grasshoppers, 60, 117
Gravitropism, 332
Gravity, plant responses to, 332
Great Salt Lake, 86
Groundwater, 317–318
  pollution of, 322
Growth, 18
Growth hormone, 196, 197
Guard cells, 154
Gullet, 95
Gymnosperms, 73
  sexual reproduction in, 159

H
Habitat, 297, 300, 328
Habits, 256
  healthy, 256–258
Hagfish, 52
Half-life, 361
Hearing, 193
Heart, 125, 126, 167, 173, 174, 199
Hemoglobin, 176
Hemophilia, 283, 284
Hepatitis B, 103
Herbivores, 117, 307, 309
Heredity, 266–270
  environment and, 280
  study of, in humans, 279–284
High blood pressure, 256
Hinge joint, 201
Homeostasis, 18
Hominids, 376–377
Homo erectus, 378, 379
Homo habilis, 378
Homologous structures, 374
Homo sapiens, 376–377, 379
Honey mushroom, 100
Hooke, Robert, 16
Hookworms, 57
Hormones, 133–134, 196–197
  stress and, 197
Horse, evolution of, 362
Horseflies, 116
Human body systems, 167–203
  circulatory, 173–177
  digestive, 168–171
  endocrine, 196–197
  excretory, 184–185
  muscular, 201–205
  nervous, 187–190
  respiratory, 181–182
  sense organs, 192–194
  skeletal, 199–201
Human Genome Project (HGP), 271
Human language, 344
Humans
  impact on ecosystems, 299–300
  study of heredity in, 279–284
Human sex-linked traits, 284
Hummingbirds, 116
Humus, 79
Hydras, 56, 117, 118, 135
Hydrochloric acid, 169
Hyphae, 99, 107
Hypothesis, 370

I
Ichthyosaur, 379
Identical twins, 279
Immune system, 243
Immunity, 244
Imprinting, 335
Impulses, 134, 189, 193, 194
Inbreeding, 283
Incomplete metamorphosis, 60
Infectious disease, 242
Influenza, 244
Infrasounds, 344
Innate behavior, 328–333
Insects, 58, 60, 116
Insight, 337
Instinct, 328
Insulin, 196
Interactions, 296
Internal fertilization, 218
Intestines, 203
Invertebrates, 56–61
  nervous systems in, 135
Investigations
  Classifying Objects, 49–50
  Comparing Cells, 10–11
  Graphing Gestation Times, 226–227
  Growing an African Violet from a Leaf, 161-162
  Growing Bread Mold, 105–106
  How Does Exercise Change Heart Rate?, 179–180

Identifying Angiosperms and Gymnosperms, 75–76
Living or Nonliving?, 29–30
Making Molds, 366
Observing Learning Patterns, 339–340
Reading Food Labels, 254–255
Studying Feeding in Hydras, 121–122
Testing the pH of Rain, 304–305
Tracing a Genetic Disease, 286–287
Involuntary muscles, 203
Iris, 192

J
Jawless fish, 52
Jellyfish, 56, 117
Jenner, Edward, 245
Joints, 200–201

K
Kangaroos, 222
Kelps, 91
Kidneys, 130, 131, 132, 176, 185, 196
Kingdom, 31, 44
Koalas, 222

L
Lampreys, 52
Landsteiner, Karl, 178
Language, human, 344
Large intestine
  digestion in, 171
Larynx, 181
Leakey, Mary and Louis, 378
Learned behavior, 335–337
Learning
  conditioning, 336–337
  insight, 337
  observational, 335
  trial-and-error, 336
Leaves
  function of, 146
  parts of, 146
Leeches, 57, 116
Leeuwenhoek, Antonie van, 28
Lens, 193
Lethal mutation, 354
Lichen, 109
Life, origin of, 210–212
Life activities
  basic, 17–19
Ligaments, 200
Light, plant responses to, 330, 333
Linnaeus, Carolus, 68
Liver, 170
Living organisms, fossils and, 363
Living things
  organization of, 296–297
  properties of, 27
Lizards, 53
Lobsters, 59, 124
Lorenz, Karl, 335
Lucy, 377
Lungs, 124, 125, 167, 176, 181, 199
Lymphocytes, 244
Lysosomes, 7

M
Mad cow disease, 14
Magnesium, 200
Magnifying glasses, 2–3
Malaria, 93
Male reproductive system, 228–229
Mammal egg, 219
Mammals, 54, 131, 222
  food for the young, 223
  number of species, 223
  offspring, 222
Mammary glands, 54, 223, 232
Marsupials, 222, 223
Mass extinctions, 361
Measles, 245
Medical technology, 171
Meiosis, 217–218, 274
Mendel, Gregor, studies of, 266–270
Menstruation, 230
Metals, 110
Metamorphosis, 52
  complete, 60, 221
  incomplete, 60, 221
Methane, 87
Microbiologists, 5, 36
Microorganisms, 32
Microscope, 2–3, 16, 70
  compound, 28
  electron, 3
Microscopic bacteria, 28
Millipedes, 58, 59
Mimosa, 332
Minerals, 15, 248, 249
  radioactive, 361
Mites, 59
Mitochondria, 6, 7
Mitosis, 215, 272
Mockingbirds, 330
Molds, 101, 102, 358
  making, 366–367
Molecules, 8
Mollusks, 58, 116, 125
Molting, 58
Monera, 31, 34, 86
Monocots, 72
Monkeys, 72
Morels, 102
Morgan, Thomas, 276
Morning glories, 333
Mosquitoes, 60, 116
Mosses, 68, 77, 78, 79
Motor neurons, 189, 190
Mouth, digestion in, 169
Movement, 17–18
Mules, 355
Mumps, 244
Muscles, 167, 200, 259
  cardiac, 173
Muscle tissue, kinds of, 203
Muscular dystrophy, 284
Muscular system, 201–205
Mushrooms, 100, 102, 103, 109
Mutations, 280
  effects of, over time, 354
  lethal, 354
  useful, 288
Mutualism, 87, 108, 109
Mycelia, 99, 101
Mycologist, 104
Mycorrhiza, 108

N
Natural selection, 371, 373
Neanderthals, 379
Nectar, 158
Nerve net, 135
Nerves, 187
  spinal, 189
Nervous system, 134, 187–190
  invertebrate, 135
  vertebrate, 136
Nest-building behavior, 328–329
Neurons
  association, 189, 190
  motor, 189, 190
  sensory, 189, 190, 192
Neurotransmitters, 134, 190
Niacin, 250
Nitrogen, 87
  adding, to soil, 322
Nitrogen cycle, 320
Nitrogen fixation, 320
Nonflowering plants, 71
Nonliving things,
  properties of, 26–27
Nonrenewable resources, 301–302
Nonvascular plants, 69
Nose, 192
Nucleus, 6, 15, 215–216
Nurse, 204
Nutrients, 15, 248–249, 313
  importance of, 15
Nutrition, 248–252
Nymphs, 221

O
Observational learning, 335
Obstetrician/gynecologist, 235
Octopuses, 58, 135
Omnivores, 307
Open circulatory systems, 125
Opossums, 222
Optic nerve, 193
Order, 44, 45
Organelles, 3, 86
Organisms, 27
  behavior of, 327–344
  changes in, 352–353
  classification of, 31–34
Organs, 4, 27
Osmosis, 96, 128
Osteoporosis, 200
Ostriches, 54
Ovary, 157, 158, 216, 230
Ovulation, 230
Oxygen, 123, 124, 146, 174, 181, 183
  importance of, 152–154
  producing, 153
  releasing, 154
Oxygen cycle, 319
Oysters, 58, 116

P

Paleontologists, 359, 363, 364
Pancreas, 170, 196
Paramecium, 92, 95
Parasites, 34, 57, 93
Parental care, 234
Pasteur, Louis, 247
Pasteurization, 36, 247
Pathogens, 242–243, 246
Pavlov, Ivan, 337
Peat, 79
Pelvis, 199
Penguins, 54
Penicillin, 88, 102
Penis, 229, 231
Peripheral nervous system, 136, 187
Peristalsis, 169, 170
Perspiration, 184
Petiole, 146
P generation, 266
Phagocyte, 243
Pharynx, 169, 181
Phenotype, 269
Phloem, 145
Phosphorus, 200, 249
Photosynthesis, 148–150, 153, 313, 318, 319, 321
Phototropism, 331
Phylum, 44, 45
Pigment, 149
Pill bugs, 59
Pistil, 157
Pituitary gland, 196
Placenta, 223, 231, 233, 236
Plague, 242
Plant kingdom, 31
Plants, 31
    cells in, 6–8, 16
    classification of, 68–69
    flowering, 74
    reproduction of, 156–159
    responses of, 330–333
    seed, 71–73, 77
    seedless, 77–79
    vascular, 77, 144–146
Plasma, 176
Platelets, 177, 258
Platypus, 222
Plesiosaur, 379
Polio, 242, 245
Pollen, 157, 158, 159
Pollination, 158
Pollution, 299–300
Populations, 297, 321, 352–353
    changes in environment, 355–356
Porpoises, 54
Potassium, 249
Potato, 145
Praying mantises, 117
Pregnancy, 231
    alcohol and, 236
    drugs and, 236
    importance of balanced diet during, 236
    smoking and, 236
Premature babies, 232
Preparator, 365
Primates, 376
Producers, 306, 319
Progesterone, 230
Properties, 26
    of living things, 27
    of nonliving things, 26–27
Prostate gland, 229
Proteins, 14, 15, 118, 168, 248
Protists, 31, 32–33, 90–93, 116, 242–243
    survival of, 95–97
Protozoans, 32, 92–93
Pseudopods, 33, 92
Puberty, 229, 234
Pulmonary artery, 181
Pulse, 173
Punnett square, 268, 269, 275, 277
Pupa, 60
Pupil, 192
Pyramid of numbers, 307

R

Racehorses, breeding, 289
Radial symmetry, 56
Radioactive minerals, 361
Rain
    acid, 299, 302
    pH of, 304–305
Rays, 52
Receptor cells, 192, 194
Recessive genes, 268, 270, 282–283, 284
Rectum, 171
Recycling, 302
Red blood cells, 176–177
Redi, Francesco, 210
Red marrow, 200
Red tide, 98
Reflex, 190, 328
Renewable resources, 301–302
Reproduction, 19, 97
    asexual, 97, 108, 156, 214–215, 216
    DNA and, 211
    of plants, 156–159
    sexual, 97, 108, 157–158, 215–218, 270, 273–277
Reproductive system
    female, 230
    male, 228–229
Reptiles, 53
Resources, 301
    nonrenewable, 301–302
    renewable, 301–302
Respiration, 17, 123, 152–153, 181–182
Respiratory system, 181–182
Retina, 193
Rhizoids, 78
Rhizomes, 77, 145
Rib cage, 199
Riboflavin, 250
Ribosomes, 6, 15
Ringworm, 102
RNA, 8, 15, 27, 242
Root hairs, 145
Roots
    function of, 144
    parts of, 145
Roundworms, 57
Rusts, 100

S

Salamanders, 52
Saliva, 169
Salivary glands, 169
Salmon, 336
Sand dollars, 61
Sanitation, 246
Saprophytes, 87
Satellites, 98
Scallops, 58
Science at Work
    dog trainer, 338
    ecologist, 303
    florist, 147
    genetic counselor, 285
    microbiologist, 5
    mycologist, 104
    nurse, 204
    obstetrician/gynecologist, 235
    preparator, 365
    registered dietetic technician, 253
    taxonomist, 35
    tree technician, 80
    veterinary assistant, 138
    zookeeper, 55
Science in Your Life
    Are bacteria helpful or harmful?, 36
    Can a fungus save your life?, 103
    Can you identify life activities?, 20
    Can you live without kidneys?, 132
    Has everything been classified?, 47
    How do dog guides and their handlers communicate?, 346
    How do ferns and mosses provide energy?, 79
    How long is 4.6 billion years?, 368
    What can smart drugs do?, 190
    What is DNA fingerprinting?, 284
    What is drug testing?, 260
    What kind of consumer are you?, 311
    What substances are harmful during pregnancies?, 236
    What's that plant?, 155
Schweitzer, Mary, 364
Scientific names, 46–47, 68
Scientific theory, 369
Scorpions, 59
Scrotum, 228
Sea cucumbers, 61
Sea lettuce, 91
Sea stars, 61
Sea turtle, 129
Sea urchins, 61
Seaweed, 91
Secreting, 118
Sediment, 358
Seedless plants, 77–79
    reproduction in, 156

Seed plants, 71–73, 77
   reproduction in, 157
Seeds, 68, 71, 78, 158
Segmented worms, 57, 135
Selective breeding, 288–289, 290
Self-pollination, 266
Semen, 229
Sense organs, 188, 192–194
Senses, 18
Sensory neurons, 189, 190, 192
Serving sizes, 252
Sex cells, 216
Sex chromosomes, 275
Sex-linked traits, 276–277
   human, 284
Sexual reproduction, 97, 108, 215–218, 270, 273–277
   in angiosperms, 157–158
   disadvantages and advantages of, 216
   in gymnosperms, 159
Sharks, 52
Shrew, 125
Sickle-cell anemia, 283
Sight, 192–193
Silica, 91
Simple carbohydrates, 13
Skates, 52
Skeletal muscles, 203
Skeletal system, 199–201
Skin, 192, 194
   in fighting disease, 243
Skinner, B. F., 336
Skull, 51, 199
Sleeping sickness, 93
Sleep movements, 333
Slugs, 58
Small intestine, 171
   digestion in, 170
Smallpox, 242, 243, 245
Smell, 194
Smoking, pregnancy and, 236
Smooth muscles, 169, 203
Smuts, 100
Snails, 58
Snakes, 53
Sneezing, 190
Sodium, 249
Sodium chloride, 12
Soil, adding nitrogen to, 322

Solution, 12
Sori, 77
Sound signals, 344
Sow bugs, 59
Spacecraft, growing plants in, 333
Species, 44, 45, 46, 352
Sperm, 156, 209, 216, 217, 218, 228, 229, 230, 231, 273, 279
Sphagnum moss, 79
Spiders, 59, 116
   behavior in, 330
Spinal cord, 136, 187, 189, 199
Spinal nerves, 189
Sponges, 56, 116, 118
Spontaneous generation, 210–211
Spores, 77–78, 99, 156
Sporozoans, 93
Squids, 58, 135
Stamens, 157
Starches, 13
Stems, 145
Stigma, 157
Stimulus, 328
St. Martin, Alexis, 172
Stoma, 146
Stomach, 171, 203
   digestion in, 169
Stomata, 146, 148, 154
Stress
   coping with, 258
   hormones and, 197
Succession, 298
Sugar, 13
Sulfur dioxide, 299
Sun
   energy from, 312–315
   and photosynthesis, 148–150
   and phototropism, 331
Sweat glands, 184
Swim bladder, 52
Symmetry
   bilateral, 57
   radial, 56
Synapse, 189

T

Tadpoles, 124
Tapeworms, 57
Taste, 194
Taste buds, 194

Taxonomist, 35
Taxonomy, 31, 43
Technology, medical, 171
Tendons, 201, 202
Tentacles, 56
Termites, 117
Territorial behavior, 330
Testes, 196, 216, 217
Testosterone, 229
Theophrastus, 68
Thiamin, 250
Threatened species, 300
Ticks, 59
Tissues, 3–4
Toads, 52
Tongue, 192
Tortoises, 53
Touch, 194
   plant responses to, 332–333
Toxins, 88, 103
Trachea, 181, 200
Traits, 160, 211, 212, 213, 278
   sex-linked, 276–277
Transfusion, 177
Tree technician, 80
Trial-and-error learning, 336
Truffles, 102
Trypanosomes, 93
Tsetse flies, 93
Tube feet, 61
Tubers, 145
Turtles, 53
Twins, 279

U

Ultrasounds, 344
Umbilical cords, 231, 233
Uranium-238, 361
Ureters, 185
Urethra, 185, 229
Urinary bladder, 185
Urine, 185
Uterus, 223, 231, 233

V

Vaccines, 244–245
Vacuoles, 6–7
Vagina, 229, 230, 233
Vascular plants, 68–69, 77
Vascular system, 146
   in plants, 144–146
Vascular tissue, 68–69

Vasopressin, 133
Veins, 175
Ventricle, 126, 174
Venus's flytrap, 332
Vertebra, 51, 199
Vertebrates, 51–54, 117
   comparing brains of, 137
   features of, 51
   nervous system in, 136
Vestigial structures, 373
Veterinary assistant, 138
Victoria (Queen), 283
Villepreux-Power, Jeanne, 120
Villi, 170–171, 176
Viruses, 31, 242–243
Visual signals, 342–343
Vitamins, 15, 36, 248, 249, 250
Voice identification, 186
Voluntary muscles, 203
Von Baer, Karl Ernst, 219

W

Wallace, Alfred, 369, 370
Warm-blooded, 54
Wastes, excreting, 131
Water, 248, 249
   importance of, 12
Water balance, 128–131
Water cycle, 317–318
Water vapor, 318
Whales, 54, 116
White blood cells, 177, 243
Worms, segmented, 57

X

X chromosomes, 275, 276, 284
X ray, 171
Xylem, 145, 148

Y

Y chromosomes, 275, 276
Yeasts, 101, 102, 108, 214

Z

Zookeeper, 55
Zygote, 156, 218, 220, 231, 273, 279

Photo and Illustration Credits

cover: © Cosmo Condina/Getty Images; p. xvi, © Jeff Hunter/The Image Bank/Getty Images; p. 3, © Antonio Mo/Taxi/Getty Images; p. 5, © Michael Rosenfeld/Stone/Getty Images; p. 13, © Rosemary Weller/Stone/Getty Images; p. 17, © Ron Kimball/Ron Kimball Studios; p. 19 (top), © Stephen Dalton/Photo Researchers, Inc.; p. 19 (middle), © Gary Meszaros/Visuals Unlimited; p. 19 (bottom), © Suzanne L. Collins & Joseph T. Collins/Photo Researchers, Inc.; p. 24, © Kevin Schafer/The Image Bank/Getty Images; p. 26, © Art Wolfe/Stone/Getty Images; p. 31, © Will Troyer/Visuals Unlimited; p. 34, © Michael Abbey/Photo Researchers, Inc.; p. 35, ©, Visuals Unlimited; p. 36, © Andrew Brookes/Corbis; p. 40, © Hank Morgan/Photo Researchers, Inc.; p. 42, © SuperStock; p. 43 (left), © Maslowski/Visuals Unlimited; p. 43 (center), © Breck P. Kent/Animals Animals; p. 43 (right), © Joe McDonald/Animals Animals; p. 46, © Darren Bennett/Animals Animals; p. 52, © J. M. Labat Jacana/Photo Researchers, Inc.; p. 53 (left), © Stuart Westmorland/Stone/Getty Images; p. 53 (right), © Art Wolfe/Stone/Getty Images; p. 54, © Will Troyer/Visuals Unlimited; p. 55, © Michael Heron/Woodfin Camp & Associates ; p. 56, © Andrew Martinez/Photo Researchers, Inc.; p. 58 (top), © Bob Cranston/Animals Animals; p. 58 (bottom), © A. Rider/Photo Researchers, Inc.; p. 59, © James H. Robinson/Animals Animals; p. 61, © Scott Smith/Animals Animals; p. 66, © Hans Strand/Corbis; p. 69, © Biophoto Associates/Science Source/Photo Researchers, Inc.; p. 71, © Alan & Linda Detrick/Photo Researchers, Inc.; p. 73 (left), © Saxon Holt/Saxon Holt Photography; p. 73 (right), © Saxon Holt/Saxon Holt Photography; p. 78, © Dave Schiefelbein/Stone/Getty Images; p. 80, © Bob Thomas/Stone/Getty Images; p. 84, © Jeffrey Lepore/Photo Researchers, Inc.; p. 86 (left), © Biophoto Associates/Photo Researchers, Inc.; p. 86 (middle), © Eric V. Grave/Photo Researchers, Inc.; p. 86 (right), © CNRI/Science Photo Library/Photo Researchers, Inc; p. 87, © Hugh Spencer/Photo Researchers, Inc.; p. 91, © Lawrence Naylor/Photo Researchers, Inc.; p. 92, © Andrew Syred/ Science Photo Library/Photo Researchers, Inc.; p. 93, © Oliver Meckers/Photo Researchers, Inc.; p. 100 (top), © Jewel Craig/Photo Researchers, Inc.; p. 100 (bottom), © Astrid & Hanns-Frieder Michler/Science Photo Library/Photo Researchers, Inc.; p. 101, © Bill Aron/PhotoEdit; p. 102, © Michael P. Gadomski/Photo Researchers, Inc.; p. 103, © Phil A. Dotson/Photo Researchers, Inc.; p. 104, © Dan Lamont/Corbis; p. 108, © Ken Eward/Photo Researchers, Inc.; p. 109, © Brock May/Photo Researchers, Inc.; p. 114, © Gary Vestal/Stone/Getty Images; p. 124, © Breck P. Kent/Animals Animals; p. 132, © Morgan/Photo Researchers, Inc.; p. 138, © Kaz Mori/The Image Bank/Getty Images; p. 142, © Steve Satushek/The Image Bank/Getty Images; p. 144, © Michael P. Gadomski/Photo Researchers, Inc.; p. 145, © Stephen J Krasemann/Photo Researchers, Inc.; p. 146, © Jerome Wexler/Photo Researchers, Inc.; p. 147, © Tony Freeman/PhotoEdit; p. 149, © Dr Kari Lounatmaa/Science Photo Library/PhotoResearchers, Inc.; p. 154, © John D. Cunningham/Visuals Unlimited; p. 156, © John D. Cunningham/Visuals Unlimited; p. 159, © Ken M. Highfill/Photo Researchers, Inc.; p. 166, © Aldo Torelli/Stone/Getty Images; p. 170, © Meckes/Ottowa/Science Photo Library/Photo Researchers, Inc.; p. 173, © Jeff Greenberg/Science Photo Library/Photo Researchers, Inc.; p. 177, © National Cancer Institute/Photo Researchers, Inc.; p. 204, © Spencer Grant/PhotoEdit; p. 208, © Dr G. Moscoso/Photo Researchers, Inc.; p. 212, © Bill Bachmann/PhotoEdit; p. 222, © Stuart Westmorland/Stone/Getty Images; p. 223, © Bonnie Sue/Photo Researchers, Inc.; p. 235, © A. Ramey/Unicorn Stock Photos; p. 240, © David Young-Wolff/PhotoEdit; P. 245, © Robert Brenner/PhotoEdit; p. 248 (left), © Felicia Martinez/PhotoEdit; p. 248 (center), © Steven Needham/Envision; p. 248 (right), © Richard Hutchings/PhotoEdit; p. 253, © Bill Aron/PhotoEdit; p. 256, © Robert Brenner/PhotoEdit; p. 264, © David Sieren/Visuals Unlimited; p. 272, © David M. Phillips/Visuals Unlimited; p. 279 (left), © Eastcott/Momatiuk/Photo Researchers, Inc.; p. 279 (right), © Tim Davis/Photo Researchers, Inc.; p. 283 (left), © Dr. Tony Brian/Science Photo Library/Photo Researchers, Inc.; p. 283 (right), © Stan Flegler/Visuals Unlimited; p. 285, © LWA-Dann Tardif/Corbis; p. 294, © Craig Aurness/Corbis; p. 301 (top), © Michael Townsend/Stone/Getty Images; p. 301 (bottom), © Tom Bean/Stone/Getty Images; p. 303, © Mark Burnett/Photo Researchers, Inc.; p. 312, © James Darell/Stone/Getty Images; p. 326, © Sanford/Agliolo/Corbis; p. 328, © Rob & Ann Simpson/Visuals Unlimited; p. 331, © David Newman/Visuals Unlimited; p. 332, © Cabisco/Visuals Unlimited; p. 336, © Benelux Press B.V./Photo Researchers, Inc.; p. 337, © Bernd Heinrich; p. 338, © Dale C. Spartas/Corbis; p. 350, © Archivo Iconografico, S.A./Corbis; p. 352, © Daniel J. Cox/Getty Images News Services p. 355 (both), © Tom & Pat Leeson; p. 358, © A.J. Copley/Visuals Unlimited; p. 363, © Tom McHugh/Photo Researchers, Inc.; p. 365, © Rich Frishman/Stone/Getty Images; p. 372 (left), © Tim Davis/Photo Researchers, Inc.; p. 372 (right), © Rob Simpson/Visuals Unlimited; p. 377, © John Reader/Science Photo Library/Photo Researchers, Inc.; p. 378, © John Reader/Science Photo Library/Photo Researchers, Inc. illustrations: John Edwards Illustration

Midterm Mastery Test

Midterm Mastery Test — Page 1

Part A Write the letter of the answer that correctly completes each sentence.

___ 1. ___ are the basic units of life.
A Organelles B Cells C Tissues D Proteins

___ 2. The life activity that gets rid of wastes is called ___.
A digestion B respiration C excretion D reproduction

___ 3. By observing the ___ of something, you can tell if it is living or nonliving.
A properties B cells C classes D organisms

___ 4. The lowest level in the classification system is ___.
A family B genus C order D species

___ 5. ___ are flowering plants.
A Angiosperms B Conifers C Gymnosperms D Ferns

___ 6. Mushrooms and yeast are examples of ___.
A toxins B organelles C fungi D algae

___ 7. ___ is the movement of water in cells from an area of high concentration to an area of low concentration.
A Osmosis B Diffusion C Vaccination D Respiration

___ 8. The ___ is the control center for the vertebrate nervous system.
A spinal cord B cerebrum C brain D nerve net

___ 9. In plants, ___ occurs when the pollen inside the tube meets the egg.
A pollination B photosynthesis C metamorphosis D fertilization

___ 10. ___ carries water and minerals from roots to other parts of a plant.
A Xylem B Phloem C Stoma D Pigment

Midterm Mastery Test — Page 2

Part B Fill in each blank with the correct word or phrase from the Word Bank.

Word Bank

cell membrane	kingdoms	sensing and responding	solution
chlorophyll	protist	sexual	vascular
circulatory	scientific name		

11. Water is an important chemical that dissolves other chemicals into a _____.

12. When plants or animals are _____, they pick up signals and then change or move.

13. There are five _____ of living things.

14. The two-word name of a species is its _____.

15. _____ plants have tubelike cells.

16. Algae belong to the _____ kingdom.

17. The _____ is the thin layer that surrounds and holds together a cell.

18. In animals, the _____ system transports oxygen throughout the body.

19. Plants with seeds usually produce offspring by _____ reproduction.

20. _____ is the green pigment in plants that captures light energy for photosynthesis.

Midterm Mastery Test — Page 3

Part C Match the terms in Column A with the descriptions in Column B. Write the letter of the correct answer on the line.

Column A		Column B
___ 21. tissues	A	invertebrates with jointed legs
___ 22. nutrients	B	bacteria
___ 23. decompose	C	contain beginning plants and stored food
___ 24. monerans	D	speed up chemical changes
___ 25. vertebrates	E	groups of similar cells that work together
___ 26. arthropods	F	plants that do not have tubelike structures
___ 27. seeds	G	animals that eat plants
___ 28. nonvascular plant	H	animals with backbones
___ 29. mutualism	I	relationship benefiting one organism and not affecting the other
___ 30. commensalism	J	all the chemicals living things need
___ 31. filter feeding	K	process in which plants make food
___ 32. enzymes	L	allow gases to enter and leave a leaf
___ 33. herbivores	M	to break down into simpler substances
___ 34. stomata	N	straining food out of water
___ 35. photosynthesis	O	two organisms living together and helping each other

Midterm Mastery Test — Page 4

Part D Answer the questions.

36. If you were looking at cells, how could you tell whether they were plant or animal cells?

37. How do monerans differ from one-celled protists?

38. Name two characteristics that can be used to identify an animal as a mammal.

39. Why are gymnosperms more like angiosperms than like ferns?

40. How do plants produce oxygen? How is the oxygen used?

Final Mastery Test

Final Mastery Test Page 1

Part A Write the letter of the answer that correctly completes each sentence.

____ 1. The life activity that releases energy from chemicals is called ____.
  A digestion   B respiration   C excretion   D reproduction

____ 2. Mushrooms and yeast are examples of ____.
  A toxins   B organelles   C fungi   D algae

____ 3. ____ is the movement of materials from an area of high concentration to an area of low concentration.
  A Osmosis   B Diffusion   C Vaccination   D Respiration

____ 4. ____ carries water and minerals from roots to other parts of a plant.
  A Xylem   B Phloem   C Stoma   D Pigment

____ 5. The ____ is the control center for the vertebrate nervous system.
  A spinal cord   B cerebrum   C brain   D nerve net

____ 6. The central nervous system has two main parts, the brain and the ____.
  A heart   B lungs   C synapses   D spinal cord

____ 7. The ____ is a part of the respiratory system.
  A brain   B trachea   C large intestine   D heart

____ 8. ____ is the joining of an egg cell and a sperm cell.
  A Adaptation   B Pollination   C Fertilization   D Reproduction

____ 9. Infectious diseases are caused by ____.
  A pathogens   B vaccines   C antibodies   D minerals

____ 10. ____ is the division of the nucleus into two new nuclei.
  A Meiosis   B Mitosis   C Heredity   D Inbreeding

Final Mastery Test Page 2

Final Mastery Test, continued

____ 11. Human ____ is the study of how humans inherit chromosomes.
  A biology   B breeding   C genetics   D engineering

____ 12. Organisms that make their own food are ____.
  A consumers   B producers   C condensers   D decomposers

____ 13. ____ is the response of plants to light.
  A Phototropism   B Imprinting   C Instinct   D Thigmotropism

____ 14. Humans and humanlike relatives from the past belong to the ____ group.
  A hominid   B rhizoid   C moneran   D vestigial

____ 15. Amino acids combine to make up ____.
  A carbohydrates   B fats   C minerals   D proteins

Part B Match the names in Column A with the descriptions in Column B. Write the letter of the correct answer on the line.

Column A
____ 16. Charles Darwin
____ 17. Ivan Pavlov
____ 18. Gregor Mendel
____ 19. Edward Jenner
____ 20. Carolus Linnaeus

Column B
A studied digestion and feeding behavior in dogs
B wrote *The Origin of Species*
C developed the system for the scientific naming of organisms that is used today
D developed the first vaccine, which was for smallpox
E made discoveries about heredity by studying pea plants

Final Mastery Test Page 3

Final Mastery Test, continued

Part C Match the terms in Column A with the descriptions in Column B. Write the letter of the correct answer on the line.

Column A
____ 21. nutrients
____ 22. photosynthesis
____ 23. artery
____ 24. tissues
____ 25. decompose
____ 26. arthropods
____ 27. phloem
____ 28. herbivores
____ 29. respiratory
____ 30. endocrine
____ 31. digestive
____ 32. adaptations
____ 33. chromosomes
____ 34. vaccines
____ 35. vitamins
____ 36. genes
____ 37. inbreeding
____ 38. biome
____ 39. community
____ 40. natural selection

Column B
A nutrients that found in small amounts in foods
B carries food from leaves to other parts of a plant
C groups of similar cells that work together
D allow organisms to live in a certain environment
E system of which glands are a part
F process in which plants make food
G animals that eat plants
H can be recessive or dominant
I blood vessel that carries blood away from the heart
J to break down into simpler substances
K an ecosystem found over a large geographic area
L contain DNA
M sexual reproduction between organisms within a small gene pool
N explains how evolution occurs
O cause the body to make antibodies against a specific pathogen
P all the chemicals living things need
Q system of which the lungs are a part
R invertebrates with jointed legs
S different populations living in one place
T system of which the small intestine is a part

Final Mastery Test Page 4

Final Mastery Test, continued

Part D Fill in each blank with the correct word or phrase from the Word Bank.

Word Bank

capillaries	genetic	motor	scientific
circulatory	homologous structures	radioactive	spontaneous
decomposers	kingdoms	reproduce	generation
food chain	membrane	responding	solution

41. The two-word name of a species is its _____ name.

42. Water dissolves other chemicals into a _____.

43. In animals, the _____ system transports oxygen throughout the body.

44. Sensing and _____ involves plants and animals picking up signals and then changing or moving.

45. There are five _____ of living things.

46. The cell _____ surrounds a cell and holds it together.

47. The _____ neurons carry impulses from the brain and spinal cord to muscles and glands.

48. Walls of the _____ are only one cell thick.

49. To survive, living organisms must _____.

50. _____ is the idea that living things can come from nonliving things.

Final Mastery Test

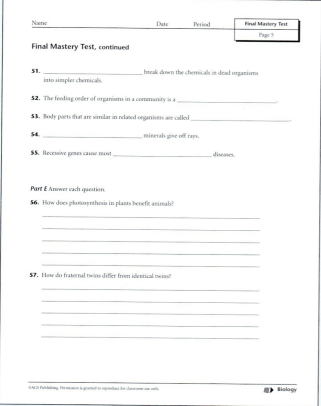

Final Mastery Test Page 5

Final Mastery Test Page 6

The lists below show how items from the Midterm and Final correlate to the chapters in the student edition.

Midterm Mastery Test

Chapter 1: 1–2, 11–12, 17, 21–22, 36

Chapter 2: 3, 13, 23–24, 37

Chapter 3: 4, 14, 25–26, 38

Chapter 4: 5, 15, 27–28, 39

Chapter 5: 6-7, 16, 29–30

Chapter 6: 8, 18, 31–33

Chapter 7: 9–10, 19–20, 34–35, 40

Final Mastery Test

Chapter 1: 1, 15, 21, 24, 42, 44, 46

Chapter 2: 2, 25, 45

Chapter 3: 26, 41

Chapter 4: 20

Chapter 6: 3, 5–6, 28, 43

Chapter 7: 4, 22, 27

Chapter 8: 7, 23, 29–31, 47–48, 60

Chapter 9: 8, 10, 33, 49–50

Chapter 10: 9, 19, 34–35, 59

Chapter 11: 11, 18, 36–37, 55, 57

Chapter 12: 12, 38–39, 51–52, 56, 58

Chapter 13: 13, 17

Chapter 14: 14, 16, 32, 40, 53–54

Teacher's Resource Library Answer Key

Workbook Activities

Workbook Activity 1—Compare and Contrast
1. a. Both enable scientists to see small things. b. Microscopes use beams of light. Electron microscopes magnify objects even more by using beams of tiny particles instead of light. 2. a. Both are made of cells. Both have functions within living things. b. Tissues are groups of cells. Organs are groups of tissues. 3. a. Both are basic units of life. b. The two kinds of cells have different functions within the body. Skin cells cover and protect your body. Muscle cells allow for movement. 4. a. Both are living things that carry out all basic life activities. b. Animals have many cells. Bacteria have only one cell. 5. a. Both are living things and are made of cells. b. Each has different kinds of cells, tissues, and organs.

Workbook Activity 2—Plant and Animal Cells
Animal cells (left side of chart): lysosomes **Both cells** (middle): mitochondria; nucleus; DNA; Golgi bodies; cell membranes; cytoplasm; vacuoles; ribosomes; endoplasmic reticulum **Plant cells** (right side of chart): cell walls; chloroplasts

Workbook Activity 3—Chemicals for Life
1. Water: dissolves other chemicals so nutrients can move from cell to cell; drinking water, tea, fruit juices, other liquids 2. Carbohydrates: provide energy for animals; sugar, fruits, vegetables, bread, cereal, pasta, rice, potatoes 3. Fats: store large amounts of energy; meat, butter, cheese, peanut butter 4. Proteins: build and repair body parts; meats, beans, nuts, eggs, cheese 5. Minerals: nutrients the body needs in small supplies to be healthy; meats, dairy products, salt, bananas 6. Vitamins: nutrients the body needs in small supplies to be healthy; vegetables, meats, grains, fruits 7. Possible vitamins are Vitamin A, found in milk, liver, eggs, green and yellow vegetables; Vitamin C, found in citrus fruits, tomatoes, strawberries, cantaloupes; Vitamin E, found in whole-grain cereals, lettuce, vegetable oils. 8. Possible minerals are iron, found in meats and eggs; potassium, found in bananas; calcium, found in dairy products. 9. Amino acids are long chains of molecules that make up proteins. There are 20 different amino acids that can be arranged in different ways to make different proteins. 10. Plants use energy from sunlight to make carbohydrates. Animals get energy from the carbohydrates that plants make.

Workbook Activity 4—Vocabulary Review
1. C 2. D 3. F 4. E 5. A 6. B 7. G 8. H 9. nucleus 10. endoplasmic reticulum 11. atom 12. cell membrane 13. chloroplast 14. mitochondrion 15. homeostasis

Workbook Activity 5—Classifying Information
1. L 2. N 3. N 4. L 5. N 6. L 7. N 8. N and L 9. organism 10. properties 11. nonliving 12. bacteria 13. organs 14. things they lack 15. properties that describe movement, behavior, or activities

Workbook Activity 6—Living Things: Vocabulary Review
1. E 2. G 3. F 4. J 5. C 6. A 7. D 8. H 9. I 10. B 11. decompose 12. algae 13. microorganisms 14. pseudopod 15. organism 16. amoeba 17. property 18. taxonomy 19. protozoan 20. bacteria

Workbook Activity 7—The Classification of Animals
1. phylum 2. class 3. genus 4. species 5. genus, species 6. taxonomy, related 7. kingdom 8. species 9. smaller group, more 10. one 11. A group of organisms that can breed with each other to produce offspring like themselves. 12. same genus 13. Common names vary, but scientific names are the same everywhere. 14. *Felis* should be capitalized and both words should be underlined or in italic type. 15. Many organisms have not yet been discovered and classified.

Workbook Activity 8—Distinguishing Vertebrates
The order of responses to 1–3 may vary. 1. They have internal skeletons. 2. They have backbones. 3. They have skulls. 4. D 5. B 6. F 7. E 8. A 9. G 10. C

Workbook Activity 9—Classifying Animals: Vocabulary Review
1. I 2. J 3. E 4. F 5. A 6. G 7. B 8. M 9. D 10. C 11. N 12. L 13. K 14. O 15. H

Workbook Activity 10—Classifying Plants
1. Aristotle 2. Linnaeus 3. genus 4. species 5. Acer 6. rubrum 7. tube 8. vessel 9. vascular 10. vein 11. stems 12. leaves 13. roots 14. nonvascular 15. moisture

Workbook Activity 11—Seed Plants: Vocabulary Review
1. angiosperms 2. seeds 3. monocots 4. dicots 5. two 6. parallel 7. gymnosperms 8. fruit 9. cone 10. conifers 11. trees 12. green 13. paper 14. ginkgo tree 15. leaves

Workbook Activity 12—Vocabulary Review
1. K 2. H 3. L 4. F 5. N 6. C 7. I 8. P 9. B 10. E 11. G
12. M 13. A 14. O 15. J 16. D 17–20. Sentences will vary.
Example sentences: 17. Moss does not have tubelike cells.
18. The cells are grouped together in vascular tissue to move food and water. 19. Moss is a nonvascular plant. 20. A rhizoid is a tiny rootlike thread in a moss plant.

Workbook Activity 13—Bacteria
1. I 2. E 3. M 4. F 5. N 6. L 7. B 8. K 9. D 10. J 11. A 12. C
13. G 14. H 15. Accept any answer that is reasonably expressed and reflects lesson content.

Workbook Activity 14—All About Protists
1. algae 2. protozoans 3. amoeba 4. oxygen 5. trypanosomes

J	I	H	A	Q	T	F	I	V	S	Q	Q	M	F	T
H	I	K	F	L	A	G	E	L	L	U	M	W	E	R
T	Y	U	I	O	G	A	S	O	O	U	R	G	N	Y
Z	X	C	V	B	N	A	Q	W	E	R	T	Y	U	P
P	R	M	S	D	F	G	E	J	K	L	Z	X	C	A
R	N	Z	Q	S	P	O	R	O	Z	O	A	N	S	N
O	F	C	V	L	L	T	X	M	B	R	O	O	B	O
T	E	O	E	I	U	I	O	P	A	S	D	X	G	S
O	K	K	Z	C	C	E	B	N	M	Q	H	O	R	O
Z	U	N	O	C	A	Q	D	F	G	W	J	X	L	M
O	C	N	B	I	M	E	W	E	T	T	Y	Y	I	E
A	A	N	D	L	G	Y	J	H	L	Z	X	G	V	S
N	M	B	W	I	R	W	Y	U	A	M	O	E	B	A
S	G	D	I	A	T	O	M	C	V	B	N	N	Q	W
E	R	T	Y	U	M	T	M	S	L	Q	S	M	H	J

Workbook Activity 15—The Diary of a Protist
1. food vacuoles 2. gullet 3. contractile vacuoles 4. osmosis 5. cell membrane 6. chloroplasts 7. flagellum 8. cilia 9. spin 10. eyespot
11. asexual 12. Because not all protists are one-celled; some types of algae, such as seaweeds, are many-celled. 13. a paramecium
14. Students' answers should include any two of the following: has gullet; has cilia; doesn't have chloroplasts; doesn't have eyespot; motion described is paramecium-type movement. 15. A pond; a paramecium lives in a body of fresh water.

Workbook Activity 16—Fungi
1. fungi 2. hyphae 3. chloroplast 4. mycelium 5. club fungi
6. mold 7. smuts 8. aflatoxin 9. destroying angel 10. athlete's foot
11. yeast 12. gills 13. preservatives 14. ringworm 15. allergy

Workbook Activity 17—Vocabulary Review
Items to be checked: 2, 3, 4, 5, 6
Items to be checked: 8, 11, 12, 13, 14, 17, 18, 19, 20, 21, 23, 24
Items to be checked: 25, 26, 27, 28, 29, 30, 32, 33, 35

Workbook Activity 18—Animals Feeding
1. A Both take in fluids. B Filter feeders strain food out of the water. Fluid feeders suck liquids out of other organisms. 2. A Both must take in food. B Herbivores eat plants and carnivores eat other animals. 3. A Both are types of digestive systems. B A gastrovascular cavity is like a sack with one opening. A digestive tract is a tube with two openings and one-way flow of food and waste. 4. water; strain (or filter) 5. piercing

Workbook Activity 19—Animal Respiratory and Circulatory Systems
1. four 2. atria 3. ventricles 4. carbon dioxide 5. left atrium 6. left ventricle 7. body 8. right ventricle 9. right atrium 10. lungs

Workbook Activity 20—Water Balance
1. lower 2. out of 3. drink 4. gills 5. higher 6. into 7. water
8. kidneys 9. gills 10. They hold in water with watertight skin, a shell, or a waxy covering.

Workbook Activity 21—Vocabulary Review
1. C 2. L 3. A 4. F 5. I 6. E 7. B 8. J 9. D 10. H 11. K 12. G
13. B, A 14. A, B 15. B, A

Workbook Activity 22—The Vascular System in Plants
1. xylem 2. phloem 3. holds plants in ground 4. absorbs water and minerals from soil 5. stores water and minerals 6. brings water and minerals to other parts of the plant 7. xylem 8. phloem 9. layer of growth 10. supports the leaves 11. transports food, water, and minerals through the plant 12. stores food 13. petiole 14. blade
15. veins 16. makes food 17. stores food 18. transports food to stems 19. allows gases to enter and leave the plant 20. Both contain xylem and phloem; both transport food, water, and minerals through the plant; both store food. Roots hold plants in the ground while stems have another function: supporting leaves so that they can receive sunlight.

Workbook Activity 23—How Plants Make Food

Across:
1. chlorophyll
4. energy
5. photosynthesis
9. glucose
10. chloroplast
14. dioxide
15. sunlight

Down:
2. oxygen
3. leaves
6. hydrogen
7. equation
8. molecule
11. pigment
12. stomata
13. food

Workbook Activity 24—How Plants Give Off Oxygen

1. releases energy 2. stores energy 3. combines with hydrogen 4. releases oxygen to make water 5. releases carbon dioxide 6. combines with hydrogen to make sugar and water 7. gives off water 8. takes in water 9. breaks down sugar 10. makes sugar 11. Particles of gases are far apart, so you can't see or hold gases. Particles in a solid are packed tightly together, so you can see and touch a solid. 12. Possible answer: A stoma has two guard cells that open and close it so that gases can move in and out. 13. Photosynthesis takes place during the day when the sun is shining, so the stomata open. They close at night when there is no light and therefore no photosynthesis. 14. When the soil and air are dry, the stomata close so the plant doesn't lose water. 15. Destroying rain forests could damage or kill many other kinds of life because their oxygen supply depends on plants' photosynthesis.

Workbook Activity 25—Vocabulary Review

1. J 2. C 3. K 4. H 5. M 6. G 7. O 8. N 9. A 10. B 11. E 12. D 13. I 14. F 15. L

Workbook Activity 26—Digestion

1. mouth 2. esophagus 3. liver 4. gallbladder 5. pancreas 6. stomach 7. large intestine 8. small intestine 9. rectum 10. anus 11. liver 12. small intestine 13. large intestine 14. stomach 15. mouth

Workbook Activity 27—In Circulation

Across:
1. capillary
5. red
6. artery
7. plasma
9. vein
10. high
11. type
12. blood
15. lungs

Down:
2. platelets
3. red blood cells
4. from
8. antibody
13. die
14. cell

Workbook Activity 28—Respiration

1. trachea 2. right bronchus 3. bronchial tubes 4. aveolus 5. bronchioles 6. diaphragm 7. pharynx 8. larynx 9. left lung 10. heart 11. diaphragm 12. larynx 13. trachea 14. bronchioles 15. pharynx

Workbook Activity 29—Excretion

1. three 2. the fatty layer 3. perspiration 4. cooling 5. the excretory system 6. two 7. blood 8. the ureters 9. the bladder 10. the urethra 11. They are both liquids containing waste material. 12. The dermis is under the epidermis. 13. Wastes contain poisons that could harm the body. 14. They are all waste materials contained in perspiration and urine. 15. Possible answer: Waste material in the body is of no use and can cause harm. The same is true of trash around the house or community.

Workbook Activity 30—The Nervous System

1. peripheral 2. impulses 3. synapse 4. reflex 5. spinal cord 6. brain 7. neuron 8. cerebrum 9. nerves 10. cerebellum

I	B	E	W	B	L	I	N	K	I	N	G	L	T
W	Z	E	N	O	B	K	C	A	B	O	H	G	L
N	A	P	E	R	I	P	H	E	R	A	L	O	A
E	L	Q	R	T	E	H	I	K	G	S	H	C	C
U	M	A	V	B	N	A	S	R	S	S	O	E	I
R	E	T	E	K	S	N	G	S	E	M	E	R	M
O	T	E	S	P	A	N	Y	S	Y	Y	F	C	E
N	S	P	T	U	H	U	L	P	K	M	L	B	H
O	N	D	B	O	M	U	S	C	L	E	S	E	C
T	I	G	O	G	P	E	Q	T	L	E	I	L	X
G	A	O	P	M	G	T	H	T	A	O	T	L	N
B	R	A	I	N	O	L	Z	X	K	L	L	U	F
S	B	H	O	U	C	E	R	E	B	R	U	M	D
B	R	N	U	D	R	O	C	L	A	N	I	P	S

Workbook Activity 31—The Senses of Sight and Hearing
1. cornea 2. lens 3. pupil 4. iris 5. optic nerve 6. retina 7. iris 8. outer ear 9. ear canal 10. tiniest bones in the body 11. auditory nerve 12. cochlea 13. inner ear 14. middle ear 15. eardrum

Workbook Activity 32—The Endocrine System
Order of answers for 1–8 may vary. 1. pancreas 2. adrenal glands 3. pituitary gland 4. insulin 5. adrenaline 6. growth hormone 7. Possible answer: pancreas, insulin 8. Possible answer: pituitary gland, growth hormone 9. adrenaline 10. stress 11. endocrine system 12. hormones 13. bloodstream 14. chemical messengers 15. feedback loop

Workbook Activity 33—Vocabulary Review
1. C 2. F 3. K 4. A 5. L 6. D 7. I 8. O 9. H 10. E 11. N 12. J 13. B 14. M 15. G 16. D 17. B 18. A 19. A 20. C

Workbook Activity 34—That's Life
1. B 2. C 3. D 4. A 5. E 6. spontaneous generation 7. diversity 8. trait 9. DNA 10. adaptations 11–15. Answers will vary. Typical answers are given. 11. Parents' DNA is passed to offspring during reproduction. 12. Frogs develop from eggs laid externally. 13. Resistance to disease is an example of an adaptation. 14. DNA provides the information that determine traits, or characteristics. 15. People once thought that rotten meat was the source for the spontaneous generation of flies.

Workbook Activity 35—Reproduction
1. Answers will vary. Typical answers are given. **A** Both mitosis and meiosis involve the division of a cell and the creation of new cells. **B** Mitosis involves one cell division and results in the creation of two new cells. Meiosis involves two cell divisions and results in the creation of four new cells. 2. a 3. d 4. a 5. d 6. a 7. d 8. both 9. asexual 10. sexual 11. sexual 12. sexual 13. asexual 14. both 15. sexual

Workbook Activity 36—Animal Development
1. 1, 5, 4, 3, 2 2. nymph 3. marsupial 4. placenta 5. uterus 6. milk 7. mammals 8. zygote 9. DNA 10. embryo 11. jobs 12. food 13. pupa 14. the period of development of a mammal, from fertilization until birth 15. In general, the larger the mammal, the longer the gestation time.

Workbook Activity 37—Vocabulary Review
1. C 2. D 3. B 4. E 5. A 6. F 7–10. Answers will vary. Typical answers are given. 7. The umbilical cord attaches a fetus to the placenta. 8. External fertilization takes place outside a woman's body, but internal fertilization takes place inside a woman's body. 9. Estrogen and progesterone are two sex hormones found in females. 10. The prostate gland produces the fluid found in semen.

Workbook Activity 38—The Body Versus Disease
1. Both are white blood cells in the body's immune system; both fight pathogens. Phagocytes engulf and destroy pathogens; lymphocytes produce antibodies that bond to pathogens and stop them from attaching to cells. 2. Both are tiny and play a role in infectious disease in the body. Pathogens cause disease, and antibodies bond with pathogens to make them harmless to cells. 3. Both are ways of preventing infectious disease. Vaccines cause the body to produce antibodies that fight pathogens inside the body. Sanitation methods destroy pathogens outside the body. 4. infectious disease 5. plague 6. pathogen 7. virus 8. polio 9. phagocytes 10. lymphocytes 11. immune system 12. vaccine 13. smallpox 14. cowpox 15. sanitation

Workbook Activity 39—Eat Right for Health
1. carbohydrates 2. proteins 3. fats 4. water 5. vitamins 6. minerals 7. nutrition 8. water 9. vitamin C 10. vitamin K 11. minerals 12. Food Guide Pyramid 13. lower 14. fats and oils 15. slice

Workbook Activity 40—Vocabulary Review
1. A 2. H 3. F 4. G 5. D 6. C 7. I 8. J 9. B 10. E 11. Pathogens 12. phagocytes; lymphocytes 13. immunity 14. nutrition 15. Drug abuse

Workbook Activity 41—Heredity
1. Rr 2. r 3. R 4. RR, Rr, rr 5. offspring with the genotype rr 6. 75 percent chance 7. Rr because two combinations result in Rr whereas only one results in RR. 8. RR and Rr 9. zero 10. 25 percent chance 11. Rr 12. Yes, the odds that an offspring will have the genotype rr is the same for each offspring. 13–15. Students' Punnett squares should be drawn according to the models on pages 268–269.

Workbook Activity 42—Chromosomes

Meiosis/Mitosis 1. Both are ways cells reproduce. **2.** In mitosis, the division of the nucleus results in two new cells that are identical to the parent cell. Mitosis occurs in the human body during growth and repair. In meiosis, the division of the nucleus occurs during the formation of sex cells. Meiosis results in four sex cells. **3.** Both were scientists who made important advances in the knowledge of genetics and heredity. **4.** Mendel experimented with pea plants; Morgan with fruit flies. Mendel showed how dominant genes and recessive genes control how parent organisms pass on their traits to offspring. Morgan proved the existence of sex-linked traits and how they are passed on. **5.** Both have chromosomes in the nucleus, made of proteins and DNA and determine an organism's traits. **6.** There are 23 chromosomes in human sex cells. There are 46 chromosomes in other human cells. **7.** Both are sex cells and have 23 chromosomes. **8.** Males produce sperm cells and females produce egg cells. **9.** Both are determined by dominant and recessive genes. **10.** Sex-linked traits are linked to an organism's sex chromosomes. **11.** gamete: sex cells; egg or sperm **12.** zygote: cell formed when an egg and sperm join **13.** carrier: an organism that carries a gene but does not show the effects of the gene **14.** X chromosome: human sex chromosome; females have two; males have one **15.** Y chromosome: human sex chromosome; males have one

Workbook Activity 43—How Heredity Is Studied in Humans

I. Kinds of Twins
  A. Identical
  B. Fraternal
II. Parts of environment that influence a person's characteristics
  A. Family
  B. Air
  C. Everything else in surroundings
III. DNA
  A. Definition: material in chromosomes that contains an organism's genes
  B. Bases
    1. Definition: molecules found in DNA used to code information
    2. Names
      a. T
      b. A
      c. C
      d. G
  C. Replication
    1. Definition: when DNA makes a copy of itself
    2. Result: two identical copies of the original DNA molecule
IV. Mutation
  A. Definition: change in the order of bases in a DNA molecule
  B. Causes
    1. X rays
    2. chemicals
V. Genetic Diseases
  A. Diabetes
    1. Description: The body does not make insulin, so people with the disease have too much sugar in the blood.
    2. Cause: Recessive genes for the disease are inherited.
  B. Sickle-cell anemia
    1. Description: Red blood cells have a sickle shape that clogs vessels; people with the disease have weakness and an irregular heartbeat.
    2. Cause: Recessive genes for the disease are inherited.
  C. Hemophilia
    1. Description: Blood does not have protein needed to make it clot. People with the disease bleed a lot when slightly injured.
    2. Cause: inbreeding

Workbook Activity 44—Vocabulary Review
1. Q 2. N 3. O 4. H 5. D 6. G 7. M 8. K 9. C 10. P 11. F
12. R 13. A 14. I 15. J 16. S 17. T 18. C 19. B 20. L

Workbook Activity 45—Ecology Crossword
Across: 1. population   Down: 1. pollution
         3. biosphere   2. acid
         5. community   4. nonrenewable
         6. biotic   6. biome
         7. resources   9. habitat
         8. endangered   10. fossil fuel
        13. interact   11. climax
        14. ecosystem   12. ecology
        15. succession

Workbook Activity 46—Food Chains and Food Webs
1. food chain 2. photosynthesis 3. first-order consumers
4. second-order consumers 5. third-order consumers 6. large
7. small 8. food webs 9. decomposers 10. energy 11. chemical energy 12. consumers 13. food chain 14. decreases 15. Sun

Workbook Activity 47—Energy Flow in an Ecosystem
1. Sun 2. life activities 3. heat 4. plant tissues 5. life activities
6. heat 7. body tissues 8. second-order consumers 9. producers; because they get unlimited energy from the Sun 10. third-order consumers; because energy is lost at each stage of the food chain and they are at the end 11. The pyramid helps show the decreasing amount of energy at each higher level of the food chain. 12. third-order consumer, or fox 13. Grasshoppers are first-order consumers, so more food is available to them than to foxes, which are third-order consumers. 14. Without the Sun, producers could not make food; there would be no food for consumers and no way for energy to pass up the food chains. 15. Possible answer: Producers use energy from the Sun to make food that is eaten by consumers.

Workbook Activity 48—Vocabulary Review
1. H 2. C 3. A 4. E 5. G 6. I 7. J 8. F 9. D 10. B 11. acid rain
12. abiotic 13. extinct 14. condenses 15. nonrenewable

Workbook Activity 49—Innate Behavior
1. courtship 2. territorial 3. phototropism 4. behavior 5. stimulus
6. heredity 7. habitat 8. instinct 9. gravitropism 10. response
11. reflex 12. nest building 13. species 14. experience 15. innate

Workbook Activity 50—Learned Behavior
1. E 2. E 3. A 4. B 5. C 6. D 7. C 8. A 9. C 10. E 11. B 12. D
13. B 14. A 15. D

Workbook Activity 51—How Animals Communicate: Vocabulary Review
Across: 1. communication   Down: 1. channel
         6. response   2. mate
         7. gravitropism   3. inherited
        11. learned   4. territorial
        12. instinct   5. observational
        13. stimulus   8. innate
        14. trial   9. behavior
        15. reflex   10. insight

Workbook Activity 52—Changes in a Population
1. evolution 2. populations 3. mutations 4. traits 5. lethal
6–10. Answers should include any five of the following: become physically separated, move to different habitats, reproduce at different times, have no attraction to each other, have infertile offspring, have different body structures that prevent mating

Workbook Activity 53—Fossils Crossword
Across: 2. radioactive   Down: 1. mass extinction
         6. DNA   3. Equus
         7. geologic time   4. half-life
        11. teeth   5. fossil
        12. paleontologist   8. time periods
        13. fishes   9. sediment
        14. wings   10. mold
        15. Schweitzer

Workbook Activity 54—Darwin's Theory of Evolution
1. fossils 2. similar 3. hypotheses 4. scientific theories
5. modification 6. natural selection 7. Both are attempts to explain a problem using logic and evidence. 8. A hypothesis is unproven, but a theory is backed up by lots of evidence from various sources.
9. Both help explain the evolutionary process. 10. Descent with modification states that evolution occurs; natural selection explains how it occurs.

Workbook Activity 55—Evolution: Vocabulary Review
1. I 2. B 3. G 4. E 5. J 6. H 7. D 8. C 9. A 10. F

Alternative Workbook Activities

Alternative Workbook Activity 1—Compare and Contrast
1. a. Both are made of cells. Both have functions within living things. **b.** Tissues are groups of cells. Organs are groups of tissues. **2. a.** Both are basic units of life. **b.** The two kinds of cells have different functions within the body. Bone cells support and protect your body. Nerve cells send and receive messages. **3. a.** Both are living things made of cells. **b.** Each has different kinds of cells, tissues, and organs that carry out the life of the organisms in different ways.

Alternative Workbook Activity 2—Plant and Animal Cells
Animal cells (left side of chart): lysosomes **Both cells** (middle): mitochondria; nucleus; DNA; cytoplasm; vacuoles; ribosomes **Plant cells** (right side of chart): cell walls; chloroplasts

Alternative Workbook Activity 3—Chemicals for Life
1. Proteins: build and repair body parts; meats, beans, nuts, eggs, cheese **2.** Fats: store large amounts of energy; meat, butter, cheese, peanut butter **3.** Water: dissolves other chemicals so nutrients can move from cell to cell; drinking water, tea, fruit juices, other liquids **4.** Carbohydrates: provide energy for animals; sugar, fruits, vegetables, bread, cereal, pasta, rice, potatoes **5.** Possible minerals are iron, found in meats and eggs; potassium, found in bananas; calcium, found in dairy products.

Alternative Workbook Activity 4—Vocabulary Review
1. B **2.** C **3.** D **4.** E **5.** A **6.** nucleus **7.** atom **8.** cell membrane **9.** chloroplast **10.** cell wall

Alternative Workbook Activity 5—Classifying Information
1. L **2.** N **3.** N **4.** N **5.** L **6.** L **7.** organism **8.** properties **9.** bacteria **10.** organs

Alternative Workbook Activity 6—Living Things: Vocabulary Review
1. B **2.** F **3.** C **4.** G **5.** D **6.** A **7.** H **8.** E **9.** algae **10.** decompose **11.** fungus **12.** microorganisms **13.** organism **14.** protozoan **15.** amoeba

Alternative Workbook Activity 7—The Classification of Animals
1. kingdom **2.** class **3.** family **4.** species **5.** genus, species **6.** Taxonomy, related **7.** higher **8.** fewer **9.** species **10.** one

Alternative Workbook Activity 8—Distinguishing Vertebrates
The order of responses to 1–3 may vary. **1.** They have internal skeletons. **2.** They have backbones. **3.** They have skulls. **4.** F **5.** B **6.** D **7.** E **8.** A **9.** G **10.** C

Alternative Workbook Activity 9—Classifying Animals: Vocabulary Review
1. I **2.** J **3.** E **4.** F **5.** A **6.** K **7.** B **8.** H **9.** D **10.** C **11.** G **12.** reptile **13.** amphibian **14.** mollusk **15.** segmented worm

Alternative Workbook Activity 10—Classifying Plants
1. Linnaeus **2.** genus **3.** species **4.** vascular **5.** vein **6.** stems **7.** leaves **8.** roots **9.** nonvascular **10.** moisture

Alternative Workbook Activity 11—Seed Plants: Terms Review
1. angiosperms **2.** seeds **3.** dicots **4.** two **5.** monocots **6.** parallel **7.** gymnosperms **8.** fruit **9.** cone **10.** conifers

Alternative Workbook Activity 12—Vocabulary Review
1. K **2.** M **3.** C **4.** G **5.** H **6.** A **7.** O **8.** E **9.** D **10.** F **11.** L **12.** B **13.** N **14.** I **15.** J

Alternative Workbook Activity 13—Bacteria
1. I **2.** E **3.** H **4.** F **5.** A **6.** D **7.** B **8.** G **9.** C **10.** Accept any answer that is reasonably expressed and reflects lesson content.

Alternative Workbook Activity 14—All About Protists
1. protist **2.** algae **3.** chloroplasts **4.** protozoans **5.** amoeba

J	I	H	A	Q	T	F	I	V	S	Q	Q	M	F	C
H	I	K	F	L	A	G	E	L	L	U	M	W	E	H
T	Y	U	I	O	G	A	S	O	O	U	R	G	N	L
Z	X	C	V	B	N	A	Q	W	E	R	T	Y	U	O
P	R	M	S	D	F	G	E	J	K	L	Z	X	C	R
R	N	Z	Q	S	P	O	R	O	Z	O	A	N	S	O
O	F	C	V	L	L	T	X	M	B	R	O	O	B	P
T	E	O	E	P	R	O	T	O	Z	O	A	N	S	L
I	K	K	Z	C	C	E	B	N	M	Q	H	O	R	A
S	U	N	O	C	A	Q	D	F	G	W	J	X	L	S
T	C	N	B	I	M	E	W	E	T	T	Y	Y	I	T
A	A	N	D	L	G	Y	J	H	L	Z	X	G	V	S
N	M	B	W	I	R	W	Y	U	A	M	O	E	B	A
S	G	D	I	A	T	O	M	C	V	B	N	N	Q	W

Alternative Workbook Activity 15—The Diary of a Protist
1. gullet 2. food vacuoles 3. osmosis 4. contractile vacuoles
5. cilia 6. eyespot 7. chloroplasts 8. flagellum 9. a paramecium
10. Students' answers should include any two of the following: has gullet; has cilia; doesn't have chloroplasts; doesn't have eyespot.

Alternative Workbook Activity 16—Fungi
1. fungi 2. penicillin 3. club fungi 4. mold 5. hyphae 6. gills
7. mycelium 8. smuts 9. yeast 10. allergy

Alternative Workbook Activity 17—Vocabulary
Items to be checked: 1, 2, 6, 7, 9
Items to be checked: 10, 11, 12, 14, 16, 17, 20, 22, 24
Items to be checked: 25, 26, 27, 29, 30

Alternative Workbook Activity 18—Animals Feeding
1. **A** Both must take in food. **B** Herbivores eat plants and carnivores eat other animals. 2. **A** Both are types of digestive systems.
B A gastrovascular cavity is like a sack with one opening. A digestive tract is a tube with two openings and one-way flow of food and waste. 3. Fluid feeders 4. water 5. gizzard

Alternative Workbook Activity 19—Animal Respiratory and Circulatory Systems
1. oxygen 2. three 3. left atrium 4. right atrium 5. lungs

Alternative Workbook Activity 20—Water Balance
1. lower 2. into 3. water 4. kidneys 5. higher 6. out of 7. drink
8. gills 9. They drink water or get fluids from their foods. 10. They hold in water with watertight skin, a shell, or a waxy coating.

Alternative Workbook Activity 21—Vocabulary Review
1. A 2. B 3. D 4. F 5. C 6. G 7. H 8. I 9. J 10. E 11. nerve net
12. circulatory 13. secrete 14. filter feeding 15. excrete

Alternative Workbook Activity 22—The Vascular System in Plants
1. phloem 2. absorbs water and minerals from soil 3. stores water and minerals 4. brings water and minerals to other parts of the plant 5. xylem 6. phloem 7. supports the leaves 8. transports food, water, and minerals through the plant 9. stores food
10. petiole 11. blade 12. veins 13. stores food 14. transports food to stems 15. allows gases to enter and leave the plant

Alternative Workbook Activity 23—How Plants Make Food
Across: 1. chlorophyll   Down: 2. oxygen
4. photosynthesis   3. leaves
8. glucose   5. hydrogen
9. chloroplast   6. equation
                             7. molecule
                             10. pigment

Alternative Workbook Activity 24—How Plants Give Off Oxygen
1. combines with hydrogen 2. releases oxygen to make water
3. releases carbon dioxide 4. combines with hydrogen to make sugar and water 5. gives off water 6. takes in water 7. breaks down sugar 8. makes sugar 9. Photosynthesis takes place during the day when the sun is shining, so the stomata open. They close at night when there is no light and therefore no photosynthesis.
10. Destroying rain forests could damage or kill many other kinds of life because their oxygen supply depends on plants' photosynthesis.

Alternative Workbook Activity 25—Vocabulary Review
1. E 2. I 3. J 4. C 5. G 6. D 7. H 8. A 9. F 10. B

Alternative Workbook Activity 26—Digestion
1. mouth 2. esophagus 3. liver 4. gallbladder 5. pancreas
6. stomach 7. large intestine 8. small intestine 9. esophagus
10. small intestine

Alternative Workbook Activity 27—In Circulation
Across: 1. capillary   Down: 2. platelets
4. artery   3. red blood cells
5. plasma   6. antibody
9. blood   7. heart
10. type   8. cardiac

Alternative Workbook Activity 28—Respiration
1. trachea 2. right bronchus 3. bronchial tubes 4. diaphragm
5. pharynx 6. larynx 7. left lung 8. heart 9. larynx 10. diaphragm

Alternative Workbook Activity 29—Excretion
1. perspiration 2. skin 3. cooling 4. kidneys 5. the ureters 6. two
7. urethra 8. bladder 9. They are both liquids produced by the body to get rid of wastes that might be harmful. 10. Perspiration comes out through thousands of sweat glands under the skin. Urine collects in the bladder and exits through a tube called the urethra.

Alternative Workbook Activity 30—The Nervous System
1. synapse **2.** reflex **3.** spinal cord **4.** brain **5.** neuron **6.** backbone **7.** brain stem **8.** muscles **9.** cerebellum **10.** cerebrum

C	M	N	E	U	R	O	N	S
E	B	U	O	B	R	I	S	P
R	R	B	S	A	E	J	Y	I
E	A	P	H	C	F	O	N	N
B	I	U	Q	K	L	W	A	A
E	N	L	T	B	E	E	P	L
L	S	T	T	O	X	M	S	C
L	T	L	A	N	R	M	E	O
U	E	E	R	E	P	P	Q	R
M	M	U	B	R	A	I	N	D
S	C	E	R	E	B	R	U	M

Alternative Workbook Activity 31—The Senses of Sight and Hearing
1. cornea **2.** lens **3.** optic nerve **4.** retina **5.** cornea **6.** ear canal **7.** auditory nerve **8.** cochlea **9.** eardrum **10.** cochlea

Alternative Workbook Activity 32—The Endocrine System
1. endocrine system **2.** hormones **3.** bloodstream **4.** chemical messengers **5.** feedback loop **6.** adrenaline **7.** stress **8.** pancreas **9.** pituitary gland **10.** adrenal glands

Alternative Workbook Activity 33—Vocabulary Review
1. C **2.** D **3.** F **4.** A **5.** H **6.** B **7.** G **8.** I **9.** J **10.** E **11.** B **12.** B **13.** B **14.** A **15.** C

Alternative Workbook Activity 34—That's Life
1. B **2.** A **3.** D **4.** C **5.** diversity **6.** DNA **7.** spontaneous generation Answers for 8–10 will vary. Typical answers are given. **8.** Resistance to disease is an example of an adaptation. **9.** DNA and a book are alike because both store information in a pattern; DNA stores it in a pattern of chemicals, and a book stores it in a pattern of letters. **10.** At one time people thought that frogs were made from mud through a process called spontaneous generation.

Alternative Workbook Activity 35—Reproduction
1. Answers will vary. Typical answers are given. **A** Both external fertilization and internal fertilization involve the fertilization of one or more eggs. **B** External fertilization takes place outside the female's body. Internal fertilization takes place inside the female's body. **2.** d **3.** a **4.** a **5.** d **6.** both **7.** sexual **8.** both **9.** sexual **10.** both **11.** sexual **12.** asexual **13.** both **14.** sexual **15.** asexual

Alternative Workbook Activity 36—Animal Development
1. 3, 1, 4, 2 **2.** kangaroo **3.** milk **4.** adult **5.** DNA **6.** placenta **7.** mammals **8.** zygote **9.** food **10.** the period of development of a mammal, from fertilization until birth

Alternative Workbook Activity 37—Vocabulary Review
1. B **2.** F **3.** D **4.** H **5.** C **6.** E **7.** G **8.** A **9–10.** Answers for 9 and 10 will vary. Typical answers are given. **9.** The vagina and fallopian tubes are female sex organs. **10.** The scrotum and prostate gland are male sex organs.

Alternative Workbook Activity 38—The Body Versus Disease
1. Both are ways of preventing infectious disease. Vaccines cause the body to produce antibodies that fight pathogens inside the body. Sanitation methods destroy pathogens outside the body. **2.** Both are white blood cells in the body's immune system; both fight pathogens. Phagocytes engulf and destroy pathogens; lymphocytes produce antibodies that bond to pathogens and stop them from attaching to cells. **3.** virus **4.** phagocytes **5.** infectious disease **6.** immune system **7.** vaccine **8.** lymphocytes **9.** sanitation **10.** pathogen

Alternative Workbook Activity 39—Eat Right for Health
1. carbohydrates **2.** fats **3.** vitamins **4.** minerals **5.** fats and oils **6.** vitamin K **7.** Thiamin **8.** minerals **9.** Food Guide Pyramid **10.** nutrients

Alternative Workbook Activity 40—Vocabulary Review
1. E **2.** B **3.** A **4.** D **5.** G **6.** C **7.** F **8.** phagocytes **9.** pathogens **10.** nutrition

Alternative Workbook Activity 41—Heredity

1. RR, Rr, rr **2.** Rr **3.** r **4.** Rr because two combinations result in Rr whereas only one results in rr. **5.** 75 percent chance **6.** Rr **7.** RR and Rr **8.** with the genotype rr **9–10.** Students' Punnett squares should be drawn according to the models on pages 268–269.

Alternative Workbook Activity 42—Chromosomes

1. Both have chromosomes in the nucleus, made of proteins and DNA and determine an organism's traits. **2.** There are 23 chromosomes in human sex cells. There are 46 chromosomes in other human cells. **3.** Both are gene combinations that determine the sex of humans. **4.** XY is a male; XX is a female. **5.** Both were scientists who made important advances in the knowledge of genetics and heredity. **6.** Mendel experimented with pea plants; Morgan with fruit flies. Mendel showed how dominant genes and recessive genes control how parent organisms pass on their traits to offspring. Morgan proved the existence of sex-linked traits and how they are passed on. **7.** Both are ways cells reproduce. **8.** In mitosis, the division of the nucleus results in two new cells that are identical to the parent cell. Mitosis occurs in the human body during growth and repair. In meiosis, the division of the nucleus occurs during the formation of sex cells. Meiosis results in four sex cells. **9.** an organism that carries a gene but does not show the effects of the gene **10.** rod-shaped bodies, made of DNA and proteins, that are in the nucleus of cells

Alternative Workbook Activity 43—How Heredity Is Studied in Humans

I. Parts of environment that influence a person's characteristics
   A. Family
   B. Air
   C. Everything else in surroundings
II. Kinds of Twins
   A. Identical
   B. Fraternal
III. DNA
   A. Definition: material in chromosomes that contains an organism's genes
   B. Bases
      1. Definition: molecules found in DNA used to code information
      2. Names
         a. T
         b. A
         c. C
         d. G
IV. Genetic Diseases
   A. Diabetes
      1. Description: The body does not make insulin, so people with the disease have too much sugar in the blood.
      2. Cause: Recessive genes for the disease are inherited.
   B. Sickle-cell anemia
      1. Description: Red blood cells have a sickle shape that clogs vessels; people with the disease have weakness and an irregular heartbeat.
      2. Cause: Recessive genes for the disease are inherited.
   C. Hemophilia
      1. Description: Blood does not have protein needed to make it clot. People with the disease bleed a lot when slightly injured.
      2. Cause: inbreeding

Alternative Workbook Activity 44—Vocabulary Review
1. F 2. A 3. T 4. D 5. J 6. P 7. G 8. B 9. R 10. M 11. Q 12. S
13. H 14. I 15. K 16. N 17. C 18. O 19. L 20. E

Alternative Workbook Activity 45—Ecology Crossword
Across: 3. population Down: 1. pollution
4. biosphere 2. fossil fuel
6. producer 5. endangered
7. community 7. climax
9. ecosystem 8. biome
10. habitat

Alternative Workbook Activity 46—Food Chains and Food Webs
1. producer 2. photosynthesis 3. second-order consumers 4. large
5. food webs 6. decomposers 7. chemical energy 8. food chain
9. decreases 10. consumers

Alternative Workbook Activity 47—Energy Flow in an Ecosystem
1. Sun 2. heat 3. plant tissues 4. life activities 5. body tissues
6. second-order consumers 7. producers; because they get unlimited energy from the Sun 8. third-order consumers; because energy is lost at each stage of the food chain and they are at the end 9. fox, bird, grasshopper, grass 10. Without the Sun, producers could not make food; there would be no food for consumers and no way for energy to pass up the food chains.

Alternative Workbook Activity 48—Vocabulary Review
1. B 2. G 3. D 4. F 5. A 6. C 7. E 8. biotic 9. acid rain
10. ecosystem

Alternative Workbook Activity 49—Innate Behavior
1. mate 2. stimulus 3. instinct 4. habitat 5. behavior
6. phototropism 7. gravitropism 8. reflex 9. territory
10. nest building

Alternative Workbook Activity 50—Learned Behavior
1. B 2. D 3. E 4. C 5. E 6. C 7. A 8. A 9. D 10. B

Alternative Workbook Activity 51—How Animals Communicate: Vocabulary Review
Across: 4. observational Down: 1. communication
6. instinct 2. error
8. stimulus 3. conditioning
9. innate 5. learned
10. gravity 7. reflex

Alternative Workbook Activity 52—Changes in a Population
1. evolution 2. populations 3. Mutations 4. lethal 5. geographic isolation 6–9. Answers should include any four of the following: become physically separated, move to different habitats, reproduce at different times, have no attraction to each other, have infertile offspring, have different body structures that prevent mating
10. They increase the chances that the individual will live to reproduce.

Alternative Workbook Activity 53—Fossils Crossword
Across: 2. radioactive Down: 1. mass extinction
3. cast 4. fossil
4. fishes 6. time periods
5. geologic time 7. sediment
9. paleontologist 8. mold
10. scientific theory

Alternative Workbook Activity 54—Darwin's Theory of Evolution
1. fossils 2. like 3. change 4. hypotheses 5. scientific theories
6. modification 7. natural selection 8. They show traces of body parts that a population possessed earlier, but which are no longer useful in its evolved state. 9. that they come from a common ancestor 10. that they evolved in different ways from a common ancestor

Alternative Workbook Activity 55—Evolution: Vocabulary Review
1. G 2. B 3. F 4. E 5. A 6. H 7. D 8. C 9. mold 10. primates

Lab Manual

Lab Manual—Safety in the Classroom
1. Answers will vary. Students should be able to describe an alternate route. They can draw it on their map in a different color. **2.** Answers will vary. In addition to a fire extinguisher and safety goggles, safety equipment might include a fire blanket and eye rinse. **3.** Answers will vary depending on the safety equipment in the classroom. **Explore Further** Diagrams will vary but should show safety equipment, electrical and gas outlets, and an emergency exit.

Lab Manual 1—Using a Microscope
1. Accept any of the following answers: The classified *e* in its natural size just fills the power field (100X). The letter *e* is easy to tell in the backward position. It is also easy to see in the upside-down image. You can easily tell which part of the *e* you are seeing. **2.** It filled the field at 100X; it was backwards; it was upside-down. **Explore Further** Based on their examination of the letter *e*, students should draw the *c* and *g* backward and upside-down.

Lab Manual 2—Using a High-Power Microscope
1. Just one part of the letter *e*. **2. a.** It appears to move to the right. **b.** It appears to move to the left. **3. a.** It appears to come toward you. **b.** It appears to move away. **Explore Further** The letter has thickness. Part of the letter comes into focus at each turn of the adjustment knob.

Lab Manual 3—Comparing Cells (Investigation 1)
1. Similarities include size and shape and presence of organelles. Differences include shape and the presence of a cell wall, large vacuole, and green chloroplasts in plant cells but not in animal cells. **2.** Differences include relative size, shape, and the absence of organelles in bacterial cells. **Explore Further** Point out that if students observe an onion cell, they will not see chloroplasts. This is because the onion is part of the root of the plant and is therefore underground.

Lab Manual 4—Living or Nonliving? (Investigation 2)
1. Answers will vary depending on the pictures chosen. **2.** Answers will vary depending on the pictures chosen. **3.** Answers will vary. Possible answers: A lawn mower gets energy from gasoline and it moves. But the lawn mower does not reproduce, develop, or grow. So, it is not a living thing. **4.** Answers may vary, but most pictures will show living things and nonliving things together. **Explore Further** Living objects to list might include other students, the teacher, plants, insects, and class pets. Possible nonliving objects could include desks, floor, walls, chalkboard, pencil, pen, clothing, shoes, backpacks, and paper.

Lab Manual 5—A Look at the Euglena
Note: You can send for euglena and other microorganism cultures from one of the biological supply houses.
Sometimes the microorganisms move so quickly that they move from the field of vision before they can be observed. You can slow them down in the following ways.
- Pull apart a piece of absorbent cotton or lens paper and put a few fibers on a slide. Add the euglena culture. The fibers will enmesh the organisms and slow them down.
- Prepare a gelatin solution by mixing one teaspoon of dry gelatin powder in a cup of cold water. Be sure the gelatin is fresh and cool or it will not work. Use a toothpick to add gelatin to the slide.
- Add a drop of methyl cellulose solution to the slide.

1. Answers will vary. The students may not be able to see all the parts labeled in the drawing. **2. a.** It has a chloroplast; it is green. **b.** It moves from place to place. **3.** Euglena belongs in the protist kingdom because it is neither a plant nor an animal, and it has a single cell. **4.** The flagellum is for movement. **Explore Further** Euglena has an eyespot that is light sensitive. It should move toward the light. This allows it to make its own food.

Lab Manual 6—Comparing Single-Celled Organisms
1. Answers will vary depending on the protist studied; for example, amoebas have undefined shapes, and paramecia and euglena are somewhat oval. **2.** Answers will vary depending on the moneran studied. For example, bacteria have three basic shapes—sphere, spiral, and rod. **3.** Yes, in the protists. **4.** They are both single-celled organisms; protists have organelles, but monerans do not. **Explore Further** Students' answers should remain the same because all single-celled protists have organelles and all single-celled monerans do not.

Lab Manual 7—Classifying Sports Equipment
1. probably two **2.** probably four **3.** A system grouping together all items related to the same sport would make the task easier. **Explore Further** Groupings may vary. Possible groups: equipment for playing and things players wear; subgroups might be catcher's equipment and other players' equipment.

Lab Manual 8—Classifying Objects (Investigation 3)

1. The names of groups will vary. Possible groups include eating utensils, fasteners, toiletry supplies, things found in a desk. **2.** The number of levels will vary, but students should be able to come up with at least three levels. **3.** Groups' systems will probably vary in their levels and names for groups. **4.** The investigation shows that objects can be classified in more than one way. Using a single system of classification eliminates confusion and duplication of naming. **Explore Further** Combining classification systems will mean that the new chart looks like neither of the original two. Encourage groups to preserve the most logical parts of each system. Emphasize that they may choose to change names of some levels completely.

Lab Manual 9—Studying Vertebrates and Invertebrates

1. a vertebrate **2.** The backbone is a series of small bones that run along the back of the fish. The bones are attached loosely to each other, so the backbone can bend. **3.** the skull, the ribs, and the bones of the tail **4.** The skull is made of bones that are firmly attached to each other. It has sockets for the eyes. **5.** no **6.** an invertebrate **7.** Vertebrates have a backbone, and invertebrates do not. **8.** the shell **Explore Further** My backbone also provides support and is made of bones that are attached in a way to provide flexibility.

Lab Manual 10—Identifying Angiosperms and Gymnosperms (Investigation 4)

1. The shape of leaves from gymnosperms such as conifers is needlelike. Angiosperms have flat, broader leaves. **2.** Monocots have long, narrow leaves with smooth edges. Dicots have broader leaves that can have lobes (rounded, protruding parts) and wavy or serrated (zigzag) edges. **3.** Some empty boxes might remain because the veins in the leaves of many gymnosperms cannot be seen, and they are not classified as monocots or dicots. **4.** Answers will vary. Students may mention size, thickness, or texture. If you picked clusters of leaves from a rose bush or a pecan or walnut tree, students may recognize that leaves may grow singly (simple) or in groups (compound). **Explore Further** Display students' collections in the classroom. Names of plants will vary depending on the collection.

Lab Manual 11—Tree Study

1–3. Answers will vary, depending on the tree specimen and students' observations. If students do not know the species of the chosen tree, they could find out by talking with people from a local nursery or by using a reference book that describes trees in your area. **Explore Further** Answers will vary; students will probably note that the temperature readings will be different and that the color, texture, and even presence of leaves on the tree may change, depending upon the season and the specimen chosen.

Lab Manual 12—Dicot and Monocot Seeds

The drawing of the lima bean should show two seed leaves, and the drawing of the corn seed should show one seed leaf.
1. The corn kernel is from a plant that is a monocot; it has one embryo leaf and one cotyledon. **2.** The lima bean is from a plant that is a dicot; it has two embryo leaves and two cotyledons. **3.** Each half of the split bean is called a cotyledon. **Explore Further** The cotyledons in lima beans store food for the embryo plant. The same stored food is nutritious to humans, who eat the beans.

Lab Manual 13—Observing Paramecia

1. The contractile vacuole should contract more frequently in distilled water than in the culture medium. **2.** The contractile vacuole should contract less frequently in salt water than in the culture medium. **3.** Water enters a paramecium more rapidly when the paramecium is in distilled water than when it is in the culture medium. Therefore, the contractile vacuoles must contract more frequently to remove excess water. Water enters a paramecium less rapidly when the paramecium is in salt water than when it is in the culture medium. Therefore, the contractile vacuoles need not contract as frequently to remove unneeded water. **Explore Further** Answers will vary, depending on the solution students choose to use.

Lab Manual 14—Studying Yeast

1. Yeast cells reproduce by budding. **2.** The culture that grew in the warm place should have the most cells. **3.** The culture that grew in the refrigerator should have the fewest cells. **4.** Results should indicate that increasing the temperature causes yeast to grow more rapidly. **Explore Further** Answers will vary, depending on the variable chosen by students.

Lab Manual 15—Growing Bread Mold (Investigation 5)

1. Students' answers should accurately reflect their observations. The dish with the moist bread kept in the dark should have the most mold. The dish with the moist bread in the light may also have significant mold growth. **2.** Students' answers should accurately reflect their observations. Both dishes with the dry bread should have little or no mold growth. **3.** The results indicate that the mold grows best in the presence of moisture and in the dark. **4.** It was important to let the dishes sit uncovered for a period of time so that mold spores from the air could fall into them. The spores could then produce mycelia that grew into the bread if it was moist. **5.** If the dark and light places were not at the same temperature, you could not conclude that differences in mold growth were due to differences in light or moisture. The difference in temperature could also be a factor. **Explore Further** If students need help choosing another environmental condition to test, suggest any of the following: temperature, growth medium (that is, using food other than bread), variable amounts of water. Suggest that students use the investigation on pages 105–106 as their model.

Lab Manual 16—Observing the Action of a Digestive Enzyme

1. Yes. A blue-black color appeared when iodine solution was added to the ground cracker and to the ground cracker + water. **2.** No. A blue-black color did not appear when iodine solution was added to any cup containing egg white. **3.** The cup containing the chewed cracker did not have a blue-black color because the digestive enzyme in saliva broke down the starch that was in the cracker. Without starch, the iodine solution could not cause the color change. **4.** Saliva contains water. It could have been the water that prevented the starch in the cracker from reacting with the iodine solution. Testing a ground cracker mixed with water ruled out this possibility. **Explore Further** The cucumber does not change color, but the banana turns blue-black. The cucumber does not contain starch but the banana does. The blue-black color gradually lightens and turns yellowish. The banana gets mushy.

Lab Manual 17—Studying Feeding in Hydras (Investigation 6)

1. Answers may vary. Typically, hydras do not respond to water vibrations; shorten their tentacles and body when touched; and bend toward the filter paper soaked in beef broth. **2.** Chemical stimuli probably trigger feeding responses. **Explore Further** If a hydra shows a feeding response when a dead, motionless water flea is nearby but not touching the hydra, then the answer to question 2 would be supported.

Lab Manual 18—Measuring Carbon Dioxide Production

1. Either the "after holding breath" or the "after exercising" flask should have needed the most drops. Actual results may vary. **2.** The "after normal breathing" flask should have needed the fewest drops. **3.** The "after holding breath" and "after exercising" flasks, in either order, should have contained the most carbon dioxide. The "after normal breathing" flask should have contained the least. **4.** Your body continues to produce carbon dioxide while you hold your breath. The carbon dioxide passes from the blood in your lungs and builds up in the air inside your lungs. When you exhale, there is more carbon dioxide than normal in your breath. **5.** Your body produces more carbon dioxide when you exercise. The extra carbon dioxide leaves the blood in your lungs. When you exhale, there is more carbon dioxide than normal in your breath. **Explore Further** Students may suggest using a kind of air pump to bubble air through water with methyl red in a flask. They could logically infer that it would have a higher concentration of carbon dioxide than outside air because everyone in the classroom is exhaling carbon dioxide, thus increasing the concentration.

Lab Manual 19—Vascular Tissue

1. Both plants were a clear and natural light green; the leaves of the celery were pale green, and the carnation flower was white. **2.** to make sure the tubes were not clogged or dried up **3.** In the carnation, the flower changed from white to a color. Colored water moved up the stem and into the leaves. **4.** The fresh cut shows the round tube openings filled with colors. **5.** The diagram will show the arrangement of the tubes in the stem. **Explore Further** The vascular tissue will look like tiny holes or dots that are part of the stem or tube.

Lab Manual 20—Sprouting Seeds

1. Paragraphs should describe what students observed about the sprouting and development of the plants. **2.** Responses should indicate that only a portion of the seeds sprouted. Paragraphs should be consistent with student graphs. **Explore Further** Students should note that more seeds will sprout, and they will sprout faster, when exposed to warmth and sunlight.

Lab Manual 21—Growing an African Violet From a Leaf (Investigation 7)

1. Students should observe that new roots grow from the stem. The new roots are white and covered with root hairs. **2.** The leaf can make food because it has chloroplasts, but it needs roots to be able to absorb nutrients from the soil. **3.** New leaves appear. **4.** This is asexual reproduction because a new plant grew from a part of the original plant. **Explore Further** Encourage students to share their results with the class.

Lab Manual 22—How Does Exercise Change Heart Rate? (Investigation 8)

1. Heart rate increases with amount of activity. **2.** When the heart rate was lowest, the heart was not beating as fast. When the heart rate was highest, the heart was beating faster than at any other time. When the heart rate was lowest, the activity level was low. **3.** The heart rate increases as activity increases because your body needs more oxygen delivered to its cells. Cells need more oxygen to release energy from glucose. The energy is needed for movement. **Explore Further** Investigations will vary, but make sure students have a clear purpose and can obtain measurable results.

Lab Manual 23—Diffusion

1. In salt water. Water (solute) moved from an area of higher concentration (the carrot cells) to an area of lower concentration (the container). As water moved out of the carrot cells, the carrot size diminished and the thread became looser. **2.** In plain water. Water moved from an area of higher concentration (the container) to one of lower concentration (the carrot cells). As water moved into the carrot cells, the carrot swelled and the thread became tighter. **3.** The string allows you to see the extent to which swelling or dehydration has occurred. It tightens or loosens as the size of the carrot changes. **4.** Liquids move from levels of low concentration to levels of higher concentration. They move to achieve balance. **Explore Further** Excretory system: water, salt, sugars; respiratory system: oxygen and carbon dioxide.

Lab Manual 24—Sensory Receptors in the Skin
1. The areas sensitive to cold have more receptors for cold. Areas sensitive to pain have more receptors for pain. **Explore Further** The hand is more sensitive than the back of the neck. Students should conclude that they can sense more stimuli on the hand than the back of the neck because the hand has a greater number of sensory receptors.

Lab Manual 25—Observing Mitosis
1. They are rectangular. **2.** Answers will depend on the slide viewed. Often pink or purple stain is used on onion root tips. **Explore Further** Answers will vary, depending on the number of cells in each phase that are visible.

Lab Manual 26—Observing Sperm and Egg Cells
1. Answers will vary. Students may write that both are single cell, but that the eggs are larger than the sperm. **2.** Its tail. **Explore Further** The round shape of the egg gives sperm a better chance of fertilizing the egg because there are no sides or corners that could inhibit or prevent fertilization.

Lab Manual 27—Graphing Gestation Times (Investigation 9)
1. the cow and the mare **2.** one month **Explore Further** The new bar graph should have a scale on the left side that represents the number of times an animal can reproduce in a year and should be numbered from 1 to 12. This is how many times each animal can reproduce in a year: rabbit, 12; cat and dog, 6; sow, 3; ewe and goat, 2 2/5; cow and mare, 1. The shorter the gestation time the more often an animal can reproduce.

Lab Manual 28—Fighting Pathogens
1. Answers should be supported by students' data. Students are likely to find that the dish with vinegar has little or no pathogenic growth, while the dish without vinegar probably shows some pathogenic growth. **2.** Answers will depend on results. Students may suggest that the acid retarded the growth of pathogens in the dish with vinegar. **Explore Further** How well an area is scrubbed and the specific household cleaner will affect the results. The effectiveness of household cleaners in preventing the growth of pathogens will vary. Students may expect the cleaners to completely eliminate the pathogens. However, some pathogens may continue to grow.

Lab Manual 29—Reading Food Labels (Investigation 10)
1–4. Answers will vary. Check to be sure answers are supported by the data. **5.** Answers will vary. Snack foods and foods high in fat usually have smaller serving sizes than foods low in fat. Make sure that answers are supported by the data. **Explore Further 1.** Answers will vary. Make sure that answers are supported by information in the lesson. **2.** Answers will vary. Make sure students give reasons for their answers.

Lab Manual 30—Comparing Calories Used in Exercise
1. jogging and swimming **2.** they use about the same number of calories per hour. **3.** Answers will vary. Students may not expect sitting and standing to use so many calories. **Explore Further** Answers will vary. Some students would prefer walking because it could be done with a partner and they could enjoy the outdoors. Others might prefer skipping rope because it burns calories more efficiently.

Lab Manual 31—Modeling Mendel's Experiments
1. Students should name their genotypes; for example, if P stands for the purple flower and p for the white flower, the parents will be Pp and Pp. The phenotypes will be purple and white. **2.** If P is dominant, students' Punnett squares should show that chances are 75% that the offspring will have purple flowers. **Explore Further** Students' answers will depend on whether their experiments with beads resulted in 3/4 of them being Pp or PP.

Lab Manual 32—Modeling Sex Determination
1–5. Answers will vary. **6.** Students should find that on each flip of the coin, there is a 50% chance of each outcome. The same applies for having a girl or a boy each time a child is born. **Explore Further** Students should recognize that the chances of having a boy or girl are 50/50 each time, regardless of how many boys or how many girls are already in a family.

Lab Manual 33—Tracing a Genetic Disease (Investigation 11)

1. A recessive gene causes albinism. Two normal parents can produce an albino child. **2.** The pedigree should show three generations: John, his parents, and his grandparents. **3.** Huntington's disease is inherited by a dominant gene. Every person with Huntington's disease has a parent who also has the disease. **4.** John probably inherited Huntington's disease from his father. **5.** Nancy is not a carrier. If she had inherited a gene for Huntington's disease, she would also have the disease because the gene is dominant. **Explore Further 1.** John's genotype is probably Hh. He received a dominant gene for Huntington's disease from his father. He received a normal gene from his mother. **2.** The Punnett square should show a cross between the genotypes Hh and hh. John's child has a 50 percent chance of inheriting Huntington's disease.

Lab Manual 34—Life in a Pond

1. Answers will vary. Sometimes microorganisms move too quickly to be observed easily. You can slow them down by adding a few cotton or lens paper fibers to the slide or by adding a drop of methyl cellulose solution. **2.** Sketches will vary. Students should include as much detail as they can. **Explore Further** Possible answer: algae, lily pads, weeds, one-celled organisms, minnows, fish, frogs, spiders, turtles, dragonflies are living parts; water and rocks are nonliving parts. Microorganisms could feed on tiny bits of plants or on each other and provide food for larger organisms. They live in the water and take oxygen from it; they might group near rocks.

Lab Manual 35—Testing the pH of Rain (Investigation 12)

1. The pH of distilled water is 7.0. The pH of rainwater will vary. It should be lower than that of distilled water. **2.** Answers will vary. Any samples with pH less than 4.9 indicate acid rain. **3.** The pH of rainwater may vary on different days because the level of pollution in the air may vary. In addition, rain comes from different areas. If it is blown in from a large city or highly industrialized area, it may be more acidic. **Explore Further** The pH of a pond, lake, or stream is likely to be different from that of rainwater. Streams and lakes fed by streams get water from other locations; they are not filled only by local rain. Natural materials, such as limestone, and the activities of organisms that live in the water also affect the pH of a body of water.

Lab Manual 36—A Food Chain

1. Protists, Beetle larvae, Dragonfly nymph, Backswimmer, Perch, and Pickerel **2.** Less energy is available to the population at each higher level of the food chain. **3.** Answers will vary but should include only organisms that might live within the same habitat. **4.** Answers will vary but should show a logical flow of energy and a logical progression through the levels of the food chain. **5.** Students should suggest that both food chains start with producers and end with third-order consumers. Populations are largest and energy availability greatest at the first level of the food chain, and both decrease at each higher level. **Explore Further** Answers will vary depending upon the habitat. Water dwellers could not survive on land, for example. Answers should show an understanding that organisms are adapted to their habitats.

Lab Manual 37—Observing Phototropism

1. Answers will vary, but students will probably write that the plants will turn toward the light. **2.** Answers will vary. **3.** phototropism **4.** The response helps plants survive by helping them get sunlight to make food. **Explore Further** Students will find that other seeds may germinate at different rates, but they react to light in a similar manner.

Lab Manual 38—Comparing Reaction Times

1. Students should find that reaction times improve across trials. **2.** Students should find that reaction times improve across trials. **3.** Students should find that they improved with practice. If they do not, students may suggest that they were tired or distracted, which increased their reaction times. **4.** The behavior is learned because it is affected by experience. **Explore Further** Students may find that the rate of improvement begins to level off or even decline. They may suggest that there is a physical limit to the speed of a person's reactions or that fatigue or loss of concentration have begun to influence the results.

Lab Manual 39—Observing Learning Patterns (Investigation 13)

1. Answers may vary and should reflect the data collected. Times should improve with practice. **2.** Answers may vary and should reflect the data collected. Students may suggest that times improved with practice because learning occurred with each try. **Explore Further 1.** Students may choose to use different colors to graph all partners on one graph. Graphs must reflect the data collected. **2.** Answers will vary. Students may find that partners are able to complete the maze in record times because they have learned the maze well. Students may find that times begin to increase because of fatigue.

Lab Manual 40—Making Molds (Investigation 14)

1. Students may notice that the specimens and molds have similar sizes, shapes, and textures. **2.** Students may point out that the specimens' color is missing from the mold. In addition, the mold does not show all sides of the specimen. **3.** Students should conclude that remains with greater depth, such as shells or bones, would produce a better mold than would flat remains, such as a leaf or feather. **Explore Further** The casts that students make should resemble the specimens that were used to make the molds.

Lab Manual 41—Adaptive Advantage

1. the red ones **2.** the green ones **3.** Answers will vary but should suggest that toothpicks with a color contrasting the background color are easier to spot than those with a color matching the background color. **4.** Those that blend in with the background have a better chance of survival than those that don't. **Explore Further** Answers will vary.

Lab Manual 42—Comparing Body Parts

1. The number and arrangement of the bones are similar. **2.** They differ in shape and function. **3.** Answers will vary. Students should realize that structure is a better indicator of relationship. Similar structures may have different functions as in the case of the bat and the alligator in the investigation. **Explore Further** Answers will vary depending on organisms chosen.

Community Connection

Completed activities will vary for each student. Community Connection activities are real-life activities that students complete outside the classroom. These activities give students practical learning and practice of the concepts taught in *Biology*. Check completed activities to see that students have followed directions, completed each step, filled in all charts and blanks, provided reasonable answers to questions, written legibly, and used appropriate science terms and proper grammar.

Self-Study Guides

Self-Study Guides outline suggested sections from the text and workbook. These assignment guides provide flexibility for individualized instruction or independent study.

Mastery Tests

Chapter 1 Mastery Test A
Part A 1. C 2. D 3. A 4. B 5. B 6. B
Part B 7. C 8. E 9. H 10. J 11. F 12. B 13. D 14. I 15. G 16. A
Part C 17. development 18. organs 19. reproduction 20. fats
Part D 21. water 22. sensing and responding 23. inside cells 24. Both are structures—organelles are inside cells and organs are made up of tissues inside the body. Both have specific functions to perform. 25. microscope

Chapter 1 Mastery Test B
Part A 1. J 2. F 3. B 4. D 5. I 6. G 7. H 8. E 9. A 10. C
Part B 11. D 12. B 13. B 14. A 15. C 16. B
Part C 17. inside cells 18. Both are structures—organelles are inside cells and organs are made up of tissues inside the body. Both have specific functions to perform. 19. sensing and responding 20. microscope 21. water
Part D 22. fats 23. reproduction 24. development 25. organs

Chapter 2 Mastery Test A
Part A 1. C 2. A 3. D 4. A 5. B
Part B 6. animal 7. monera 8. fungi 9. protist 10. plant
Part C 11. C 12. F 13. A 14. I 15. B 16. H 17. D 18. E 19. J 20. G
Part D 21. according to their similarities 22. All are means by which protozoans move. 23. They lack organelles. 24. one-celled organisms with some animal traits 25. They decompose matter, help blood clot, give plants nitrogen, and help make foods.

Chapter 2 Mastery Test B
Part A 1. B 2. A 3. C 4. A 5. D
Part B 6. fungi 7. animal 8. protist 9. monera 10. plant
Part C 11. J 12. G 13. C 14. F 15. A 16. I 17. B 18. H 19. D 20. E
Part D 21. All are means by which protozoans move. 22. according to their similarities 23. one-celled organisms with some animal traits 24. They decompose matter, help blood clot, give plants nitrogen, and help make foods. 25. They lack organelles.

Chapter 3 Mastery Test A
Part A 1. B 2. C 3. D 4. A 5. B
Part B 6. C 7. D 8. B 9. E 10. A
Part C 11. bony fish 12. mammals 13. sponges 14. segmented worms 15. echinoderms
Part D 16. hair on bodies, lungs, and mammary glands 17. *Ursus horribilis* 18. complete metamorphosis 19. reptiles 20. Many live early life in water and later life on land.

Chapter 3 Mastery Test B
Part A 1. B 2. E 3. A 4. D 5. C
Part B 6. mammals 7. bony fish 8. echinoderms 9. segmented worms 10. sponges
Part C 11. A 12. C 13. B 14. D 15. B
Part D 16. complete metamorphosis 17. hair on bodies, lungs, and mammary glands 18. *Ursus horribilis* 19. Many live early life in water and later life on land. 20. reptiles

Chapter 4 Mastery Test A
Part A 1. B 2. C 3. A 4. C 5. B
Part B 6. Ferns are vascular; mosses are nonvascular. Ferns have well-developed roots, leaves, and stems; mosses do not. 7. Linnaeus developed a method for classifying plants and animals. Our classification system today is based on that method. 8. Conifers and ginkgo trees are both gymnosperms.
Part C 9. K 10. J 11. C 12. B 13. L 14. D 15. A 16. I 17. H 18. E 19. G 20. F
Part D 21. vascular 22. seed 23. peat 24. dicot 25. spores

Chapter 4 Mastery Test B
Part A 1. A 2. H 3. L 4. C 5. I 6. B 7. J 8. G 9. E 10. F 11. K 12. D
Part B 13. seed 14. dicot 15. spores 16. peat 17. vascular
Part C 18. A 19. C 20. B 21. B 22. C
Part D 23. Linnaeus developed a method for classifying plants and animals. Our classification system today is based on that method. 24. Conifers and ginkgo trees are both gymnosperms. 25. Ferns are vascular; mosses are nonvascular. Ferns have well-developed roots, leaves, and stems; mosses do not.

Chapter 5 Mastery Test A
Part A 1. D 2. B 3. A 4. B 5. B
Part B 6. B 7. E 8. J 9. G 10. H 11. C 12. F 13. A 14. I 15. D 16. M 17. K 18. L
Part C 19. Fungi get food by extending their hyphae into a food source such as dead animal or plant matter, bread, or living plant tissue. 20. Possible answers: Bacteria help break down dead plant or animal tissue; they change nitrogen into ammonia for plants to use; they produce drugs for people; they process foods people eat, such as yogurt and sauerkraut. 21. Protist 22. An amoeba extends parts of its cell called "pseudopods," which push the cell along. 23. cell membrane
Part D 24. Both are produced by molds. Penicillin is an antibiotic used to treat bacterial infections and cyclosporine is used to keep the body from rejecting transplanted organs. 25. Both are plantlike protists with chloroplasts, so they can make their own food. Euglena is a single-celled alga that uses a flagellum to move from place to place. Kelp is a multicelled alga.

Chapter 5 Mastery Test B
Part A 1. M 2. B 3. E 4. J 5. G 6. H 7. C 8. K 9. F 10. A 11. L 12. I 13. D
Part B 14. B 15. B 16. D 17. B 18. A
Part C 19. Both are plantlike protists with chloroplasts, so they can make their own food. Euglena is a single-celled alga that uses a flagellum to move from place to place. Kelp is a multicelled alga. 20. Both are produced by molds. Penicillin is an antibiotic used to treat bacterial infections and cyclosporine is used to keep the body from rejecting transplanted organs.
Part D 21. cell membrane 22. Protist 23. Fungi get food by extending their hyphae into a food source such as dead animal or plant matter, bread, or living plant tissue. 24. An amoeba extends parts of its cell called "pseudopods," which push the cell along. 25. Possible answers: Bacteria help break down dead plant or animal tissue; they change nitrogen into ammonia for plants to use; they produce drugs for people; they process foods people eat, such as yogurt and sauerkraut.

Chapter 6 Mastery Test A
Part A 1. A 2. D 3. A 4. B 5. C
Part B 6. invertebrate nervous system 7. simple brain 8. nerve clusters 9. vertebrate nervous system 10. brain 11. spinal cord 12. nerves
Part C 13. E 14. I 15. B 16. G 17. A 18. C 19. F 20. H 21. J 22. D
Part D 23. They take in oxygen, give off carbon dioxide, and excrete extra salt. 24. Hormones secreted by endocrine glands are carried by the blood; these chemical signals affect only certain cells and are for slow responses. Nerve impulses sent by the nervous system travel rapidly along nerve cells and move from cell to cell by neurotransmitters and are for instant responses. 25. They have a very thin body wall so all cells are reachable. Gas exchange and food transport can occur at the cell level.

Chapter 6 Mastery Test B
Part A 1. invertebrate nervous system 2. simple brain 3. nerve clusters 4. vertebrate nervous system 5. brain 6. spinal cord 7. nerves
Part B 8. B 9. G 10. A 11. C 12. F 13. H 14. J 15. I 16. D 17. E
Part C 18. C 19. B 20. D 21. A 22. A
Part D 23. They have a very thin body wall so all cells are reachable. Gas exchange and food transport can occur at the cell level. 24. They take in oxygen, give off carbon dioxide, and excrete extra salt. 25. Hormones secreted by endocrine glands are carried by the blood; these chemical signals affect only certain cells and are for slow responses. Nerve impulses sent by the nervous system travel rapidly along nerve cells and move from cell to cell by neurotransmitters and are for instant responses.

Chapter 7 Mastery Test A
Part A 1. D 2. B 3. C 4. D 5. A
Part B 6. I 7. C 8. F 9. G 10. E 11. B 12. A 13. J 14. D 15. H
Part C 16. Germination is when a seed starts to grow into a new plant. In order to germinate, the temperature and amount of water must be right. 17. When insects and birds drink nectar from flowers, pollen sticks to their bodies. They carry the pollen to the pistil of other flowers or the same flower. 18. Seed plants can reproduce asexually from a piece of a plant called a cutting. The leaf or stem can grow roots and become a new plant. 19. Gymnosperms do not have flowers. They do not have fruits like angiosperms; the uncovered seeds are under the cones' scales. Examples of gymnosperms include evergreens such as spruce, fir, and pine trees. 20. Gymnosperms' reproductive organs are in cones. During reproduction, male cones release millions of pollen grains, and some reach female cones.

Chapter 7 Mastery Test B
Part A 1. C 2. H 3. I 4. J 5. D 6. A 7. E 8. B 9. F 10. G
Part B 11. C 12. D 13. A 14. B 15. D
Part C 16. When insects and birds drink nectar from flowers, pollen sticks to their bodies. They carry the pollen to the pistil of other flowers or the same flower. 17. Gymnosperms do not have flowers. They do not have fruits like angiosperms; the uncovered seeds are under the cones' scales. Examples of gymnosperms include evergreens such as spruce, fir, and pine trees. 18. Gymnosperms' reproductive organs are in cones. During reproduction, male cones release millions of pollen grains, and some reach female cones. 19. Germination is when a seed starts to grow into a new plant. In order to germinate, the temperature and amount of water must be right. 20. Seed plants can reproduce asexually from a piece of a plant called a cutting. The leaf or stem can grow roots and become a new plant.

Chapter 8 Mastery Test A
Part A 1. D 2. A 3. D 4. C 5. C
Part B 6. C 7. F 8. A 9. H 10. D 11. B 12. E 13. G
Part C 14. spinal cord 15. motor neurons 16. joints 17. sense 18. nerve
Part D Possible answers: 19. Enzymes in the mouth and stomach break food down to a liquid called chyme. Nutrients in the chyme pass through the walls of the small intestine and are carried by the blood to the cells. 20. Cartilage is a thick, smooth tissue that is softer than bone. Most of our skeleton, including our backbone and skull, is made of bone. We have cartilage in our nose and ears.

Chapter 8 Mastery Test B
Part A 1. A 2. D 3. H 4. E 5. C 6. G 7. F 8. B
Part B 9. A 10. C 11. D 12. C 13. D
Part C 14. joints 15. nerve 16. sense 17. motor neurons 18. spinal cord
Part D Possible answers: 19. Cartilage is a thick, smooth tissue that is softer than bone. Most of our skeleton, including our backbone and skull, is made of bone. We have cartilage in our nose and ears. 20. Enzymes in the mouth and stomach break food down to a liquid called chyme. Nutrients in the chyme pass through the walls of the small intestine and are carried by the blood to the cells.

Chapter 9 Mastery Test A
Part A 1. C 2. D 3. A 4. B 5. C 6. B 7. D 8. B 9. D 10. A
Part B 11. female 12. male 13. both 14. male 15. female 16. female 17. male 18. both 19. male 20. female
Part C 21. Sexual 22. identical 23. mitosis 24. Asexual 25. Meiosis
Part D 26. so the population can continue to exist 27. puberty 28. the belief that living things come from nonliving things 29. DNA 30. cells taking on different jobs

Chapter 9 Mastery Test B
Part A 1. female 2. female 3. male 4. male 5. both 6. female 7. male 8. both 9. female 10. male
Part B 11. A 12. C 13. B 14. A 15. B 16. D 17. D 18. B 19. C 20. D
Part C 21. mitosis 22. Sexual 23. Meiosis 24. identical 25. Asexual
Part D 26. puberty 27. cells taking on different jobs 28. DNA 29. the belief that living things come from nonliving thing 30. so the population can continue to exist

Chapter 10 Mastery Test A
Part A 1. A 2. B 3. D 4. D 5. B
Part B 6. I 7. F 8. H 9. C 10. G 11. D 12. A 13. E 14. J 15. B
Part C 16. Fats, Oils, & Sweets 17. Milk, Yogurt, and Cheese Group 18. Vegetable Group 19. Meat, Poultry, Fish, Dry Beans, Eggs, & Nuts Group 20. Fruit Group 21. Bread, Cereal, Rice, and Pasta Group
Part D 22. Sanitation methods provide clean surfaces that limit the growth of pathogens and so stop pathogens from entering your body. 23. Lymphocytes make antibodies that protect against the disease; phagocytes surround and destroy the pathogens. 24. Control of serving size helps present overeating and weight gain. 25. Stress can lead to headaches, grinding teeth, depression, and damage to the immune system.

Chapter 10 Mastery Test B
Part A 1. A 2. C 3. B 4. F 5. G 6. D 7. E 8. I 9. J 10. H
Part B 11. D 12. D 13. A 14. B 15. B
Part C 16. Lymphocytes make antibodies that protect against the disease; phagocytes surround and destroy the pathogens. 17. Sanitation methods provide clean surfaces that limit the growth of pathogens and so stop pathogens from entering your body. 18. Stress can lead to headaches, grinding teeth, depression, and damage to the immune system. 19. Control of serving size helps present overeating and weight gain.
Part D 20. Fats, Oils, & Sweets 21. Milk, Yogurt, and Cheese Group 22. Vegetable Group 23. Meat, Poultry, Fish, Dry Beans, Eggs, & Nuts Group 24. Fruit Group 25. Bread, Cereal, Rice, and Pasta Group

Answer Key

Chapter 11 Mastery Test A
Part A 1. F 2. I 3. H 4. A 5. E 6. G 7. D 8. J 9. B 10. C
Part B 11. B 12. C 13. D 14. B 15. B
Part C 16. Mendel experimented with pea plants. He showed that recessive and dominant genes determine the traits offspring inherit from their parents. 17. Sex cells have 23 chromosomes; other cells have 46. 18. Examples include breeding sheep for short legs, breeding cows to produce more milk and chickens to lay more eggs, and great racehorses.

19.

	Q	q
Q	QQ	Qq
Q	QQ	Qq

The Punnett square represents crosses between the traits of organisms.

20. Possible answers: diabetes causes too much sugar in the blood; sickle cell anemia causes deformed red blood cells that cause weakness and heart problems; hemophilia is a disease in which a person's blood fails to clot and can result in excessive bleeding from a slight injury.

Chapter 11 Mastery Test B
Part A 1. B 2. D 3. B 4. B 5. C
Part B 6. J 7. E 8. C 9. G 10. F 11. H 12. B 13. A 14. D 15. I
Part C 16. Examples include breeding sheep for short legs, breeding cows to produce more milk and chickens to lay more eggs, and great racehorses. 17. Possible answers: diabetes causes too much sugar in the blood; sickle cell anemia causes deformed red blood cells that cause weakness and heart problems; hemophilia is a disease in which a person's blood fails to clot and can result in excessive bleeding from a slight injury.

18.

	Q	q
Q	QQ	Qq
Q	QQ	Qq

The Punnett square represents crosses between the traits of organisms.

19. Mendel experimented with pea plants. He showed that recessive and dominant genes determine the traits offspring inherit from their parents. 20. Sex cells have 23 chromosomes; other cells have 46.

Chapter 12 Mastery Test A
Part A 1. C 2. B 3. C 4. D 5. B
Part B 6. D 7. C 8. A 9. E 10. J 11. H 12. I 13. B 14. G 15. F
Part C 16. 2, 1, 3 17. 1, 4, 3, 2 18. 2, 3, 1
Part D 19. the study of interactions among living things and nonliving parts of an environment 20. to break down dead organisms into chemicals plants can use 21. climax community 22. all the linked food chains in a community 23. They cannot be replaced by nature. 24. They give off carbon dioxide gas during respiration or when they decompose. 25. They have more energy available to them than organisms higher up the food chain.

Chapter 12 Mastery Test B
Part A 1. G 2. I 3. E 4. A 5. H 6. D 7. F 8. C 9. B 10. J
Part B 11. B 12. D 13. C 14. B 15. C
Part C 16. They have more energy available to them than organisms higher up the food chain. 17. They give off carbon dioxide gas during respiration or when they decompose. 18. climax community 19. the study of interactions among living things and nonliving parts of an environment 20. They cannot be replaced by nature. 21. to break down dead organisms into chemicals plants can use 22. all the linked food chains in a community
Part D 23. 1, 4, 3, 2 24. 2, 1, 3 25. 2, 3, 1

Chapter 13 Mastery Test A
Part A 1. C 2. A 3. B 4. A 5. D
Part B 6. G 7. D 8. I 9. B 10. J 11. C 12. H 13. E 14. A 15. F
Part C 16. Answers will vary. 17. Insight depends upon remembering and using previous experience. 18. They are all innate behaviors. 19. Examples will vary. 20. Language allows people to talk about the past and the future and things that are not present.

Chapter 13 Mastery Test B
Part A 1. I 2. A 3. C 4. J 5. F 6. B 7. E 8. D 9. H 10. G
Part B 11. A 12. C 13. D 14. B 15. A
Part C 16. Language allows people to talk about the past and the future and things that are not present. 17. They are all innate behaviors. 18. Answers will vary. 19. Insight depends upon remembering and using previous experience. 20. Examples will vary.

Chapter 14 Mastery Test A
Part A 1. B 2. A 3. C 4. D 5. B
Part B 6. D 7. A 8. E 9. B 10. C
Part C 11. modification 12. population 13. gametes 14. Fossils 15. half-life
Part D 16. an improved chance for survival because of a characteristic that makes an organism more successful 17. They suggest a common ancestor in which a version of the structure first appeared; the organisms with homologous structures evolved in slightly different directions. 18. They can compare locations in rock and use dating techniques to see which fossils are older and which are newer. They can see how groups of organisms changed over time and when extinction occurred. 19. Cro-Magnons lived more recently and were more advanced than Neanderthals. They also resemble us in more ways than Neanderthals do. For example, their brain was about the same size as that of modern humans. 20. They were tested by many scientists using large quantities of data. Evidence has overwhelmingly supported his ideas, so they are widely accepted as true.

Chapter 14 Mastery Test B
Part A 1. E 2. C 3. D 4. A 5. B
Part B 6. A 7. B 8. D 9. C 10. B
Part C 11. They can compare locations in rock and use dating techniques to see which fossils are older and which are newer. They can see how groups of organisms changed over time and when extinction occurred. 12. They suggest a common ancestor in which a version of the structure first appeared; the organisms with homologous structures evolved in slightly different directions. 13. an improved chance for survival because of a characteristic that makes an organism more successful 14. They were tested by many scientists using large quantities of data. Evidence has overwhelmingly supported his ideas, so they are widely accepted as true. 15. Cro-Magnons lived more recently and were more advanced than Neanderthals. They also resemble us in more ways than Neanderthals do. For example, their brain was about the same size as that of modern humans.
Part D 16. Fossils 17. population 18. half-life 19. gametes 20. modification

Midterm Mastery Test
Part A 1. B 2. C 3. A 4. D 5. A 6. C 7. A 8. C 9. D 10. A
Part B 11. solution 12. sensing and responding 13. kingdoms
14. scientific name 15. Vascular 16. protist 17. cell membrane
18. circulatory 19. sexual 20. chlorophyll
Part C 21. E 22. J 23. M 24. B 25. H 26. A 27. C 28. F 29. O
30. I 31. N 32. D 33. G 34. L 35. K
Part D 36. Plant cells have cell walls for structure, chloroplasts for photosynthesis, and one large vacuole. Animal cells have mitochondria to produce energy, lysosomes to break down substances, and many small vacuoles. 37. Monerans do not have organelles in their cells, but one-celled protists do. 38. A mammal feeds its young with milk from mammary glands and has hair on its body. 39. Angiosperms and gymnosperms are both seed plants whereas ferns produce spores for reproduction rather than seeds.
40. Plants produce oxygen during photosynthesis when water is broken into hydrogen and oxygen. Plants use some of the oxygen in cellular respiration and release the rest into the air.

Final Mastery Test
Part A 1. A 2. C 3. A 4. A 5. C 6. D 7. B 8. C 9. A 10. B 11. C
12. B 13. A 14. A 15. D
Part B 16. B 17. A 18. E 19. D 20. C
Part C 21. P 22. F 23. I 24. C 25. J 26. R 27. B 28. G 29. Q
30. E 31. T 32. D 33. L 34. O 35. A 36. H 37. M 38. K 39. S
40. N
Part D 41. scientific 42. solution 43. circulatory 44. responding
45. kingdoms 46. membrane 47. motor 48. capillaries
49. reproduce 50. spontaneous generation 51. decomposers
52. food chain 53. homologous structures 54. radioactive
55. genetic
Part E 56. Photosynthesis produces food that is a source of energy for animals, and it releases oxygen, which animals need for respiration. 57. Fraternal twins develop from two different zygotes and do not have identical genes. Identical twins develop from a single zygote that separates completely; they have identical genes.
58. Dinosaurs are extinct because they no longer exist on Earth.
59. The food guide pyramid is a guide for good nutrition that can help you choose the right foods and quantities of foods to eat.
60. All are functions of the human skeletal system.

Materials List for Biology Lab Manual

Note: You will want to have enough of the following materials for each student in your class. Students can do some of these activities in pairs or groups.

Chapter 1

Lab Manual 1: Using a Microscope
microscope, microscope slide, eyedropper, lowercase e from the classified ad section of the newspaper, coverslip

Lab Manual 2: Using a High-Power Microscope
compound microscope, microscope slide, eyedropper, coverslip, microscope, lowercase e from the classified ad section of the newspaper

Lab Manual 3: Comparing Cells
Investigation 1
prepared slides of animal, plant, and bacterial cells; light microscope

Chapter 2

Lab Manual 4: Living or Nonliving?
Investigation 2
5 pictures from a magazine or book, numbered 1 - 5

Lab Manual 5: A Look at the Euglena
100 power microscope, depression slides, medicine dropper, coverslips, euglena culture

Lab Manual 6: Comparing Single-Celled Organisms
high-power microscope, prepared slide of single-celled moneran, prepared slight of single-celled protist

Chapter 3

Lab Manual 7: Classifying Sports Equipment
baseball, baseball bat, golf ball, golf club, racquet ball, racquetball racquet, tennis ball, tennis racquet

Lab Manual 8: Classifying Objects
Investigation 3
assortment of objects found in a classroom

Lab Manual 9: Studying Vertebrates and Invertebrates
small, cooked, bony fish; paper towels, paring knife, tweezers, cooked clam

Chapter 4

Lab Manual 10: Identifying Angiosperms and Gymnosperms
Investigation 4
5 different kinds of leaves

Lab Manual 11: Tree Study
meat thermometer, air thermometer, tape measure, trowel (small garden shovel), hand lens, paper, crayons

Lab Manual 12: Dicot and Monocot Seeds
presoaked lima beans and corn seeds, hand lens, knife, paper towels, stereomicroscope (if available)

Chapter 5

Lab Manual 13: Observing Paramecia
safety glasses, medicine droppers, methyl cellulose, microscope slide, paramecium culture, coverslip, compound light microscope, stopwatch, distilled water, paper towel, salt water

Lab Manual 14: Studying Yeast
safety glasses, 3 test tubes, package of dry yeast, 3 medicine droppers, maple or corn syrup, refrigerator, thermometer, microscope slide, iodine solution, cover slip, compound light microscope

Lab Manual 15: Growing Bread Mold
Investigation 5
slice of dried white bread, masking tape, table knife, stereomicroscope, 4 petri dishes with lids

Chapter 6

Lab Manual 16: Observing the Action of a Digestive Enzyme
mortar and pestle, soda cracker, 4 small paper cups, 2 medicine droppers, iodine solution, cooked egg white

Lab Manual 17: Studying Feeding in Hydras
Investigation 6
microscope well slide, hydra culture, 2 medicine droppers, stereomicroscope, water flea culture, filter paper, scissors, forceps, beef broth, dish labeled "used hydras"

Lab Manual 18: Measuring Carbon Dioxide Production
safety glasses, 100 mL graduated cylinder, 4 flasks (250 mL), 0.1% methyl red, 2 medicine droppers, plastic wrap, grease pencil, drinking straw, 0.04% sodium hydroxide

Chapter 7

Lab Manual 19: Vascular Tissue
1 celery stalk with leaves, 1 white carnation, 2 regular water glasses, red or blue food coloring, knife or scalpel

Lab Manual 20: Sprouting Seeds
100 bean seeds, 12- inch x 12-inch piece of towel, 5-6 pages of newspaper, rubber bands, water

Lab Manual 21: Growing an African Violet from a Leaf
Investigation 7
African violet plant, water, 2 paper cups, aluminum foil, potting soil

Chapter 8

Lab Manual 22: How Does Exercise Change Heart Rate?
Investigation 8
watch or clock with a second hand, graph paper

Lab Manual 23: Diffusion
2 carrots, 4 beakers or jars, salt, 60 cm thread or thin string, metric ruler, labels, marking pens, water, scalpel or knife, stirring rod

Lab Manual 24: Sensory Receptors in the Skin
outlines of each partner's left hand, 3 thin nails, cup of ice, cup of hot water

Chapter 9

Lab Manual 25: Observing Mitosis
compound light microscope, prepared slide of an onion root tip

Lab Manual 26: Observing Sperm Cells and Egg Cells
compound light microscope, prepared slide of sea star egg cells, prepared slide of sea star sperm cells

Lab Manual 27: Graphing Gestation Times
Investigation 9
pencil or colored pencils, graph paper

Chapter 10

Lab Manual 28: Fighting Pathogens
2 disposable nutrient agar dishes, marker, 2 sterile cotton swabs, sterile distilled water, distilled white vinegar, spoon, tape

Lab Manual 29: Reading Food Labels
Investigation 10
8 food labels

Lab Manual 30: Comparing Calories
pencil or colored pencils

Materials List for Biology Lab Manual

Chapter 11

Lab Manual 31: Modeling Mendel's Experiments
2 purple beads, 2 white beads

Lab Manual 32: Modeling Sex Determination
a nickel, a dime, masking tape, marking pen

Lab Manual 33: Tracing a Genetic Disease
Investigation 11
paper, pencil

Chapter 12

Lab Manual 34: Life in a Pond
hand lens, depression slides, cover slips, jar of pond water, microscope, medicine dropper

Lab Manual 35: Testing the pH of Rain
Investigation 12
3 small trash cans, 3 new plastic trash bags, pH paper, pH scale, distilled water

Lab Manual 36: A Food Chain
encyclopedia, dictionary

Chapter 13

Lab Manual 37: Observing Phototropism
scissors, milk carton, potting soil, radish seeds, shoe box, metric ruler

Lab Manual 38: Comparing Reaction Times
meterstick

Lab Manual 39: Observing Learning Patterns
Investigation 13
large sheet of unlined paper, drinking straws, safety glasses, scissors, transparent tape, blindfold, stopwatch or a watch with a second hand

Chapter 14

Lab Manual 40: Making Molds
Investigation 14
seashell, leaf with thick beins and a petiole, petroleum jelly, disposable gloves, safety glasses, plaster mixture, small containers such as aluminum pie pans or disposable plastic bowls

Lab Manual 41: Adaptive Advantage
100 green toothpicks, 100 red toothpicks, stopwatch with a second hand, green construction paper, red construction paper

Lab Manual 42: Comparing Body Parts
AGS *Biology* textbook

Some Suppliers of Science Education Materials

Carolina Biological Supply Company
700 York Road
Burlington, NC 27215
800-334-5551
Fax: 800-222-7112
www.carolina.com

Fisher Science Education
4500 Turnberry Drive
Hanover Park, PA 60133
800-955-1177
Fax: 800-955-0740
www.fisheredu.com

NASCO
901 Janesville Avenue
Fort Atkinson, WI 53538
800-558-9595
Fax: 920-563-8296
www.nascofa.com

Sargent-Welch
P.O. Box 5229
Buffalo Grove, IL 60089-5229
800-727-4368
Fax: 800-676-2540
www.Sargentwelch.com

National Science Teachers Association (NSTA)
1840 Wilson Blvd.
Arlington, VA 22201
703-243-7100
nsta.org
suppliers.nsta.org for supplier list

Teacher Questionnaire

Attention Teachers! As publishers of *Biology*, we would like your help in making this textbook more valuable to you. Please take a few minutes to fill out this survey. Your feedback will help us to better serve you and your students.

1. What is your position and major area of responsibility? _____
   _____
   _____

2. Briefly describe your setting:
   ____ regular education ____ special education ____ adult basic education
   ____ community college ____ university ____ other _____

3. The enrollment in your classroom includes students with the following (check all that apply):
   ____ at-risk for failure ____ low reading ability ____ behavior problems
   ____ learning disabilities ____ ESL ____ other _____

4. Grade level of your students: _____

5. Racial/ethnic groups represented in your classes (check all that apply):
   ____ African-American ____ Asian ____ Caucasian ____ Hispanic
   ____ Native American ____ Other

6. School Location:
   ____ urban ____ suburban ____ rural ____ other _____

7. What reaction did your students have to the materials? (Include comments about the cover design, lesson format, illustrations, etc.)
   _____
   _____
   _____

8. What features in the student text helped your students the most?
   _____
   _____
   _____

9. What features in the student text helped your students the least? Please include suggestions for changing these to make the text more relevant.

   _____
   _____

10. How did you use the Teacher's Edition and support materials, and what features did you find to be the most helpful?

    _____
    _____

11. What activity from the program did your students benefit from the most? Please briefly explain.

    _____
    _____
    _____

12. Optional: Share an activity that you used to teach the materials in your classroom that enhanced the learning and motivation of your students.

    _____
    _____
    _____

Several activities will be selected to be included in future editions. Please include your name, address, and phone number so we may contact you for permission and possible payment to use the material.

Thank you!

▼ fold in thirds and tape shut at the top ▼

BUSINESS REPLY MAIL
FIRST-CLASS MAIL   PERMIT NO.12   CIRCLE PINES MN

POSTAGE WILL BE PAID BY ADDRESSEE

AGS Publishing   ATTN: Marketing Support
4201 WOODLAND ROAD
PO BOX 99
CIRCLE PINES MN 55014-9911

NO POSTAGE
NECESSARY
IF MAILED
IN THE
UNITED STATES

Name: _____
School: _____
Address: _____
City/State/ZIP: _____
Phone: _____